PYTHON TOOLS FOR SCIENTISTS

PYTHON TOOLS
FOR SCIENTISTS

An Introduction to Using Anaconda, JupyterLab, and Python's Scientific Libraries

by Lee Vaughan

no starch press

San Francisco

Printed in the United States of America

First printing

26 25 24 23 22   1 2 3 4 5

ISBN-13: 978-1-7185-0266-6 (print)
ISBN-13: 978-1-7185-0267-3 (ebook)

Publisher: William Pollock
Managing Editor: Jill Franklin
Production Manager: Sabrina Plomitallo-González
Production Editor: Katrina Horlbeck Olsen
Developmental Editor: Frances Saux
Cover Illustrator: Gina Redman
Interior Design: Octopod Studios
Technical Reviewer: John Mayhew
Production Services: Octal Publishing, LLC

For information on distribution, bulk sales, corporate sales, or translations, please contact No Starch Press, Inc. directly at info@nostarch.com or:

No Starch Press, Inc.
245 8th Street, San Francisco, CA 94103
phone: 1.415.863.9900
www.nostarch.com

Library of Congress Control Number: 2022942882

This book is dedicated to the worldwide army of open source software developers. I am immensely grateful for your hard work and the immeasurable good it produces.

About the Author

Lee Vaughan is a programmer, educator, and author of *Impractical Python Projects* (No Starch Press, 2019) and *Real-World Python* (No Starch Press, 2021). As an executive-level scientist at ExxonMobil, he constructed and reviewed computer models, developed and tested software, and trained geoscientists and engineers. His books are dedicated to helping self-learners develop and hone their Python skills and have fun doing it!

About the Technical Reviewer

John Mayhew is a geoscientist with an extensive background in mathematics, data analysis, and scientific computing. He is a co-founder of the nonprofit organization Land of Jershon and currently serves on its board of directors and as the CEO. He has also established a charitable giving consultantship, East Gate Advocates, designed to connect donors with nonprofit projects.

BRIEF CONTENTS

Acknowledgments xxi

Introduction xxiii

PART I: SETTING UP YOUR SCIENTIFIC CODING ENVIRONMENT 1

Chapter 1: Installing and Launching Anaconda 7

Chapter 2: Keeping Organized with Conda Environments 21

Chapter 3: Simple Scripting in the Jupyter Qt Console 49

Chapter 4: Serious Scripting with Spyder 61

Chapter 5: Jupyter Notebook: An Interactive Journal for Computational Research 93

Chapter 6: JupyterLab: Your Center for Science 139

PART II: A PYTHON PRIMER 173

Chapter 7: Integers, Floats, and Strings 175

Chapter 8: Variables 201

Chapter 9: The Container Data Types 219

Chapter 10: Flow Control 257

Chapter 11: Functions and Modules 283

Chapter 12: Files and Folders 315

Chapter 13: Object-Oriented programming 347

Chapter 14: Documenting Your Work 377

PART III: THE ANACONDA ECOSYSTEM 397

Chapter 15: The Scientific Libraries 399

Chapter 16: The InfoVis, SciVis, and Dashboarding Libraries 419

Chapter 17: The GeoVis Libraries 457

PART IV: THE ESSENTIAL LIBRARIES 489

Chapter 18: NumPy: Numerical Python 491

Chapter 19: Demystifying Matplotlib 537

Chapter 20: pandas, seaborn, and scikit-learn 583

Chapter 21: Managing Dates and Times with Python and pandas 625

Appendix: Answers to the "Test Your Knowledge" Challenges 665

Index 685

CONTENTS IN DETAIL

ACKNOWLEDGMENTS   **XXII**

INTRODUCTION   **XXIII**

Why Python?. xxv
Navigating This Book xxvii
      Part I: Setting Up Your Scientific Coding Environment xxviii
      Part II: A Python Primer xxix
      Part III: The Anaconda Ecosystem xxx
      Part IV: The Essential Libraries. xxx
      Appendix. xxxi
Updates and Errata xxxi
Leaving Reviews. xxxi

PART I: SETTING UP YOUR SCIENTIFIC CODING ENVIRONMENT   1

1
INSTALLING AND LAUNCHING ANACONDA   7

About Anaconda 7
Installing Anaconda on Windows 9
Installing Anaconda on macOS 11
Installing Anaconda on Linux. 12
Getting to Know Anaconda Navigator 13
      Launching Navigator. 13
      The Home Tab 13
      The Environments Tab 15
      The Learning Tab. 17
      The Community Tab. 17
      File Menu. 18
Summary 19

2
KEEPING ORGANIZED WITH CONDA ENVIRONMENTS   21

Understanding Conda Environments. 22
Working with Conda Environments Using Navigator 24
      Launching Navigator. 25
      Creating a New Environment 25
      Managing Packages 27
      Duplicating Environments 33
      Backing Up Environments 33
      Removing Environments 34

Working with Conda Environments Using the Command Line Interface 34
    Launching the Command Line Interface 34
    Creating a New Environment 36
    Specifying an Environment's Location 37
    Managing Packages 39
    Duplicating and Sharing Environments. 44
    Restoring Environments 46
    Removing Environments 47
    Cleaning the Package Cache 48
Summary 48

3
SIMPLE SCRIPTING IN THE JUPYTER QT CONSOLE   **49**

Installing seaborn. 50
Installing and Launching the Jupyter Qt Console Using Navigator 51
Installing and Launching the Jupyter Qt Console Using the CLI 52
The Qt Console Controls 53
    Choosing a Syntax Style 53
    Using Keyboard Shortcuts 54
    Using Tabs and Kernels 55
    Printing and Saving 56
    Multiline Editing 58
Summary 59

4
SERIOUS SCRIPTING WITH SPYDER   **61**

Installing and Launching Spyder with Anaconda Navigator. 62
Installing and Launching Spyder Using the CLI 63
Launching Spyder from the Start Menu 64
Configuring the Spyder Interface 64
Using Spyder with Environments and Packages 66
    The Naive Approach. 66
    The Modular Approach 66
Using Project Files and Folders. 68
    Creating a Project in a New Directory 69
    Creating a Project in an Existing Directory 70
    Using the Project Pane 72
The Help Pane 72
The IPython Console 74
    Using the Console for Output and Plotting 75
    Using Kernels with the Console 76
    Clearing the Namespace 76
    The History Pane. 77
    Special Consoles 77
The Editor Pane 78
    Writing a Program Using the Editor 78
    Defining Code Cells 81
    Setting the Run Configuration 83
    Autocompleting Text 84
The Code Analysis Pane 85

The Variable Explorer Pane 86
The Profiler Pane 89
The Debugger Pane 90
Summary 91

5
JUPYTER NOTEBOOK: AN INTERACTIVE JOURNAL FOR
COMPUTATIONAL RESEARCH   93

Installing Jupyter Notebook 94
    The Naive Approach 94
    The Modular Approach 96
Your First Jupyter Notebook 97
    Creating Dedicated Project Folders 98
    Navigating the Notebook Dashboard and User Interface 100
    Naming a Notebook 101
    Adding Text with a Markdown Cell 102
    Adding Code and Making Plots with a Code Cell 104
    Working with Output Cells 106
    Adding an Image with a Markdown Cell 107
    Saving the Notebook 109
    Closing the Notebook 109
Getting Help 109
    Keyboard Shortcuts 110
    The Command Palette 112
Using Notebook Extensions 113
    Installing Extensions 113
    Enabling Extensions 113
Working with Widgets 115
    Installing ipywidgets 115
    Creating Widgets with Interact 116
    Creating Widgets with Interactive 118
    Manually Creating Widgets 119
    Handling Events 120
    Customizing Widgets 121
    Embedding Widgets in Other Formats 122
Sharing Notebooks 122
    Checking and Running Notebooks with the Kernel Menu 123
    Downloading Notebooks 123
    Sharing Notebooks via GitHub and Gist 125
    Sharing Notebooks via Jupyter Notebook Viewer 128
    Sharing Notebooks via Binder 129
    Other Sharing Options 131
    Trusting Notebooks 131
Turning Notebooks into Slideshows 132
    Installing the RISE Extension 132
    Creating a Slideshow 133
    Using Speaker Notes 136
Summary 136

6

JUPYTERLAB: YOUR CENTER FOR SCIENCE 139

When to Use JupyterLab Instead of Notebook? 140
Installing JupyterLab 140
    The Naive Approach 140
    The Modular Approach 142
Building a 3D Astronomical Simulation 144
    Using Dedicated Project Folders 144
    The JupyterLab Interface 145
    The Menu Bar 146
    The Left Sidebar 147
    Creating a New Notebook 148
    Naming the Notebook 149
    Using Markdown Cells 149
    Adding Code and Making Plots 150
    Adding a Console 152
    Displaying an Image File 153
    Exploring the Simulation 154
    Opening Multiple Notebooks 156
    Saving the Workspace 156
    Clearing the Workspace 157
    Closing the Workspace 157
Taking Advantage of the JupyterLab Interface 157
    Creating Synchronized Views 158
    Copying Cells Between Notebooks 158
    Staying Focused by Using Single Document Mode 160
Using the Text Editor 161
    Running a Script in a Terminal 162
    Running a Script in a Notebook 163
    Simultaneously Writing and Documenting Code 164
Using JupyterLab Extensions 165
    Installing and Managing Extensions with the Extension Manager 166
    Installing and Managing Extensions Using the CLI 169
    Installing ipywidgets for JupyterLab 170
    Creating Custom Extensions 171
Sharing 171
Summary 171

PART II: A PYTHON PRIMER 173

7

INTEGERS, FLOATS, AND STRINGS 175

Mathematical Expressions 176
    Mathematical Operators 176
    The Assignment Operator 177
    Augmented Assignment Operators 178
    Precedence 178
    The math Module 179

Error Messages 182
Data Types 184
    Accessing the Data Type 185
    Integers 185
    Floats 186
    Strings 189
Summary 200

8
VARIABLES   201

Variables Have Identities 202
Assigning Variables 203
    Using Expressions 204
    Operator Overloading 204
    Using Functions 204
    Chained Assignment and Internment 205
    Using f-Strings 206
Naming Variables 206
    Reserved Keywords 207
    Variables Are Case Sensitive 209
    Best Practices for Naming Variables 209
    Managing Dynamic Typing Issues 211
    Handling Insignificant Variables 212
Getting User Input 213
Using Comparison Operators 214
Summary 217

9
THE CONTAINER DATA TYPES   219

Tuples 220
    Creating Tuples 221
    Converting Other Types to Tuples 221
    Working with Tuples 222
Lists 229
    Creating Lists 230
    Working with Lists 230
Sets 239
    Creating Sets 239
    Working with Sets 241
    Creating Frozensets 245
Dictionaries 246
    Creating Dictionaries 247
    Combining Two Sequences into a Dictionary 248
    Creating Empty Dictionaries and Values 249
    Working with Dictionaries 249
Summary 256

10
FLOW CONTROL

257

The if Statement 258
     Working with Code Blocks 259
     Using the else and elif Clauses 260
     Using Ternary Expressions 262
     Using Boolean Operators 263
Loops 264
     The while Statement 265
     The for Statement 267
     Loop Control Statements 269
     Replacing Loops with Comprehensions 271
Handling Exceptions 274
     Using try and except 274
     Forcing Exceptions with the raise Keyword 276
     Ignoring Errors 277
Tracing Execution with Logging 278
Summary 281

11
FUNCTIONS AND MODULES

283

Defining Functions 284
     Using Parameters and Arguments 285
     Positional and Keyword Arguments 286
     Using Default Values 287
     Returning Values 289
     Naming Functions 290
     Built-in Functions 290
Functions and the Flow of Execution 292
     Using Namespaces and Scopes 293
     Using Global Variables 294
     Using a main() Function 295
Advanced Function Topics 297
     Recursion 297
     Designing Functions 298
     Lambda Functions 299
     Generators 300
Modules 303
     Importing Modules 304
     Inspecting Modules 306
     Writing Your Own Modules 307
     Naming Modules 310
     Writing Modules That Work in Stand-Alone Mode 310
     Built-in Modules 311
Summary 313

12
FILES AND FOLDERS 315

Creating a New Spyder Project 316
Working with Directory Paths. 316
    The Operating System Module 317
    Absolute vs. Relative Paths 319
    The pathlib Module 320
    The Shell Utilities Module 322
Working with Text Files 325
    Reading a Text File 325
    Closing Files Using the with Statement 329
    Writing to a Text File 330
    Reading and Writing Text Files Using pathlib 332
Working with Complex Data 333
    Pickling Data 334
    Shelving Pickled Data 336
    Storing Data with JSON 339
    Catching Exceptions When Opening Files 342
    Other Storage Solutions 343
Summary 344

13
OBJECT-ORIENTED PROGRAMMING 347

When to Use OOP 348
Creating a New Spyder Project 348
Defining the Frigate Class 349
    Defining Instance Methods 352
    Instantiating Objects and Calling Instance Methods 353
Defining a Guided-Missile Frigate Class Using Inheritance 355
    Instantiating a New Guided-Missile Frigate Object 357
    Using the super() Function for Inheritance 358
Objects Within Objects: Defining the Fleet Class 359
Reducing Code Redundancy with Dataclasses 361
    Using Decorators 362
    Defining the Ship Class 364
    Identifying Friend or Foe with Fields and Post-Init Processing. 370
    Optimizing Dataclasses with __slots__ 372
Making a Class Module 373
Summary 375

14
DOCUMENTING YOUR WORK 377

Comments 378
    Single-Line Comments 379
    Multiline Comments 380
    Inline Comments 380
    Commenting-Out Code 381

Docstrings 382
    Documenting Modules 384
    Documenting Classes. 386
    Documenting Functions and Methods. 387
    Keeping Docstrings Up to Date with doctest 388
    Checking Docstrings in the Spyder Code Analysis Pane. 391
Summary 396

PART III: THE ANACONDA ECOSYSTEM   397

15
THE SCIENTIFIC LIBRARIES   399

The SciPy Stack 400
    NumPy 401
    SciPy 401
    SymPy 402
    pandas 403
A General Machine Learning Library: scikit-learn 404
The Deep Learning Frameworks 406
    TensorFlow 407
    Keras. 407
    PyTorch 408
The Computer Vision Libraries 409
    OpenCV 409
    scikit-image 410
    PIL/Pillow. 410
The Natural Language Processing Libraries 411
    NLTK 411
    spaCy 412
The Helper Libraries 413
    Requests. 413
    Beautiful Soup 414
    Regex 415
    Dask 416
Summary 418

16
THE INFOVIS, SCIVIS, AND DASHBOARDING LIBRARIES   419

InfoVis and SciVis Libraries 420
    Matplotlib 422
    seaborn 424
    The pandas Plotting API 428
    Altair 429
    Bokeh 430
    Plotly 431
    HoloViews 436
    Datashader 441
    Mayavi and ParaView. 443

Dashboards 445
    Dash 446
    Streamlit 447
    Voilà 448
    Panel 449
Choosing a Plotting Library 450
    Size of Dataset 451
    Types of Plots 452
    Format 452
    Versatility 453
    Maturity 454
    Making the Final Choice 455
Summary 456

17
THE GEOVIS LIBRARIES

457

The Geospatial Libraries 458
    GeoPandas 460
    Cartopy 464
    Geoplot 465
    Plotly 467
    folium 470
    ipyleaflet 473
    GeoViews: The HoloViz Approach 476
    KeplerGL 479
    pydeck 481
    Bokeh 484
Choosing a GeoVis Library 484
Summary 487

PART IV: THE ESSENTIAL LIBRARIES

489

18
NUMPY: NUMERICAL PYTHON

491

Introducing the Array 492
    Describing Arrays Using Dimension and Shape 492
    Creating Arrays 494
    Accessing Array Attributes 504
    Indexing and Slicing Arrays 506
Manipulating Arrays 518
    Shaping and Transposing 518
    Joining Arrays 521
    Splitting Arrays 522
Doing Math Using Arrays 524
    Vectorization 524
    Broadcasting 526
    The Matrix Dot Product 527
    Incrementing and Decrementing Arrays 528
    Using NumPy Functions 529

Reading and Writing Array Data 533
Summary 536

19
DEMYSTIFYING MATPLOTLIB   537

Anatomy of a Plot 538
The pyplot and Object-Oriented Approaches. 539
Using the pyplot Approach 540
        Creating and Manipulating Plots with pyplot Methods 542
        Working with Subplots. 545
        Building Multipanel Displays Using GridSpec 549
Using the Object-Oriented Style 555
        Creating and Manipulating Plots with the Object-Oriented Style 557
        Working with Subplots. 561
        Building Multipanel Displays Using GridSpec. 564
        Insetting Plots 567
        Plotting in 3D 568
        Animating Plots 569
Styling Plots. 573
        Changing Runtime Configuration Parameters 574
        Creating and Using a Style File. 576
        Applying Style Sheets 578
Summary 580

20
PANDAS, SEABORN, AND SCIKIT-LEARN   583

Introducing the pandas Series and DataFrame. 584
        The Series Data Structure 584
        The DataFrame Data Structure 585
The Palmer Penguins Project 586
        The Project Outline 587
        Setting Up the Project 587
        Importing Packages and Setting Up the Display 589
        Loading the Dataset. 589
        Displaying the DataFrame and Renaming Columns 590
        Checking for Duplicates. 592
        Handling Missing Values 592
        Exploring the Dataset. 596
        Predicting Penguin Species Using K-Nearest Neighbors 609
Summary 623

21
MANAGING DATES AND TIMES WITH PYTHON AND PANDAS   625

Python datetime Module 626
        Getting the Current Date and Time 626
        Assigning Timestamps and Calculating Time Delta. 627
        Formatting Dates and Times 628
        Converting Strings to Dates and Times. 630
        Plotting with datetime Objects. 632
        Creating Naive vs. Aware Objects 633

Time Series and Date Functionality with pandas 636
    Parsing Time Series Information 637
    Creating Date Ranges 640
    Creating Periods 642
    Creating Time Deltas 644
    Shifting Dates with Offsets 645
    Indexing and Slicing Time Series 646
    Resampling Time Series 647
Summary 663

APPENDIX
ANSWERS TO THE "TEST YOUR KNOWLEDGE" CHALLENGES   **665**

INDEX   **685**

ACKNOWLEDGMENTS

Thanks to Bill Pollock, founder and president of No Starch Press, for letting me write yet another book. Thanks also to Frances Saux for sticking with me through two whole books and providing the best editing money can buy. To Gina Redman, Jill Franklin, and Octopod Studios for another spectacular cover illustration. To Sarah De Vos for marketing assistance, Katrina Horlbeck Olsen for production editing, and the rest of the staff at No Starch Press who work tirelessly to produce "the finest in geek entertainment."

Special thanks to Anaconda Inc. co-founder and CEO, Peter Wang, for his vision to "empower the whole world with data literacy." To James Bednar, director of custom services at Anaconda, for his invaluable time, guidance, and advice with respect to the data visualization chapters. Thanks also to Anaconda data scientist Albert DeFusco for technical assistance in setting up the Anaconda distribution and for useful discussions around project management best practices and understanding the relationships among products like Anaconda Cloud and Nucleus.

Thanks to John Mayhew for his thorough technical review and helpful suggestions. Two heads really are better than one! Thanks also to Mike Driscoll, content writer at Real Python, for advice on the Jupyter Notebook and JupyterLab chapters.

Finally, extra special thanks to ExxonMobil geological modeler Andy Maas for his frank and frustrated discussions on Python's large selection of plotting and coding tools. Although others shared his concerns, he directly inspired this book. Hopefully, I've added some clarity to these issues.

INTRODUCTION

This book is for scientists and budding scientists who want to use the Python programming language in their work. It teaches the basics of Python and shows the easiest and most popular way to gain access to Python's universe of scientific libraries, the preferred method for documenting work, and how to keep various projects separate and secure.

As a mature, open source, and easy-to-learn language, Python has an enormous user base and a welcoming community eager to help you develop your skills. This user base has contributed to a rich set of tools and supporting libraries (collections of precompiled routines) for scientific endeavors such as data science, machine learning, language processing, robotics, computer vision, and more. As a result, Python has become one of the most important scientific computing languages in academia and industry.

Popularity, however, comes with a price. The Python ecosystem is growing into an impenetrable jungle. In fact, this book sprang from conversations with scientific colleagues in the corporate world. New to Python, they were frustrated, stressed, and suffering from *paralysis by analysis*. At every turn, they felt they had to make critical and difficult decisions such as which library to use to draw a chart and which text editor to use to write their programs. They didn't have the time or inclination to learn multiple tools, so they wanted to choose the option with the fewest repercussions down the road.

This book is designed to address those concerns. Its goal is to help you get started with scientific computing as quickly and painlessly as possible. Think of it as a machete for hacking through the dense jungle of Python distributions, tools, and libraries (Figure 1).

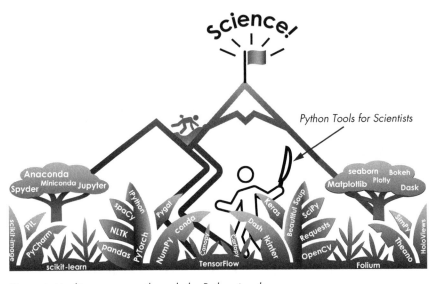

Figure 1: Hacking your way through the Python jungle

To reach this goal, I'll help you make some decisions. As everyone's needs are unique, these won't always be perfect, but they should represent sensible, "no regret" choices that will position you to customize your setup later, after you have more experience.

To begin, you'll use the free *Anaconda Distribution* of Python. As the most popular Python distribution platform, it has more than 30 million users worldwide. Provided by Anaconda, Inc. (*https://www.anaconda.com/*), it's the platform of choice for Python data science. Anaconda will make it easy to install Python, set up your computing environment, and keep it organized and up to date over time.

Please note that this book is intended for scientists who write scripts for their own personal use or for that of their team. It's not intended for professional software developers or engineers working on *enterprise* software. It also addresses only free, open source software. Your place of work may use proprietary or commercial libraries that supersede those listed here.

Finally, this book won't show you how to *do* science, or data analysis, or whatever your job entails. It won't teach you how to use your operating system, and it won't provide detailed instructions on how to use every important scientific library. Each of these requires large, dedicated volumes, which you can readily find in bookstores or online. Rather, this book will introduce you to basic tools and libraries useful across a wide range of scientific disciplines, help you to install them, and help you to get started using them. And, hopefully, it will take a lot of the stress out of setting up and using Python for science.

Why Python?

Because you're reading this book, you've probably already made up your mind about using Python. If you're still mulling it over, however, let's look at some reasons why you might want to choose Python for scientific programming. Otherwise, feel free to skip to the next section, "Navigating This Book" on page xxvii.

Python's design philosophy stresses simplicity, readability, and flexibility. These priorities make it a useful language for all stages of research and scientific endeavors, including general computing, design of experiments, building device interfaces, connecting and controlling multiple hardware/software tools, heavy-duty number crunching, and data analysis and visualization. Let's take a look at some of the key features of Python and why they are great selling points for science:

Free and open source: Python is *open source*, which means that the original source code is freely available and may be redistributed and modified by anyone. It is continuously developed by a team of volunteers and managed by the nonprofit Python Software Foundation (*www .python.org/*). A strong point of open source software is that it's *hardened*; that is, scrubbed of bugs and other problems by a large, involved user base. In addition, these users often publish and share their code so that the entire community has access to the latest techniques. On the downside, open source software can be more vulnerable to malicious users, less user friendly, and more poorly documented and supported than commercial alternatives.

High level: Python is a *high-level* programming language. This means that significant areas of the computing system, such as memory management, are automated and hidden from view. As a result, Python's syntax is very readable by humans, making it easy to learn and use.

Interpretive: Python is an *interpretive* language, which means it executes instructions immediately—similar to applying a calculation in a spreadsheet—without the need to compile the code. This gives you instant feedback, makes Python highly interactive, and helps you to catch errors as soon as they occur. It does slow the language down, however, compared to compiled languages such as Java and C++.

Platform neutral: Python runs on Windows, macOS, and Linux/Unix, and apps are available for Android and iOS.

Widespread support and shared learning: Millions of developers provide a strong support system to Python. Thanks to this large community, all the major Python products include online documentation, and you can easily find help and guidance through both free and fee-based online support sites and tutorials. Likewise, the number of Python-related print and ebooks has exploded in recent years and cover a wide range of subjects for beginners through advanced users.

Python's helpful user base is important, as the key to programming lies not in memorizing all the commands, but in *understanding what you want to do.* You will spend as much time in online search engines as you will in Python, and knowing how to construct a *task-specific* question (such as "How do I post text on an image in OpenCV?") will become an essential skill (Figure 2).

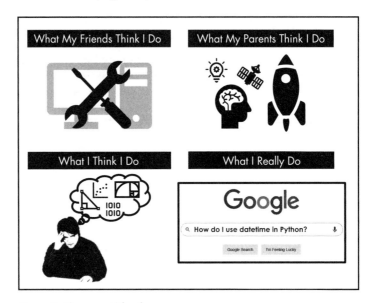

Figure 2: The secret life of programmers

Among the more popular support sites is Stack Overflow (*https:// stackoverflow.com/*). In many cases, you'll find that your query has already been answered. If not, be sure to take the tour (*https://stackoverflow.com/ tour/*) and visit the *Asking* section (*https://stackoverflow.com/help/asking/*) to review the proper way to post questions.

You can also find sites dedicated to the use of Python in specific sciences. For example, *Practical Python for Astronomers* (*https://python4 astronomers.github.io/*) is a useful site for astronomers, and *Analytics Vidhya* (*https://www.analyticsvidhya.com/*) is designed for data scientists.

Batteries included: A motto of Python is "batteries included," which means that it comes with all the possible parts required for full usability. In addition to a large *standard library* of useful tools, Python can be easily upgraded from a wide selection of third-party libraries. These are Python programs written and tested by experts in a field that you can apply in your own work. Some examples include OpenCV, used to work with image and video data; TensorFlow, used for machine learning projects; and Matplotlib, used for generating charts and diagrams. These libraries will greatly reduce the amount of code that you need to write to conduct experiments, analyze and visualize data, design simulations, and complete your projects.

Scalable: Python can easily handle the large datasets commonly used in science and engineering. Your main limitations will be the processing speed and memory of your computer. For comparison, Microsoft Excel spreadsheets have speed and stability issues with as few as tens of thousands of datapoints. Complex Excel projects become fragile as the number of spreadsheets grow, resulting in errors that are difficult to recognize, find, and fix.

Python supports both procedural and object-oriented programming that will help you write clear, logical code for both small- and large-scale projects. Python will also catch errors for you as soon as they occur.

Flexible: Python can handle multiple data formats and can run instrumentation and sensors for scientific experiments and data gathering. As a "glue" language, it's easy to integrate with lower-level languages such as C, C++, and FORTRAN, and it's useful for connecting multiple scripts or systems, including databases and web services. The large number of third-party libraries available makes Python extendable to many tasks.

Navigating This Book

This book is designed for both true beginners and those familiar with Python but not Anaconda or some of the various scientific libraries. It's designed to be "one-stop shopping" that will get you up and running with enough knowledge to begin working with data and writing your own programs.

True beginners who want a quick start learning Python should first read the chapters shown boxed in Figure 3, and then return to Part I to finish Chapters 5 and 6.

Part I: Setting Up Your Scientific Coding Environment

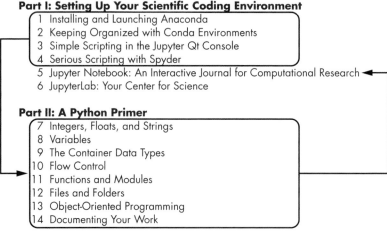

1 Installing and Launching Anaconda
2 Keeping Organized with Conda Environments
3 Simple Scripting in the Jupyter Qt Console
4 Serious Scripting with Spyder
5 Jupyter Notebook: An Interactive Journal for Computational Research
6 JupyterLab: Your Center for Science

Part II: A Python Primer

7 Integers, Floats, and Strings
8 Variables
9 The Container Data Types
10 Flow Control
11 Functions and Modules
12 Files and Folders
13 Object-Oriented Programming
14 Documenting Your Work

Part III: The Anaconda Ecosystem

15 The Scientific Libraries
16 The InfoVis, SciVis, and Dashboarding Libraries
17 The GeoVis Libraries

Part IV: The Essential Libraries

18 Numpy: Numerical Python
19 Demystifying Matplotlib
20 pandas, seaborn, and scikit-learn
21 Managing Dates and Times with Python and Pandas
Appendix: Answers to the "Test Your Knowledge" Challenges

Figure 3: The fast track to learning Python

More experienced users might want to skip around (for example, omitting the Python primer). With that in mind, here's a short synopsis of the book's contents.

Part I: Setting Up Your Scientific Coding Environment

Part I provides instructions on how to install, launch, and navigate Anaconda, and how to use the conda package manager, an open source package and environment management system that runs on Windows, macOS, and Linux. In addition, you'll be introduced to the world of shells, interpreters, text editors, notebooks, and integrated development environments (IDEs), including when and why you need them. Part I includes the following chapters:

Chapter 1, Installing and Launching Anaconda: How to install Anaconda on Windows, macOS, and Linux, followed by a tour of the Anaconda Navigator graphical user interface (GUI) and the alternative terminal-based command prompt.

Chapter 2, Keeping Organized with Conda Environments: Introduces the concept of virtual environments that let you isolate your projects and use different versions of Python and its scientific libraries. You'll set up your first *conda environment*, a directory that contains a specific version of Python, into which you'll install a specific collection of conda packages. This will allow you to keep your projects organized and prevent any conflicts among different versions of Python and/or the various libraries.

Chapter 3, Simple Scripting in the Jupyter Qt Console: Introduces the *Jupyter (IPython) Qt console*, a lightweight interface useful for interactive coding, quick concept testing, and data exploration.

Chapter 4, Serious Scripting with Spyder: Introduces *Spyder*, the *Scientific Python Development Environment* included with Anaconda. Spyder was designed by scientists, engineers, and data analysts, and provides the advanced editing, analysis, debugging, and profiling functionality of a comprehensive development tool with the data exploration, interactive execution, deep inspection, and visualization capabilities of a scientific application. If you're completely new to Python, skip down to Part II, where you'll use this tool and the Qt Console to learn the basics of Python.

Chapter 5, Jupyter Notebook: An Interactive Journal for Computational Research: Introduces the *Jupyter (IPython) Notebook*, a web-based interactive computing platform that combines live code, equations, descriptive text, interactive visualizations, and other types of media. Programs written in Jupyter can be extensively documented in-place and turned into publishable articles, interactive dashboards, and presentation-quality slideshows.

Chapter 6, JupyterLab: Your Center for Science: Introduces *JupyterLab*, a web-based interactive development environment for Jupyter notebooks, code, and data. JupyterLab's flexible interface can be configured to support a wide range of workflows in data science, scientific computing, and machine learning. In fact, you may spend most of your scientific computing "life" here, especially if you're a data scientist.

Part II: A Python Primer

Part II is a quick introduction to the Python programming language. If you're already familiar with the basics, you can skip this part and just use it as a reference when needed. Part II includes the following chapters:

Chapter 7, Integers, Floats, and Strings: Introduces some of Python's basic data types, operators, and error messages.

Chapter 8, Variables: Introduces variables and variable naming conventions.

Chapter 9, The Container Data Types: Introduces Python's tuple, list, set, and dictionary data types.

Chapter 10, Flow Control: Introduces flow-control statements, line structure, and methods for handling exceptions (errors).

Chapter 11, Functions and Modules: Introduces important concepts like abstraction and encapsulation, used to make programs easier to read and maintain.

Chapter 12, Files and Folders: Introduces modules and functions for working with files, folders, and directory paths.

Chapter 13, Object-Oriented Programming: Introduces the basics of object-oriented programming (OOP), used to make programs easier to maintain and update.

Chapter 14, Documenting Your Work: Presents best practices for in-code documentation.

Part III: The Anaconda Ecosystem

Part III introduces the Anaconda Python ecosystem and includes high-level summaries of many important scientific and visualization libraries, such as NumPy, pandas, and Matplotlib, and how to choose among the many options available. Part III includes the following chapters:

Chapter 15, The Scientific Libraries: Overviews of the core scientific libraries grouped by function, such as data analysis, machine learning, language processing, computer vision, deep learning, and so on. Guidelines are provided for choosing among competing libraries, along with a discussion of methods and libraries for dealing with very large datasets.

Chapter 16, The InfoVis, SciVis, and Dashboarding Libraries: Overviews of the most important libraries used to plot statistical and 3-D data and generate dashboards. Guidelines are provided for choosing among competing libraries.

Chapter 17, The GeoVis Libraries: Overviews of the most important libraries used to plot geospatial data. Guidelines are provided for choosing among competing libraries.

Part IV: The Essential Libraries

Part IV introduces you to the basics of working with NumPy, Matplotlib, and pandas—the "Big Three" of Python scientific libraries. These libraries are important and wildly popular ones on which many others are based. Part IV includes the following chapters:

Chapter 18, NumPy: Numerical Python: Introduces *NumPy*, the module used for mathematical calculations in Python. Many useful scientific libraries such as pandas and Matplotlib are built on NumPy. This section covers some of its key concepts and base functionality.

Chapter 19, Demystifying Matplotlib: Covers the basics of Matplotlib, the granddaddy of plotting in Python, including some of its more confusing aspects.

Chapter 20, pandas, seaborn, and scikit-learn: Introduces *pandas*, the Python library designed for data loading, manipulation, and analysis. It offers data structures and operations for manipulating numerical tables and time series and includes data visualization functionality. This chapter is built around a machine learning classification problem that also involves seaborn, used for easier Matplotlib plotting, and scikit-learn, used for building predictive models.

Chapter 21, Managing Dates and Times with Python and Pandas: Addresses methods for working with dates and times in both native Python and pandas.

Appendix

The appendix presents answers to the "Test Your Knowledge" challenges throughout the book.

Updates and Errata

This book will likely have multiple printings, and you can check for any updates or corrections at *https://www.nostarch.com/python-tools-scientists*. In the event you find any typos or errors, please report them to *errata@nostarch.com*. Be sure to include the book's title and the page numbers affected (ebook readers should mention the chapter and the subsection).

As Python, Anaconda, and the scientific libraries are constantly evolving, I provide links to their official sites where appropriate so that you can always find the most up-to-date information regarding these products.

Leaving Reviews

If you find this book helpful, please take the time to leave an online review, even if it's just a ranking with stars. Your unbiased opinion will help other users navigate the increasingly crowded market of Python programming books.

PART I

SETTING UP YOUR SCIENTIFIC CODING ENVIRONMENT

In Part I, you'll create a scientific coding environment to build upon for years to come. You'll start by installing *Anaconda,* a distribution of Python that works on Windows, macOS, and Linux and provides access to the science libraries we'll use in this book. You'll then learn to use the conda package and environment manager to keep your projects organized and up to date. After that, you'll familiarize yourself with the popular coding tools Jupyter Qt console, Spyder, Jupyter Notebook, and JupyterLab.

These coding tools help you write code, run code, and review the output, and are summarized in Table I-1. If you're unsure of the meaning of any of the terminology in the table, see the "Terminology" sidebar.

Table I-1: Coding Tool Summaries

Qt Console	Notebook	Spyder	JupyterLab
https://qtconsole .readthedocs.io/	https://jupyter -notebook.readthe docs.io/	https://docs.spyder -ide.org/current/ index.html	https://jupyterlab .readthedocs.io/en/ stable/
A terminal-like GUI application with enhancements such as inline figures, syntax highlight-ing, media output, session export, and more. Good for learning and testing.	A web-based com-puting platform that allows you to interact with code, widgets, equa-tions, narrative text, plots, images, and video in a single document.	A scientific IDE with features for editing, analysis, debugging, and profiling, plus data exploration, inter-active execution, introspection, and visualization.	An IDE-like interface for Jupyter that lets users work with Jupyter notebooks, text editors, and custom components in a flexible, inte-grated, and exten-sible manner.

The Jupyter Qt console lets you execute commands inside windows called IPython interpreters and immediately displays the results. You can use this console to interact with and visualize data. It's also great for learn-ing Python.

The famous Jupyter Notebook is a web application that allows you to create and share documents that contain live code, equations, visualiza-tions, and narrative text. It's a wildly popular tool for data science that lets you do everything from exploring and cleaning data to producing polished and interactive reports, presentations, and dashboards. Using the cloud-based *JupyterHub*, you can serve Jupyter notebooks to multiple users such as a class of students or a scientific research group.

Spyder and JupyterLab are *integrated development environments (IDEs)*. An IDE is an application that provides programmers with a set of tools for software development. For example, an IDE might include tools for debug-ging software and timing how long the code, or parts of the code, take to run. IDEs are built to work with specific application platforms and remove barriers involved in the development life cycle. They are generally used for more heavy-duty programming than is normally done in consoles or note-books. *JupyterLab*, the next-generation user interface for Anaconda's Project Jupyter, combines the classic Jupyter Notebook with a user interface that offers an IDE-like experience. It will someday replace Jupyter Notebook.

These coding tools are products of *Interactive Python (IPython)*, a com-mand shell used for interactive computing. (A *command shell* exposes the operating system's services to a program or human user.) IPython is still evolving, and in 2015 the project split so that the language-agnostic parts, such as the notebook format, Qt console, web applications, message proto-col, and so on, were put in the Jupyter project.

The name *Jupyter* references the Julia, Python, and R languages, though the project supports more than 40 languages. After the split, some terms changed. Most notably, IPython Notebook became Jupyter Notebook. There is also some overlap in the functionality of IPython products. This can cause confusion, especially given the volume of online articles and tutorials that reference the old terminology. If you're interested in the history of IPython and Jupyter Notebook, check out the datacamp blog post "IPython or Jupyter?" at *https://www.datacamp.com/community/blog/ipython-jupyter/*.

TERMINOLOGY

The following are some important terms that we'll be using in Part I.

Debugging

A multistep process for finding, isolating, and resolving problems that prevent proper program operation, known as *bugs*. Debugging is usually performed with a program called, appropriately, a *debugger*. Debuggers run the problem program under controlled conditions in a step-by-step mode to track its operations. This typically involves running or halting the program at specific points, skipping over certain parts, displaying memory content, showing the position of errors that cause the program to crash, and so on.

Extensible

Extensibility is a principle used in software engineering and systems design that indicates whether a tool provides for future growth. JupyterLab, for example, is designed as an extensible environment. JupyterLab *extensions* are add-ons that provide new interactive features to the JupyterLab interface. For instance, *JupyterLab LaTeX* is an extension that lets you live-edit LaTeX documents, *JupyterLab Plotly* is an extension for rendering Plotly charts, and *JupyterLab System Monitor* lets you monitor your own resource usage, such as memory and CPU time. You can even write custom plug-ins for your own projects.

IDE

An IDE is a coding tool that integrates other specialized utilities into a single programming environment. Among these specialized tools are a text editor, a debugger, functions for autocompleting code, functions for highlighting mistakes, file managers, project managers, a performance profiler, a deployment tool, a compiler, and so on. By combining common software-writing tools into a single application, IDEs increase programmer productivity and make it easier to manage big projects with lots of interrelated scripts. The downside is that IDEs can be *heavy*, meaning they can take up a lot of system resources. They can also be a bit intense for beginners and those who need to write only relatively simple scripts.

(continued)

Introspection

The ability to determine the type of an object and check its properties at run-time. In Python, an *object* is a code feature that has attributes and methods; you'll learn more about these in Chapter 13. Code introspection dynamically examines these objects and provides information about them. When introspection is available, hovering the cursor over an object in your code will launch a pop-up window listing the type of object as well as useful tips about using it.

Kernel

The computational engine at the core of an operating system. It is always resident in memory, which means that the operating system is not permitted to swap it out to a storage device. The kernel manages disks, tasks, and memory and acts as a bridge between applications and the data processing performed at the hardware level.

Profiling

An analysis that measures the amount of time or memory required for a program, or a program's components, to run. Profiling information can optimize code and improve its performance. IDEs, such as Spyder, come with profiling tools built in.

Qt

Pronounced *cute*, this is a widget ("Windows gadget") toolkit for creating graphical user interfaces and cross-platform applications that run on Windows, macOS, Linux, and Android.

Terminal

In modern usage, *terminal* refers to a *terminal emulator* rather than actual hardware such as a monitor and keyboard. Emulators provide a text-based interface at which to enter commands and may also be referred to as a *command line interface (CLI)*, *command prompt*, *console*, or *shell*. The major operating systems all come with some type of terminal. Windows includes the Command Prompt executable, *cmd.exe*, for running Disk Operating System (DOS) commands and to connect to other servers. macOS ships with the aptly named *Terminal*, which you can use to run Unix commands within the operating system or to access other machines using the Zsh or Z shell. Unix normally includes a program called *xterm*, which can run *Bash* or other Unix shells.

Terminals are not very user friendly, but they allow access to information and software that sometimes is available only on a central computer, such as a File Transfer Protocol (FTP) server. Manipulating thousands of files and folders in the operating system is also easier in a terminal than in a graphics window. You can automate and expedite workflows on your computer, saving you time and aggravation. Additionally, you can run Python programs from a terminal as well as a lot of Anaconda operations (as an alternative to performing them with the Anaconda Navigator GUI). Best of all, knowing how to use a terminal will greatly impress your colleagues.

After you finish Chapter 4 in Part I, you can proceed to Part II, "A Python Primer," for an introduction to Python programming. If you're comfortable with Python, complete Part I and go straight to Part III, "The Anaconda Ecosystem," to learn more about the essential packages for scientific computing.

1

INSTALLING AND LAUNCHING ANACONDA

Anaconda, the world's most popular data science platform, provides access to a large collection of commonly used science libraries. This chapter walks you through the Anaconda installation process for Windows, macOS, and Linux. To verify the installation, you'll launch Navigator, the GUI interface for Anaconda, and take a quick tour of its features.

About Anaconda

Among other features, Anaconda includes tools to help you write code and work with datasets; the Python language itself; collections of prewritten programs called *packages*; the Navigator GUI; and *Nucleus*, a community learning and sharing resource. Much of this content, summarized in Figure 1-1, is created and maintained by other organizations and distributed through Anaconda.

Figure 1-1: The key components of Anaconda

If you're new to programming, you might be unfamiliar with the concept of packages. Packages are collections of *modules*, which are single programs that perform tasks that other programs can use. For example, a module might load an image and convert it from color to grayscale. Another module might resize or crop the image. Several of these image-manipulation modules might be grouped together into a package, and groups of packages form a *library* (Figure 1-2). The OpenCV computer vision library, for example, includes packages that do simple image manipulations, others that work with streaming video, and others still that perform machine learning tasks like detecting human faces.

Figure 1-2: The definitions of modules, packages, and libraries

Unfortunately, the terms *module*, *package*, and *library* are used interchangeably so often that they might as well refer to the same thing. To make matters worse, *package* may also refer to a unit of distribution, sharable with a community, that can contain a library, an executable, or both. So, you shouldn't get too hung up on the definitions.

Many of the scientific packages that ship with Anaconda require numerous *dependencies* (specific versions of other supporting packages) to run. They might require a specific version of Python, as well. To keep the various Python installations and other packages from interacting and breaking, and to keep them up to date, Anaconda uses a binary package and environment manager called *conda*. You can use conda to install thousands of packages from the Anaconda public repository. There are also tens of thousands of packages from community channels such as conda-forge. These are in addition to several hundred packages that are automatically installed with Anaconda.

Conda will make sure that all necessary dependencies are installed with each library, saving you considerable trouble. It will also alert you if you are missing a dependency. Lastly, to prevent various packages from conflicting, conda lets you create *conda environments*, which are secure, isolated laboratories for your science projects. Packages in a conda environment will not interfere with packages in other locations, and when you share an environment, you can be sure that all the necessary packages are included. You'll learn how to create conda environments in Chapter 2.

When you download Anaconda, you get access to *Anaconda.org*, a package management system that makes it easy to find, access, store, and share public notebooks, environments, databases, and packages in both conda and the Python Package Index (PyPI). You can use it to share your work collaboratively on the cloud or search and download popular Python packages and notebooks. You can also build new conda packages using conda-build and then upload them to the cloud to share with others (or to access them from anywhere).

Anaconda is developed and maintained by Anaconda, Inc. In addition to the free Anaconda Distribution (previously called *Anaconda Individual Edition*) that we'll be using, the company also provides commercial versions. You can find the official documentation for all editions at *https:// docs.anaconda.com/anacondaorg/*. Anaconda is also a distribution of the R programming language, and conda provides package, dependency, and environment management for languages such as Ruby, Lua, Scala, Java, JavaScript, C/C++, FORTRAN, and more. In this book, however, we'll focus solely on its use with Python.

You'll need about 5GB of free hard drive space to install Anaconda. Otherwise, you'll need to install Miniconda, a minimal installation that requires around 400MB and comes with Python but not the other preinstalled libraries. There is also no need to uninstall any existing Python installations or packages prior to installing Anaconda.

In the event that you encounter problems, see the troubleshooting guide at *https://docs.anaconda.com/anaconda/user-guide/troubleshooting/* and the FAQ at *https://docs.anaconda.com/anaconda/user-guide/faq/*. If you encounter any divergence in the instructions, defer to those in the installation wizard.

Installing Anaconda on Windows

You can find the official Windows-specific installation instructions at *https:// docs.anaconda.com/anaconda/install/windows/*. Step 1 is to download the Anaconda Installer. You might need to choose between the 32- or 64-bit installer. Unless you have a very dated computer, you'll want to click the 64-bit option. If you're unsure, you should be able to verify your system type by navigating to **Settings ▸ System ▸ About**.

Clicking an installer downloads an *.exe* file into your *Downloads* folder (this can take a few minutes to complete). At this point, you have the option of checking the integrity of the installer using the SHA-256 *checksum*, which is a mathematical algorithm that checks files for corruption. Comparing a newly generated checksum against one generated ahead of time lets you detect errors introduced during data transmission. If you choose to run the checksum, see the instructions at *https://docs.anaconda.com/anaconda/install/hashes/*.

To start the installation, right-click the downloaded *.exe* file and choose the **Run as Administrator** option from the pop-up window. As Administrator, you'll have permission to install Anaconda anywhere you want on your system. The installer will ask you for permission to make changes to your computer. Click **Yes**. The setup wizard should now appear. Click **Next** and then agree to the license.

The next window asks you to choose the installation type. Select the recommended **Just Me** option and then click **Next**. Next, you're asked to choose an installation location. The installer will suggest a folder on the *C:* drive under your username. Note that this path should contain only 7-bit ASCII characters (numbers, letters, and certain symbols) and no spaces. Make a note of this default location and then click **Next**.

In the Advanced Installation Options window, register Anaconda *as default Python* and don't add it to PATH. This is the recommended approach. It just means that you'll need to open Anaconda Navigator or the Anaconda Command Prompt using the Start menu. By selecting the environment variable checkbox Add Anaconda3 to my PATH, you'll be able to use Anaconda in the command prompt; however, this can cause problems down the road. Also, you can always add Anaconda to your PATH later. Click **Install** to continue. When the installation is complete, click **Next**.

After the installation window closes, you might be presented with the option to install the PyCharm or DataSpell IDE. If so, ignore it and click **Next**. We'll be using the Spyder IDE, which comes preinstalled on Anaconda.

The installation should now be complete. In the final window, check the tutorial boxes if you want to view these later, and then click **Finish**. At this point, a window might open, welcoming you to Anaconda and inviting you to register for Anaconda Nucleus. You should also see an *Anaconda3* folder in your Start Menu (Figure 1-3). This folder should contain a number of items, such as Navigator and prompts, which are terminals for entering text commands. You might also see icons for launching Jupyter and Spyder.

To verify that Anaconda loaded correctly, click the Windows **Start** button, navigate to the Anaconda3 app, and then launch Anaconda Navigator from the drop-down menu. You can also enter `anaconda-navigator` in the Anaconda Prompt terminal. This window doesn't always automatically pop up, so be sure to check the taskbar at the bottom of your screen.

To see detailed information about your Anaconda distribution and Python version, type `conda info` in Anaconda Prompt.

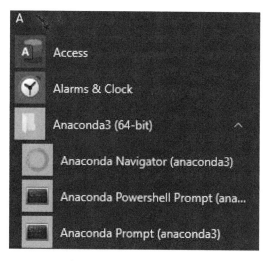

Figure 1-3: The Anaconda3 program folder on the
Windows Start menu

Installing Anaconda on macOS

You can install Anaconda Individual Edition on macOS using either a graphical setup wizard or through the command line. You can find instructions
for both at *https://docs.anaconda.com/anaconda/install/mac-os/*. Choose the
installer for your version of the operating system by scrolling to the bottom
of the download page. When the download completes, you have the option
of verifying the data's integrity using the SHA-256 checksum algorithm
(see the section "Installing Anaconda on Windows" on page 9). Then,
double-click the downloaded file and click **Continue** to launch the installation process.

You'll be taken through the obligatory Introduction, Read Me, and
License screens. The Important Information box in the Read Me screen
will include specific instructions in the event that you want to deviate from
any of the recommended default choices. When you finish with these screens,
click the **Install** button to install Anaconda in your *~/opt* directory. This
is the recommended location, though you have the option of changing it
using the Change Install Location button.

On the next screen, choose **Install for me only** and then click **Continue**.
You might now have the option to install the PyCharm or DataSpell IDE.
We will be using the Spyder IDE that comes preinstalled with Anaconda,
so skip this step by clicking **Continue**. At this point you should see a screen
indicating a successful installation. I highly recommend taking the time to
look at the quick start guide and tutorial.

To end the installation process, click **Close**.

To verify installation, click **Launchpad** and then select **Anaconda Navigator**. Alternatively, use CMD-SPACE to open Spotlight Search and then enter Navigator to open the program. You can also see detailed information on the installed Anaconda distribution and Python version by visiting the Mac terminal and entering conda info.

Installing Anaconda on Linux

Because there are so many flavors of Linux, I strongly recommend you visit the official Anaconda installation instructions at *https://docs.anaconda.com/anaconda/install/linux/*. If you are running Linux on an IBM PowerPC or Power ISA computer, see *https://docs.anaconda.com/anaconda/install/linux -power8/*. These sites will help you install the dependencies that you'll need to use GUI packages with your particular Linux distribution. The instructions presented in this section are for the x86 architecture.

Linux has no graphical installation option for installing Anaconda, so you'll need to use the command line for most of the process. To begin, scroll to the bottom of the download page and click the installer for your system. When the download completes, you have the option of verifying the data integrity using the SHA-256 checksum algorithm (see the section "Installing Anaconda on Windows" on page 9). Open a terminal and enter the following:

```
sha256sum /path/filename
```

Then, enter the following to begin installation:

```
bash ~/Downloads/Anaconda4-202x.xx-Linux-x86_64.sh
```

Note the date in the *.sh* filename above. This should be set to the name of the file you downloaded. If you did not download the installer to your *Downloads* directory, replace ~/Downloads/ with the correct path.

At the installer prompt, click **Enter** to view the license terms and then click **Yes** to agree. Next, the installer will prompt you to click **Enter** to accept the default install location, which is recommended, or specify an alternate installation directory. If you accept the default, the installer displays the following:

```
PREFIX=/home/<user>/anaconda<2 or 3>
```

It then continues the installation, which may take a few minutes to complete. When the installer asks, "Do you wish the installer to initialize Anaconda3 by running conda init?" the recommended answer is "yes." If for some reason you decide to say "no," see the instructions and FAQ on the installation website.

When the installer finishes, you'll see a message, thanking you for installing Anaconda. Ignore the link for installing the PyCharm or DataSpell IDE, as we'll be using the Spyder IDE that comes preinstalled.

For the installation to take effect, you'll need to either close and open your terminal window or enter the command `source ~/.bashrc`. To control whether each shell session has the base environment activated by default, run `conda config –set auto_activate_base True`. If base activation is not desired, set this to `False`. In general, you will want to use the base environment as the default.

To verify installation, open a terminal and type `conda list`. If Anaconda is working correctly, this will display a list of all installed packages and their version numbers. You can also enter `anaconda-navigator` to open Navigator.

Getting to Know Anaconda Navigator

Anaconda Navigator is a desktop GUI. It provides a friendly point-and-click alternative to opening a command prompt or terminal and using typed commands to manipulate Anaconda. You can use Navigator to launch applications, search for packages on *Anaconda.org* or in a local Anaconda repository, manage conda environments, channels, and packages, and access a huge volume of training material. It works on Windows, macOS, and Linux.

Launching Navigator

On Windows, the installer will create a Start menu shortcut for Navigator. For Linux or macOS with Anaconda installed via the **.sh* installer (as we did previously), open a terminal and enter `anaconda-navigator`. If you used the GUI (*.pkg*) installer on macOS, click the Navigator icon in Launchpad.

The Home Tab

Navigator opens with a window similar to the one shown in Figure 1-4. The app tiles, such as for Jupyter Notebook and Spyder, may be arranged differently in your view.

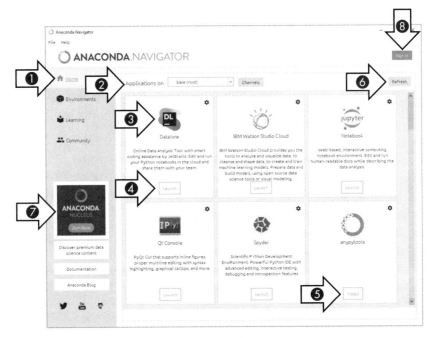

Figure 1-4: The Anaconda Navigator Home tab

The initial window you see is the Home tab **❶**. Additional pages are listed below Home and include Environments, Learning, and Community. When you launch Navigator, you'll start in the base (root) environment **❷**. Environments are just folders or directories used to isolate and manage packages. The base environment is the folder in which Anaconda is installed, such as *C:\Users\<your_username>\anaconda3* on Windows.

The scrollable main screen is filled with square tiles for applications such as Datalore, Spyder, Command Prompts, and more **❸**. Each tile contains a logo icon, the name of the app, a description of the app, and either a Launch **❹** or Install **❺** button depending on its current state. The "gear" icon in the upper-right corner of each tile also lets you install the app as well as update, remove, or install a specific version. The nice thing about Anaconda is that when it installs an app, it automatically finds and installs all the dependencies (other packages) that the app needs to run, and it shows you a list of these in a pop-up window.

If you install a package or tool from the Anaconda Prompt command line interface, the Navigator Home tab might not reflect the change. To ensure this tab is always up to date, you can click the Refresh button in the upper right **❻**.

At the lower left of the Home tab, you might see a link for Anaconda Nucleus **❼**. You can join here, or sign into an existing account using the button in the upper-right corner **❽**. Note that this button may be named either "Sign in" or "Connect." You need to sign in only if you're going to be accessing Anaconda Nucleus for sharing projects over the cloud or if you're accessing repositories such as *Anaconda.org*.

The Environments Tab

Now let's take a look at the Environments tab (Figure 1-5). To open it, click the **Environments** link ❶ beneath Home. Here, you'll be able to manage conda environments and install and uninstall libraries from Anaconda, conda forge, and other sites. We'll go into the details of this in Chapter 2.

Figure 1-5: Anaconda Navigator Environments tab

At this point, you should only see the base (root) environment ❷. The other environments shown, such as "Levy," "golden_spiral," and "penguins," are ones I've created previously using the Create button ❸ at the bottom of the screen. Note that there are additional buttons for cloning, importing, and removing environments. Newer versions may show an additional button for backing up environments to the cloud.

Only one environment can be active at a time. Clicking an environment link deactivates the current environment (such as "base") and activates the one you've clicked (such as "penguins"). It can take a few seconds for the screen to update. The right half of the screen will now show you a list of the packages installed in that environment, along with a description and version number. Also note that you can change environments using the applications in the drop-down menu on the Home tab.

If you click the Installed drop-down menu, you'll see choices for Not installed, Updatable, Selected, and All ❹. At the bottom of the screen, you'll see how many packages are currently installed and available ❺. For the base environment, the packages preinstalled by Anaconda may change slightly over time, so the number you see might be different.

NOTE *You can also see which packages come preinstalled with Anaconda by going to https://docs.anaconda.com/anaconda/packages/pkg-docs/. You'll need to know your operating system and Python version.*

When you select **Not installed**, you'll see a list of packages available from Anaconda but not currently installed in the selected environment. To see packages available from other sources, such as conda-forge, simply click the **Channels** button ❻ and select or add a new channel (Figure 1-6). A *channel* is just the path that conda takes to look for packages. Other options for working with packages include updating the packages list for the enabled channels (Update index) and searching for a package.

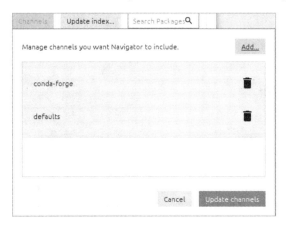

Figure 1-6: The Channels drop-down menu lets you add, update, and delete channels.

To remove a package from the active environment, click the checkbox next to the package (Figure 1-7). This opens a menu that offers choices such as marking a package for removal or installing a specific version number, which opens another menu.

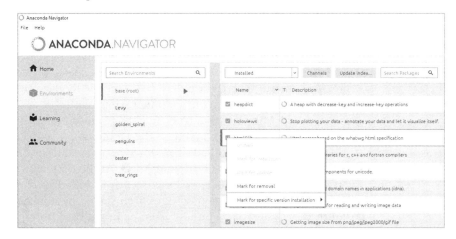

Figure 1-7: Marking a package for an action

We talk more about managing packages in the next chapter. You can also visit the Anaconda documentation for more on this subject (*https://docs .anaconda.com/anaconda/navigator/tutorials/manage-packages/*).

The Learning Tab

On the Learning tab (Figure 1-8), you can discover more about Navigator, the Anaconda platform, and open data science. To open it, click the **Learning** link beneath Home ❶.

Figure 1-8: The Anaconda Navigator Learning tab

Click the Documentation, Training, Webinars, or Video buttons ❷ to see related tile items ❸. You can turn on all the categories at once. To turn off a highlighted category, just click it again. Clicking a tile item button will open it in a browser window ❹. The button choices are Read, View, and Explore.

The Community Tab

On the Community tab (Figure 1-9) you can learn more about events, free support forums, and social networking relating to Navigator. To open it, click the **Community** link beneath Home ❶.

Figure 1-9: The Anaconda Navigator Community tab

Clicking the Events, Forum, or Social buttons ❷ changes the displayed tiles. Depending on the type of tile, you can Learn More ❸, Explore ❹, or Engage ❺. Clicking a tile button opens it in a browser window.

File Menu

The File menu in the upper-left corner of the Navigator screen includes options to let you set preferences (Figure 1-10) and quit the program. Users of macOS will see additional options in the Preferences menu, including Services, for linking to your computer's system preferences menu; Hide Anaconda-Navigator, for hiding the Navigator window; Hide Others, to hide all window except Navigator; and Show All, for showing all windows. For a detailed explanation of the Preferences menu options, see *https://docs .anaconda.com/anaconda/navigator/overview/*.

The Quit option shuts down Navigator and releases the memory resources used by Anaconda.

This completes the overview of Anaconda Navigator. You can find more information in the official documentation at *https://docs.anaconda.com/ anaconda/navigator/*. In the next chapter, we'll use Navigator, along with the command line interface, to set up conda environments that keep your projects separate, safe, and organized.

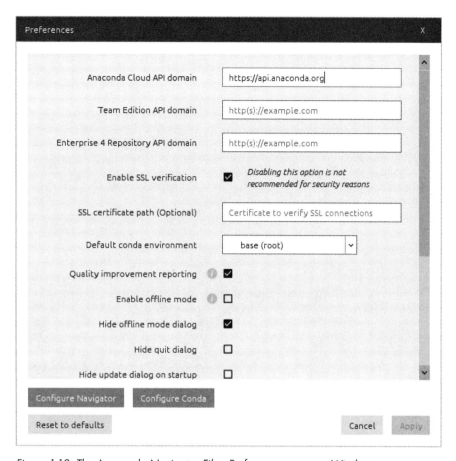

Figure 1-10: The Anaconda Navigator File ▸ Preferences menu on Windows

Summary

With Anaconda installed on your computer, you now have easy access to Python and its ecosystem of thousands of useful packages. You're also part of the Anaconda community, with storage options, lots of learning opportunities, and the ability to upload and share packages you've built yourself. Lastly, you've become familiar with the Navigator interface, letting you run Anaconda with point-and-click convenience.

2

KEEPING ORGANIZED WITH CONDA ENVIRONMENTS

Each of your Python projects should have its own conda environment. Conda environments let you use any version of any package you want, including Python, without the risk of compatibility conflicts. You can organize your packages based on project needs rather than cluttering your base directory with unnecessary packages. And you can share your environments with others, making it possible for them to perfectly reproduce your projects.

Anaconda Navigator, introduced in the previous chapter, provides an easy point-and-click interface for managing environments and packages. For even more control, conda lets you perform similar tasks using text commands in Anaconda Prompt (for Windows) or in a terminal (for macOS or Linux).

In this chapter, we'll use both Navigator and conda to create conda environments, install packages, manage the packages, remove the environment, and more. Before we begin, let's take a closer look at why a conda environment is useful.

Understanding Conda Environments

You can think of conda environments as separate Python installations. The *conda environment manager*, represented by the cargo ship in Figure 2-1, treats each environment much like a secure shipping container. Each "container" can have its own version of both Python and any other packages you need to run for a specific project. These containers are nothing more than dedicated directories in your computer's directory tree.

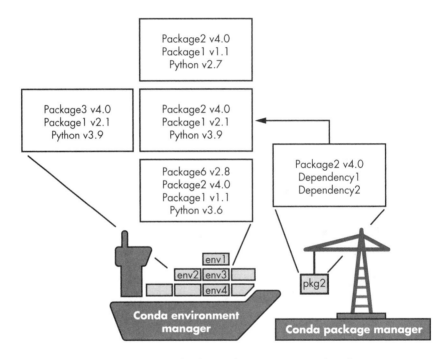

Figure 2-1: A conceptual diagram for the conda environment and package managers

As shown in Figure 2-1, you can have different versions of Python and different versions of the same libraries loaded on your computer. If they're in separate environments, they'll be isolated and won't conflict with one another. This is important because you might inherit legacy projects that run only with older versions of some packages.

The *conda package manager*, represented by the crane in Figure 2-1, finds and installs packages into your environments. Think of each package as a separate item packed in a shipping container like that heavy box of *National Geographic* magazines you should've recycled years ago.

The package manager ensures that you have the latest stable version of a package or of a version that you specify. It also finds and loads all the *dependencies* the main package needs to run at the correctly matched versions. A dependency is just another Python package that provides supporting functionality. For example, Matplotlib (for plotting) and pandas (for data analysis) are both built on NumPy (Numerical Python) and won't run without it. For this reason, it's best to install all the packages that you're going to need for a project at the same time, if possible, to avoid dependency conflicts.

If you're worried that installing packages in each conda environment is poor space management, set your mind to rest. No copies are created. Conda downloads packages into a package cache, and each environment links to the appropriate packages in this cache.

By default, this package cache is in the *pkgs* directory of your Anaconda distribution. To find it, open Anaconda Prompt or a terminal (see the instructions in Chapter 1) and enter `conda info`. Depending on your operating system, you should find the package cache at *C:\Users\<username>\anaconda3\pkgs* (Windows), *~/opt/anaconda3* (macOS), or */home/<username>/anaconda3/pkgs* (Linux).

Of course, *<username>* here refers to your personal username. The macOS location shown is for the graphical installation. If you installed Anaconda using the shell, you can find it at */Users/<username>/anaconda3*. In any case, the `conda info` command will reveal its location.

NOTE *By default, each user has their own package cache that's not shared with anyone else. It's possible to set up a* shared *package cache to save disk space and reduce installation times. If you want to share packages among multiple users, see the instructions at* https://docs.anaconda.com/anaconda/user-guide/tasks/shared-pkg-cache/.

You can also use the `conda info` command (or `conda info --envs`) to show where your conda environments are stored. In Windows, for example, the default location is *C:\Users\<username>\anaconda3\envs*.

The base environment is created by default when you install Anaconda, and it includes a Python installation and core system libraries and dependencies of conda. As a general guideline, *avoid installing additional packages into your base environment*. If you need to install additional packages for a new project, first create a new conda environment.

CONDA AND PIP

You'll occasionally encounter a package that can't be installed with conda. In this case, you'll need to do so using the Python *package management system (pip)*. Conda and pip work similarly with two exceptions. First, pip works only with Python, whereas conda works with multiple languages. Second, pip installs packages from the *Python Package Index* (*https://pypi.org/*), otherwise known as *PyPI*, whereas conda installs packages from the *Anaconda repository* (*https://repo.anaconda.com/*) and *Anaconda.org* (*https://anaconda.org/*). You can also install packages from PyPI in an active conda environment using pip. For your convenience, conda will automatically install a copy of pip in each new environment you create.

Unfortunately, issues can arise when conda and pip are used together to create an environment, especially when the tools are used back to back multiple times, establishing a state that can be difficult to reproduce. Most of these issues stem from that fact that conda, like other package managers, has limited abilities to control packages that it did not install. When using conda and pip together, here are the general guidelines:

- Install packages needing pip only after installing packages available through conda.
- Don't run pip in the root environment.
- Re-create the conda environment from scratch if changes are needed.
- Store conda and pip requirements in an environment (text) file.

For more details on this issue, see *https://www.anaconda.com/blog/using-pip-in-a-conda-environment/*. For more on pip, see *https://packaging.python.org/guides/installing-using-pip-and-virtual-environments/#creating-a-virtual-environment/*. We'll look at creating a requirements text file later in this chapter.

Working with Conda Environments Using Navigator

Setting up your first conda environment is easy. In the sections that follow, we'll use the Anaconda Navigator GUI to work with conda environments. Later in this chapter, we'll use conda in Anaconda Prompt (or a terminal) to do the same things. Anaconda Prompt and Navigator were introduced in Chapter 1.

Launching Navigator

In Windows, go to the Start menu and click the Anaconda Navigator desktop app. In macOS, open Launchpad and then click the Anaconda-Navigator icon. In Linux, open a terminal window and enter `anaconda-navigator`.

When Navigator starts, it automatically checks for a new version. If you see an Update Application message box asking you if you would like to update Navigator, click **Yes**. For a review of the Navigator interface, see Chapter 1.

Creating a New Environment

In Navigator, select the **Environments** tab and then click the **Create** button. This opens the Create New Environment dialog (Figure 2-2). Because this is your first environment, name it *my_first_env*.

Figure 2-2: The Navigator Create New Environment dialog

Note the Location information in Figure 2-2. By default, conda environments are stored in the *envs* folder within your Anaconda installation. For this reason, you must give each environment a unique name when using Navigator. It's also possible to create environments in other locations using the command line interface. We'll look at this option later in the section "Specifying an Environment's Location" on page 37.

The first package installed is Python. By default, this is the same version of Python you used when you downloaded and installed Anaconda. If you want to install a different version, you can use the pull-down menu to select it.

Click **Create**. In a minute or so, you should see the new environment on the Environments tab. You should now have two environments, *base (root)* and *my_first_env*. The arrow to the right of the name indicates that *my_first _env* is now the active environment (Figure 2-3). *Active* means that this is the environment in which you are now working, and any packages you load will be put in this folder. Clicking a name in the list activates that name and deactivates any other environments.

Figure 2-3: The newly created active environment (my_first_env) on the Navigator Environments tab

Also on the Environments tab is a listing of packages installed in *my_first_env* and their version numbers (Figure 2-4). At the bottom of the window, you can see that 12 packages were installed. These are all packages associated with Python. Over time, the number of packages may change, so you may see a different number.

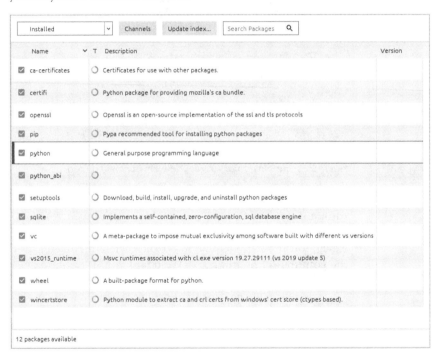

Figure 2-4: The list of initially installed packages on the Navigator Environments tab

Congratulations, you just created your first conda environment! You can start using Python right away. But if you need additional packages, such as pandas and NumPy, you must install them in this environment. So let's get to it.

Managing Packages

After you create an environment, you can use the Environments tab to see which packages are installed, check for available packages, find a specific package and install it, and update and remove packages.

Finding and Installing Packages

To find an installed package, activate the environment you want to search by clicking its name (see Figure 2-3). If the list of installed packages in the pane on the right is long and you don't want to scroll, start typing the name of the package in the Search Packages box. This will reduce the number of packages displayed until only the package you want remains.

To find a package that is not installed, change the selection of packages displayed in the right pane by clicking the drop-down menu above it and selecting **Not installed** (see Figure 2-5).

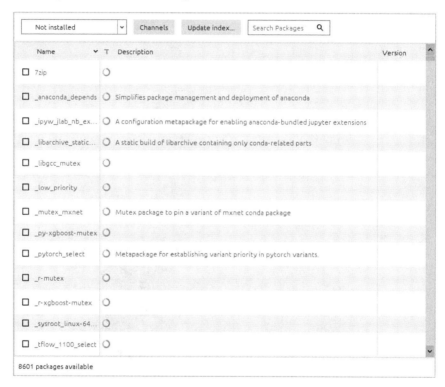

Figure 2-5: The list of available but uninstalled packages on the Navigator Environments tab

As shown in the lower left of Figure 2-5, there are currently 8,601 packages automatically available after you create the new environment (this number may change over time, so the one you see might be different). To see more packages, you can add a channel using the Channels button on the Environments tab.

Click **Channels** to open a dialog (Figure 2-6). Then, enter **conda-forge** for access to the conda-forge community channel. This channel is made up of thousands of contributors who provide packages for a wide range of software (for more information, see *https://conda-forge.org/docs/user/introduction .html/*).

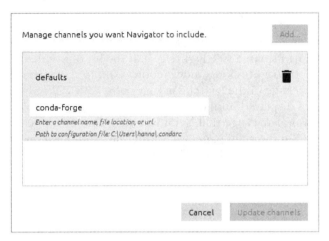

Figure 2-6: Adding conda-forge using the Channels dialog

Press ENTER and then click the **Update channels** button to add conda-forge (Figure 2-7).

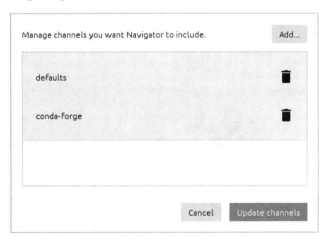

Figure 2-7: Updating channels with the Channels dialog

The pane on the right side of the Environments tab should now refresh to show that you have tens of thousands of packages available. You can remove channels by clicking the corresponding trash cans in the dialog (see Figure 2-7).

NOTE *If a package you want isn't available from Anaconda, you can try installing it from the Python Package Index (PyPI.org/) using pip, which conda installs by default in conda environments (see the "Conda and PIP" sidebar on page 24).*

Remember that we wanted to add NumPy and pandas. Because NumPy is a requirement for pandas, it's included in the pandas dependencies list. Consequently, you need to install only pandas. Enter **pandas** in the search box at the top of the right pane (Figure 2-8). Then, click the checkbox next to the pandas package and click **Apply** at the lower right. To install multiple packages at the same time, click each of the corresponding checkboxes prior to clicking Apply.

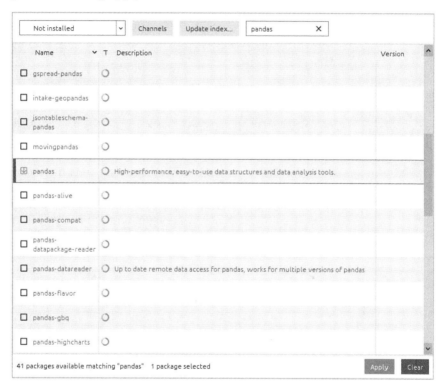

Figure 2-8: Finding and installing the pandas package on the Environments tab

A new dialog opens and, after a few moments, displays a list of packages on which pandas is dependent (Figure 2-9). As you can see, NumPy is among them. Click the **Apply** button to complete the installation of pandas.

If you switch to the Installed list, the number of installed packages will have increased, and the list will include both pandas and NumPy. Be aware that you might need to clear the Search Packages box to see the full list.

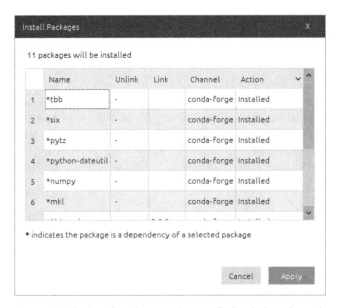

Figure 2-9: The list of packages to be installed including dependencies

You might notice that some of the major libraries appear to be duplicated in the Not installed list. For example, you can choose between "matplotlib" and "matplotlib-base" (Figure 2-10). The "-base" options tend to be lighter versions for when a package, like Matplotlib, is used by other packages as a dependency. As a result, it might not be fully functional; thus, you should not install this "-base" version when installing packages like Matplotlib or NumPy. This way, you can be sure that everything will work with no surprises.

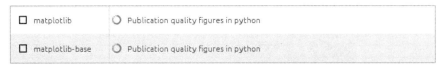

Figure 2-10: There are two choices for the matplotlib library in the list of uninstalled packages.

Updating and Removing Packages

Over time, newer versions of installed packages may become available. To check for these, select the **Updatable** filter at the top of the right pane of the Environments tab (Figure 2-11). The list you see might not exactly match the one shown.

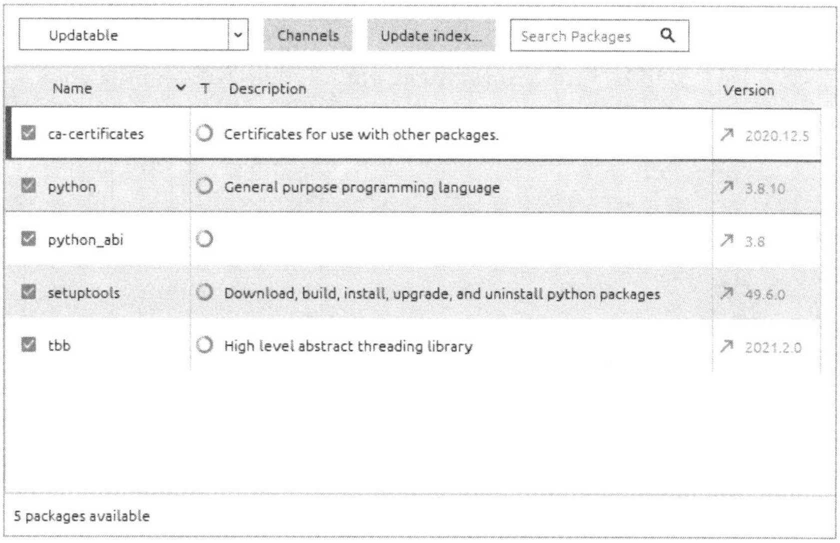

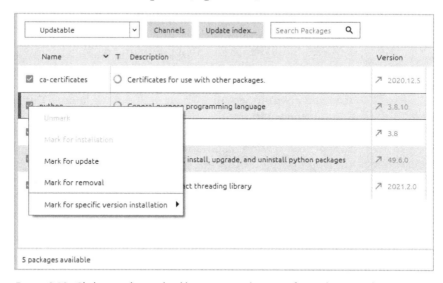

Figure 2-11: The right pane of the Environments tab, showing installed packages with available updates

In this example, Python is out of date, so let's update it to the current version. If your version is already current, try updating another package in the Updatable list.

First, click the checkbox next to Python and then, from the pop-up menu, select **Mark for update** (Figure 2-12).

Figure 2-12: Click a package checkbox to open the menu for updating and removing packages.

Click the **Apply** button at the lower right. This will open the Update Packages window, showing you which packages will be modified and which will be installed (Figure 2-13).

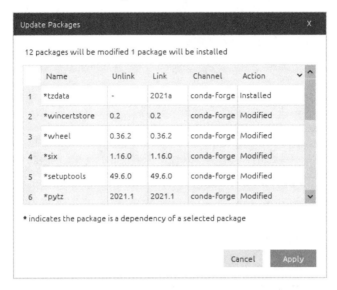

Figure 2-13: The Update Packages window for updating Python

Click **Apply** to continue. After a few minutes, Python will vanish from the Updatable list. Change the filter to **Installed** and you will see that the version of Python has changed. The Python version in the base (root) environment has not changed since all the changes you are making are to the *active* conda environment only, which is *my_first_env*.

You should be careful about updating packages for no reason, as other packages may be dependent on an older version. In the event you do break your environment in this way, it's not the end of the world; you can restore it using an environment file, which we'll discuss later in this chapter.

If you want to install a *specific version* of a package, click the checkbox by the installed package name and use the menu in Figure 2-12. Click **Mark for specific version installation** and then select the version number from the pop-up window that appears. Click **Apply** to launch the installation.

If the list of updatable packages is long and you don't want to click each checkbox, you can use the command line for efficiency. On the Environments tab, click the triangular arrow beside the active environment's name (see Figure 2-3). Then, select **Open Terminal** and enter the following:

```
conda update --all
```

You'll be shown a list of packages to be updated and asked to proceed or not. Later in this chapter, when we look at the command line interface, we'll talk about this command in more detail. We'll also discuss ways to lock or freeze a package so that it doesn't update.

To remove a package from an environment, click its checkbox, select the **Mark for removal option** (see Figure 2-12), and then click **Apply**. This will remove the package, *including its dependencies*. That last part is important. If you remove pandas from *my_first_env*, you will also remove NumPy! To prevent this, you need to explicitly install NumPy before installing pandas.

Duplicating Environments

The Clone and Import buttons at the bottom of the Environments pane (Figure 2-14) let you make an exact copy of an environment and create a new environment from a specifications file, respectively. To clone an environment, you first must activate that environment by clicking its name. When using Import, you'll be prompted to name the new environment and point to a specifications file. We'll look at creating a new environment from a file in more depth in the section "Duplicating and Sharing Environments" on page 44.

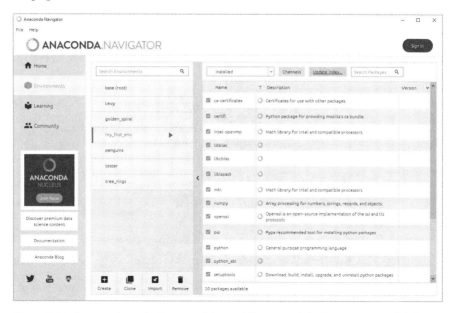

Figure 2-14: Buttons along the bottom of the middle pane of the Environments tab help you to manage conda environments.

Backing Up Environments

In newer releases of Anaconda Navigator, you might see a Backup button beside the Import button. This button lets you back up your environments to the cloud and import them back. You can use this to checkpoint your work, keep a copy for disaster recovery, or migrate from machine to machine. You'll need to have an Anaconda Nucleus account. For details, visit *https://www.anaconda.com/blog/keeping-your-conda-environments-safe-and -secure-with-your-anaconda-nucleus-account/*.

Removing Environments

To delete a conda environment, first click the name of the environment that you want to remove and then click the **Remove** button (with the trash can icon) at the bottom of the Environments tab (Figure 2-14). A pop-up window will then show you the location of the environment and request confirmation.

It's good practice to make an environment file before deleting an environment so that you can restore it if necessary. We go over how to do this in a later section.

Also be aware that environments are folders, and any data that you store in that folder will be deleted when you remove the environment. You should keep data in a separate folder or folders.

Working with Conda Environments Using the Command Line Interface

You can also work with conda environments in the *command line interface*, or *CLI* (pronounced *Clie*). Advanced users might prefer the control offered by this text-based interface to the point-and-click functionality offered by Navigator.

Launching the Command Line Interface

To begin in Windows, use the Start menu to launch Anaconda Prompt; in macOS or Linux, open a terminal window. In the CLI, the conda command is the primary interface for managing environments and installations of various packages. Like Navigator, you can use it to do the following:

- Query and search the Anaconda package index and current Anaconda installation
- Create and manage conda environments
- Install and update packages into existing conda environments

As with Navigator, you should start by either creating a new conda environment or activating an existing one. Table 2-1 lists some of the more useful single-line conda commands for working with environments. These commands let you reproduce and augment the functionality in Navigator. You'll need to replace words in all uppercase with a specific name. For example, for ENVNAME, you should substitute the actual name of your environment, such as *my_first_env*. You can also abbreviate many command options that begin with two dashes (--) to a single dash plus the first letter of the option. In other words, you can use -n instead of --name, and -e instead of --envs. We'll look at these commands in more detail in the sections that follow.

Table 2-1: Useful conda Commands for Working with Environments

Command	Description
`conda help`	Display an explanation of conda positional arguments
`conda info`	Verify installations, version numbers, directory locations
`conda update --name base conda`	Update conda to the current version
`conda create --name ENVNAME python`	Create a new environment and install Python
`conda create --name ENVNAME python=3.x`	Create a new environment with a specific Python version
`conda create --prefix path\ENVNAME`	Create a new environment at a named disk location
`conda activate ENVNAME`	Activate the named environment
`conda activate path\to\environment-dir`	Activate an environment at a named disk location
`conda deactivate`	Deactivate the current environment
`conda list`	List all packages and versions in the active environment
`conda list --name ENVNAME`	List all packages and versions in a named environment
`conda list --revisions`	List the versions of an active environment
`conda install -n ENVNAME --revision REVNUM`	Restore an environment to a previous version
`conda remove --name ENVNAME --all`	Delete a deactivated environment
`conda create --clone ENVNAME --name NEWENV`	Make an exact copy of an environment
`conda env export --name ENVNAME > envname.yml`	Export an environment to a readable YAML file
`conda env create --file ENVNAME.yml`	Create an environment from a YAML file
`conda list --explicit > pkgs.txt`	Export environment with exact versions for one OS
`conda create --name NEWENV --file pkgs.txt`	Create environment based on exact package versions

For a complete list of commands, see the "conda cheat sheet" at *https://docs.conda.io/projects/conda/en/4.6.0/_downloads/52a95608c49671267e40c689e0bc00ca/conda-cheatsheet.pdf.*

NOTE *This chapter assumes that you followed the instructions in Chapter 1 for installing Anaconda. Doing so will ensure that Anaconda is correctly added with respect to your PATH, the environment variable that specifies a set of directories where executable programs are located on your computer. This is important for using conda commands in the terminal with macOS and Linux.*

Creating a New Environment

Let's create a new conda environment named *my_second_env*, given that we've already used Navigator to create *my_first_env*. In the Anaconda Prompt window or terminal, enter the following:

```
conda create --name my_second_env python
```

This will create a new environment with the current version of Python. Enter y when asked if you want to proceed (and continue to do this throughout the chapter).

NOTE   *You can disable the verification prompt by adding the --yes or -y flag to the end of a command. This is helpful if you are automating processes, but you should avoid it with day-to-day work to lessen the chance of error.*

If you want to install a particular version of Python, say 3.9, you can use this command (but don't run it now):

```
conda create --name my_second_env python=3.9
```

This command is subtle. Because we used a *single* equal sign (=) when assigning the Python version, the result is the *latest* version in the Python 3.9 tree (such as Python 3.9.4). To get *exactly* Python 3.9, you must use a double equal sign (==) when assigning the version number.

To install multiple packages when creating an environment, list them after the Python installation (don't do this now, either):

```
conda create --name my_second_env python numpy pandas
```

To activate the new environment, enter the following:

```
conda activate my_second_env
```

Next, let's check that the environment was created and is active:

```
conda env list
```

This will produce the list shown in Figure 2-15. The asterisk (*) marks the active environment. You can also see *my_first_env*, which we created with Navigator in the previous section, as well as environments that I created earlier, some of which we'll use later in this book.

So that you're always cognizant of which environment is active, the command prompt now includes the name of the environment (first line in Figure 2-15).

To see the list of the packages currently installed in the environment, enter `conda list`. This returns the package names, versions, build, and channel information. To see the contents of a *non-active* environment, such as *my_first_env*, use `conda list -n my_first_env`. Remember, -n is just shorthand for --name.

```
 ■ Anaconda Prompt (anaconda3)                                    —    □    ✕

(my_second_env) C:\Users\hanna>conda env list
# conda environments:
#
base                                 C:\Users\hanna\anaconda3
Levy                                 C:\Users\hanna\anaconda3\envs\Levy
golden_spiral                        C:\Users\hanna\anaconda3\envs\golden_spiral
my_first_env                         C:\Users\hanna\anaconda3\envs\my_first_env
my_second_env                     *  C:\Users\hanna\anaconda3\envs\my_second_env
penguins                             C:\Users\hanna\anaconda3\envs\penguins
tester                               C:\Users\hanna\anaconda3\envs\tester
tree_rings                           C:\Users\hanna\anaconda3\envs\tree_rings
```

Figure 2-15: The output of the conda env list *command in the Anaconda Prompt window*

Specifying an Environment's Location

The conda environments that you create are stored by default in the *envs* folder beneath your Anaconda installation. For example, on my Windows machine, the environment we just created is stored at *C:\Users\hanna\ anaconda3\envs\my_second_env*. (My wife, Hannah, set up the computer; hence, she's listed as the user.)

It's possible, however, to store the environment elsewhere. This lets you place the conda environment in a project folder and consistently name it something like *conda_env* (Figure 2-16).

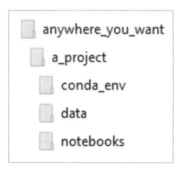

Figure 2-16: An example directory tree for storing a conda environment outside of the default location

To create a conda environment outside of the default *envs* folder, replace the --name (-n) flag with --prefix (-p):

```
conda create -p D:\Documents_on_D\anywhere_you_want\a_project\conda_env
```

To activate the environment, run the following:

```
conda activate D:\Documents_on_D\anywhere_you_want\a_project\conda_env
```

Placing your conda environment within a project directory provides several benefits. First, you can immediately determine whether a project uses an isolated environment. Second, it makes your project self-contained, as opposed to having the environment, the data, and things like Jupyter notebooks stored in different, unrelated locations. Third, you can use the same name, such as *conda_env*, for all your environments, making them instantly recognizable to anyone.

Like environments in the default location, your new environment will show up when using the `conda env list` and `conda info -e` commands, though it won't have an official name, such as *my_first_env* or *base* (see the * in Figure 2-17).

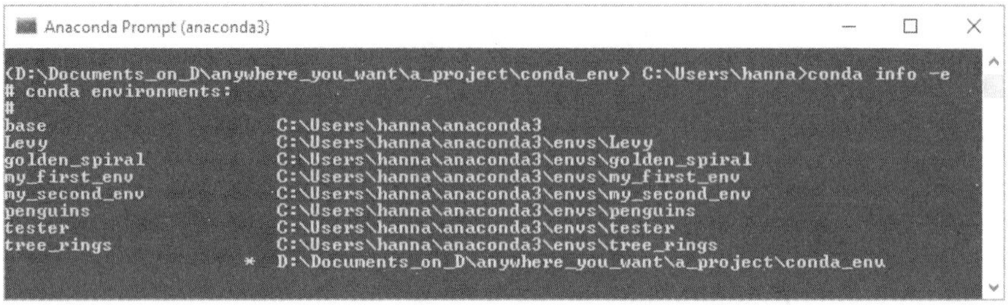

Figure 2-17: The output of the `conda info -e` command showing the active environment on the D:\ drive

Not surprisingly, there are some drawbacks to specifying an install path other than the default location when creating conda environments. For instance, conda won't be able to find your environment with the `--name` flag. For example, to list the contents of *my_first_env*, located in the default location, you can simply enter the following:

```
conda list -n my_first_env
```

For environments in other locations, you must use the `--prefix` flag plus the full path:

```
conda list -p D:\Documents_on_D\anywhere_you_want\a_project\conda_env
```

Another issue is that your command prompt is now prefixed with the active environment's *absolute path* rather than the environment's *name*. This can make for some long and unwieldy prompts, as seen in the first line in Figure 2-17.

You can force conda to always use the environment name for the prompt by modifying the env_prompt setting in the *.condarc* file. This is the *conda configuration file*, an optional runtime configuration file that allows advanced users to configure various aspects of conda, such as which channels are searched for packages. You can read about it in the documentation at *https://conda.io/projects/conda/en/latest/user-guide/configuration/index.html*.

If you want to alter (or create) a *.condarc* file to shorten the long prefix in your shell prompt, use this command:

```
conda config --set env_prompt '({name})'
```

Now you will see only the environment name in the prompt, regardless of where the environment is stored. This won't be very enlightening if you use the generic *conda_env* moniker, and you run the risk of getting confused and working in the wrong environment. For this reason, you might want to stick with the long prefix format or append each environment name with the project name, such as *conda_env_penguins* or *conda_env_covid*.

Managing Packages

After you create an environment, you can use conda to check for all available packages, find a specific package and install it, and update and remove packages. As noted in the "Launching Navigator" section, you should install all the packages you need for a project at the same time to ensure that there are no dependency conflicts.

Table 2-2 lists some useful conda commands for working with packages. The table mainly shows commands for working *within* active environments, as this is considered a best practice. You'll need to replace words in all uppercase with a specific name.

Table 2-2: Useful conda Commands for Working with Packages

Command	Description
`conda search PKGNAME`	Search for a package in currently configured channels
`conda search PKGNAME=3.9`	Search for a specific version in configured channels
`conda search PKGNAME --info`	Get detailed info on a package including its dependencies
`conda install PKGNAME`	Install current version of a package in an active environment
`conda install PKGNAME=3.4.2`	Install specific package version in an active environment
`conda install PKG1 PKG2 PKG3`	Install multiple packages in an active environment
`conda install -c CHANNELNAME PKGNAME=3.4.2`	Install specific version from named channel in an active environment
`conda uninstall PKGNAME`	Remove a package from an active environment
`conda update PKGNAME`	Update a specific package in an active environment
`conda update --all`	Update all updatable packages in an active environment
`conda list`	List all packages in an active environment
`conda list anaconda`	Show version number of installed Anaconda distribution
`conda clean --all`	Remove unused cached files including unused packages
`conda config --show`	Examine the conda configuration file
`PKGNAME --version`	Show the version number of an installed package

For a complete list of commands, see the "conda Cheat Sheet" at *https:// docs.conda.io/projects/conda/en/4.6.0/_downloads/52a95608c49671267e40c689e0b c00ca/conda-cheatsheet.pdf.*

Installing Packages

The recommended way to install packages using conda is *from within an active environment.* Alternatively, you can install packages from outside an environment by using --name or the --prefix flag with a directory path. This approach isn't advisable. Not only is it more trouble, but you run the risk of installing packages in the wrong environment.

To demonstrate how to find and install packages using conda, let's add to *my_second_env* two packages, Matplotlib (for plotting) and pillow (used for working with images). First, activate the environment:

```
conda activate my_second_env
```

It's preferable to specify the version of each package when installing. This will help you to explicitly capture what's in your environment in the event that you want to rebuild or share your project in the future. Because we have no need to use an old version of Matplotlib or pillow, let's search for the package to see its current version number:

```
conda search matplotlib
```

This returns a long list of all the available versions of Matplotlib, shown truncated for brevity in the example that follows. The far-right column represents the channel information. Of course, the version numbers will change through time, so you'll see a different list:

```
--snip--
matplotlib          3.3.4    py39haa95532_0   pkgs/main
matplotlib          3.3.4    py39hcbf5309_0   conda-forge
matplotlib          3.4.1    py37h03978a9_0   conda-forge
matplotlib          3.4.1    py38haa244fe_0   conda-forge
matplotlib          3.4.1    py39hcbf5309_0   conda-forge
matplotlib          3.4.2    py37h03978a9_0   conda-forge
matplotlib          3.4.2    py38haa244fe_0   conda-forge
matplotlib          3.4.2    py39hcbf5309_0   conda-forge
```

The *pkgs/main* channel is the top priority channel in conda's *defaults* channel, which is set by default to the Anaconda Repository. In this example, notice that the defaults channel has Matplotlib 3.3.4, whereas the conda-forge channel has Matplotlib 3.4.2.

Packages on conda-forge may be more up to date than those on the defaults channel, and you can find packages on conda-forge that aren't in defaults. With the defaults channel, however, you can be certain that the available packages have been checked for compatibility, thus making it the "safest" alternative.

If you don't specify a channel, Anaconda will automatically use the channel at the *top* of the channels configuration list in your *.condarc* file. To see your channels list, enter:

```
conda config --show channels
```

which produces this output:

```
channels:
  - conda-forge
  - defaults
```

As configured in this example, Anaconda will look for packages in the conda-forge channel first.

If the package you're looking for is in the highest priority channel, it will be installed, *even if more up-to-date versions are available in the next channel in the list.* In this case, if you install Matplotlib without specifying a version or channel, you'll end up with the most up-to-date version available, as conda-forge has top priority.

When I repeat the previous exercise for the pillow package, I see that both channels use the same version (8.2.0), so the channel doesn't matter. Now, let's install both packages together in *my_second_env*, specifying the latest versions (use the version numbers shown here or update them to whatever is current for you):

```
conda install matplotlib=3.4.2 pillow=8.2.0
```

Now let's verify the installation:

```
conda list
```

You should see the correct versions for the packages as well as the conda-forge source channel. The defaults channel, however, will appear as a blank in the "Channel" column.

If you want conda to install the newest version of a package in *any* listed channel, you can turn off the channel priority order using this command:

```
conda config --set channel_priority: false
```

You can force conda to use a specific channel by using the --channel flag and the name of the channel, like this (for the defaults channel):

```
conda install -c defaults matplotlib=3.3.4
```

To get the most up-to-date version available on that channel, you can leave off the version number, though this is not advised.

To change the membership and ordering in the list of channels in your configuration file, you can use flags like --remove, --append, and --prepend. Generally, you'll want the defaults channel on top, so let's move it up by first removing it and then adding it back:

```
conda config --remove channels defaults
conda config --prepend channels defaults
```

You can add your own channel by signing up at https://anaconda.org/ *and uploading your own conda packages.*

If you can't find a package you need through Anaconda, try the Python Package Index (*https://pypi.org/*). For more on this resource, see the "Conda and PIP" sidebar on page 24. When you install a package using pip and then use the conda list command, the channel designation for that package will be "pypi."

Finally, if you want to install a base package or set of packages in every environment you create, you can edit your configuration file to automatically add them. For example, to always install the highest version of Python by default, run the following:

```
conda config --add create_default_packages python
```

Now every time you create a new conda environment, Python will be included by default. If you do a lot of data science work, you'll probably want to add NumPy, pandas, and a plotting library, as well. You can review the default packages list by entering this:

```
conda config --show
```

To remove a package from the default packages list, use the --remove flag in place of --add. You can also override this option at the command prompt with the --no-default-packages flag.

For more options for editing your configuration file, enter **conda config --help**. For more information on installing packages and managing channels, go to *https://docs.conda.io/* and search for "installing with conda" and "managing channels," respectively.

Updating and Removing Packages

Over time, newer versions of installed packages may become available. The following commands will help you to keep your environment up to date.

First, be sure that conda is up to date by running the following (from anywhere):

```
conda update -n base conda
```

To check whether an update is available for a specific package, such as pip, in an active environment, enter:

```
conda update pip
```

If updates are available, you'll be shown the new package information, such as its version, build, memory requirements, and channel, and you'll be prompted to accept or decline the update.

To update all the packages in an active environment to the current version, enter:

```
conda update --all
```

To update a non-active environment, enter the following, where *ENV_NAME* is the name of the environment.

```
conda update -n ENV_NAME --all
```

Even though the update command tries to make everything as new as it can, it might not be able to upgrade all packages to the latest versions. If there are conflicting constraints in your environment, Anaconda might use an older version of some packages to satisfy dependency constraints when updating.

With great power comes great responsibility. Be careful about updating the Anaconda package itself, as upgrades to this metapackage are released less frequently than those for other packages. Thus, you can unknowingly downgrade packages with the update. Also never try to manage an exact set of packages in the base (root) environment. The latter is the job of specific conda environments.

For more on these topics, see *https://www.anaconda.com/blog/keeping -anaconda-date/, https://docs.anaconda.com/anaconda/install/update-version/,* and *https://www.anaconda.com/blog/whats-in-a-name-clarifying-the-anaconda -metapackage/.*

NOTE *It's possible to prevent some packages from updating by creating an exceptions list and saving it as a file named* pinned.txt *in an environment's* conda-meta *directory. You can learn more at "Preventing packages from updating (pinning)" at* https:// docs.conda.io/.

To *remove* a package, such as Matplotlib, from an active environment, enter:

```
conda remove matplotlib
```

To remove multiple packages at the same time, list them one after another. Let's do this now for *my_second_env*:

```
conda remove matplotlib pillow
```

To remove the same package from a non-active environment, provide the name of the environment using the --name (-n) flag:

```
conda remove -n ENV_NAME matplotlib
```

Remember, working with non-active environments in this way is discouraged due to the increased chance for error. Whether you work with Navigator or conda, it's shockingly easy to lose track of which environment you're working in and cause all kinds of mayhem.

To verify the results of updating and removing packages, use the **conda list** command in the active environment.

Duplicating and Sharing Environments

You can exactly duplicate an environment by either cloning it or using a special file that lists its contents, which makes it easy to share environments with others, archive them, or restore deleted versions.

Cloning Environments

The simplest way to duplicate an environment is to use the --clone flag. For example, to produce an exact copy of *my_second_env* called *my_third_env*, use the following:

```
conda create --name my_third_env --clone my_second_env
```

To verify the results, enter:

```
conda env list
```

Using an Environment File

You can also duplicate an environment by recording its contents. An *environment file* is a text file that lists all the packages and versions that are installed in an environment, including those installed using pip. This helps you both restore an environment and share it with others.

The environment file is written in *YAML (.yml)*, a human-readable data-serialization format used for data storage. YAML originally meant "Yet Another Markup Language" but now stands for "YAML Ain't Markup Language" to stress that it's more than just a document markup tool.

To generate an environment file, you must activate and then export the environment. Here's how to make a file for *my_second_env*:

```
conda activate my_second_env
conda env export > environment.yml
```

You can name the file any valid filename, such as *my_second_env.yml*, but be careful because an existing file with the same name will be overwritten.

By default, this file is written to the user directory. For my Windows setup, this is *C:\Users\hanna*. Here are the file contents (specific versions and dates have been replaced with *x*, as these values are time dependent and your output may differ):

```
name: my_second_env
channels:
  - conda-forge
  - defaults
dependencies:
  - ca-certificates=202x.xx.x=h5b45459_0
  - certifi=202x.xx.x=py39hcbf5309_1
  - openssl=1.1.1k=h8ffe710_0
  - pip=21.1.x=pyhd8ed1ab_0
  - python=3.x.x=h7840368_0_cpython
  - python_abi=3.x=1_cp39
  - setuptools=49.x.x=py39hcbf5309_3
  - sqlite=3.xx.x=h8ffe710_0
  - tzdata=202x=he74cb21_0
  - vc=14.c=hb210afc_4
  - vs20xx_runtime=14.28.29325=h5e1d092_4
  - wheel=0.xx.x=pyhd3deb0d_0
  - wincertstore=0.x=py39hcbf5309_1006
prefix: C:\Users\hanna\anaconda3\envs\my_second_env
```

You can now email this file to a coworker, and they can perfectly reproduce your environment. If they use a different operating system, you can use the --from-history flag to generate a file that will work across platforms:

```
conda env export --from-history > environment.yml
```

Here's how the new environment file looks:

```
name: my_second_env
channels:
  - conda-forge
  - defaults
dependencies:
  - python
prefix: C:\Users\hanna\anaconda3\envs\my_second_env
```

In this case, the environment file includes only packages that you've explicitly asked for, like Python, without their dependencies. Solving for dependencies can introduce packages that might not be compatible across platforms, so they are not included.

Remember when I said that it was best practice to specify a version number when installing a package, even if you wanted to take the most recent version? Well, look at -python in the last environment file listing: *there's no version number*. When you use the history flag, the environment file includes *exactly what you asked for*. By not specifying a version, you told conda to install the current version of Python. If someone uses your file after the release of a new version of Python, not only will they not reproduce your environment (assuming you haven't updated it), but they also won't *know* they haven't reproduced it!

After you have an *environment.yml* file, you can use it to re-create an environment. For example, a coworker could duplicate *my_second_env* by entering this command:

```
conda env create -n my_second_env -f \directory\path\to\environment.yml
```

You can also add packages in the file to another environment, by providing the environment name, represented here by *ENV_NAME*:

```
conda env update -n ENV_NAME -f \directory\path\to\environment.yml
```

For more on environment files, including how to manually produce them, visit *https://conda.io/projects/conda/en/latest/user-guide/tasks/manage-environments.html#sharing-an-environment/*.

Using a Specifications File

If your environment does not include packages installed using pip, you can also use a *specifications file* to reproduce a conda environment on the same operating system. To create a specification file, activate an environment, such as *my_second_env*, and enter the following command:

```
conda list --explicit > exp_spec_list.txt
```

This produces the following output, truncated for brevity:

```
# This file may be used to create an environment using:
# $ conda create --name <env> --file <this file>
# platform: win-64
@EXPLICIT
https://conda.anaconda.org/conda-forge/win-64/ca-certificates-202x.xx.x-h5b45459_0.tar.bz2
https://conda.anaconda.org/conda-forge/noarch/tzdata-202xx-he74cb21_0.tar.bz2
--snip--
```

To re-create *my_second_env* using this text file, run the following:

```
conda create -n my_second_env -f \directory\path\to\exp_spec_list.txt
```

Note that the --explicit flag ensures that the targeted platform is annotated in the file, in this case, # platform: win-64 in the third line.

Restoring Environments

Because conda keeps a history of all the changes made to an environment, you can always roll back to a previous version. To see the list of available versions, first activate the environment and then enter the following:

```
conda list --revisions
```

In the list of revisions, a plus sign (+) before a package name means that it was added, a minus sign (-) means that it was uninstalled, and no symbol before the name means that it was updated.

To restore the environment to a previous version, such as rev 3, use this command:

```
conda install --revision 3
```

Alternatively, enter the following:

```
conda install --rev 3
```

If you restore to an older revision, this revision will get its own number, so you can still restore back to an earlier one. For example, if the revisions list shows eight revisions, and you restore to revision 6, when you regenerate the revisions list, you'll see nine revisions. Revision 9 will be identical to revision 6.

Removing Environments

To delete a conda environment, you first must deactivate it by running the following:

```
conda deactivate
```

Then, to remove the deactivated environment, run this command, substituting the name of the environment for *ENVNAME*:

```
conda remove -n ENVNAME --all
```

Alternatively, you can run the following:

```
conda env remove -n ENVNAME --all
```

To verify the removal, run:

```
conda env list
```

You can also use the info command to verify this:

```
conda info -e
```

The removed environment should be absent from the environments list.

Remember, for environments outside of Anaconda's *envs* folder, you'll need to include the directory path:

```
conda remove -p PATH\ENVNAME --all
```

Cleaning the Package Cache

Over time, as you create and remove environments and install and uninstall packages, your *anaconda3* folder will consume more and more disk space. You can recover some of this space by cleaning the package cache. As discussed in "Understanding Conda Environments" on page 22, this is the folder that holds all your installed packages.

To clean the package cache, run the `conda clean` command from any environment. To get a preview of the files it flags for removal, you can make a dry run:

```
conda clean --all --dry-run
```

To commit, use:

```
conda clean --all
```

This will remove the index cache, unused cache packages (packages that are no longer linked to any environment), tarballs (files that combine and compress multiple files), and lock files from under the *pkgs* directory. Windows users will want to reboot after running this command.

For more options when running `conda clean`, see *https://docs.conda.io/projects/conda/en/latest/commands/clean.html*.

Summary

Every Python project should have its own conda environment to keep your work organized, isolated, up to date, reproducible, and sharable. Although Anaconda Navigator provides easy point-and-click manipulation of environments, you'll want to learn some command line interface commands for complete control.

3

SIMPLE SCRIPTING IN THE JUPYTER QT CONSOLE

The *Jupyter Qt console* is a lightweight application that blends the simplicity of a terminal with features possible only in a GUI, such as viewing inline figures. It's designed for quickly testing ideas, exploring datasets, and working through tutorials rather than extended interactive use.

When I say that the Qt console is "lightweight," I mean that it has a small memory footprint and doesn't burden your CPU. Likewise, it doesn't overwhelm users with a bewildering number of controls and options. The interface is clean and sparse (Figure 3-1), much like the interactive shell that ships with Python, but with many improvements. These include line numbers, the ability to open multiple tabs, the support of rich media output (such as images, video, audio, and interactive elements), command history retrieval across sessions, proper multiline editing with syntax highlighting, session export, and more.

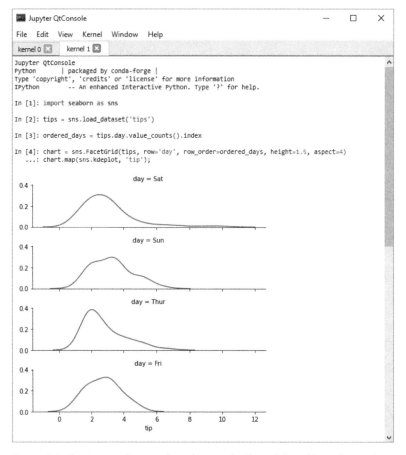

Figure 3-1: The Jupyter Qt console with two tabs (kernel 0 and kernel 1) and an inline figure

The sections that follow provide a broad introduction to the Qt console. For a more in-depth study, you can find the official documentation at *https://qtconsole.readthedocs.io/*.

Installing seaborn

If you want to reproduce the plots shown in this chapter, you'll need to install the seaborn data visualization library. Use the *my_first_env* conda environment created in the last chapter.

Open Anaconda Prompt (in Windows) or a terminal (in macOS and Linux) and then enter the following:

```
conda activate my_first_env

conda install seaborn
```

You'll also need knowledge of multiline editing, which you can find in the "Multiline Editing" section later in this chapter.

Installing and Launching the Jupyter Qt Console Using Navigator

There are two ways to install the Jupyter Qt console using Anaconda Navigator. If you have trouble with the first method, proceed to the second.

The easiest way is to use the Qt Console tile on the Home tab. First activate the environment by selecting its name in the **Applications on** pull-down menu near the top of the Home tab (Figure 3-2). In this example, we're using *my_first_env*, created in the last chapter. Next, click the **Install** button on the Qt Console app tile. You might need to scroll down the Home tab to find the tile.

Figure 3-2: The Anaconda Navigator Home tab showing the active environment (my_first_env) and the Qt Console tile

NOTE   *Ignore the package named* jupyter console. *This version of the console is purely terminal based and doesn't involve Qt for graphics.*

After a few moments, the Install button should change to a Launch button. Click this to launch the console. Note that, even though the tile on the Home tab has an IPython (IP[y]) icon, the console window name is Qt Console.

If for some reason you don't see the installation tile on the Home tab, click the **Environments** tab, switch the view to **Not installed**, search for **qtconsole** in the **Search Packages** box, and then click the button beside **qtconsole** in the list (Figure 3-3).

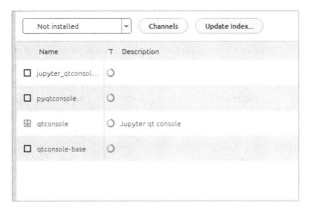

Figure 3-3: Installing the Jupyter Qt console through the Environments tab

Next, click the **Apply** button at the bottom of the screen and then click **Apply** in the pop-up window that opens. You should now see the Qt Console tile on your Home tab (Figure 3-3). If not, try clicking the **Refresh** button in the upper-right corner of the Home tab.

By installing Qt console in each environment, you'll be able to import and use other packages in that environment.

Installing and Launching the Jupyter Qt Console Using the CLI

To install the Jupyter Qt console in a new environment using the CLI rather than Anaconda Navigator, first open Anaconda Prompt (in Windows) or a terminal (in macOS and Linux) and activate the conda environment. Let's do this for *my_first_env*, created in the last chapter:

```
conda activate my_first_env
```

Next, use conda to install the console:

```
conda install qtconsole
```

Notice PyQt in the list of packages to be installed. This library enables the use of graphics in the same window as code and explains the "Qt" in Jupyter Qt console. Enter Y when prompted to complete the installation.

To start the program, enter:

```
jupyter qtconsole
```

If the console doesn't appear automatically, check your task bar. Later, if you want to update the app, enter the following:

```
conda update qtconsole
```

NOTE *If you have both Navigator and Anaconda Prompt (or the Terminal) open at the same time, and are working between them, you'll need to click the Refresh button on the Navigator Home screen after using conda to install or remove packages. This will update the Install and Launch buttons on the Navigator app tiles to the correct state.*

The Qt Console Controls

The Qt console is *interactive*, which means that it behaves like an electronic calculator. Any instructions you enter will be run immediately. In fact, you can use the console as a calculator:

```
In [1]: 5 * 2 + (10 / 2)
Out[1]: 15.0
```

Notice how the console marks input versus output and includes line numbers. Although you can't see it in a black and white book, the Qt console also uses different colors to distinguish among keywords, comments, error messages, and so on. Called *syntax highlighting*, this color-coding helps you to visually categorize your code.

You can also choose a light or dark background. Indeed, this would be a good time to play with some of the screen configuration choices to see which one you like best.

Choosing a Syntax Style

At the top of the Jupyter Qt console window, click **View ▸ Syntax Style**. You'll see a list of around 36 style types, including the popular emacs, vim, and vs styles. Choose one and then enter the code you see in Figure 3-4, which will let you see some of the theme's color choices. This book uses the default syntax style unless noted otherwise.

To compare styles, open new tabs using **File ▸ New Tab with New kernel**. Then use **Window ▸ Rename Current Tab** to name the tab for the style currently on display (such as "Monokai" in Figure 3-4). You can copy the code from one tab to the next to see the highlighting changes.

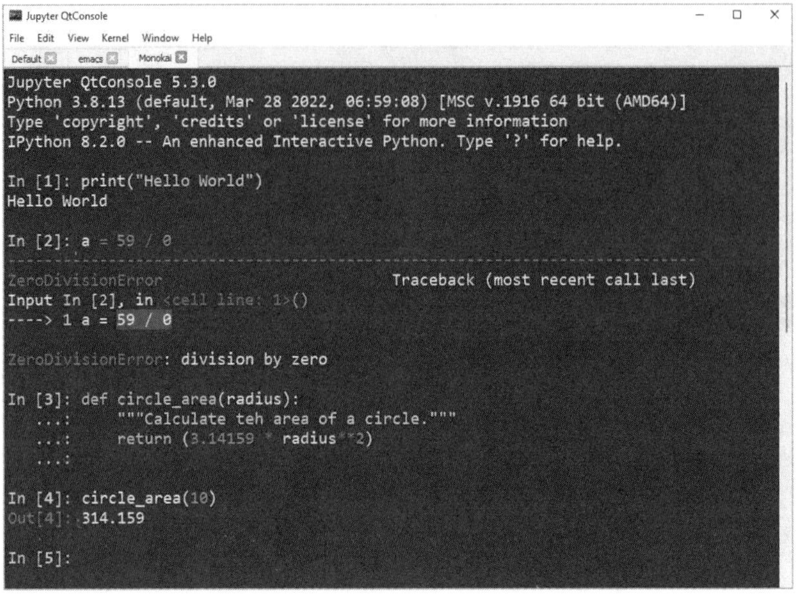

Figure 3-4: The Monokai syntax style

If you start Jupyter Qt console from the command line, you can specify a style at the same time. For example, to choose Monokai, enter the
following:

```
jupyter qtconsole --style monokai
```

Of course, there's no need to specify the `default` style.

It's even possible to configure the console and set your own style (see
"Colors and Highlighting" and "Fonts" at *https://qtconsole.readthedocs.io/_/
downloads/en/stable/pdf/*).

Using Keyboard Shortcuts

Jupyter Qt console supports keyboard shortcuts, or *keybindings*, including
the familiar CTRL-C and CTRL-V for copy and paste, respectively (Table
3-1). You can launch a list of the keybindings by clicking **Help ▸ Show
QtConsole help**. To exit help, use the ESC key.

Table 3-1: Some Common Keybindings Available in Jupyter Qt Console

Keybinding	Description
CTRL-C	Copy highlighted text to clipboard without prompts
CTRL-SHIFT-C	Copy highlighted text to clipboard with prompts
CTRL-V	Paste text from clipboard
CTRL-Z	Undo

Keybinding	Description
CTRL-SHIFT-Z	Redo
CTRL-S	Save to HTML/XHTML
CTRL-L	Clear terminal
CTRL-A	Go to beginning of line
CTRL-E	Go to end of line
CTRL-U	Delete from cursor to the beginning of the line
CTRL-K	Delete from cursor to the end of the line
CTRL-P	Previous line (like up arrow)
CTRL-N	Next line (like down arrow)
CTRL-F	Forward (like right arrow)
CTRL-B	Back (like left arrow)
CTRL-D	Delete next character or exit if input is empty
ALT-D	Delete next word
ALT-BACKSPACE	Delete previous word
CTRL-.	Force kernel to restart
CTRL-+	Increase font size
CTRL-hyphen	Decrease font size
CTRL-T	Open new tab with new kernel
CTRL-SHIFT-P	Print
F11	Toggle full screen mode
CTRL-R	Rename current tab
ALT-R	Rename window

Among the more useful shortcuts are the up and down arrow keys. These let you cycle through lines you've already entered to use them again.

Using Tabs and Kernels

Jupyter Qt console supports multiple tabs, which you can open from the **File** menu. You must select a kernel option, which is the active "computational engine" that executes the code. There are three choices:

New tab with new kernel   Opens a new tab with a new IPython kernel.

New tab with same kernel   Creates a child of a parent kernel loaded on a particular tab. Objects initialized on the parent tab will be accessible in both tabs.

New tab with existing kernel   Opens a new tab and lets you choose from kernels other than IPython.

Printing and Saving

If you're one of those "old-school" people who likes to print programs to paper and edit them with a red pen, you'll like the **File ▸ Print** command, which will produce a hardcopy of your code as it appears in the console.

You can save the Qt console session as an HTML or XHTML file using **File ▸ Save to HTML/XHTML**. If you have any inline figures or images, you can choose to write them to an external PNG file. PNG images can be either saved in an external folder or inlined to create a larger but more portable file. In Windows, the external folder, named *ipython_files*, is stored beneath the HTML file location.

With the XHTML option, your figures will be inlined as SVG files. To switch the format of inline figures from the default PNG format to SVG, see "Saving and Printing" at *https://qtconsole.readthedocs.io/*.

Although the Qt console is meant for interactive work, it's possible to copy code from the saved HTML/XHTML file or from an external text editor into the console to run it again. You'll need to strip out any output lines, however, and you'll lose the line number formatting (compare Figure 3-5 to Figure 3-1).

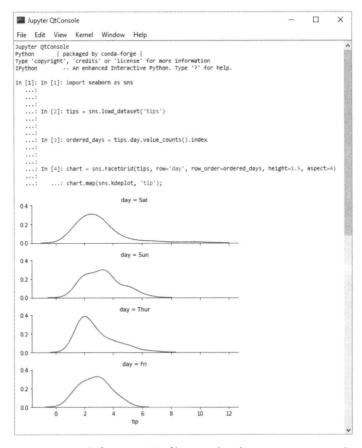

Figure 3-5: Code from an HTML file copied and run again in a new Qt console session

You can also use the %load magic command to take any script, such as a text file or existing Python file, and paste its contents as your next input in the Qt console. You then can edit it or execute it as is.

A magic command *is a special IPython enhancement added over normal Python code that facilitates common tasks such as loading a file. Line magic commands, as used in the Qt console, are prefixed by the percent symbol (%).*

To see how the %load command works, open the text editor for your platform and enter this:

```
print()
print("This is just a test.")
```

The print() function is a built-in Python routine (mini-program) that prints output to the screen. We look at functions in greater detail in Chapter 11.

Save this somewhere as *test.py* or *test.txt*. In the Qt console, type **%load** plus the path to your file, as with my example in Figure 3-6. Press ENTER to load the file, and again to execute the code.

Figure 3-6: Using the %load magic command to load and execute code from a text file

The %load command can also load code from other sources such as a URL.

A subset of commonly used magic commands are listed in Table 3-2. You can read more about them at *https://ipython.readthedocs.io/en/stable/interactive/magics.html.*

As you can see from this example, you don't need a fancy tool to write a Python program; a simple notepad application will suffice. But you can do better. Text editors dedicated to programming such as Emacs, Vim, IDLE, Notepad++, Sublime Text, and many others have built-in functionality that helps you code much more efficiently. We look at Spyder's text editor in the next chapter.

Table 3-2: Common Line Magic Commands

Command	Description
%cd	Change the current working directory
%cls (or %clear)	Clear the screen
%conda	Run the conda package manager within the current kernel
%load	Load code into current frontend
%lsmagic	List the currently available magic functions (ESC to exit)
%matplotlib qt	Display Matplotlib plots in interactive Qt window versus inline
%pprint	Toggle pretty printing on/off
%precision	Set floating-point precision for pretty printing
%pwd	Return the current working directory path
%quickref	Display reference material for magic functions (ESC to exit)
%reset	Remove all variables from the session memory
%timeit	Time the execution of a Python statement or expression
%MAGIC?	Adding a "?" behind a magic command displays its docstring

Multiline Editing

Multiline editing is a useful feature that's not available in terminals but is supported by Qt console. It lets you enter multiple lines without executing them by using CTRL-ENTER in place of ENTER.

NOTE *This book uses Windows conventions. macOS users should substitute the COMMAND key for CONTROL, and the OPTION key for ALT when using keyboard shortcuts.*

If you look closely at the code in Figure 3-1, you'll notice that Line 4 looks odd. The second line is not numbered; instead, it's three dots precede it:

```
In [4]: chart = sns.FacetGrid(tips, row='day', ...
   ...: chart.map(sns.kdeplot, 'tip');
```

Because I pressed CTRL-ENTER after typing Line 4, the line didn't execute. As a result, I was able to fully define the chart before drawing it. Had I entered and executed each of these lines independently, I would've gotten the unacceptable results shown in Figure 3-7.

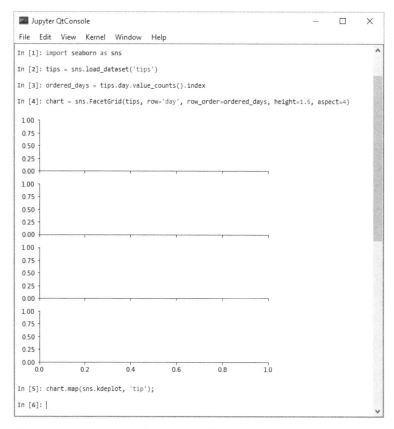

Figure 3-7: Executing each line individually causes the chart to plot without data

Additionally, at any point in a multiline block, you can force its execution (without having to go to the bottom) by using SHIFT-ENTER.

Multiline editing is a convenient feature that distinguishes Jupyter Qt console from more basic interpreters. It's wonderful for short code snippets, but as your programs become longer, you'll want the efficiency and persistence of a true text editor.

Summary

The Jupyter Qt console is a lightweight application for writing code snippets, quickly exploring datasets and testing ideas, and working through coding tutorials. For writing large persistent programs, you'll want to use other coding tools such as Jupyter Notebook, JupyterLab, or Spyder.

4

SERIOUS SCRIPTING
WITH SPYDER

The *Scientific Python Development IDE (Spyder)* is an open source interactive development environment designed *by* scientists *for* scientists. It integrates numerous specialized tools, such as a text editor, debugger, profiler, linter, and console, into a comprehensive tool for software development.

Spyder is built for heavy-duty work and consequently has a larger system footprint and more complicated interface (Figure 4-1) than the Jupyter Qt console covered in the previous chapter. But that doesn't mean you can't use Spyder for small tasks. It includes both a console, for executing ad hoc code, and a text editor for writing persistent, easily editable scripts of any size. We use Spyder in Part II of this book, which provides a primer to Python programming in the event you need to learn Python or refresh certain concepts.

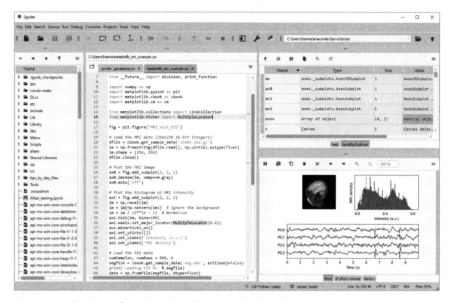

Figure 4-1: The Scientific Python Development IDE (Spyder)

In general, if you plan on writing elaborate programs or developing applications, you'll want to use Spyder or a similar IDE.

Installing and Launching Spyder with Anaconda Navigator

Spyder comes preinstalled in your *base* environment. To install it in a different environment using Anaconda Navigator, first activate the environment by selecting its name in the **Applications on** pull-down menu near the top of the Home tab (Figure 4-2). In this example, we are using *my_first_env* that we created in Chapter 2. Next, click the **Install** button on the Spyder app tile to install it. You might need to scroll down the Home tab to find the tile.

After a few minutes, the Install button should change to a Launch button. Click it to start Spyder. Remember, if you want to install a specific version of Spyder, click the gear icon at the upper right of the tile to see a listing of available version numbers.

For more information on installing Spyder, see the Installation Guide at *https://docs.spyder-ide.org/current/installation.html*.

Figure 4-2: The Anaconda Navigator Home tab showing the active environment (my_first _env) and the Spyder tile

Installing and Launching Spyder Using the CLI

Spyder comes preinstalled in your *base* environment. To install it in a new environment using conda, first open Anaconda Prompt (in Windows) or a terminal (in macOS and Linux) and activate the conda environment. Let's do this for *my_first_env* by entering the following:

```
conda activate my_first_env
```

Next, use conda to install spyder:

```
conda install spyder
```

To install specific versions, such as 5.0.3, enter:

```
conda install spyder=5.0.3
```

To launch Spyder from the command line, enter:

```
spyder
```

For more information on installing Spyder, see the installation guide at *https://docs.spyder-ide.org/current/installation.html.*

Launching Spyder from the Start Menu

On most platforms, the official documentation recommends launching Spyder from Anaconda Navigator. In Windows, however, the recommended method is to launch Spyder from the Start menu (Figure 4-3).

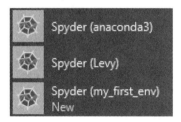

Figure 4-3: Spyder installations in the Windows Start menu under Anaconda3

There, you should see a listing of all your Spyder installations and the environments in which they're loaded, under the *Anaconda3* folder.

Configuring the Spyder Interface

Figure 4-4 shows the Spyder interface with the major panes and toolbars labeled. Note that I've changed its appearance from the "factory settings" view to facilitate this walkthrough and make it easier to see in a black-and-white book. Don't be intimidated by all the controls and panes. Spyder can be as easy or as difficult as you want to make it.

So that you can more easily follow along, let's configure your screen to look closer to that shown in Figure 4-4. First, set the syntax highlighting theme in the Preferences window by either clicking **Tools ▶ Preferences** from the top toolbar in Windows and Linux; **Python/Spyder ▶ Preferences** on macOS; or the wrench icon on the main toolbar near the top of the screen (Figure 4-4).

Find the Syntax highlighting theme menu, choose the **Spyder** option, and then click **Ok**. This sets the background to white (use **Spyder Dark** if you have sensitive eyes). Note that you have many color choices for highlighting code, just as you did with the Jupyter Qt console in the previous chapter.

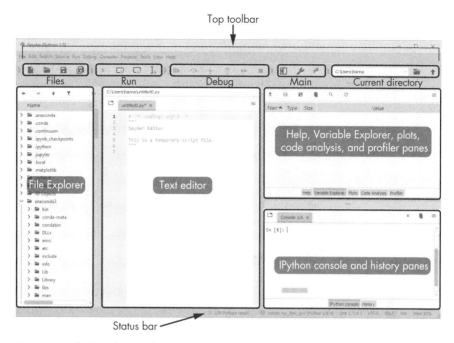

Figure 4-4: The Spyder interface with key components labeled

Now, let's move the File Explorer pane to the left side of the screen. From the toolbar at the top of the interface, click **View ▸ Unlock panes and toolbars**. This lets you drag them around just like moving windows on your desktop. In the upper-right pane, find the gray tab labeled **Files** and click it. The upper-right pane should now show a File Explorer window. Grab the top of it and drag it to the left side of the interface, as shown in Figure 4-4. You can grab the sides of the panes to resize them.

From the top toolbar, click **Run ▸ Run profiler**, followed by **Source ▸ Run code analysis**. These should automatically appear as tabs in the upper-right pane, as in Figure 4-4. Using the profiler, you can measure your code's runtime, whereas code analysis checks for style violations and potential bugs.

To save this or any layout, on the top toolbar choose **View ▸ Window layouts ▸ Save current layout** and give the layout a unique name. This becomes the default layout when you start Spyder. To choose another layout, look for it under **View ▸ Window layouts**.

As you can see, Spyder is highly configurable. You can undock panes by dragging them outside of Spyder. You can turn panes and toolbars on and off using **View ▸ Panes** and **View ▸ Toolbars**, respectively. With time, your interface will evolve and become uniquely yours. Just remember to save that window layout!

In the sections that follow, we'll look at how to use Spyder with different environments, set up Spyder projects, and use Spyder's panes and toolbars. Other good references are the Spyder home page (*https://www.spyder-ide.org/*), documentation (*http://docs.spyder-ide.org/current/index.html*), and frequently asked questions (*https://docs.spyder-ide.org/5/faq.html*).

Using Spyder with Environments and Packages

Spyder is a package like any other and must be installed in *some* conda environment. This means that you'll get an error if you try to import and use a package that's not in the same environment as Spyder. To manage this issue, let's look at the easy but resource-heavy *naive approach* and the lighter but more involved *modular approach*.

The Naive Approach

The simplest solution to using Spyder with environments is to install Spyder directly into each conda environment and run it from there, as we did in the previous installation examples. This works with all Spyder versions and should require no extra configuration after the IDE is installed. Unfortunately, it results in multiple installations to manage and isn't as flexible or configurable as other alternatives.

For example, suppose that you start a new project with a new environment in January and install the current version of Spyder into that environment. Six months later, in July, you start an additional project and load Spyder into that project's new environment. This version of Spyder might possibly be newer than the one you installed in January. At this point, your *pkgs* folder has two separate Spyder installations taking up space. If you don't need to keep older versions, one option is to run `conda update spyder` in your environments, to bring them all up to the current version, and then run `conda clean -all` to remove any versions not linked to an environment.

You may find the naive approach a suitable solution if you don't plan to use Spyder a lot, if you won't be working on a lot of projects at the same time, or if your system isn't severely resource constrained. It certainly fits the *science first, programming second* mindset. Otherwise, check out the modular approach in the next section.

The Modular Approach

Another way to work with existing environments is to install Spyder in one location and then change its default Python interpreter. The interpreter is the *python.exe* file that resides in each conda environment folder. Depending on your system, you might see it called *python.exe, pythonw.exe, python,* or *pythonw.*

With the modular approach, you install Spyder only once and put it in a dedicated environment (let's call it *spyder_env*). This way, you can update it separately from other packages and avoid conflicts. You can perform either a minimal install of Spyder or a full install that includes all of Spyder's optional dependencies for full functionality.

Let's use the command line to create the dedicated environment and perform a full install, adding packages like NumPy, pandas, and so on:

```
conda create -n spyder_env spyder numpy scipy pandas matplotlib sympy cython
```

From now on, you will start Spyder from this dedicated environment.

To allow the Spyder package in *spyder_env* to import and use packages in another environment, you must install the lightweight *spyder_kernels* package into the other environment, using either Navigator or conda. For example, we have not installed Spyder in *my_second_env*, created in Chapter 2. To use Spyder there, activate that environment and run the install like this:

```
conda activate my_second_env
conda install spyder_kernels
```

Now, you can point your Spyder application, running in *spyder_env*, to the interpreter in *my_second_env* so that it can find and use the packages installed in *my_second_env*.

To change the Python interpreter in Spyder, click the name of the current environment in the **Status** bar (see Figure 4-4) and then select **Change default environment in Preferences** (you can also use the "wrench" icon in the main toolbar). From the Preferences dialog, select **Python interpreter** and then click the radio button next to **Use the following Python interpreter** (Figure 4-5). Choose the environment from the drop-down list or use the text box (or the Select file icon to the right of the text box) to provide the path to the Python interpreter that you want to use.

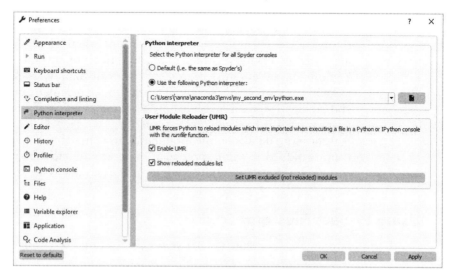

Figure 4-5: Changing the Python interpreter using the Preferences dialog

Click **OK** to change the interpreter and then click **Consoles ▸ Restart kernel** on the top toolbar for the changes to take effect. The environment name on the Status bar should change from *spyder_env* to *my_second_env* (Figure 4-6). Now, Spyder can find and import packages from the selected environment, no matter the location of the Spyder package.

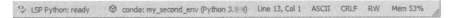

Figure 4-6: The Spyder Status bar showing the name of the source environment for the Python interpreter

Note that if you change the interpreter to an environment that does not have either the Spyder or spyder-kernels package installed, you'll get an error message in the console when you try to restart it. Likewise, if you try to start a new console, you'll get the informative message shown in Figure 4-7.

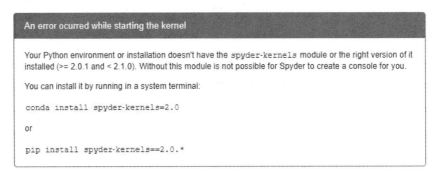

Figure 4-7: Starting a new console in an environment without the spyder-kernels package produces a useful error message.

As you can imagine, using the modular approach with multiple environments can become tedious, and you can lose track of which environment you're working in. Some of Spyder's functionality, like the Variable Explorer, might not work correctly for specific data types. And if you need to lock down a specific version of Spyder in a particular project, you could end up having to run multiple Spyder installations anyway so that other projects use the most current version.

For more details on the modular approach, see the Spyder development team's guide for working with environments and packages at *https://github .com/spyder-ide/spyder/wiki/Working-with-packages-and-environments-in-Spyder/*.

Using Project Files and Folders

Spyder lets you create special *project files* to store everything you do. These help you to stay organized and let you reload projects later to seamlessly continue your work. Projects are managed (opened, closed, created, and so on) using the **Projects** menu on Spyder's top toolbar.

Creating a Project in a New Directory

To create a new project file as a new directory, on the top toolbar, click **Projects ▸ New Project**. This opens the Create New Project dialog shown in Figure 4-8. Name the new project *my_spyder_proj*, choose a disk location, and then click **Create**.

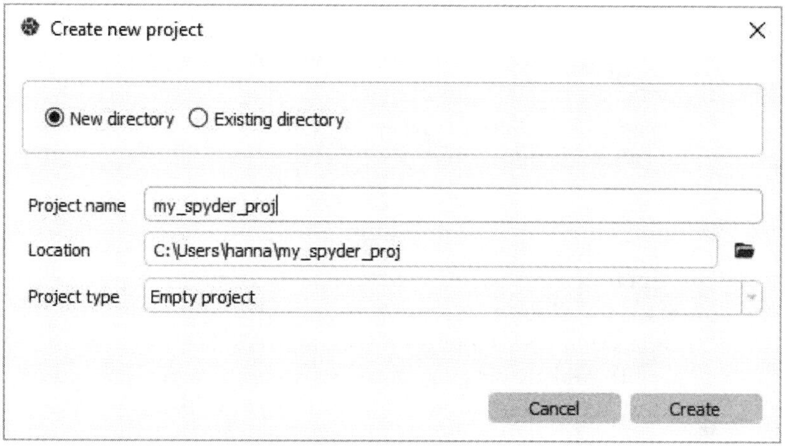

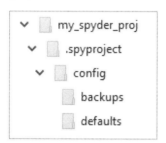

Figure 4-8: The Create New Project dialog

This creates the directory structure shown in Figure 4-9. In addition to the folders shown, Spyder will create eight files to help it manage your project.

Figure 4-9: The initial directory structure after creating a new project with Spyder

To keep your project organized, you can add additional folders to *my_spyder_proj*. Ideally, these will use standardized names that are clear and concise so that you can easily work among projects and share them with others. Let's do this now as an example. If you already have your own system, feel free to use that.

In Spyder's File Explorer pane, right-click *my_spyder_proj* and then select **New ▸ Folder** from the pop-up menu. Add the folders shown below the existing *.spyproject* folder, as shown in Figure 4-10.

Figure 4-10: The new project folders displayed in the File Explorer pane

In the naming format shown, *code* is for your Python code; *data* is for data files such as Excel spreadsheets, *.csv* files, images, and so on; *documents* is for text documents, such as reports; *output* is for things like figures and tables that your code produces; and *misc* is for everything else.

To make your project truly self-contained, I recommend including your conda environment, with its list of Python packages, in the project folder. To do this, create the Spyder project in an existing directory, as described in the next section.

Creating a Project in an Existing Directory

There are times when you'll want to create your Spyder project in an *existing* directory. A good example of this is when you want to include your conda environment in the project so that this important folder is bundled together with your other project files, allowing you to easily share or archive the project.

When stored in a Spyder project, the environment folder should be named something like *env* or *conda_env*. If you're working multiple projects, you'll want to append a project name—abbreviated if long—such as *env_PROJ_NAME*. This way, you can identify the correct Spyder installation when launching from the Windows **Start** menu. Recall that creating an environment outside of the default *pkgs* folder has some minor drawbacks, so you'll want to review "Specifying an Environment's Location" on page 37 before committing.

To include the conda environment folder in your Spyder project, we'll create both the project and environment folder using the command line. Let's name the project folder *spyder_proj_w_env* and use conda to simultaneously create both folders. In this example, I'm placing it under my *C:\Users\ hanna* folder in Windows, but you can put it anywhere you want.

NOTE *The subsequent instructions follow the naive approach, described in "Using Spyder with Environments and Packages," earlier in this chapter. If you're using the modular approach, you need only to install the spyder-kernels package in your project's conda environment. After that, start Spyder from its own dedicated environment and then change its Python interpreter to your project's conda environment.*

To begin, if Spyder is currently running, use **File ▸ Quit** on the top toolbar to exit it. Next, open Anaconda Prompt (in Windows) or a terminal (in macOS and Linux) and enter the following:

```
conda create -p C:\Users\hanna\spyder_proj_w_env\conda_env python=3.9 spyder=5.0.3
```

Remember, -p is short for --prefix, which lets you include a directory path. We've also installed Python and Spyder at the same time, specifying the version number as recommended. This represents a minimal installation of Spyder. To install all of Spyder's optional dependencies for full functionality, you can append these package names *after* Spyder in the previous command (I've omitted version numbers for brevity):

```
numpy scipy pandas matplotlib sympy cython
```

Now, activate the new environment and start Spyder by entering the following two lines, substituting the path to your environment:

```
conda activate C:\Users\hanna\spyder_proj_w_env\conda_env
spyder
```

At this point, you can create a new project by selecting **Projects ▸ New Project** from Spyder's top toolbar. Only this time, select **Existing directory**, leave the project name blank, and set the location to the path to the new project folder, *spyder_proj_w_env*, as shown in Figure 4-11.

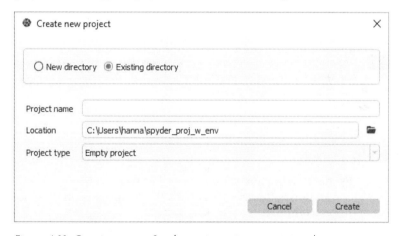

Figure 4-11: Creating a new Spyder project using an existing directory

You can now add the additional folders for code, data, and so on, as we did in the previous section. At this point, you'll have a self-contained project (Figure 4-12).

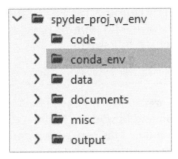

Figure 4-12: The new Spyder project with embedded conda environment (conda_env folder)

Again, you can use any file organization system you prefer, but I would strongly advise against dumping everything straight into the project folder. This will create a confusing mess, especially with large projects.

Using the Project Pane

When it comes to working with your project folders, you have several choices. Figure 4-10 was taken from Spyder's File Explorer pane. If you'd rather see just your project folders when using Spyder, open the Project pane by clicking **View ▶ Panes ▶ Project** from the top toolbar. To close the File Explorer pane, use the "hamburger" icon in the upper-right corner of the pane or use **View ▶ Panes** from the top toolbar and then deselect the pane.

 *You can also view your project folder from your operating system's file explorer while in Spyder. From either the Project or File Explorer pane, right-click the project folder and then select **Show in folder**.*

The Help Pane

Spyder's Help pane is useful whether you're a beginner or an experienced programmer. To activate it, click the **Help** tab along the bottom of the upper-right pane in Figure 4-4.

When you start Spyder for the first time, you'll see a message in the Help pane asking you to read a short introductory tutorial (Figure 4-13). I highly recommend it, but if you want to wait, you'll be able to read it later using the Help menu on the top toolbar.

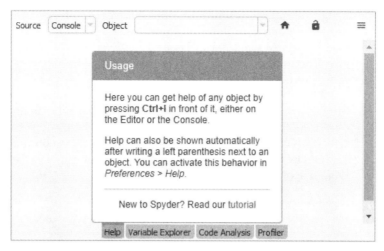

Figure 4-13: The Help pane at initial startup

In addition to the Introduction tour, the toolbar Help menu provides access to the longer Spyder tutorial that displays in the Help pane (Figure 4-14). You can also watch videos, access both the Spyder and IPython documentation, see a summary of keyboard shortcuts, and more.

Figure 4-14: The Help menu and the Spyder tutorial displayed in the Help pane

If you have the Help pane open while you code, it can find, render, and display documentation for any object with a *docstring* (descriptive text summary), including modules, classes, functions, and methods. This lets you access documentation directly from Spyder, without having to interrupt your workflow and look elsewhere.

The Source menu at the top of the Help pane lets you select between the Editor and the IPython console (Figure 4-15). Manually clicking an object, such as the print() function in Figure 4-15, and then pressing CTRL-I (CMD-I in macOS) will display information on that item. You can get help by manually entering an object's name (such as "print") in the **Object** textbox at the top of the pane.

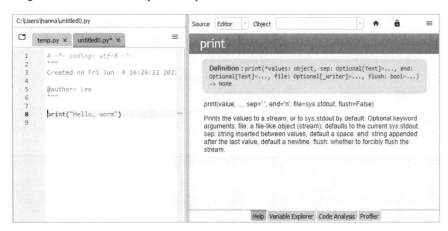

Figure 4-15: Help output for the print() function used in the Editor, invoked using CTRL-I in Windows

To enable automatic help for both the Editor and Console, first click the wrench icon on the main toolbar (see Figure 4-4), and then select **Help** and click the radio buttons for Editor and IPython console under Automatic connections. It then can be turned on and off using the "lock" icon at the top of the Help pane. When on, simply typing a left parenthesis character (() after a function or method name will show its associated help document.

You can also access summary help for objects by hovering over them in the Editor. Clicking the hover pop-up will open the full documentation in the Help pane. Just be sure that the Source menu is set to "Editor."

Finally, the "hamburger" icon at the upper right of the Help pane lets you toggle features in the display mode, such as rich or plain text, dock and undock the pane, close the window, and so on.

The IPython Console

The IPython console, located in the lower-right pane in Figure 4-4, represents a direct connection to Python that lets you run code interactively. We reviewed most of its functionality in Chapter 3, so I won't repeat that here.

With Spyder, you can open multiple consoles, restart the kernel, clear the namespace, view a history log, undock the window, and perform similar tasks. You can select some or all these options by clicking the named tab at the top of the IPython console pane, by using the "hamburger" icon in the upper-right corner of the pane, or by clicking **Consoles** on the top toolbar. You also get full GUI integration with the enhanced Spyder Debugger and the Variable Explorer, which we'll look at in later sections.

Using the Console for Output and Plotting

When you use Spyder's text editor, any text-based output will appear in the console. Likewise, any Matplotlib-based graphics will display either in the console, as you saw in Chapter 3, or in Spyder's Plots pane. The Plots pane is the default location, but you can force graphics to display *within* the console by opening the **Plots** pane, clicking the "hamburger" icon in the upper-right corner, and then deselecting **Mute inline plotting**. You can also control the graphics display from the top toolbar by choosing **Tools ▸ Preferences ▸ IPython console ▸ Graphics** and then selecting from the Graphics backend menu (Figure 4-16).

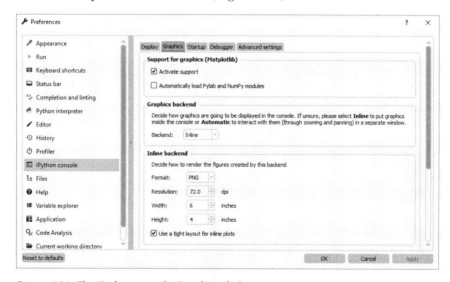

Figure 4-16: The IPython console Graphics dialog

Displaying graphics in the console is a good choice if you want to save a record of an interactive session. If you need to interact with the plot, however, such as zooming, configuring subplots, manipulating files, and saving them using different formats, you'll want to open it in a new window. You can do this by adding the magic command `%matplotlib qt` after the imports at the top of your program.

NOTE   *Some types of graphics can't be displayed within Spyder but will instead open in a browser or an external native window. These include web-based graphics and Turtle and TKinter windows.*

Using Kernels with the Console

The Python kernel is a computational engine that executes the code. You have several options for working with kernels in the console, including starting new kernels and interrupting running kernels. These are accessible from either Consoles on the top toolbar, the named console tab, or the "hamburger" icon on the IPython console pane.

You can also connect to external local and remote kernels (including those managed by Jupyter Notebook or Qt console) through the Consoles menu. To learn more, see *https://docs.spyder-ide.org/5/panes/ipythonconsole.html*.

Clearing the Namespace

The Python kernel keeps track of objects such as variables and functions that you use when coding. This collection of objects, defined in the console at any given time, is called the *namespace*. To keep the namespace from becoming cluttered, Spyder allows you to clear it whenever you want.

Let's look at an example. The left pane in Figure 4-17 is the text editor, and the right pane is the console. You can use both to write code. In the editor, I set x = 5 and then pressed F5 to run the program. Because I didn't include a print() function, nothing seems to happen, but internally, Python has assigned the x variable a value of 5.

Figure 4-17: Both the text editor (left) and console (right) share the same x value

Now I decide to stop and test a coding idea in the console. I want to use an x value of 10, but I forget to type that in. Instead, I immediately multiply x by 10 and get an output of 50 (Figure 4-17). Normally, this would raise an error, as I haven't named x yet, but because I did this earlier in the editor (in what I think is a separate program), x is already in the namespace. From my perspective, this result is unexpected.

Debugging this in a tiny snippet is easy, but imagine that you're working with longer, more complicated programs. A single occurrence of the x variable might be buried in 200 lines of code. Even with small programs, a common error is to run the program, delete something important, and then not notice the mistake as the program runs correctly, because the deleted object is resident in memory.

These persistent objects are easily forgotten and can come from numerous sources including previously executed code, interactive work in the console, or convenience imports of libraries (Spyder may do some of those convenience imports automatically). To remove these objects and clear the namespace without restarting the kernel, you can click **Remove all variables** under the Consoles menu on the top toolbar, or under the Console tab in the Console pane. You can also remove all the variables by entering the following in the console:

```
%reset
```

If you want to view the objects defined in the global namespace of a session, use:

```
dir()
```

Note that even after removing all variables, a dozen or so built-in objects will remain. The namespace will never be completely empty.

As a rule, whenever you finish coding a program, you should check that it runs independently by first removing all variables or starting a new kernel.

The History Pane

The History pane (Figure 4-4) contains a timestamped record of all the commands and code that you've run in a console. You can use this log to retrace your steps and reproduce your work. It won't show output or messages, however, and if you run a program in the Editor pane, it will show only that the file was *run*, not what commands were executed. And no matter how many consoles you have open, there'll be only one History pane. All the commands from the various consoles will be listed in the order in which they were executed, with no indication as to the source console.

You can copy commands from the History pane and paste them in both the console and the editor. Currently, only 1,000 lines of history can be shown in the pane and there's no way to clear the history. The list of commands are stored in *history.py* in the *.spyder-py3* directory in your user home folder (such as *C:/Users/<username>* on Windows, */Users/<username>* for macOS, and */home/<username>* on GNU/Linux).

Special Consoles

In addition to the IPython console, Spyder supports several *special consoles* that you can launch from either Consoles on the top toolbar or by using the "hamburger" icon on the IPython console pane. For example, the Cython console lets you use Cython (a superset of the Python language) to speed up your code and call C functions directly from Python. The SymPy console enables the creation and display of symbolic math expressions. You can also activate symbolic math usage through **Preferences ▶ IPython console ▶ Advanced Settings ▶ Use symbolic math**, assuming you have the SymPy package installed. For more on this, click **Help ▶ Spyder tutorial** on the top toolbar.

The Editor Pane

The text editor (Figure 4-4) is the heart and soul of Spyder. Whereas a console is basically a "scratch pad" designed for throw-away, interactive scripting with little to no persistence, Spyder's Editor pane lets you create programs that you can save and run (or edit) later. You can think of it as a word processor with coding-friendly features like syntax highlighting, real-time code and style analysis, on-demand completion, common keyboard shortcuts, horizontal and vertical splitting, and more.

Writing a Program Using the Editor

To take the editor for a test drive, use either the command line or Navigator to activate the *spyder_proj_w_env* environment that you made in "Creating a Project in an Existing Directory" earlier in the chapter To try out plotting in the IDE, install the NumPy and Matplotlib packages in the active environment using either Navigator or the command line. In the command line, this looks like the following:

```
conda install matplotlib numpy -y
```

The -y (short for --yes) just confirms the installation command at execution so that you don't have do it manually during the process. I show this for your convenience, but it's always safer to manually confirm installations and removals. This gives you another opportunity to confirm that the correct environment is activated and that conda isn't having to downgrade an existing package, due to some dependency.

Next, launch Spyder from the same environment, using either the Windows Start menu (be sure to pick the icon with the proper environment name), Anaconda Navigator, or the command line. You're now ready to write your first program with the editor.

To evaluate Spyder's plotting capability, let's use the "Stem Plot" example from the Matplotlib gallery (*https://matplotlib.org/stable/gallery/index.html*). Start a new file by clicking **File ▶ New file** on the top toolbar or by using CTRL-N. You'll see a new "untitled" tab appear in the Editor pane.

Delete the boilerplate text in the editor pane and type in the code that follows. Unlike the console, it's okay to press ENTER to add new lines. In script mode, your code is executed later using special commands. If you're a complete beginner, don't worry about the code details; for now, focus on how the Editor pane works:

```
"""Stem plot example from matplotlib gallery"""

import matplotlib.pyplot as plt
import numpy as np

x = np.linspace(0.1, 2 * np.pi, 41)
```

```
y = np.exp(np.sin(x))

plt.stem(x, y)

print("\nThis is a stem plot.")
```

Although it's possible to run this code now, let's save it first using **File ▸ Save as** on the top toolbar. Name the file *stem_plot.py* and save it in the *code* folder of your Spyder project.

NOTE   *The text editor supports many keyboard shortcuts. To see a list, on the top toolbar, click **Help ▸ Shortcuts summary**. To search for a specific shortcut, click **Tools ▸ Preferences ▸ Keyboard shortcuts**, also on the top toolbar.*

To execute the code, you have several choices. You can use the "play" arrow on the left side of the Run toolbar (Figure 4-4), click inside the Editor pane and press CTRL-ENTER, or press F5 (or FN-F5, depending on your keyboard) from within the Editor pane.

If this is the first time you've run a program in Spyder, you'll be asked to choose a run configuration (Figure 4-18). Choose the default selection, **Execute in current console**. We'll talk about what this means in "Setting the Run Configuration" later in the chapter.

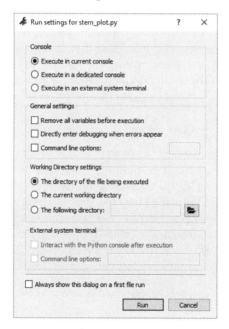

Figure 4-18: The Run settings dialog

You should now see the results shown in Figure 4-19.

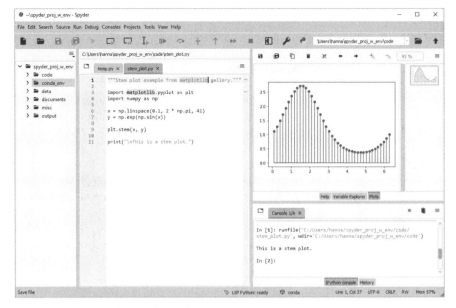

Figure 4-19: The Spyder interface after executing the stem_plot.py program

One thing to note in Figure 4-19 is that the text output, "This is a stem plot," appears in the console window. This is a cleaner outcome compared to programming in the console, where your code, along with any output—including error messages—are displayed together in the same window.

After you run a program in the editor, it's "known" to the console. This means that it remembers things like named variables and defined functions. To see an example, in the IPython console, enter the following and then press ENTER:

```
plt.stem(x, y)
```

This should regenerate the stem plot in Figure 4-19.

Behavior such as this is useful when developing and debugging complex programs and when exploring large datasets that you don't want to load more than once. It can also lead to unexpected results, as discussed earlier in "Clearing the Namespace" earlier in the chapter.

Now, let's look at another way to display a plot. In the Editor pane, add the %matplotlib qt magic command below the imports, as follows:

```
"""Stem plot example from matplotlib gallery"""

import matplotlib.pyplot as plt
import numpy as np
%matplotlib qt
--snip--
```

Save the file as *stem_plot.py* and then use CTRL-ENTER to run the program. In this case, the plot displays in an external window (check your task bar for an icon if nothing pops up). This Qt window has a toolbar with more options than are available inline (Figure 4-20).

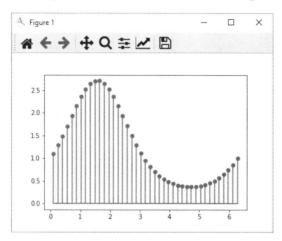

Figure 4-20: The stem plot displayed in an external Qt window

This is a good time to revisit the global namespace. Close the Qt window. Now, delete the %matplotlib qt magic command from the file and save it by pressing CTRL-S. Whether you rerun the file from the editor or use plt.stem(x, y) in the console as we did earlier, the Qt window pops up again. Even though you removed it from the file, the old command is still persistent in memory.

To restore the inline plotting, restart the kernel or run the magic command %matplotlib inline, either in the console or through the editor.

Defining Code Cells

In the previous example, you typed a complete program into the Editor pane and then ran it. You can also run it one line at a time, or run a block of connected lines, referred to as a "code cell."

To look at an example, create a new file, only this time, use the rectangular "New file" icon on the left side of the Files toolbar (Figure 4-4). Name the file *temperature_converter.py* and save it in the *code* folder.

In this example, assume that you're always having to convert temperature measurements from one scale to another and you want to put several conversion equations in one file, for convenience. You don't want to run the whole program every time, because you're normally performing just one conversion, such as Fahrenheit to Celsius, or Celsius to Kelvin. In this case, code cells are a handy solution.

Enter the following code into the new file and save it. Use #%% to separate the code into discrete cells:

```
#%% Convert Temperature: Fahrenheit to Celsius
degrees_f = 0
degrees_c = (degrees_f - 32) * 5 / 9
print(f"{degrees_f} F = {degrees_c} C")

#%% Convert Temperature: Celsius to Fahrenheit
degrees_c = -40
degrees_f = (degrees_c * 9 / 5) + 32
print(f"{degrees_c} C = {degrees_f} F")

#%% Convert Temperature Celsius to Kelvin
degrees_c = 0
degrees_k = degrees_c + 273.15
print(f"{degrees_c} C = {degrees_k} K")
```

Adding a description to the right of the separator not only documents what the cell does, it names that cell in the Outline pane. To activate this pane, go to the top toolbar and click **View ▸ Panes ▸ Outline**. Figure 4-21 shows the Editor and Outline panes together.

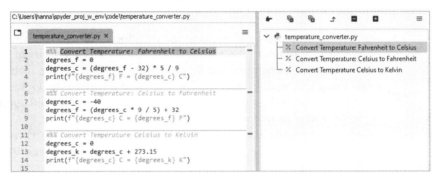

Figure 4-21: The Editor pane (left) and Outline pane (right) for the temperature_converter .py program

In the Editor pane in Figure 4-21, notice how the horizontal lines divide the script into cells starting with the #%% separators. In the Outline pane, the descriptions of the cells appear in order, from top to bottom. If you click a description, the corresponding cell in the Editor pane is highlighted and activated. You can also highlight a cell by clicking in it in the text editor.

To see the options for running cells, hover your cursor over the icons on the Run toolbar (Figure 4-4). This will also reveal the keyboard shortcuts. To run just the middle cell that converts Celsius to Fahrenheit, click in the cell, and then press CTRL-ENTER or click the **Run current cell** icon from the toolbar. You can also run a selection of code or a single line using an icon or the F9 key.

The ability to run selected cells or single lines is useful when designing and debugging programs. It's also handy for, say, changing a plot's parameters and evaluating the results without reloading all the input data. You can also use this to update part of a program without rerunning all the code, but keep in mind that the console will "remember" only the last thing that was run. If this becomes confusing, you'll want to refresh the console by restarting the kernel or removing all variables from the namespace.

Setting the Run Configuration

When you run a program in the editor for the first time, either using **Run ▸ Run** from the toolbar or by pressing F5, a dialog will open and ask you to choose the method for executing the file (see Figure 4-18). You'll have three choices:

- Execute in current console (the default)
- Execute in dedicated console
- Execute in an external system console

We'll look at the first two in more detail next, but the recommendation for beginners is to use the default option, **Execute in current console**, and then verify that completed code executes independently. This requires clearing the namespace by removing all variables or restarting the kernel prior to checking the program.

Don't worry about getting locked into a decision. You can change the run configuration at any time by selecting **Run ▸ Configuration per file** from the top toolbar.

Executing in the Current Console

When a file is executed in the current console, you can continue to interact with the console after the file runs. This lets you inspect and interact with any objects created during execution. This is a useful feature for incremental coding, testing, and debugging. As you saw in "Writing a Program Using the Editor," it lets you call commands and functions from the console without executing the file again.

This comes at a price, however. Objects can persist in the global namespace from before execution of the code (see "Clearing the Namespace" on page 76). One way to ensure that your code does not depend on existing but transitory objects in the namespace is to execute the file in a new console, as described next.

Executing in a Dedicated Console

Choosing the **Execute in a dedicated console** option means that a new IPython console is opened every time you execute code in the editor. With this option, you can be sure that there are no persistent global objects polluting the namespace, such as undefined functions, unnamed variables, or unimported packages. It's a safe option, but it provides a bit less flexibility for interacting with your code. It can also generate a lot of console tabs to manage. So, if you're aware of namespace issues, the **Executing in the current console** option is preferred.

Autocompleting Text

To save you keystrokes, both the text editor and console support the use of *autocompletion* using the TAB key. For example, enter the following long variable name in the editor:

```
supercalifragilisticexpialidocious = 'wonderful'
```

Now, slowly start typing it again and watch what happens.

When you start typing the name of an object such as a command, function, variable, class, and so on, the editor will present you with a list of objects that start with those letters (Figure 4-22). In the console, you must press TAB to display the list.

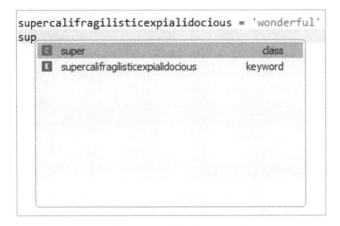

Figure 4-22: Using the Autocomplete functionality

If the name is unique, or at the top of the list, you can press the TAB or ENTER key immediately and Spyder will fill out the rest of the name. If there are multiple choices, you can either keep typing until only the name you want remains and then press TAB or ENTER; use the arrow keys to select the correct name and press TAB or ENTER; or double-click the correct name with the mouse. You can change the number of characters that you need to type to see the list of suggested completions by going to **Tools ▶ Preferences ▶ Completion and Linting** on the top toolbar.

Autocompletion is a great feature because it supports writing "Pythonic" code that's easy to read. With autocompletion, you can use highly descriptive variable and function names, like `photoshpere_temperature_in_celsius` or `step_2_apply_Gaussian_blur()` without incurring repetitive strain injuries.

The Code Analysis Pane

Python has certain guidelines for writing code that the community is expected to follow. The goal is to produce Pythonic code that others can easily pickup and understand. We examine these guidelines later in Part II. For now, know that *linters* are tools that review your code and provide feedback on where you might have violated a guideline. Spyder uses the best-in-class *Pylint* linter in its Code Analysis pane.

Code analysis will help you to improve your code by detecting style issues, bad practices, and potential bugs. You should not consider a program complete—or ready to post on an online help site—until you've run it through a linter.

Let's look at how this works. Use **Projects ▸ Open Project** from the top toolbar to open the *spyder_proj_w_env* project that you made earlier in "Creating a Project in an Existing Directory." Then, open the *stem_plot.py* file in the editor using **File ▸ Open** from the top toolbar. We made this file previously in "Writing a Program Using the Editor." Next, open the Code Analysis pane by clicking in the Editor pane and pressing F8, or by using **Source ▸ Run Code Analysis** on the top toolbar. You should get the results shown in Figure 4-23.

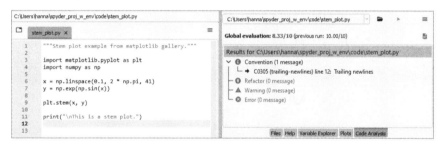

Figure 4-23: The Editor pane (left) and Code Analysis pane (right) for the stem_plot.py program

At the top of the Code Analysis pane on the right, you can see that the code was given a high evaluation score of 8.33 out of 10. The only violation was for having trailing newlines (blank lines) at the end of the program.

You can customize the code analysis by clicking **Tools ▸ Preferences** and then selecting from the **General**, **Linting**, **Code style and formatting**, and **Docstring style** tabs. There's lots to choose from, including ignoring certain errors and warnings, changing the tool used to format code, picking the convention used to lint docstrings, underlining errors and warnings, and more.

You can also suppress messages by adding specific comments to your code. For example, an expectation is that most variables in the global space represent *constants* and should be named using all caps. In short programs, you might choose to ignore this by inserting the following comment at the top of your file:

```
# pylint: disable=invalid-name
```

To find the proper message name, such as "trailing-newlines," check the results in the Code Analysis pane (see Figure 4-23).

For more on code analysis, see the "Panes in Depth" section of the Spyder documentation (*https://docs.spyder-ide.org/*). For more on the Python style guide, see *https://pep8.org/*.

The Variable Explorer Pane

The Variable Explorer pane lets you view and edit variables generated during the execution of a program in the text editor, or those entered directly in the IPython console. These are the namespace contents of the current IPython console session, and you can use the Variable Explorer to inspect, add, remove, and edit their values through a variety of GUI-based editors.

Let's try it out. First, on the top toolbar, click **Consoles ▸ Restart kernel** to start a new IPython console session. This will delete any old variables that might be persistent in memory. Now, in the upper-right pane, click the **Variable Explorer** tab or, on the top toolbar, click **View ▸ Panes ▸ Variable Explorer**.

In the IPython console, enter the following:

```
In [1]: import numpy as np

In [2]: an_array = np.random.randn(10, 5)

In [3]: a_list = ['talc', 'gypsum', 'calcite']

In [4]: a_dictionary = {'gold': 'Au', 'silver': 'Ag'}

In [5]: a_sum = 1 + 2 + 3

In [6]: a_float = 10 / 3

In [7]: a_string = "latchstring"
```

Each time you press ENTER, the Variable Explorer pane should update until it looks like Figure 4-24.

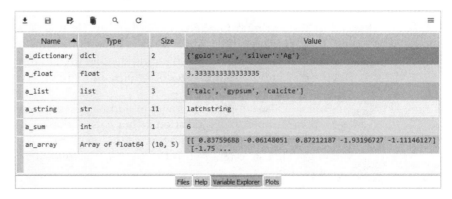

Figure 4-24: The Variable Explorer pane

The pane shows you the name of the variable; its type, such as an integer, string, dictionary, and so on; its size; and its value. Right-clicking objects in the Variable Explorer displays options to plot and analyze these further. The pane supports editing lists, strings, dictionaries, NumPy arrays, pandas DataFrames, pandas Series, Pillow images, and more, letting you plot and visualize them with one click. For example, although the 10-row-by-5-column NumPy array is too large to show in the Value column, if you double-click it, an Object Viewer window appears that lets you view the array and manipulate its contents (Figure 4-25).

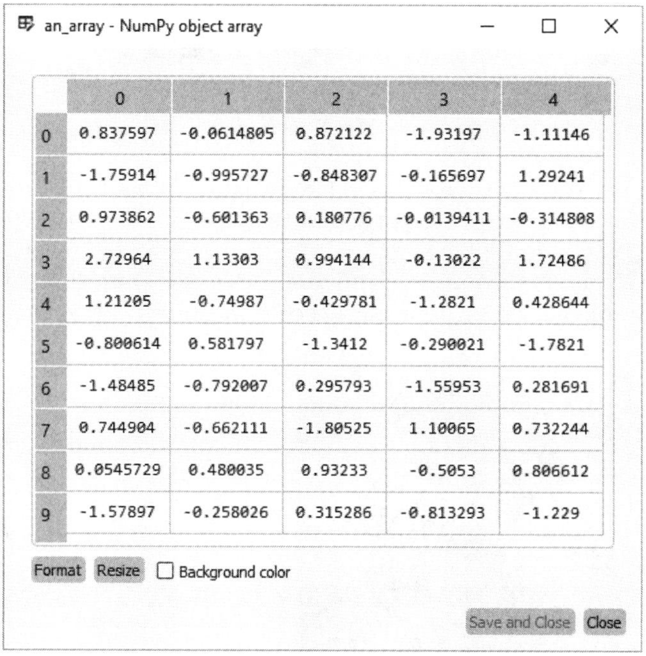

	0	1	2	3	4
0	0.837597	-0.0614805	0.872122	-1.93197	-1.11146
1	-1.75914	-0.995727	-0.848307	-0.165697	1.29241
2	0.973862	-0.601363	0.180776	-0.0139411	-0.314808
3	2.72964	1.13303	0.994144	-0.13022	1.72486
4	1.21205	-0.74987	-0.429781	-1.2821	0.428644
5	-0.800614	0.581797	-1.3412	-0.290021	-1.7821
6	-1.48485	-0.792007	0.295793	-1.55953	0.281691
7	0.744904	-0.662111	-1.80525	1.10065	0.732244
8	0.0545729	0.480035	0.93233	-0.5053	0.806612
9	-1.57897	-0.258026	0.315286	-0.813293	-1.229

Figure 4-25: Object Viewer display of the an_array object

Likewise, double-clicking anywhere within the list object's row in the Variable Explorer pane will launch an Object Viewer (Figure 4-26). By right-clicking in a row in the Object Viewer, you can perform operations such as inserting a row and adding a new item, like "fluorite."

If you use the `a_list` variable again in the current session, it will contain the new item, "fluorite." You can also use the Variable Explorer's toolbar to save the current session's data as a *.spydata* file, which you can load later to recover all the variables stored. However, be aware that changing the value of an object in an Object Viewer doesn't alter your code. If you rerun the code that generated the `a_list` variable, whether from a file or the console, it won't contain "fluorite."

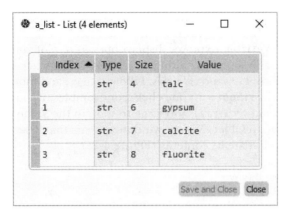

Figure 4-26: Object Viewer displaying a list object

You can filter the items in the Variable Explorer by clicking the "hamburger" icon at the upper right of the pane. If an item can be plotted, you can generate a plot of its values, appropriate to its data type, by right-clicking the object. For example, right-click the an_array object and then choose **Show image**. This will produce a color heatmap of the array (Figure 4-27).

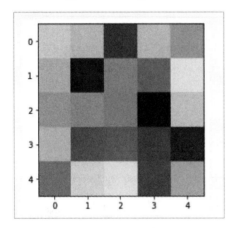

Figure 4-27: A heatmap of the an_array object

One thing that Variable Explorer won't do is let you see "local" variables defined within functions (we look at functions in Chapter 11). If you define a function using the following code, you won't be able to see the var1 and var2 variables in the pane.

```
In [1]: def a_function():
    ...:     var1 = 42
    ...:     var2 = "spam"
    ...:
```

The Variable Explorer lets you keep track of your program's global variables. It helps you to develop and test programs by permitting the inspection and editing of variables in a friendly GUI format. To read more about the options available, see the Spyder documentation at *https://docs .spyder-ide.org/5/panes/variableexplorer.html*.

The Profiler Pane

The Profiler helps you to optimize your code by measuring the runtime and number of calls for every function or method called in a file. You can use it to identify bottlenecks and quantitatively measure performance improvements after you make changes.

Let's look at an example of how this works. On the top toolbar, open a new file in the editor by clicking **File ▸ New file**. Save this file in the *code* folder of your *spyder_proj_w_env* project (or anywhere else you want) as *hoot.py*. Now, enter the following code:

```
def search_list(my_iterable):
    if 'hoot' in my_iterable:
        print("Hooty hoot!")

def search_set(my_iterable):
    if 'hoot' in my_iterable:
        print("Hooty hoot!")

my_list = [i for i in range(1000)]
my_list[998] = 'hoot'

my_set = set(my_list)

search_list(my_list)
search_set(my_set)
```

In this example, we defined two functions, search_list and search_set, that are identical in all but name. We're going to use the Profiler to prove that it's a lot faster to search for an item in a Python *set* versus a Python *list*, so we need to distinguish between the functions (we look at sets and lists in more detail in Chapter 9).

We next created a list with the numbers 0 to 999 (Python starts counting at 0, not 1) and replaced the next-to-last item (index 998) with "hoot." We then made a set from this list, named my_set. Now we call each function and pass them either the list or set, as appropriate (*pass* means that we enter the name of our list or set in the function's parentheses). When each function reaches the "hoot" item in the list or set, it immediately prints "Hooty hoot!" in the console.

Run the file by clicking in the Editor pane and pressing F5. You should see Hooty hoot! display twice in the IPython console.

To see how long each function took to run, click **Run ▸ Run profiler** from the top toolbar. This launches the Profiler pane and displays the run statistics (Figure 4-28).

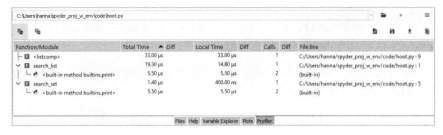

Function/Module	Total Time	Diff	Local Time	Diff	Calls	Diff	File:line
⊢ ▣ \<listcomp>	33.00 µs		33.00 µs		1		C:/Users/hanna/spyder_proj_w_env/code/hoot.py : 9
∨ ▣ search_list	19.30 µs		14.80 µs		1		C:/Users/hanna/spyder_proj_w_env/code/hoot.py : 1
┊ └ ◆ \<built-in method builtins.print>	5.50 µs		5.50 µs		2		(built-in)
∨ ▣ search_set	1.40 µs		400.00 ns		1		C:/Users/hanna/spyder_proj_w_env/code/hoot.py : 5
└ ◆ \<built-in method builtins.print>	5.50 µs		5.50 µs		2		(built-in)

Files Help Variable Explorer Plots Profiler

Figure 4-28: The Profiler pane

The Total Time column shows the time taken by the specified item and every function called by it (indented underneath it). The Local Time column counts only the time spent in a particular callable object's own scope. Based on the Local Time, the list object took 14.8 microseconds to run versus only 400 nanoseconds for the set. Because the two functions were identical except for their inputs, we can surmise that a hashable set is a better data type than a list for doing membership searches.

Note that you can select files by using the text box at the top of the Profiler pane and run them with the green "play" arrow to the right of the box (Figure 4-28). Other options include showing the program's output, saving the profiling data, loading profiling data for comparison, and clearing comparisons.

To learn more about the Profiler, including the option to measure the memory usage of your code, see *https://docs.spyder-ide.org/5/panes/profiler.html*.

The Debugger Pane

Debugging is the process of detecting and removing errors ("bugs") in code that can cause it to crash, return incorrect results, or otherwise behave unexpectedly. Python automatically produces error messages that can help you determine what part of your code is failing.

For a more sophisticated approach, Spyder integrates the enhanced ipdb debugger that's part of the Python standard library. With the debugger tool, you can walk through your code line by line checking for problems.

Going into the details of the debugger is beyond the scope of this book, and you'll probably write a lot of code without needing it. If you're curious, however, you can get a good overview at *https://docs.spyder-ide.org/5/panes/debugging.html*, and there are many online tutorials and videos for the debugger that use real-life coding examples.

Summary

Spyder is powerful enough for full-time developers, so there's a lot here we haven't covered. But despite its sophistication, it's easy for a beginner to pick up and use, and its editor and IPython console are great if you're looking only to knock off short scripts. Although much of your scientific programming will probably be performed in Jupyter Notebook, covered next, there are many coding tasks for which Spyder is more appropriate, and you'll be glad to have it in your repertoire.

If you're new to Python and want to start learning the language right now, you can skip ahead to Part II, "A Python Primer." When you finish, don't forget to circle back to Part I and check out the chapters on Jupyter Notebook and JupyterLab (Chapters 5 and 6).

5

JUPYTER NOTEBOOK: AN INTERACTIVE JOURNAL FOR COMPUTATIONAL RESEARCH

The classic Jupyter Notebook is the world's most popular tool for data science. As a savable web-based application, Notebook lets you capture the entire computational process, from loading and exploring data to developing and executing code, and even documenting and presenting the results. It's no wonder Notebook has become the default environment for code-based research.

To paraphrase James Bednar, director of custom services at Anaconda, *notebooks tell stories*. They're designed to capture and convey a code-based narrative that has a linear flow and is composed of small, human-digestible steps. They can include documentation that concisely and precisely explains what's going on. This helps scientists, researchers, developers, and students generate *reproducible* code-based research.

Like a personal science journal, a Jupyter notebook can serve as a complete record of a computational session. To make your work more understandable and repeatable, you can interleave inputs and outputs with narrative text, mathematical formulas, images, links, and more. You can also share your notebooks directly or turn them into interactive slideshows or dashboards.

In this chapter, we delve into the details of Jupyter Notebook using the classic version. In the next chapter, we'll look at the newer implementation in JupyterLab, the next-generation interface for Project Jupyter. Except for some slight rearrangement of the menus, the newer version works the same and uses the same file formats as classic Notebook. In fact, the two can be run side by side on the same computer, and JupyterLab even comes with a button for launching the classic version.

NOTE *In the pages that follow,* Jupyter Notebook *or* Notebook *(uppercase "N") refers to the* application, *whereas* Jupyter notebook *or* notebook *(lowercase "n") refers to an actual notebook file generated by the application. These files have a* .ipynb *extension, which is short for "IPython notebook."*

To supplement this chapter, you can find a quick start guide at *https:// jupyter-notebook-beginner-guide.readthedocs.io/en/latest/* and the full documentation at *https://jupyter-notebook.readthedocs.io/en/stable/notebook.html*.

Installing Jupyter Notebook

Jupyter Notebook is an open source package that comes preinstalled in Anaconda's *base* environment. However, it's not a good idea to work on projects in *base*, as that can get messy. To keep your project packages organized, safe, and sharable, they need to be in dedicated conda environments.

To use Jupyter Notebook with conda environments, you have two main options. You can install Jupyter Notebook directly in each conda environment, or you can link each environment to the Notebook installation in the *base* environment. To mimic what we did with Spyder in Chapter 4, let's call the first option the *naive* approach and the second the *modular* approach. Although the modular approach is generally recommended, if a project needs to lock down a specific version of Notebook, you'll need to use the naive approach.

The Naive Approach

The naive approach is to install Jupyter Notebook directly in each of your conda environments. Notebook can then import and use any packages installed in the same environment.

This is the simplest approach, but it can become resource intensive over time as your *pkgs* folder becomes populated with different versions of Notebook. You might also struggle to keep all of the installations up to date and might not be able to see or switch to other environments from within Notebook.

Installing and Launching Jupyter Notebook Using Anaconda Navigator

To install Jupyter Notebook in a new environment using Anaconda Navigator, first launch Navigator using the Start menu in Windows, or Launchpad in macOS, or by entering `anaconda-navigator` in a terminal in Linux. Then, in the Applications on pull-down menu near the top of the Home tab (Figure 5-1), activate the environment by selecting its name. In this example, we're using *my_first_env*, created in Chapter 2. If you skipped this step in Chapter 2, see "Creating a New Environment" on page 36.

Figure 5-1: The Anaconda Navigator Home tab showing the active environment (my_first _env) and the Notebook tile

Next, find the **Jupyter Notebook** app tile and then click the **Install** button. You might need to scroll down the Home tab to find the tile. This will install the most current version of Notebook available from the top channel in your Channels listing, located near the top of the Home tab. If you want to install a specific version of Jupyter Notebook, click the "gear" icon at the upper right of the Notebook tile to see a listing of available version numbers (see Figure 5-1).

After a few moments, the Install button should change to a Launch button. This button starts a local web server on your computer that displays the Jupyter dashboard. Because it's running locally, you don't need an active internet connection. You'll need to leave Navigator open, however, because it's running the local server for Notebook that lets you interact with your web browser.

Installing and Launching Jupyter Notebook Using the CLI

To install Jupyter Notebook in a new environment using conda, first open Anaconda Prompt (in Windows) or a terminal (in macOS and Linux) and activate the conda environment. Let's do this for *my_second_env* that we created in Chapter 2.

If you skipped this step in Chapter 2, create the environment using the following:

```
conda create --name my_second_env
```

Now, activate the environment by entering:

```
conda activate my_second_env
```

Next, use conda to install Notebook:

```
conda install notebook
```

To install a specific version, such as 6.4.1, you would use this:

```
conda install notebook=6.4.1
```

To launch Notebook, enter:

```
jupyter notebook
```

This starts a local web server on your computer that displays the Jupyter dashboard. Because it's running locally, you don't need an active internet connection. You'll need to leave your Prompt window or terminal open, however, because it's running the local server for Notebook that lets you interact with your web browser.

NOTE *There are conda packages named* notebook *and* jupyter. *The notebook package is the classic Jupyter Notebook application. The larger jupyter package bundles Jupyter Notebook, Qt console, and IPython kernel.*

The Modular Approach

The modular approach links each conda environment back to the Notebook package that was loaded in the *base* environment when you installed Anaconda. This approach is resource efficient, lets you easily keep the Notebook package up to date, and lets you see and choose among different environments from the same Notebook.

You can use the modular approach with either Navigator or the CLI. For simplicity, let's use the CLI. Open Anaconda Prompt (in Windows) or a terminal (in macOS or Linux) and enter the following to create a new environment named *my_jupe_env*:

```
conda create --name my_jupe_env
```

Enter **y** when prompted to accept the installation. Next, activate the new environment:

```
conda activate my_jupe_env
```

To link this environment with the Jupyter Notebook installation in the *base* environment, use the following:

```
conda install ipykernel
```

Because we're using the *ipykernel* package, we don't need to explicitly install Python in the environment. However, if you do need to use a specific version of Python in your project, you'll need to install it in the environment.

Now, deactivate *my_jupe_env*, which returns you to *base*, and then install the *nb_conda_kernels* package (you'll need to do this only once):

```
conda deactivate
conda install nb_conda_kernels
```

The nb_conda_kernels package enables a Jupyter instance in an environment to automatically recognize any other environment that has the ipykernel package installed. It's this combination of nb_conda_kernels in the *base* environment and ipykernel in other conda environments that allows you to use a single installation of Jupyter Notebook.

To start Notebook from *base* you'll need to enter this:

```
jupyter notebook
```

This starts a local web server on your computer that displays the Jupyter dashboard. Because it's running locally, you don't need an active internet connection. You'll need to leave your Prompt window or terminal open, however, because it's running the local server for Notebook that lets you interact with your web browser.

Your First Jupyter Notebook

To begin, let's work through an example. In this case, we'll use a notebook to summarize the eruption cycle of the famous Old Faithful geyser in Yellowstone National Park. We'll load some data, prepare it, plot it, and then add a decorative image.

If you launched Notebook in the previous sections, your browser opened a dashboard page like the one in Figure 5-2. Shut it down now using the **Quit** button in the upper-right corner of the page, and then close the browser tab. If Navigator is open, close it by selecting **File ▸ Quit**.

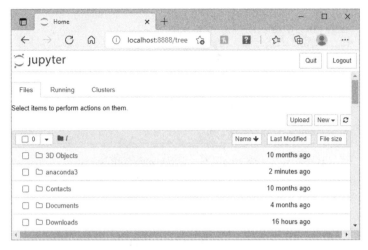

Figure 5-2: The Jupyter dashboard appears in your browser when you start Jupyter Notebook.

Going forward, we'll use the modular approach described in the previous section, so be sure to install the nb_conda_kernels package in your *base* environment if you haven't done so already. To do this with the CLI, open Anaconda Prompt (Windows) or a terminal (macOS or Linux) and then activate *base*:

```
conda activate base
```

Then enter the following:

```
conda install nb_conda_kernels
```

The notebook package ships with Anaconda, so it's already in the *base* environment.

Creating Dedicated Project Folders

Jupyter notebooks are saved to the folder from which you started the application. This means notebooks will tend to accumulate in your home or user directory. In addition, Anaconda uses dedicated folders to keep track of your installed packages and conda environments (see Chapter 2). Although Anaconda is designed to work smoothly with this structure and help you navigate it, not everyone wants their project files scattered around their directory tree. As we discussed in the previous chapter, there are multiple benefits to keeping all your project files together in a single folder.

For this project, let's store the conda environment and Jupyter notebooks in a folder named *my_nb_proj*, short for "my notebook project." I'll create this in my user directory in Windows, and I suggest that you use a similar location on your system. Although you can do this through Anaconda Navigator, the command line is more succinct, so we'll use that going forward.

To make the directories for the project, open Anaconda Prompt (in Windows) or a terminal (in macOS or Linux) and enter the following (using your own directory path):

```
mkdir C:\Users\hanna\my_nb_proj
mkdir C:\Users\hanna\my_nb_proj\notebooks
mkdir C:\Users\hanna\my_nb_proj\data
```

This makes a *my_nb_proj* directory with *notebooks* and *data* subdirectories. Next, create a conda environment named *my_nb_proj_env* in the project directory, activate it, and install some libraries (substitute your own path where needed):

```
conda create --prefix C:\Users\hanna\my_nb_proj\my_nb_proj_env
conda activate C:\Users\hanna\my_nb_proj\my_nb_proj_env
conda install ipykernel pandas seaborn
```

As described previously, the ipykernel package lets you use the Jupyter Notebook application in the *base* environment. The pandas package is Python's primary data analysis library, and seaborn is a plotting library that includes some useful datasets. (We look at these libraries in more detail later in the book.)

At this point, your project directory structure should look like Figure 5-3. Of course, with a real project, you might include additional folders for specific types of data, non-notebook scripts, miscellaneous items, and more.

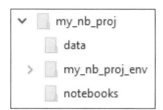

Figure 5-3: Directory structure for my_nb_proj

Jupyter Notebook likes to save to its current directory. The first time you save a file, it's easiest if you start Notebook from within that folder. Afterward, you can launch Notebook from anywhere and still access the file. To launch Notebook in your new *notebooks* folder, first activate the *base* environment (where Jupyter Notebook is installed), and then use the cd command to change directories:

```
conda activate base
cd C:\Users\hanna\my_nb_proj\notebooks
```

Because this folder is already in my user directory, I could have also used the relative path:

```
cd my_nb_proj\notebooks
```

To start Notebook, enter:

```
jupyter notebook
```

You should now see the Jupyter dashboard in your browser.

Navigating the Notebook Dashboard and User Interface

The Jupyter Notebook dashboard, also called the *Home page*, opens with an intuitive file explorer tab (see Figure 5-4). This tab displays notebook documents and other files in the directory from which you launched Notebook, known as the *current directory*. When you click a file or folder, you're presented with standard options like duplicating, renaming, deleting, and so on. The dashboard also helps you to create new notebooks, exit the application, and manage currently running Jupyter processes and clusters used for parallel processing.

Because we launched Notebook from the empty *notebooks* folder, no files or folders are visible. Let's fix that by creating a new notebook. Start by clicking the **New** button at the upper right of the Files tab to open a drop-down menu, as shown in Figure 5-4.

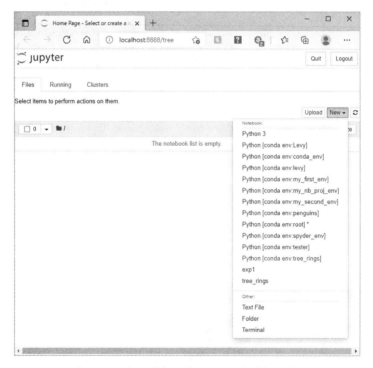

Figure 5-4: Choosing a kernel from the Jupyter dashboard New menu

The menu offers you a choice of kernels in the various conda environments you've created, including ones not in the default *envs* folder. It's able to do this thanks to the nb_conda_kernels package installed in *base* and the ipykernel package installed in each environment. At the bottom of the list, you have other choices including creating a new text file, folder, or terminal.

To activate the kernel in your *my_nb_proj* environment, select **Python [conda env:my_nb_proj_env]** from the list. This opens the notebook's user interface (UI). The notebook UI is where you interactively build your notebook document. Its primary components are the menu bar, toolbar, and cells (Figure 5-5). I encourage you to take the quick interactive tour of these components by clicking **Help ▶ User Interface Tour** in the menu.

Figure 5-5: The notebook user interface

At the right side of the menu bar, you'll see the active kernel and conda environment (*Python[conda env:my_nb_proj_env]*). If this is not the name you expect, you're using packages from a different environment, which might not contain packages you need or their correct versions.

The modular nature of Jupyter Notebook is the key to its success. It's built of blocks, called *cells*, that can contain either code or "text" (such as headers, bulleted lists, images, and hyperlinks). Code cells can be run independently or all at once, and each has its own output area. This lets you break your computational problem into pieces and organize related ideas into cells. When you get a cell (or cells) working properly, you can move on. This is convenient for interactive exploration and is especially useful for long-running processes that you need to run only once per session.

Naming a Notebook

Let's learn about the UI components and workflow by actively creating a notebook. First, give your new notebook a name by clicking **Untitled**, located just above the menu bar, entering **geyser** in the text box, and then clicking **Rename** (Figure 5-6).

Figure 5-6: Renaming a notebook

At this point, you should see a new file and folder appear in your project's *notebooks* folder. The *geyser.ipynb* file is the notebook document. This is just a plaintext JSON file saved with a *.ipynb* extension. The *.ipynb_check points* folder contains the *geyser-checkpoint.ipynb* file, which lets you restore your notebook back to a previous version.

You'll also see the notebook file appear in your dashboard (Figure 5-7). If you click the box next to its name, you'll launch a menu bar with options for working with the file, such as moving, renaming, and deleting it (the "trash can" icon).

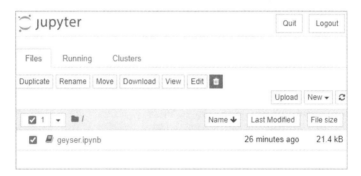

Figure 5-7: The Jupyter dashboard with the file actions menu active

To open the geyser notebook in the future, just click the filename in the dashboard. The "book" icon will turn green to indicate an actively running notebook, and you can use the drop-down menu just above the filename to filter all running notebooks. Note that you can't access notebook files that are outside (above) the root of the directory tree shown in the dashboard. The root directory is the directory from which you started Notebook.

Adding Text with a Markdown Cell

Now let's provide a descriptive header for the notebook, to let people know what it's about and to cite the source of the geyser data. Click in the first cell, labeled In [ ]: on the left side. Next, on the toolbar, change the cell type from **Code** to **Markdown**, as depicted in Figure 5-8.

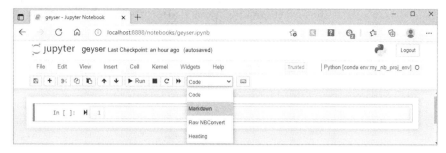

Figure 5-8: Changing the cell type using the Toolbar

Markdown (*https://daringfireball.net/projects/markdown/*), a superset of the HTML markup language, lets you add explanatory text to your notebook. You can style this text in multiple ways, including text size, bold, italics, and strike-through. You can change colors, use style sheets, make lists, and add hyperlinks. You can even drag and drop images and videos into a Markdown cell.

Some commonly used Markdown styles are listed in Table 5-1. Insert your own text for words in ALL-CAPS. To see additional styles, search for "Markdown cells" at *https://jupyter-notebook.readthedocs.io/*.

Table 5-1: Common Markdown Styles

Style syntax	Description
# YOUR TEXT	Heading size; # (largest) → ##### (smallest)
YOUR TEXT	Makes text **bold**
YOUR TEXT	Makes text *italic*
~~YOUR TEXT~~	Strikes through ~~text~~
- YOUR TEXT	Creates a bulleted list (also accepts + and *)
YOUR TEXT	Changes text to the specified color (in JupyterLab)
[Text](URL)	Inserts a hyperlink to a website URL
 	Inserts an image using a filename or URL address. You can also drag and drop an image into the cell.

NOTE *The Heading choice in the style menu has been deprecated and no longer works. Notebook will direct you to the Markdown option for creating headers.*

To make a header for your notebook, click in the cell and enter the following:

```
## Old Faithful geyser eruption dataset
### (Weisberg (2005) in *Applied Linear Regression*)
```

Be sure to include a space after the hash marks. Your cell should look like Figure 5-9.

Figure 5-9: Creating a header using a Markdown cell

To run the cell, click the **Run** button on the toolbar or, on your keyboard, press SHIFT-ENTER. Your cell should look like Figure 5-10. Notice that "Applied Linear Regression" is in italics. To go back and edit the cell again, just double-click it.

Figure 5-10: The formatted header

NOTE *SHIFT-ENTER executes a cell and advances the cursor to the next cell, creating a new cell if necessary. CTRL-ENTER executes the current cell but does not advance to the next one.*

Adding Code and Making Plots with a Code Cell

Notebook supports in-browser code editing, and it includes features found in Spyder, like automatic syntax highlighting, indentation, and tab completion/introspection. In other words, you can execute code from the browser, and see the output of the computations, including plots and images, in dedicated output cells within the notebook.

To begin coding, click in the new code cell and enter the following:

```
%matplotlib inline
import pandas as pd
```

```
import seaborn as sns

df = sns.load_dataset('geyser')  # Times are in minutes.
display(df.head())
df = df.rename(columns={'kind': 'eruption_cycle'})
```

This time, use CTRL-ENTER to run the cell. You may have noticed that this is the opposite of the Jupyter Qt console, in which you execute code by pressing ENTER and add multiple lines without execution using CTRL-ENTER.

To add another cell in a different way, from the menu bar click **Insert ▸ Insert Cell Below**. Click in this new cell and enter and run the following code to make a "violin plot":

```
sns.violinplot(x=df.eruption_cycle, y=df.duration, inner=None);
```

The semicolon at the end of the line prevents Notebook from displaying the *textural* information about the plot object. Your notebook should now look like Figure 5-11.

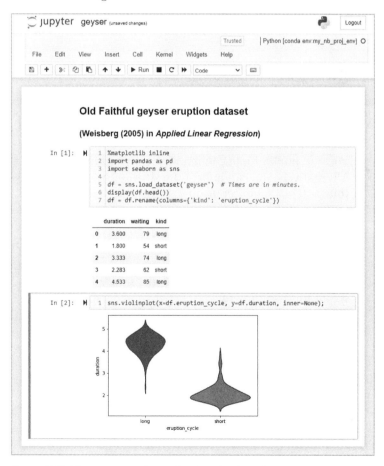

Figure 5-11: The geyser.ipynb *notebook with inline plot*

Something important just happened. In the first cell, you imported packages, loaded the seaborn "geysers" dataset as a pandas DataFrame, peeked at the first five lines of the DataFrame (`df.head()`), and then changed the name of one of the columns to something more meaningful. In the second cell, you plotted the DataFrame.

The key here is that you isolated the (potentially) time-consuming steps of data loading and preparation in their own cell. If you import packages and load the dataset in the first cell (labeled `In [1]:`), you're then free to "play" with the data in subsequent cells. There's no reason to wait for data to load with each execution. You saw a similar cellular approach in Spyder (see "Defining Code Cells" on page 81).

Another thing worth noting is that you used a magic command to make Matplotlib plot *inside* the notebook (you didn't need to explicitly import the Matplotlib library, because it's a dependency of seaborn). You can also add simple interactivity to the plot by using `%matplotlib notebook`, though this can slow down rendering. Magic commands were first introduced in Chapter 3. To see the list of magics, including *cell* magics, run `%lsmagic` in a cell. Cell magics are preceded by *two* percent signs (`%%`).

The plot itself shows that Old Faithful has a short and long eruption cycle. The longer you wait, the longer the eruption tends to last, so as a tourist, your patience is rewarded.

Working with Output Cells

By default, Notebook shows only the output of the last command in a code cell. Depending on the circumstance, you can get around this by using either the `print()` or `display()` functions. In the previous section, we used `display()` to show the head (first few rows) of the DataFrame. It also works to put multiple commands, separated by commas, on the same line. Alternatively, you can import the IPython InteractiveShell at the start of your notebook and set its interactivity option to "all":

```
from IPython.core.interactiveshell import InteractiveShell
InteractiveShell.ast_node_interactivity = "all"
```

Besides `all`, other `InteractiveShell` options include `none`, `last`, `last_expr`, and `last_expr_or_assign` (where "expr" stands for "expression" and "assign" stands for "assignment").

For more control over output cells, use the Cell menu (Figure 5-12). The Current Outputs and All Output options let you hide output, clear output, or toggle scrolling for a single cell or the entire notebook, respectively.

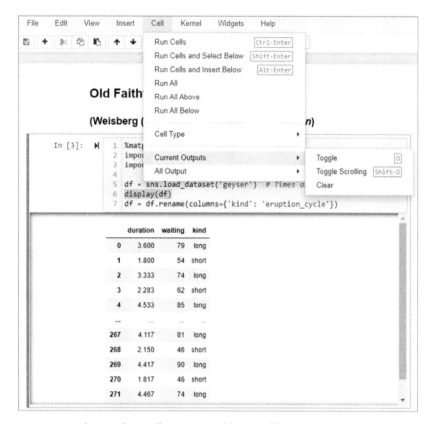

Figure 5-12: The Toggle Scrolling option adds a scrollbar to an output cell

The scrolling option is helpful if you're displaying an object that's too large for the output cell. Figure 5-12 displays the full DataFrame in the geyser notebook (using `display(df)`), and toggling on the scrollbar allows you to view the last lines of the DataFrame.

Adding an Image with a Markdown Cell

To finish off the notebook, let's add Jim Peaco's ariel view of Old Faithful, available from the National Park Service image gallery. Insert a new cell and change its type to **Markdown**. Enter and run the following code, which references the image's web address:

```
![title](https://www.nps.gov/npgallery/GetAsset/393757C9-1DD8-B71B
-0BEF06BE19C76D4D/proxy/hires)
```

Assuming that you have an active internet connection, this produces the output in Figure 5-13.

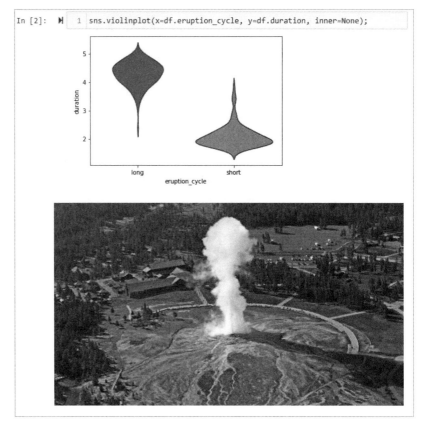

Figure 5-13: Bottom half of notebook with National Park Service image of Old Faithful

If you want control over the size of the image, use this command:

```
<img src=https://www.nps.gov/npgallery/GetAsset/393757C9-1DD8-B71B
-0BEF06BE19C76D4D/proxy/hires width="250">
```

If you don't want to worry about broken links, unreliable internet connections, or keeping track of external files, you can embed images in notebooks using **Edit ▸ Insert Image** from the menu bar, by pasting from your clipboard, or by dragging and dropping the image into the Markdown cell. Embedded images make your notebook more portable but have the disadvantage of increasing its file size and making code revisions less friendly.

Markdown cells, and notebooks in general, make it easy to include code, equations, and graphics in formatted documents. In fact, many online articles, such as those on *Medium* and *Towards Data Science*, are created using Jupyter Notebook.

Saving the Notebook

Notebook automatically saves after a set period, usually 120 seconds. You can override this by running the %autosave n magic command in a cell, where n is the number of seconds and n=0 disables autosaving. This applies only to an individual notebook and for the current session. You'll need to run the cell containing the magic command every time you open the notebook for it to take effect. For instructions on how to globally change autosave settings for all notebooks, search online for the autosavetime Jupyter extension (we'll look at using extensions later in this chapter).

To manually save your notebook at any time, use either the **Save** icon on the toolbar, the keyboard shortcut CTRL-S, or **File ▸ Save and Checkpoint** from the menu bar.

Each time you manually save your notebook, you create a *checkpoint* file in a folder named *.ipynb_checkpoints*, located in the same folder as the initial *.ipynb* file. You can reset your notebook to the checkpoint version by clicking **File ▸ Revert to Checkpoint** from the menu and then clicking on the date stamp for the last checkpoint.

Checkpoints are important because autosaving updates only the *.ipynb* file. This lets you safely work for a while without saving manually. If you find you've gone down a blind alley or made some mistake, you can always restore back to an earlier copy using the checkpoint file.

Closing the Notebook

To properly close your notebook, from the menu bar select **File ▸ Close and Halt**. Next, in the dashboard, press the **Quit** button, and then close the window.

If you are logged in to another server, as opposed to working locally, you'll want to log out using either the **Logout** button at the upper right of the notebook or at the upper right of the Jupyter dashboard.

Getting Help

The Help menu, though very intuitive, is useful enough to warrant a mention. In addition to the Notebook interface tour and documentation, it provides handy links to the documentation of many useful libraries like Python, NumPy, pandas, Matplotlib, and more (Figure 5-14).

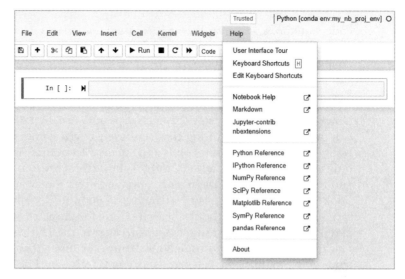

Figure 5-14: The Help menu

Keyboard Shortcuts

You can also bring up a list of keyboard shortcuts for both *command* mode and *edit* mode. You may have noticed that cell borders start out blue and then switch to green when you click inside them. *Blue* cells indicate that you're in command mode; *green* cells designate edit mode.

In command mode, the whole notebook is selected. In edit mode, the focus is on a single cell. Although there's some overlap, keyboard shortcuts for the command mode (Table 5-2) help you manipulate *cells*, while shortcuts for the edit mode (Table 5-3) help you work with *text in cells*.

Table 5-2: Selected Command Mode Keyboard Shortcuts

Shortcut	Description
H	Show all keyboard shortcuts
ENTER	Enter cell Edit mode
SHIFT-ENTER	Run cell and select cell below
CTRL-ENTER	Run selected cell
F	Find and replace
Y	Change cell mode to Code
M	Change cell mode to Markdown
1 through 6	Change cell to heading mode (1 = Largest; 6 = Smallest)
UP	Select cell above
DOWN	Select cell below
A	Insert cell above

Shortcut	Description
B	Insert cell below
X	Cut selected cell
C	Copy selected cell
V	Paste cells below
SHIFT-V	Paste cells above
D, D	Delete selected cell
Z	Undo cell deletion
SHIFT-M	Merge selected cells, or current cell with one below if only one selected
S (or CTRL-S)	Save and Checkpoint
L	Toggle line numbers
O	Toggle output of selected cells
I, I	Interrupt the kernel
SPACE	Scroll notebook down
SHIFT-SPACE	Scroll notebook up

Table 5-3: Selected Edit Mode Keyboard Shortcuts

Shortcut	Description
CTRL-M (or ESC)	Enter Command mode
UP	Move cursor up
DOWN	Move cursor down
CTRL-UP	Go to cell start
CTRL-DOWN	Go to cell end
CTRL-LEFT	Move one word left
CTRL-RIGHT	Move one word right
CTRL-]	Indent
CTRL-[	Dedent
CTRL-/	Toggle comment
CTRL-D	Delete whole line
CTRL-A	Select all
CTRL-Z	Undo
CTRL-Y	Redo
CTRL-BACKSPACE	Delete word before
CTRL-DELETE	Delete word after
SHIFT-ENTER	Run cell and select below

(continued)

Table 5-3: Selected Edit Mode Keyboard Shortcuts *(continued)*

Shortcut	Description
CTRL-ENTER	Run selected cells
CTRL-SHIFT-hyphen	Split cell at cursor
INSERT	Toggle overwrite flag
CTRL-S	Save and Checkpoint

To see the complete list of available shortcuts, click **Help ▶ Keyboard Shortcuts** or, on your keyboard, press H while in command mode. If these shortcuts aren't enough, you can customize the command mode shortcuts from within the Notebook application itself, using the **Edit keyboard Shortcuts** item. A dialog will guide you through the process of adding custom keyboard shortcuts. Afterward, the keyboard shortcut set from within Notebook will be saved to your configuration file.

The Command Palette

You can hover your cursor over items in the toolbar (Figure 5-5) to reveal their purpose. These are straightforward except perhaps for the Command Palette icon, shaped like a keyboard.

In Jupyter Notebook and JupyterLab, all user actions are processed through a centralized command system. These include the menu bar, context menus, keyboard shortcuts, and so on. For your convenience, the command palette provides a keyboard-driven way to search for and run these commands (Figure 5-15).

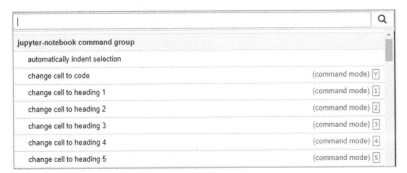

Figure 5-15: A portion of the Command Palette

You can also open the command palette using P in command mode and CTRL-SHIFT-P in edit mode. To exit the command palette, press the ESC key.

Using Notebook Extensions

You can expand the functionality of the Notebook environment by using extensions written in JavaScript. These modules, referred to as *nbextensions*, are basically add-ons or plug-ins that do things like autocomplete code, hide coding cells, spellcheck Markdown cells, create a table of contents, open a "scratchpad" cell for isolated experimentation, and more. You can also write your own custom extensions. To see the complete list of available extensions, visit *https://jupyter-contrib-nbextensions.readthedocs.io/en/latest/nbextensions.html*.

NOTE   *Classic Notebook extensions won't work in the JupyterLab version, which has its own set of extensions. You can read about these in the next chapter.*

Installing Extensions

The *jupyter_contrib_nbextensions* package is a collection of community-contributed nbextensions. To load these extensions locally in your browser, you need to install it in your *base* environment (if you're using the modular approach) or your project environment (if using the naive approach). For example, to install in *base* using the CLI, first activate the environment with this command:

```
conda activate base
```

Then, enter the following:

```
conda install -c conda-forge jupyter_contrib_nbextensions
```

Finally, install the JavaScript and CSS files to a location where Notebook can find them:

```
jupyter contrib nbextension install --user
```

CSS (short for *Cascading Style Sheets*) describe how the HTML elements in notebooks are displayed. The --user flag installs into the user's home Jupyter directories. Alternatively, using the --system flag will install into system-wide Jupyter directories.

After you've confirmed installation, restart the notebook server.

Enabling Extensions

You should now have an Nbextensions tab on the Jupyter Home page with a list of selectable nbextensions, as illustrated in Figure 5-16.

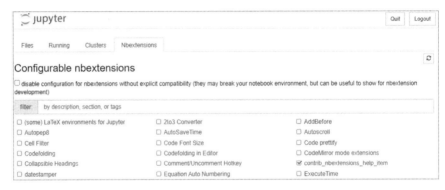

Figure 5-16: The new Nbextensions tab (shown truncated) on the Jupyter dashboard

Clicking an extension name launches its README file. For example, if you click the **Tree Filter** nbextension and scroll down, you'll see a description of what it does and a demonstration of how to use it (Figure 5-17).

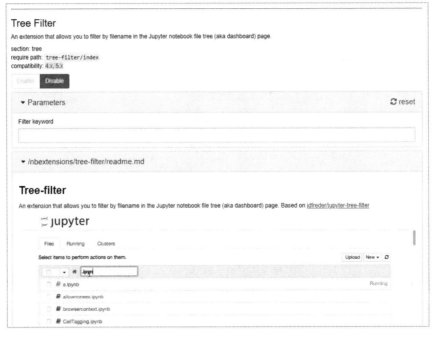

Figure 5-17: Clicking an nbextension name launches its descriptive README file.

Clicking the checkbox beside an extension turns on that extension. You can also enable and disable nbextensions from the CLI (where *<extension _name>* represents the name of the extension):

```
jupyter nbextension enable <extension_name>
```

and:

```
jupyter nbextension disable <extension_name>
```

To learn more about the jupyter_contrib_nbextension package, visit *https://jupyter-contrib-nbextensions.readthedocs.io/.* To find the latest extensions, search online for "useful Jupyter Notebook extensions."

Working with Widgets

Widgets, short for "Windows Gadgets," are interactive objects such as sliders, radio buttons, drop-down menus, checkboxes, and the like. Widgets let you build a GUI for your notebook, making it easier to explore data, set up simulations, accept user input, and so on.

In this section, we'll use the *ipywidgets* extension to create widgets. Some examples are shown in Figure 5-18. For a full list of widgets, along with their configurable parameters, visit the documentation at *https://ipywidgets.readthedocs.io/en/stable/examples/Widget%20List.html.*

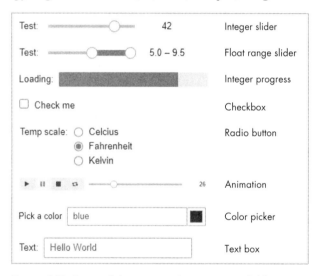

Figure 5-18: Some of the many widget types available in ipywidgets

We won't cover every type of widget here, but we'll cover enough for you to feel confident exploring on your own

Installing ipywidgets

The ipywidgets extension can be installed like any package. The instructions that follow will use the CLI, but you can easily duplicate them using the Navigator GUI.

If you're using the naive approach, wherein the notebook package is installed in your conda environment, open Anaconda Prompt (Windows) or a terminal (macOS and Linux), activate the target environment, and then enter the following:

```
conda install -c conda-forge ipywidgets
```

If you're using the modular approach, wherein your Jupyter Notebook and the IPython kernel are installed in different environments, you'll need to also install the *widgetsnbextension* package in the environment containing the Jupyter Notebook server. The widgetsnbextension package configures the classic Jupyter Notebook to display and use widgets. Let's do this now for the *base* and *my_nb_proj_env* environments (you'll need to substitute your path to *my_nb_proj_env*):

```
conda install -n base -c conda-forge widgetsnbextension
conda install -p C:\Users\hanna\my_nb_proj\my_nb_proj_env -c conda-forge ipywidgets
```

With the ipywidget package installed, you can easily create widgets either manually or by using the interact or interactive classes.

Creating Widgets with Interact

The ipywidgets.interact class helps you generate widgets for exploring and interacting with data. Let's try it out in a new notebook.

Open Anaconda Prompt (in Windows) or a terminal (in macOS or Linux). You should be in the *base* environment (if not, enter **conda activate base**). Because we're saving a new notebook, navigate to your *my_nb_proj\ notebooks* directory before launching Jupyter Notebook:

```
cd C:\Users\hanna\my_nb_proj\notebooks
jupyter notebook
```

From the Jupyter Dashboard, select **New ▶ Python [conda env:my_nb _proj_env]**. When the untitled notebook appears in your browser, rename it *widgets* and save it.

In the first cell, enter and run the following code:

```
import numpy as np
import matplotlib.pyplot as plt
❶ from ipywidgets import interact

x = np.linspace(0, 6)

def sine_wave(w=1.0):
    plt.plot(x, np.sin(w * x))
    plt.show()

❷ interact(sine_wave);
```

With this code, you import NumPy, Matplotlib, and the interact class from ipywidgets ❶. You then use NumPy's linspace() method to return an array of evenly spaced numbers over the specified interval (0-6), and assign it to the x variable (we look at NumPy in detail in Chapter 18). Next, you define a short function that multiplies the x values by the sine of x times a scalar named w and plots the results. Finally, you call interact() and pass it the sine_wave() function ❷. This produces the slider widget shown in Figure 5-19. Sliding the control nob redefines the value of w and automatically calls the sine_wave() function to update the plot.

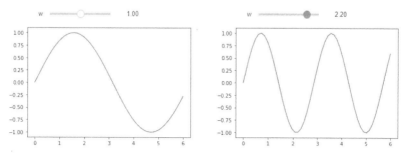

Figure 5-19: Changing the slider interactively updates the sine wave plot.

Note that you didn't need to specify a slider widget. Ipywidgets detected that we passed the sine_wave() function a floating-point value (w=1.0) and knew to use a float slider. Had we passed an integer, it would've made an integer slider.

You can also use interact() as a Python *decorator*. We covered decorators in Chapter 13; these are functions used to enhance the behavior of another function. To use interact() as a decorator, insert a new cell at the bottom of your *widgets* notebook and run the following code:

❶
```
@interact(w=1.0)
def sine_wave(w):
    plt.plot(x, np.sin(w * x))
    plt.show()
```

You must run the proceeding cell for this to work, as we aren't reimporting the libraries or reassigning the x value. When you run the current cell, the decorator ❶ calls the sine wave function for you.

In these examples, interact() tries to update the plot as you move the slider, sometimes introducing latency in the display. To prevent interact() from immediately updating, you can instead import the interact_manual() method from ipywidgets and use it to call the sine_wave() function. In this case, the plot won't update until you stop moving the slider and press the **Run Interact** button (Figure 5-20).

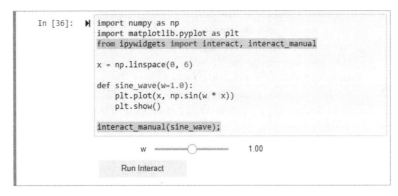

```
In [36]:  ▶  import numpy as np
             import matplotlib.pyplot as plt
             from ipywidgets import interact, interact_manual

             x = np.linspace(0, 6)

             def sine_wave(w=1.0):
                 plt.plot(x, np.sin(w * x))
                 plt.show()

             interact_manual(sine_wave);
```

Figure 5-20: The interact_manual() method produces a button for manually running interact.

As you've seen, interact determines the type of widget to produce based on the input. If you pass it a Boolean, such as x=True, it will produce a checkbox. A string, like x='Hello, World!', produces a textbox. Passing a list or dictionary will generate a drop-down menu. For example, insert a new cell at the bottom of your *widgets* notebook and run the following code:

```
def languages(descriptor):
    return descriptor

options = {'The King': 'Python', 'Not bad': 'Julia', 'Up and Coming': 'Go'}
interact(languages, descriptor=options);
```

You should get the output shown in Figure 5-21.

Figure 5-21: An interact-generated drop-down menu

The interact class abstracts away a lot of decisions, so it's easy to use. For more control, you'll want to try the interactive class or manually generate the widgets.

Creating Widgets with Interactive

The ipywidgets.interactive class gives you access to information that is bound to the widget, such as its keyword arguments and result. Unlike with interact(), you'll need to explicitly show the widget on the screen using the display() method.

Let's look at an example. Insert a new cell at the bottom of your widgets notebook and enter the following code:

```
from ipywidgets import interactive

def my_function(x):
    return x

widget = interactive(my_function, x=5)
display(widget)
```

Run the cell and move the slider to a value of 8. Now, insert a cell below and run this code:

```
print(widget.result)
```

The output should be 8. This lets you use the widget's result in subsequent code rather than just viewing the result.

Manually Creating Widgets

The interact and interactive classes make creating widgets almost automatic. But if you want more control over the process, you can create them "manually" by specifying which widget you want. You'll be able to define the layout and style, name the widgets, link them together, get events, and more.

Let's work an example. Start by inserting a new cell at the bottom of your *widgets.ipynb* notebook, and then run this code:

```
import ipywidgets as widgets

slider = widgets.IntSlider(value=0,
                           min=0,
                           max=20,
                           step=2,
                           description='A Slider',
                           orientation='horizontal')
display(slider)
```

This produces the integer slider bar in Figure 5-22.

Figure 5-22: A named integer slider bar

By building the widget directly, you're able to specify additional parameters like the displayed name (description) and orientation (horizontal or vertical). To see all the available parameters, add display(slider.keys) to the current cell or, in a new cell, run slider.keys. You can find example use cases at the documentation link cited earlier.

A slider bar by itself isn't much use, but as with the `interactive` class, you have access to the slider value, in this case through the `.value` attribute. In a new cell, run this code:

```
print(f"Slider value = {slider.value}")
```

This should produce the following output:

```
Slider value = 0
```

Handling Events

A user interacting with a widget creates an *event*. For example, a *click* event occurs when you press a button widget. When you handle an event, you tell your program what to do with the results. This usually involves writing an "event handler" function.

To capture output and ensure that it's displayed, you must send it to an `Output` widget or put the information you want to display into an `HTML` widget. Let's look at an `Output` example.

Insert a new cell at the bottom of your widgets notebook and then run the following:

```
  import ipywidgets as widgets
❶ from IPython.display import clear_output

  button1 = widgets.Button(description='Python')
  button2 = widgets.Button(description='Go')
  button3 = widgets.Button(description='Rust')
❷ output = widgets.Output()

  print("Pick your favorite language:")
  display(button1, button2, button3, output)

❸ def event_handler(button):
      with output:
          clear_output()
          print("Your favorite language is {}".format(button.description))

❹ button1.on_click(event_handler)
  button2.on_click(event_handler)
  button3.on_click(event_handler)
```

This produces the output in Figure 5-23. Clicking a button prints out the button's description (name).

Figure 5-23: Handling a button click event

Each time you click a button, the output will hang around in the output cell, so import the clear_output() method from IPython ❶. This method will let you start fresh each time a button is clicked. Next, make three button widgets and an output widget to display the results ❷.

To handle the button click event, define a function called event_handler() that takes a button object as an argument ❸. Using the output widget, first clear the display to remove the output of any previous button clicks, and then print the name of the clicked button. Finally, for each button, use the Button widget's on_click() method and pass it the event handler function ❹. This binds the function to the button click event.

Customizing Widgets

The widgets provided by ipywidgets are attractive out of the box, but you can modify them if you want. The widget layout attribute gives you control over things like the widget size, borders, alignment, and position. You can also arrange widgets in gridded patterns.

In the events-handling code example from the previous section, add the following line above the button1 variable assignment and change the button1 code, as indicated here:

```
layout = widgets.Layout(width='300px', height='50px', border='solid 2px')
button1 = widgets.Button(description='Python', layout=layout)
```

Run the cell and you should get the output shown in Figure 5-24. As you might have guessed, px stands for pixel.

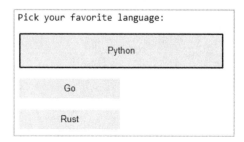

Figure 5-24: The new layout for the Python button (button1)

Conveniently, many widgets let you use predefined styles. In the previous example, change the button1 assignment as follows and then run the cell:

```
button1 = widgets.Button(description='Python', button_style='danger')
```

The Python button should turn red. Other predefined button style choices are primary (blue), success (green), info (teal), and warning (orange).

If you need even more control, the style attribute exposes non-layout-related styling attributes of widgets. The properties of this attribute are widget specific; you can list them by using the keys property. For example, for button1 in the previous example, you would run button1.style.keys.

Suppose that you want the Python button to be pink, a color not available in the predefined styles. In this case, you'd first change button1 back to its original state and then set its background color using the style attribute:

```
button1 = widgets.Button(description='Python')
button1.style.button_color = 'pink'
```

These examples are just a small taste of what you can do. To see more options, visit the documentation at *https://ipywidgets.readthedocs.io/en/latest/examples/Widget%20Styling.html*.

Embedding Widgets in Other Formats

The Notebook menu bar provides a Widgets option for embedding interactive widgets into static web pages, Sphinx documentation (the familiar "Read the Docs" web pages), and HTML-converted notebooks on the *nbviewer* web app. Following are the menu items:

Save Notebook Widget State   Saves with the current widget state as metadata, allowing the notebook file to be rendered with rendered widgets.

Clear Notebook Widget State   Deletes the saved state (you'll need to restart the kernel).

Download Widget State   Triggers the downloading of a JSON file containing the serialized state of all the widget models currently in use.

Embed Widgets   Provides a dialog containing an HTML page, which embeds the current widgets. To support custom widgets, it uses the RequireJS embedder.

To learn more about embedding, visit *https://ipywidgets.readthedocs.io/en/latest/embedding.html#*.

Sharing Notebooks

Scientific work is rarely done in isolation. You'll need a way to share your notebooks. In some cases, you'll want to share an *executable* version; for example, for coworkers who will run and modify the notebook (think coders). In other cases, you'll want to share a static copy of an *executed* notebook that contains all the generated plots and outputs (think non-coders). This latter group might include stakeholders who don't want to install Notebook, deal with its data or package dependencies, or wait for long-running notebooks to complete.

Notebooks are saved in JSON format and need to be rendered to be readable. In the following sections, we'll talk about some of the methods for downloading and sharing notebooks.

Checking and Running Notebooks with the Kernel Menu

A problem with notebooks is that cells can be run out of order, they can be deleted, and there's no guarantee that the correct execution order is easily repeatable. And as we saw in previous chapters on the Jupyter Qt console and Spyder, imports and variable assignments that are resident in memory can cause confusion and unintended consequences.

Consequently, before sharing your work, it is strongly recommended that you click **Restart & Run All** from the **Kernel** menu (Figure 5-25). If errors occur, fix the first one, repeat the command, and move to the next.

Figure 5-25: The Notebook Kernel menu

Another useful menu item is Interrupt. This is handy for long-running notebooks, in the event that you forgot to change a parameter or you recognize some error in the code or input.

Downloading Notebooks

Notebooks are automatically saved as interactive **.ipynb* files. You can email these directly to a colleague who uses Notebook. Alternatively, the **File ▸ Download as** command lets you save your notebook in many different formats (Figure 5-26). Some of these formats, like PDF via LaTeX, require certain packages be installed (if they're not, don't worry, you'll get an error message informing you of what's needed). Among the more important formats are HTML and Python.

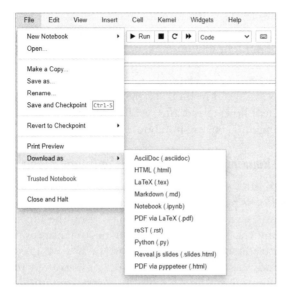

Figure 5-26: The Notebook File menu

The Python option saves your notebook as a text file with a *.py* extension in your downloads folder. You can then run this file as a Python script in a console or an IDE like Spyder. Markdown cells, cell numbers, and other non-code material is commented out by using the # symbol.

For example, if you download the geyser notebook from the previous section using Python (*.py*), you'll get the following script:

```python
#!/usr/bin/env python
# coding: utf-8

#  ## Old Faithful geyser eruption dataset
#  ### (Weisberg (2005) in *Applied Linear Regression*)

# In[1]:
get_ipython().run_line_magic('matplotlib', 'inline')
import pandas as pd
import seaborn as sns

df = sns.load_dataset('geyser')  # Times are in minutes.
display(df.head())
df = df.rename(columns={'kind': 'eruption_cycle'})

# In[2]:
sns.violinplot(x=df.eruption_cycle, y=df.duration, inner=None);

# ![title](https://www.nps.gov/npgallery/GetAsset/393757C9-1DD8-B71B-0BEF06BE19C76D4D/proxy/
hires)
```

If you open this *geyser.py* file in Spyder and run it, you'll see the tabular DataFrame output and the violin plot, but not the header or the image of Old Faithful.

You can also export your notebook from the command line using the *nbconvert* tool. This tool already powers the **Download as** menu option, but when used in the CLI (invoked as `jupyter nbconvert`), you can conveniently convert a batch of notebook files to another format with a single command. To learn more, visit *https://nbconvert.readthedocs.io/*.

After you've downloaded your notebook in the appropriate format, you'll still need to share it. Email is one option, but for collaborative work, you'll need to include any external data files that you used. And if your notebook uses third-party packages, you'll want to share an environment or requirements file (see Chapter 2) so that those with whom you're sharing can set up an identical environment. In the following sections, we'll look at some convenient ways to share notebooks via third-party websites.

Sharing Notebooks via GitHub and Gist

An easy and flexible way to deploy notebooks is to put them in a *code repository*. These sites store source code archives, provide version control to track changes, and have both public and private components. Although there are many free hosts to choose from, the most popular is *GitHub*.

GitHub, Inc., a subsidiary of Microsoft, is a provider of internet hosting for software development and version control using the *Git* program. Git lets you store the notebooks you want to share in a folder on your computer, which you can think of as a *local* repository. To make this folder function as a repository, Git also stores snapshots (records of the state of versions at a specific point in time) and metadata in a hidden folder named *.git*. This lets it keep track of contents and changes to the files. You can also include supporting data files and folders.

The GitHub website lets you host clones of these Git repositories online for the purpose of sharing, performing collaborative work, and providing a safe backup. You can include a *README.md* file to describe what's in the repository. Other users can download your notebooks to run and edit them. They can upload their changes using Git's version control system, which ensures your original work isn't overwritten.

To see an example repository, follow this link: *https://github.com/rlvaugh/ Impractical_Python_Projects/*. Be sure to scroll down to see the README file.

You can run Git from the command line, but if you're new to the process, or just want to share notebooks, I recommend using the *GitHub Desktop* GUI. The Desktop website (*https://docs.github.com/en/desktop/*) will walk you through the steps for setting up a free account and creating your first repository. In addition, a quick online search will reveal many excellent tutorials for using GitHub.

Alternatively, if you want a fast, easy, and lightweight option for sharing a notebook, you can use *GitHub Gist*. Gist is basically a tool for sharing text, and because notebooks are saved in JSON format, they qualify. Gist is a simple solution for when you don't need a big repository, yet you still get Git's version control system. In fact, a *gist* is a Git repository, with full commit history, differences (diffs), fork and clone options, and more.

When you create a gist, you have the option to add multiple files but with limits. For example, to add an Excel spreadsheet, you'll need to save it as a comma-separated values (CSV) file. Likewise, you can't add image files. Nor can you add directories. So, if your project is data heavy, you'll probably want to create a full GitHub repository using GitHub Desktop or Git with the command line.

Given that our *geyser.ipynb* notebook is simple and stand-alone, let's add it to Gist. First, go to the website at *https://gist.github.com/*. If you already have a GitHub account, click **Sign in** on the right side of the Gist banner (Figure 5-27). Otherwise, click **Sign up** to create a free account.

Figure 5-27: The GitHub Gist startup banner

After you sign in, click the plus sign (+) on the banner (Figure 5-28) to create a gist.

Figure 5-28: The GitHub Gist banner after sign-in

In the next window, you'll see a large blank area for adding text (Figure 5-29). You'll also be prompted to add a filename with extension. Enter **geyser.ipynb** and then, in the **Gist description** box, enter **Old Faithful eruption notebook**.

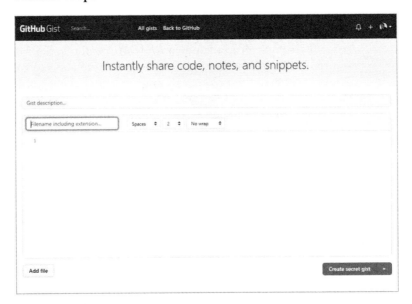

Figure 5-29: The Gist creation page

Now for the fun part. In the file explorer for your operating system, navigate to the *geyser.ipynb* file that we built earlier and drag and drop it into the blank text box area in the Gist creation page. You should see the JSON text file for your notebook (Figure 5-30).

Figure 5-30: The results of dragging and dropping the notebook into Gist

Finish by clicking the *arrow* on the green button in the lower right to see the save options. You can create a private (secret) or public gist (Figure 5-31). With the secret option, only people who know your URL can see the contents. Let's keep this between us, so click **Create secret gist**.

Figure 5-31: The options for creating a gist

NOTE   *If someone guesses or accidentally discovers the URL to a secret gist, they'll be able to see it. For better security, you'll need to use GitHub Desktop or Git to create a private repository (see "Creating Your First Repository Using GitHub Desktop" at https:// docs.github.com/).*

After a few seconds, you should see your notebook fully rendered as a static HTML file. Output such as the violin plot will show up only if the notebook was executed prior to saving.

If you scroll down, you'll see a box for adding comments. If you scroll to the top, you'll see icons for actions such as deleting or editing files, as illustrated in Figure 5-32. If you click Edit, the notebook will revert to the JSON format (Figure 5-30). Although it's possible to edit this text and change the notebook, I doubt you'll want to. Let's look at some other options.

Figure 5-32: Options for working with the gist

By clicking the **Download ZIP** button, users can download your gist as a folder on their machine, where they can edit and run the notebook. The Embed pull-down menu provides options for embedding the gist in a website (such as a blog post), copying a sharable link, or cloning the gist. The embed option works in any text field that supports JavaScript. To the left of the **Download ZIP** button, there's an icon for saving the gist to your computer and using it in GitHub Desktop.

NOTE *If your main reason for sharing notebooks on GitHub is to work collaboratively on the notebook's content, you should clear the output from your notebook before adding it to the repository. This will make it easier to track changes to the code. To learn more, visit* https://mg.readthedocs.io/git-jupyter.html.

To see the full documentation for Gist, visit *https://docs.github.com/en/ github/writing-on-github/editing-and-sharing-content-with-gists/creating-gists/.* There's also a notebook extension, called *gist-it*, for creating gists (*https:// jupyter-contrib-nbextensions.readthedocs.io/en/latest/nbextensions/gist_it/readme .html*).

With your notebooks in a repository, you'll have additional options for distributing them. Let's look at some of the most popular.

Sharing Notebooks via Jupyter Notebook Viewer

Jupyter nbviewer, or *Notebook Viewer,* is a free service for rendering GitHub-hosted notebooks online. It's useful for when the GitHub rendering engine has difficulty, such as with large notebooks, using a mobile device, or using some JavaScript-based libraries. Colleagues and stakeholders can use nbviewer to view inputs and outputs, but to execute code, they must download the notebook to a local Jupyter installation.

To use nbviewer, you simply launch the website (*https://nbviewer.jupyter .org/*) and paste the notebook's URL into a text box (Figure 5-33). This renders the notebook as a static HTML web page and gives you a stable link to that page that you can share with others. This link will remain active so long as the notebook location in the GitHub repository doesn't change.

Figure 5-33: The nbviewer web application

The application also supports browsing collections of notebooks and rendering notebooks in other formats, such as slides and scripts. To share multiple notebooks, first place them all in the same repository. Then, point nbviewer to the repository's address, and it will automatically create a navigable index for users.

You can test nbviewer using the gist we made in the previous section. Just copy the link using either the Embed menu or the clipboard icon (see Figure 5-32) and then paste it into nbviewer.

Sharing Notebooks via Binder

Binder (*https://mybinder.org*) is a free website designed for use with *public* repositories such as on GitHub. Binder lets you run notebooks stored in these static repositories by building a *Docker image* of the repository. A Docker image is a combination of a filesystem and parameters (see *https:// www.docker.com/*). When you share your notebook, via a URL, Binder provides both your code and all the software needed to run it. The user doesn't need to download or install anything.

Binder's base environment is barebones. If your project uses any third-party packages, such as Matplotlib or NumPy, your GitHub repository should include either an *environment.yml* or a *requirements.txt* file. These files list your project's package requirements (see the "Duplicating and Sharing Environments" on page 44). Binder will read the file and include any packages in the Docker image. It will update this image if you commit new changes to GitHub.

After the image is built, you can use the Binder URL to share your notebook. Binder uses a free *JupyterHub* (*https://jupyter.org/hub*) server to host your repository. JupyterHub is an open source service that allows institutions to share notebooks across large pools of users. With the public IP address that you provide, users can interact with your code and environment within a live JupyterHub instance.

Figure 5-34 shows the Binder start-up screen. I highly recommend viewing the "Zero-to-Binder" beginner tutorial by clicking the Python link visible at the top of this figure. Additional guidance (not shown) is included at the bottom of the home page.

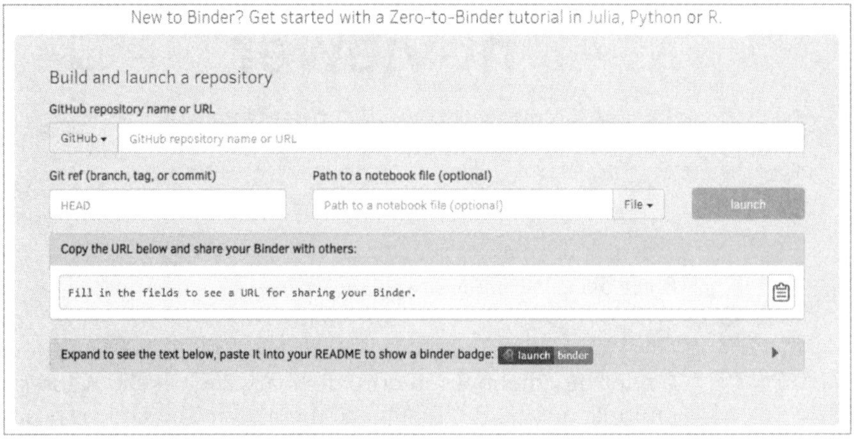

Figure 5-34: The Binder online form for sharing interactive notebooks

Users can execute your notebooks, so you'll need to provide any data dependencies. If these data files require 10MB or less of memory, the simplest solution is to add them directly into your GitHub repository. Remember, this must be a public repository for Binder to access it, so you don't want to include any sensitive information. And you need to keep in mind Binder downloads data only when the Docker image is built, not when the Binder link is clicked. Images are rebuilt only when there is a new commit to the repository.

For data sizes between 10MB and a few hundred megabytes, you need to add a file called *postBuild* to your repository. This file is a shell script that is executed as part of the Docker image construction and is executed only once when a new image is built. To learn more, see the documentation at *https://mybinder.readthedocs.io/en/latest/using/config_files .html#postbuild-run-code-after-installing-the-environment/*.

It's impractical to place large files in your GitHub repository or include them directly in the image that Binder builds. You're better off using a library specific to the data format to stream the data as you're using it. Alternatively, you can download it on demand as part of your code. For security reasons, outgoing traffic is restricted to HTTP or GitHub connections only, so you can't use FTP sites to fetch data using Binder.

If a user changes your notebook through Binder, they will not be able to save or push changes to the GitHub repository. To save changes, they will need to download the notebook to their computer by clicking **File ▸ Download as ▸ Notebook (.ipynb)**.

Because of its data limits, saving issues, and lack of version control, Binder is best for viewing and running notebooks. To collaboratively *develop* notebooks, Git with GitHub is preferred.

Other Sharing Options

Other options for sharing notebooks include—but are not limited to—*Jovian* (*https://jovian.ai/docs/*), *Google Colaboratory* (*https://colab.research.google .com/notebooks/intro.ipynb/*), and *Microsoft Azure Notebooks* (*https://notebooks .azure.com/*). These options tend to require more setup than those we discussed previously and might not play well with GitHub. All require you have an account, and Jovian needs to be locally installed. The notebook interface will look a bit different in the Google and Microsoft options.

Colab lets users collaborate and run code that exploits Google's cloud resources. This includes using free GPUs, saving documents to Google Drive, and running the *TensorFlow* machine learning library directly in the browser. In fact, Google has a "Seedbank" repository of example deep learning notebooks that you can open and run with the click of a button.

Jovian permits cell-level commenting and discussion to aid collaboration. Azure helps you to create interactive presentations from your notebooks and share them easily, though this is simple to do, regardless.

Finally, if you want total control over who can access your notebooks and how they're used, you can set up your own *JupyterHub* multiuser *Hub*. This lets you offer notebook servers to a class of students, a corporate data science workgroup, a scientific research project, and so on.

To use JupyterHub, you need a Unix server (typically Linux) running somewhere that is accessible to your users on a network. This can require configuring a public server, something best done by an IT team to ensure security issues are properly addressed. To learn more, visit *https://jupyterhub .readthedocs.io/en/latest/* and *https://jupyter-server.readthedocs.io/en/latest/ operators/public-server.html*.

NOTE *If all you need is remote access to your personal machine, you can set up a public server with a single user by following the instructions at* https://jupyter-notebook .readthedocs.io/en/stable/public_server.html.

Trusting Notebooks

If you're running Notebook locally on your own computer, your notebook is as secure as your computer. But if you're accessing a notebook remotely, sharing your notebooks, or creating a server for multiple users, the potential for hackers to exploit the notebook increases.

The problem is that a notebook includes output that exists in a context that can execute code (via JavaScript). Ideally, code should not execute just because a user opens a notebook, especially code that they didn't write. But after a user decides to execute code in the notebook, it should be considered trusted, regardless of what it does.

To address this, the Jupyter developers have implemented security models designed to prevent execution of untrusted code without explicit user input. To ensure that a notebook is "trustworthy," whenever it's executed and saved, a signature is computed from a digest of the notebook's contents along with a secret key. This is stored in a database, writable only by the current user. By default, here's where this database is located:

- *%APPDATA%/jupyter/nbsignatures.db* in Windows
- *~/Library/Jupyter/nbsignatures.db* in macOS
- *~/.local/share/jupyter/nbsignatures.db* in Linux

Each signature represents a series of outputs, which were produced by code that the user executed. As stated previously, any output generated and saved during an interactive session is considered trustworthy.

When a user opens a notebook, the server computes its signature. If it finds it in the user's database, any HTML and JavaScript output will be trusted. Otherwise, it's untrusted.

When collaborating on a notebook, other users will have different keys, so the notebook will be in an untrusted state when shared to them. There are three recommended methods for managing this situation:

- Rerun the notebook after opening (not always viable and you should trust the sender).
- Explicitly trust notebooks via **File ▸ Trusted notebook** (see Figure 5-26) or, at the CLI, run `jupyter trust /path/to/notebook.ipynb`. These methods load the notebook, compute a new signature, and add that signature to the user's database
- Share a "notebook signatures database" and use a configuration dedicated to the project.

For detailed instructions on the last approach, along with more information on notebook and server security, see the documentation at *https://jupyter-notebook.readthedocs.io/en/stable/security.html*.

Turning Notebooks into Slideshows

When you complete your project, you can present the results directly from your notebook by turning it into a slideshow. This works much like Microsoft PowerPoint, with the notable difference being that you can run code live for a dynamic and immersive experience. Let's work through an example using the modular approach, whereby you run Jupyter Notebook from the *base* environment.

Installing the RISE Extension

To enable interactive coding in the slideshow, you'll need to install the *Reveal.js – Jupyter/IPython Slideshow Extension (RISE)*. First, shut down any currently running Jupyter notebooks. Next, open Anaconda Prompt (on Windows) or a terminal (on macOS or Linux) and run the following in the *base* environment:

```
conda install -c conda-forge rise
```

Now, Notebook can find this extension and display it on the nbextensions tab on the Dashboard page. Make sure that you install RISE in the conda environment in which you installed Notebook.

Creating a Slideshow

Let's create a new notebook that we can use to demonstrate slideshow capabilities. Because we're saving a new file, we'll launch Notebook from the *notebooks* folder, created previously.

Activate the *base* environment (where Jupyter Notebook is installed). Next, use the cd command and your personal path to open your *notebooks* directory and then start Notebook:

```
conda activate base
cd C:\Users\hanna\my_nb_proj\notebooks
jupyter notebook
```

When the Notebook dashboard opens in your browser, click the **New** button at the upper right of the Files tab to open a drop-down menu (Figure 5-4). To activate the kernel in your *my_nb_proj* environment, select **Python [conda env:my_nb_proj_env]** from the list. Remember, this lets us start Notebook from the *base* environment and then work in *another* environment.

When the blank notebook appears, click **Untitled** near the top of the window, rename the new notebook **slideshow** and save it. You should also see the new RISE icon on the far-right side of the toolbar (Figure 5-35).

RISE icon

Figure 5-35: The RISE icon at the end of the toolbar

From the top menu, click **View ▶ Cell Toolbar ▶ Slideshow**. The first empty cell in your notebook should now include a drop-down menu for selecting the slide type, as illustrated in Figure 5-36.

Figure 5-36: An empty notebook cell in slideshow mode. Note the menu for selecting the slide type on the right.

This menu gives you the six options, described in Table 5-4. The most used are Slide, Skip, and Notes.

Table 5-4: The Slide Type Menu

Slide type	Description
Slide	Start a new slide. When presenting, use the left and right arrow keys to switch slides.
Sub-slide	Create a subpage with transition animation. Use up and down arrow keys to switch.
Fragment	Create a hidden part of a slide transitioned to using the spacebar.
Skip	Indicate that the selected slide should be skipped and not shown. Useful for hiding code that does not generate an in-show visualization.
Notes	Indicate that the selected slide represents speaker notes.
-	Indicate that the current cell should behave like the previous cell.

Now, let's make a short slideshow about *logarithmic spirals*, a common shape found in nature (Figure 5-37).

Hurricane   Galaxy   Nautilus   Pine cone

Figure 5-37: Some examples of the logarithmic spiral in nature

Start by making a title slide. In the first cell, set the **Slide Type** menu to **Slide**. Then, using the top toolbar, change the cell type to **Markdown** and enter the following:

```
# Spira mirabilis:  The Miraculous Spiral
```

On your keyboard, press CTRL-ENTER to exit Markdown mode.

Insert a new cell beneath the title, set its types to **Slide** and **Markdown**, as before, and then enter the following:

```
### - Why does a hurricane look like a galaxy? Or the chambers in a nautilus
shell resemble the swirls in a pinecone?

### - Growth in nature is a geometric progression, and spirals that increase
geometrically are *logarithmic*.

### - Logarithmic spirals can be plotted using Python with the polar equation:
## $r = ae^{b\theta}$
*Where:
r = radius
a is the scaling factor (size of spiral)
b is the growth factor that controls the "openness"
$\theta$ controls length of spiral*
```

Press CTRL-ENTER to execute the code.

Insert a new cell and set its types to **Slide** and **Code**. Next, enter the following code, which applies the polar equation and generates interactive slider bars. These sliders let you assess the impact of the a, b, and θ parameters. Don't worry about all the details for now; we'll go over the NumPy and Matplotlib libraries later in the book.

```
import numpy as np
import matplotlib.pyplot as plt
from ipywidgets import interact

def log_spiral(a=1, b=0.2, t=4):
    theta_radians = np.arange(0, t * np.pi, 0.1)
    radii = [a * np.exp(b * rad) for rad in theta_radians]
    plt.polar(theta_radians, radii, 'o', c='black')

interact(log_spiral);
```

Next, insert a new cell below the previous cell and set its types to **Notes** and **Markdown**. This cell can prompt you on the meaning of the parameters while you're describing the dynamic plot to your audience. Enter the following and then press CTRL-ENTER to execute:

```
#### a: (scaling factor) controls size
#### b: (growth factor) controls openness
#### t: (theta) controls length
```

Cells representing "slide notes" must come immediately after the cell with which they're associated.

Finish the presentation with a new cell whose types are set to **Slide** and **Markdown**. Enter the following and then press CTRL-ENTER:

```
# The End
```

To launch the slideshow, first save the notebook, next click in the top cell, and then click the **RISE** button (Figure 5-35). To operate the slideshow, use the keyboard shortcuts in Table 5-5. For a full list of shortcuts, including ones for operating a virtual chalkboard, click the **?** icon visible at the lower left of each slide.

Table 5-5: Selected RISE Keyboard Shortcuts

Shortcut	Result
ALT-R	Enter or exit RISE (slideshow mode)
SPACE	Move forward to next slide
SHIFT-SPACE	Move back to previous slide
SHIFT-ENTER	Evaluate and select next cell if visible
HOME/END	Jump to start/end
T	Open speaker notes window

Use the spacebar to navigate to the code cell. If you haven't executed this cell already, do so now by pressing either CTRL-ENTER or SHIFT-ENTER. You may need to manually adjust the window for the plot to fit correctly. Move the sliders slowly to see how the parameters affect the plot. You can't do this in PowerPoint!

Using Speaker Notes

Slideshow comes with a Speaker Notes window that can help you run the presentation. It shows the current slide, upcoming slide, speaker notes, and current and elapsed time (Figure 5-38). You can have this window open on your laptop screen while you're projecting the slideshow. To enter this mode, press the T key while in the slideshow.

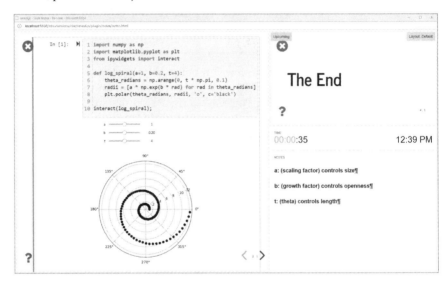

Figure 5-38: The RISE Speaker Notes window

For remote meetings, Binder (discussed in "Sharing Notebooks via Binder" on page 129) lets you host a live slideshow session for free within a browser, and users won't need Python or Jupyter to view it. For non-live viewing, the slideshow can be exported to a single HTML file.

There's a lot more to slideshows than I can cover here. The Jupyter Project documentation doesn't include much on the slideshow mode, but you can find plenty of detailed tutorials and tips by searching online for "Jupyter Notebook slideshow."

Summary

Jupyter Notebook is wildly popular for good reason; it's useful, easy, and fun! By allowing you to store all your analysis and commentary in one place, notebooks make it simple to document your work, share it, present it, and quickly pick up where you left off.

Despite this, don't become *too* enamored with notebooks, because they're not perfect. Their cellular organization encourages pollution of the global namespace, discourages writing reusable functions and classes, and makes source control and unit testing difficult. Those are some of the reasons why we learned about Spyder in Chapter 4, and why we'll look at JupyterLab next. With knowledge of Notebook, Spyder, and JupyterLab, you'll always be ready to apply the best tool to the task at hand.

6

JUPYTERLAB: YOUR CENTER FOR SCIENCE

JupyterLab is a web-based interface for Project Jupyter. It combines every scientific computing component you'll need into a single interactive and collaborative environment. Its flexible layout lets you efficiently execute complete computational workflows, from loading data to generating a final report. Its architecture is also hackable, meaning it's extensible and open to developers.

If you've been reading this book from the beginning, you've already been introduced to multiple JupyterLab components, such as a console (Chapter 3), a text editor (Chapter 4), and Jupyter Notebook (Chapter 5). So, we won't retrace our steps here; instead, we'll focus on the new interface and a few other features JupyterLab brings to the table.

NOTE   *JupyterLab is under active development, and new features are being added regularly. To ensure that you're up to date on the current state of JupyterLab, be sure to check out the full documentation at* https://jupyterlab.readthedocs.io/.

When to Use JupyterLab Instead of Notebook?

The JupyterLab developers created the new interface based on the results of a 2015 user experience survey that indicated the need for more flexibility and integration in Jupyter Notebook. Users wanted easy access not only to notebooks but also to text editors, terminals, data viewers, debuggers, Markdown editors, CSV editors, a file browser, IPython cluster manager, and so on.

JupyterLab not only provides these tools, it lets you share a kernel among them. In a single browser window, you can work with a notebook on one tab, edit a related data file on another, check resources or processes in a terminal, test concepts in a console, easily find and open files in a file manager, change the display language, and more. And like Notebook, it's free and open source.

Whereas Jupyter Notebook is great for data exploration, incremental software development, and documentation, JupyterLab permits more serious software development by providing many features found in traditional IDEs. But if you love Notebook, don't worry; JupyterLab is basically a new frontend that exists on top of the existing Jupyter architecture. It uses the same server and file format as the classic Jupyter Notebook, so it's fully compatible with your existing notebooks. In fact, you can run the classic Notebook app and JupyterLab side by side on the same computer.

Installing JupyterLab

As we discussed in Chapter 2, it's best to have a dedicated conda environment for each of your projects. To work with these environments with JupyterLab, you have two main options: you can either install JupyterLab directly in each conda environment, or you can link each environment to the JupyterLab installation in the *base* environment. We'll call the first option the *naive* approach, and the second the *modular* approach. Although the modular approach is generally recommended, if a project needs to lockdown a specific version of JupyterLab, you'll want to use the naive approach.

The Naive Approach

With the naive approach, you install JupyterLab directly in a conda environment. You then can import and use any packages installed in the same environment. This is the simplest approach, but it can become resource intensive over time as your *pkgs* folder becomes populated with different versions of JupyterLab.

Installing and Launching JupyterLab Using Anaconda Navigator

To install JupyterLab in a new environment using Anaconda Navigator, first launch Navigator using the Start menu in Windows, or Launchpad in macOS, or by entering anaconda-navigator in a terminal in Linux. Then, activate the environment by selecting its name in the **Applications on** drop-down menu near the top of the Home tab (Figure 6-1). In this example, we're using the *base* environment.

Figure 6-1: The Anaconda Navigator Home tab showing the active environment (base, or root) and the JupyterLab tile

Next, find the JupyterLab app tile and click the **Install** button. You might need to scroll down the Home tab to find the tile. If you are unable to find the tile, then install JupyterLab using the CLI, as described in the next section.

NOTE *If you see a Launch button rather than an Install button, JupyterLab now comes pre-installed with Anaconda.*

The Install button will install the most current version of JupyterLab available from the top channel in your Channels listing, located near the top of the Home tab. If you want to install a specific version, click the "gear" icon at the upper right of the JupyterLab tile (see Figure 6-1) to see a listing of available version numbers.

If you need the absolute most current version of JupyterLab, make sure the conda-forge channel is at the top of your channels list. Packages in the defaults channel might be slightly older, but as compensation, they have passed the most rigorous compatibility testing. For more on using channels see Chapter 2.

After a few moments, the Install button should change to a Launch button. This button starts a local web server on your computer that displays the JupyterLab interface. Because it's running locally, you don't need an active internet connection; however, you will need to leave Navigator running.

Installing and Launching JupyterLab Using the CLI

To install JupyterLab in a new environment using conda, first open Anaconda Prompt (in Windows) or a terminal (in macOS and Linux) and activate the conda environment. Let's do this for *my_second_env* that we created in Chapter 2. If you skipped this step in Chapter 2, create the environment by doing the following:

```
conda create --name my_second_env
```

Now, activate the environment:

```
conda activate my_second_env
```

Next, use conda to install JupyterLab:

```
conda install -c conda-forge jupyterlab
```

To install a specific version, such as 3.1.4, you would use the following:

```
conda install -c conda-forge jupyterlab=3.1.4
```

To launch JupyterLab from the command line, enter:

```
jupyter lab
```

This starts a local web server on your computer that displays the JupyterLab interface. Because it's running locally, you don't need an active internet connection. You'll need to leave your Prompt window or terminal open, however, as it's running the local server for Notebook that lets you interact with your web browser.

The Modular Approach

With the modular approach, you link each conda environment back to the JupyterLab package in your *base* environment. This approach is resource efficient. It also lets you easily keep the package up to date and choose among different environments from the same instance of JupyterLab.

You can use the modular approach with either Navigator or the CLI. For simplicity, let's use the CLI. Open Anaconda Prompt (in Windows) or a terminal (in macOS or Linux) and enter the following to create a new environment named *my_lab_env*:

```
conda create --name my_lab_env
```

Enter **y** when prompted to accept the installation. Next, activate the new environment:

```
conda activate my_lab_env
```

To link this environment with the JupyterLab installation in the *base* environment, enter the following:

```
conda install ipykernel
```

Thanks to ipykernel, we didn't need to explicitly install Python in the environment. However, if you do need to use a *specific* version of Python in your project, you'll want to explicitly install it in the environment.

Now, deactivate *my_lab_env*, which returns you to *base*:

```
conda deactivate
```

If JupyterLab is already installed in *base*, you can skip the next step. Otherwise, install JupyterLab using this command:

```
conda install -c conda-forge jupyterlab
```

To install a specific version, such as 3.1.4, you would use the following:

```
conda install -c conda-forge jupyterlab=3.1.4
```

Next, install the nb_conda_kernels package to *base*. You'll need to do this only once, so if you worked through Chapter 5, it should already be installed (you can check by running `conda list nb_conda_kernels` after activating the environment):

```
conda install nb_conda_kernels
```

The nb_conda_kernels package enables a Jupyter instance in an environment to automatically recognize any other environment that has the ipykernel package installed. It's this combination of nb_conda_kernels in the *base* environment and ipykernel in other conda environments that allows you to use a single installation of JupyterLab.

To launch JupyterLab from the active *base* environment enter the following:

```
jupyter lab
```

This launches a local web server on your computer that displays the JupyterLab interface. Because it's running locally, you don't need an active internet connection. You'll need to leave your Prompt window or terminal open, however, as it's running a local server that lets you interact with your web browser.

Building a 3D Astronomical Simulation

It's time to start working with JupyterLab! In this example, we'll use Jupyter-Lab to build a 3D simulation of an astronomical oddity: a *globular cluster*. Globular clusters are spherical collections of stars that orbit most spiral galaxies such as our Milky Way. They are among the oldest features in a galaxy and can contain millions of tightly packed stars.

Let's start off fresh to avoid confusion. If you started JupyterLab in the previous sections, go to the browser page it opened and shut it down by clicking **File ▸ Shut Down**. If Navigator is open, close it by selecting **File ▸ Quit**.

Going forward, we'll use the modular approach, so be sure to install JupyterLab and the nb_conda_kernels package in your *base* environment, as described in the previous section.

Using Dedicated Project Folders

Anaconda uses dedicated folders to keep track of your installed packages and conda environments (see Chapter 2). Although Anaconda is designed to work smoothly with this structure and help you navigate it, not everyone wants their project files scattered around their directory tree. As we discussed in Chapter 4, there are multiple benefits to keeping all of your project's files and folders within a single master folder.

Let's work through an example in which we store the conda environment and Jupyter notebooks in a folder named *my_jlab_proj*, short for "my JupyterLab project." I'll create this in my user directory in Windows (*C:\ Users\hanna*), and I suggest you use a similar location on your system.

NOTE   *The root directory for JupyterLab's file browser (that is, the highest directory in the hierarchy) is the directory from which you launched JupyterLab. This is usually your home directory that holds the* anaconda3 *folder. As a result, you won't be able to access files or folders above this directory structure within JupyterLab.*

Although you can create directories and environments with Anaconda Navigator, the command line is more succinct, so we'll use that going forward. To make the directories for the project, open Anaconda Prompt (in Windows) or a terminal (in macOS or Linux) and enter the following (using your own directory path up to *\my_jlab_proj*):

```
mkdir C:\Users\hanna\my_jlab_proj
mkdir C:\Users\hanna\my_jlab_proj\notebooks
mkdir C:\Users\hanna\my_jlab_proj\data
```

This makes a *my_jlab_proj* directory with notebooks and data subdirectories. Next, create a conda environment named *my_jlab_proj_env* under the project directory, activate it, and install some libraries (substitute your own path where needed):

```
conda create --prefix C:\Users\hanna\my_jlab_proj\my_jlab_proj_env
conda activate C:\Users\hanna\my_jlab_proj\my_jlab_proj_env
conda install ipykernel matplotlib
```

As described previously, the ipykernel package lets you use a single JupyterLab application installed in the *base* environment. The Matplotlib package is Python's primary plotting library. It includes the NumPy (Numerical Python) package as a dependency. We explore these libraries in more detail in later chapters of this book.

At this point, your project directory structure should look like Figure 6-2. Of course, with a real project, you might include additional folders for specific types of data, non-notebook scripts, miscellaneous items, and more.

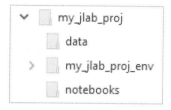

Figure 6-2: Directory structure for my_jlab_proj

To launch JupyterLab, first return to the base environment:

```
conda deactivate
```

Then, enter the following:

```
jupyter lab
```

The JupyterLab Interface

When you launch JupyterLab, a new tab should appear in your browser with a file manager along the left side and a Launcher tab in the main work area (Figure 6-3). If for some reason you don't see the Launcher pane, on the menu bar at the top, select **File ▸ New Launcher**.

The default view in Figure 6-3 is just a starting point. Indeed, JupyterLab's building blocks are so flexible and customizable that there's no such thing as a standard view, though there are some common features.

JupyterLab sessions reside in a *workspace* that contains the *state* of JupyterLab; that is, the files that are currently open, the layout of the application areas and tabs, and so on. The workspace consists of a *main work area*, or *Launcher* pane, containing tabs of documents and activities; a *menu bar*; and a collapsible *left sidebar*. The left sidebar contains a file browser and icons for the list of open tabs and running kernels and terminals, a table of contents, and an extensions manager.

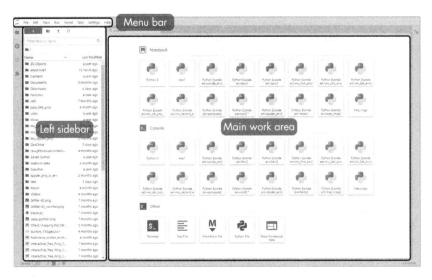

Figure 6-3: The JupyterLab workspace with major components labeled

Within the Launcher pane, you'll see sections for Notebooks and Consoles. Within them are tiles for kernels in your various conda environments (your view will differ from Figure 6-3, as I've set up some environments on my own). You'll also see an Other section from within which you can open a terminal, text file, Markdown file, Python file, or a contextual help page.

The Menu Bar

The menu bar at the top of JupyterLab (Figure 6-3) offers top-level menus that expose available actions along with their keyboard shortcuts. These are specific to which tab is active in the main work area; unavailable actions will be visible but grayed out (half intensity). For convenience, some actions are duplicated in the left sidebar. Following are the default menus:

File   Actions related to files and folders, including shutting down and logging off

Edit   Actions related to editing documents and working with notebook cells

View   Actions to alter JupyterLab's appearance and open the Command Palette

Run   Actions for running code in notebooks and consoles

Kernel   Actions for managing kernels

Tabs   Actions for working with tabs, plus a listing of open tabs

Settings   Settings for themes, languages, key maps, font sizes, and more

Help   Links for JupyterLab help, plus a launcher for Classic Jupyter Notebook

JupyterLab extensions can also create new top-level menus in the menu bar. These will be specific to the extension.

The Left Sidebar

The left sidebar provides access to commonly used tabs, such as a file browser, a list of open tabs and running terminals and kernels, a table-of-contents generator, and a manager for third-party extensions, as illustrated in Figure 6-4.

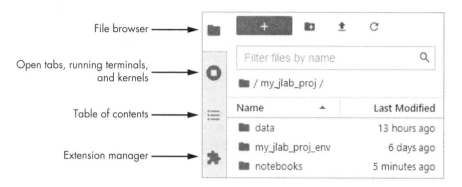

Figure 6-4: The left sidebar with the file browser active

When you close a notebook, code console, or terminal, the underlying kernel or terminal running on the server continues to run. This enables you to perform long-running actions and return later. The Running panel (Figure 6-5) lets you reopen the document linked to a given terminal or kernel. You can also shut down any open kernels or terminals.

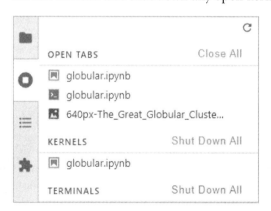

Figure 6-5: The Running Terminals and Kernels panel

The table-of-contents extension, now built in to JupyterLab, makes it easy to see and navigate the structure of a document. The table is automatically generated in the left sidebar when you have a notebook, Markdown,

LaTeX, or Python file open. Each listed section is hyperlinked to the actual section within the document. You can number headings, collapse sections, and navigate into the file.

The tool uses the headings from your Markdown cells to generate the table of contents. The Toggle Auto-Numbering option, visible when you have a file open, will go through a notebook and number the sections and subsections as designated by the headings. This lets you move big sections around without having to go through the document and renumber them.

The extensions manager helps you to manage any third-party extensions you have installed. We'll talk more about extensions, later.

The left sidebar is collapsible. Just click the icon for the active panel or select **View ▸ Show Left Sidebar** from the menu bar to toggle it off.

Creating a New Notebook

Let's make a new Jupyter notebook in the *notebooks* folder to hold our globular cluster code and output. In the file manager view, also called the *left sidebar*, navigate to the folder and open it. Then, in the Notebook section in the Launcher pane, find and click the tile labeled Python[conda env:my_jlab_ proj_env](if the tile labels are truncated, hover your cursor over the tile to see the complete name). This opens a new untitled notebook that uses the kernel in the specified environment (Figure 6-6).

Figure 6-6: A new untitled notebook in the JupyterLab work area

Note that the tab for the notebook is marked with a colored top border (blue by default). The work area permits only one current activity, and this lets you know which tab is active.

If you read Chapter 5, you probably recognize the notebook interface, though there are a few changes relative to classic Notebook. The icons and menu choices along the top of the cell (the toolbar beneath the Untitled .ipynb tab), are more streamlined and simplified, and they share functionality with the more fully featured menu bar that runs along the top of the interface. Take a moment to hover over the toolbar icons, and then click the main menu items such as File, Edit, and Run to see the available options. These should be familiar to you from Chapter 5.

Naming the Notebook

Now, let's rename the notebook. You can do this in one of several ways. You can use **File ▸ Rename Notebook** from the main menu. Alternatively, you can open contextual menus by right-clicking the **Untitled.ipynb** tab or by right-clicking the filename in the file browser and selecting **Rename** (Figure 6-7).

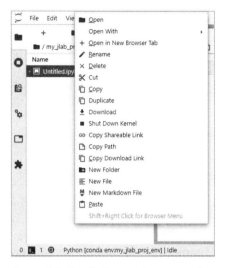

Figure 6-7: The file browser's contextual menu for working with files

JupyterLab comes with a lot of convenient contextual menus. Just about anything that's clickable, including the blank area under notebook cells, has a menu available.

Use the context menu shown in Figure 6-7 to name the notebook **globular.ipynb**. The notebook tab name should also change.

Using Markdown Cells

To make a descriptive header, click in the first cell and use the toolbar at the top of the notebook (Figure 6-8) to change the cell type from **Code** to **Markdown**.

Figure 6-8: The Notebook toolbar

Now, enter the following and press CTRL-ENTER to run the cell:

```
## Simulate a Globular Star Cluster with Matplotlib
```

For more on Markdown, see "Adding Text with a Markdown Cell" on page 102.

Adding Code and Making Plots

You could easily run the simulator code in a single cell, but for the sake of the narrative, let's spread it over multiple cells. Creating modular programs like this has its advantages. For example, you can isolate the imports and data loads in the first cell, so you don't need to rerun them every time you make a change in subsequent cells.

Start by adding a new cell using the "+" from the notebook toolbar (Figure 6-8). New cells are automatically code cells, so you're ready to start coding. The first step is to import the libraries needed to build the simulation:

```
%matplotlib inline
import numpy as np
from matplotlib import pyplot as plt
plt.style.use('dark_background')
```

This code starts with a magic command that makes Matplotlib plot *inline*. This means that it will plot to an output cell *within* the notebook. The next two lines import NumPy and Matplotlib. The final line selects Matplotlib's dark theme for plots, so our white stars will have a black backdrop. Execute the cell by pressing SHIFT-ENTER, which runs the cell and adds a new code cell below, or click the triangular "play" icon (▶) in the toolbar (Figure 6-8).

Now define a generic function that creates a list of x, y, z coordinates arrayed in a spherical volume. In the new cell, enter the following:

```
def spherical_coords(num_pts, radius):
    """Return list of uniformly distributed points in a sphere."""
❶  position_list = []
    for _ in range(num_pts):
❷      coords = np.random.normal(0, 1, 3)
        coords *= radius
        position_list.append(list(coords))
    return position_list
```

The function takes as arguments the number of points (num_pts) and the radius of the sphere (radius). This determines the size of the cluster and how many stars it contains. You then create an empty list ❶ to hold the coordinates and loop through the number of points, each time drawing three random values from a normal distribution with a mean of 0 and a standard deviation of 1 ❷. These three values will represent the x, y, z coordinates of a star. Multiplying the coordinates by the radius stretches or shrinks the size of the cluster. At the end of each loop, you append the coordinates to the list and end the function by returning the list.

Run the cell by pressing SHIFT-ENTER to add a new cell at the bottom of the notebook.

Now, create a globular cluster and plot it. In the new cell, enter the following code:

```
rim_radius = 1
num_rim_stars = 3000
rim_stars = spherical_coords(num_rim_stars, rim_radius)
❶ core_stars = spherical_coords(int(num_rim_stars/4), rim_radius/2.5)

❷ fig, ax = plt.subplots(1, 1, subplot_kw={'projection':'3d'})
ax.axis('off')
❸ ax.scatter(*zip(*core_stars), s=0.5, c='white')
ax.scatter(*zip(*rim_stars), s=0.1, c='white')
❹ ax.set_xlim(-(rim_radius * 4), (rim_radius * 4))
ax.set_ylim(-(rim_radius * 4), (rim_radius * 4))
ax.set_zlim(-(rim_radius * 3), (rim_radius * 3))
ax.set_aspect('auto')
```

The "rim" variables represent the radius and number of stars for the full cluster. Generate the coordinates by calling your function. Then, call it again to generate coordinates for stars in the densely packed core region at the center of the cluster ❶. Notice how you can alter the input arguments as you pass them to the function, by dividing them by a scaling factor and ensuring that the number of stars variable remains an integer. You can play with these scalers to change the appearance of the core region.

Time to plot the stars. Don't worry about Matplotlib's arcane syntax for now; we'll go into this in more detail later in the book. Basically, plots, referred to as Axes (ax for short), reside in Figure (fig) objects that serve as containers ❷. To make a single 3D ax object you call the plt.subplots() method and set the projection type to 3d. Then, turn off the x-, y-, and z-axes of the plot; we want our cluster to float in the blackness of space.

To post the star points, call the scatter() method twice: once for the rim stars, and once for the core ❸. This lets you specify different point sizes for the two regions. The scatter() method expects x, y, z points, but the data is currently a list of lists, with each point's coordinates in its own list:

```
[[-1.3416146295620397, 0.24853387721205472, -1.3228171973149565],
 [-0.23230429303889005, 0.04705622148151854, 0.7578767084376479]...]
```

To extract these coordinates, we'll use Python's built-in zip() function in conjunction with its *splat* (*) operator that unpacks multiple variables. Finish by setting the axis limits so their aspect ratio is equal and they're big enough to hold the cluster ❹. By relating the limits to the rim_radius variable, rather than specifying an absolute size, the plot will automatically adjust if you change the radius value.

Press CTRL-ENTER to run the cell and generate the plot without adding a new cell. Your finished notebook should look like Figure 6-9.

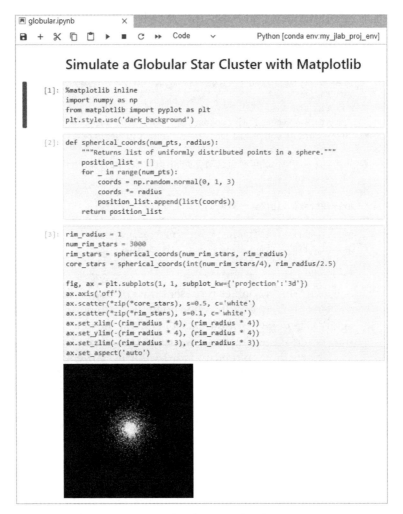

```
🖼 globular.ipynb          ✕

🖫  +  ✂ 🗗 🗎 ▶ ■ C ⏭    Code    ⌄              Python [conda env:my_jlab_proj_env]

        Simulate a Globular Star Cluster with Matplotlib

[1]: %matplotlib inline
     import numpy as np
     from matplotlib import pyplot as plt
     plt.style.use('dark_background')

[2]: def spherical_coords(num_pts, radius):
         """Returns list of uniformly distributed points in a sphere."""
         position_list = []
         for _ in range(num_pts):
             coords = np.random.normal(0, 1, 3)
             coords *= radius
             position_list.append(list(coords))
         return position_list

[3]: rim_radius = 1
     num_rim_stars = 3000
     rim_stars = spherical_coords(num_rim_stars, rim_radius)
     core_stars = spherical_coords(int(num_rim_stars/4), rim_radius/2.5)

     fig, ax = plt.subplots(1, 1, subplot_kw={'projection':'3d'})
     ax.axis('off')
     ax.scatter(*zip(*core_stars), s=0.5, c='white')
     ax.scatter(*zip(*rim_stars), s=0.1, c='white')
     ax.set_xlim(-(rim_radius * 4), (rim_radius * 4))
     ax.set_ylim(-(rim_radius * 4), (rim_radius * 4))
     ax.set_zlim(-(rim_radius * 3), (rim_radius * 3))
     ax.set_aspect('auto')
```

Figure 6-9: The completed globular cluster notebook

To save your work, on the toolbar, click the floppy disk icon, or use CTRL-S.

Adding a Console

Everything we've done to this point you could accomplish in the classic Jupyter Notebook app. Now let's see what JupyterLab can bring to the table, namely the ability to work with multiple tabs connected to the same kernel.

As you work with code, especially code that you inherit from teammates, you'll want to investigate data types, list contents, function returns, and so on. Normally, investigating side issues would clutter up your notebook. But JupyterLab lets you open multiple tabs *and* connect these tabs to the running kernel. This allows you to perform exploratory work outside of your notebook yet still within the workspace.

To open a console connected to the current kernel, right-click in any cell and then, on the context menu that opens, select **New Console for Notebook**. A console should appear beneath your notebook, as depicted in Figure 6-10.

Figure 6-10: A new console linked to the globular.ipynb notebook

To see the format of the coordinates in the rim_stars list, place your cursor in the empty box at the bottom of the console, enter the following, and then run it by pressing SHIFT-ENTER:

```
print(rim_stars[:3])
```

This displays the first three lines of the list:

```
[[0.9223767036280706, -1.0746823426729988, 0.30774451034833233],
 [0.25440816717656933, 0.21302429871155004, 0.7991568645529153],
 [-0.922974317327836, 0.49065537767349343, 0.5170958730770349]]
```

You can see that you're dealing with a list of lists, and each nested list holds three float values, representing x, y, and z coordinates. Because the notebook and console share the same kernel, as soon as you run the notebook, any imports, variable assignments, function definitions, and so on become resident in memory and accessible to the console. You can even copy all of cell [3] into the console, tweak the parameters, and plot the results there, leaving your notebook untouched.

To keep the console uncluttered, open its contextual menu and select **Clear Console Cells**.

Displaying an Image File

What if you want to compare your output to a photograph of a globular cluster, to help you tweak the input variables for a realistic-looking simulation? You could always add an image to a Markdown cell, but you might need to scroll down to see it, and you'll have to remember to delete it later. To avoid that aggravation, you can display the image in a separate JupyterLab window.

To begin, go to the Wikimedia Commons internet site (*https://commons .wikimedia.org/*) and search for "The Great Globular Cluster in Hercules – M13." Save or download the image to your *my_jlab_proj\data* folder. I used the 640-pixel resolution available at *https://upload.wikimedia.org/wikipedia/ commons/thumb/6/6f/The_Great_Globular_Cluster_in_Hercules_-_M13.jpg/ 640px-The_Great_Globular_Cluster_in_Hercules_-_M13.jpg*.

Back in JupyterLab, navigate to the image in the file browser and open it by right-clicking the filename and then selecting **Open**, or by double-clicking it. Next, drag and stack both the new image pane and the console onto the right side of the screen to produce the layout shown in Figure 6-11.

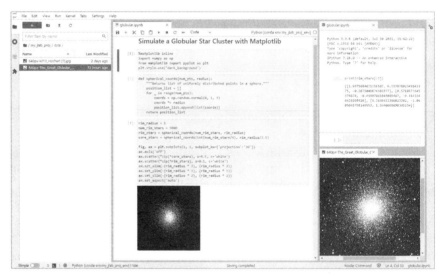

Figure 6-11: Our final workspace with file browser (left), notebook (center), console (upper right), and .jpg image (lower right)

A workspace much like this, with the file browser, notebook, and console, is a nice setup for a beginner.

Exploring the Simulation

You can change the appearance of the simulation by opening it in an external window, changing the background color, adding gridlines, and so on. To explore the simulation in 3D, change the first line in cell [1] to this:

```
%matplotlib qt
```

Then, select **Run ▸ Run All Cells** from the main menu. This opens an external Qt window that will let you spin the cluster around to view it from all sides. Check your task bar if the window doesn't appear on its own.

If you want to see the plot's 3D grid, it's best to use a negative image. First, find and comment-out the following two lines using the CTRL / or CMD / shortcut:

```
# plt.style.use('dark_background')
# ax.axis('off')
```

Then, change the star color to black:

```
ax.scatter(*zip(*core_stars), s=0.5, c='black')
ax.scatter(*zip(*rim_stars), s=0.1, c='black')
```

Save the notebook as *globular_black.ipynb* and run all cells. You might need to restart the kernel to clear the dark background plot style. If so, from the menu bar select **Kernel ▸ Restart Kernel and Run All Cells**. You should get the plot shown in Figure 6-12.

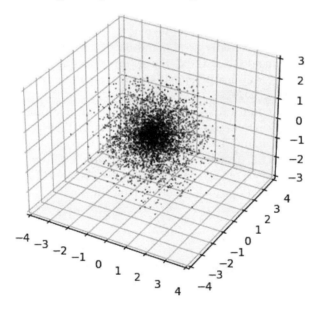

Figure 6-12: The "black" cluster simulation with grid lines

NOTE *You can use the* jupyterlab-matplotlib *extension to interact with plots within a notebook output cell. We'll look at JupyterLab extensions later in this chapter.*

Opening Multiple Notebooks

One advantage of JupyterLab is that it lets you work on multiple Notebook projects at once. Let's assume that you want to make an edit to the *geyser .ipynb* file that you made in Chapter 5. With JupyterLab, you can navigate to the notebook in the file manager and double-click it to open a new tab (Figure 6-13).

Figure 6-13: Two notebooks open in the same browser window

You now have two notebooks open in the same browser window, and they use different kernels, as indicated in the upper-right corner of each notebook.

Saving the Workspace

Documents within a workspace, such as Jupyter notebooks and text files, can be saved using standard commands like CTRL-S, **File ▸ Save Notebook**, and so forth. In addition, the *layout* of your workspace (that is, the tabs you have open, their arrangement, and their content) can be saved as a **.jupyterlab-workspace* file.

If you plan on using your current layout multiple times, or if you plan on having multiple project-dependent layouts, you'll want to give each workspace a unique name. To store this layout file in your project folder, go to the JupyterLab file browser and ensure that you're in the *my_jlab_proj* folder. Next, use the New Folder icon (a folder with a "+" in it) to create a folder named *workspaces* (Figure 6-14). Now open this folder.

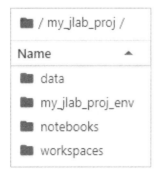

Figure 6-14: Adding the workspaces *folder*

To preserve the current state of JupyterLab, on the menu bar, select **File ▸ Save Current Workspace As**. A pop-up window will ask you for a name, in the following format:

my_jlab_proj/new-workspace.jupyterlab-workspace

Change the *new-workspace* text to *globular* and then click **Save**:

my_jlab_proj/globular.jupyterlab-workspace

To restore to a saved workspace, just open the *.jupyterlab-workspace* file.

Clearing the Workspace

To clear the contents of a workspace, use the reset URL parameter. The example here shows the general format:

```
http(s)://<server:port>/<lab-location>/lab/workspaces/<workspace-name>?reset
```

For example, to reset our globular workspace, in your browser's address bar, use reset, as shown here:

```
http://localhost:8888/lab/workspaces/globular?reset
```

This will configure your workspace similar to that shown in Figure 6-3. You can still restore the previous layout if you saved it in a *.jupyterlab-workspace* file, as described in the previous section.

For more on managing workspaces, visit *https://jupyterlab.readthedocs.io/en/stable/user/urls.html*.

Closing the Workspace

As with Jupyter Notebook, simply closing the browser tab does not stop JupyterLab. To completely shut it down, on the menu bar, use **File ▸ Shut Down**. If you are logged in to another server rather than working locally, you can log out using **File ▸ Log Out**.

NOTE *Be aware that some service providers, such as universities, might have specific logout procedures for their servers. Not following these protocols can waste allocated time resources and can result in unexpected usage fees.*

Taking Advantage of the JupyterLab Interface

Hopefully, the previous example gave you an appreciation for the JupyterLab interface. In the sections that follow, we'll take a closer look at some of its myriad components and controls. Many of these are self-explanatory, so we'll focus on the most useful and less intuitive ones.

As you saw in the globular cluster exercise, the work area lets you tie tools together in a customized layout. It also brings some nice features to Jupyter notebooks, including the ability to rearrange a notebook by dragging and dropping cells, dragging cells between notebooks to copy content, and creating multiple synchronized views of the same notebook.

Creating Synchronized Views

Let's look at the last case on synchronized views. It's not uncommon to want to look at both the top and bottom of a long notebook at the same time, or to scroll down to see interactive output. To manage this, JupyterLab lets you open the same notebook more than once.

To see how this works, in the globular cluster session, click in the *globular* notebook and then select **File ▸ New View for Notebook**. Next, arrange the layout so that the two notebooks are side by side. Then, shorten your browser window so that you can't see the entire notebook along with its output, mimicking a long notebook. In the left-hand notebook, scroll up to see the code. In the right-hand notebook, scroll down to see the plot, as in Figure 6-15. Now, rerun the cells in the first notebook. The plot on the right should update.

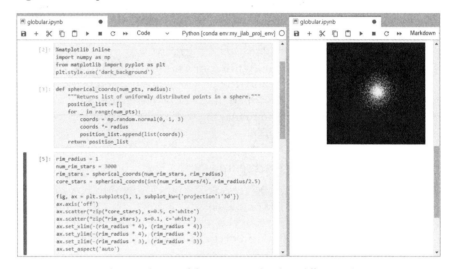

Figure 6-15: A synchronized view of the same notebook in different tabs

Alternatively, you can move the output cell into a new pane. Simply open a context menu in the output cell containing the globular cluster simulation and then select **Create New View for Output** (Figure 6-16). You then can drag it wherever you want in the workspace.

If you use sliders or other widgets to interactively change parameters and update the visualization, these will be included in the new view. This lets you create pseudo-dashboards within your workspace.

Copying Cells Between Notebooks

To drag and copy cells between notebooks, open a new notebook using **File ▸ New ▸ Notebook**. Drag the new notebook beside the *globular* notebook. From the *globular* notebook, click your cursor on a cell *number* (such as [1]:) and drag it into the new untitled notebook. You should see results similar to that shown in Figure 6-17.

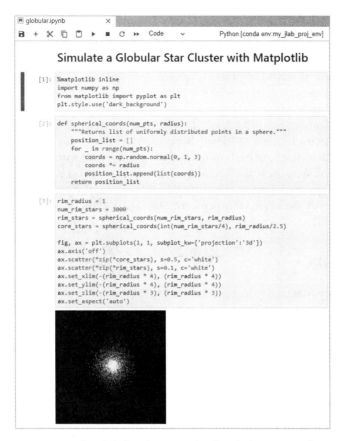

Figure 6-16: The globular cluster notebook with the output cell in a separate pane

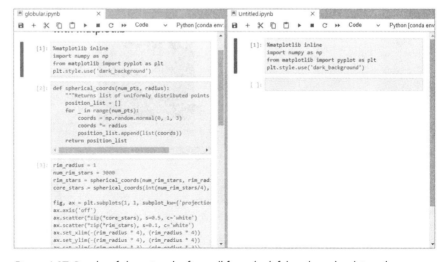

Figure 6-17: Results of dragging the first cell from the left-hand notebook into the right-hand notebook

Staying Focused by Using Single Document Mode

A nice thing about classic Jupyter Notebook is that you can focus on a task without the app "getting in your way." The JupyterLab developers took note of this and included a setting that lets you concentrate on a single document or activity without having to close all the other tabs in the main work area.

To toggle on this setting, activate a tab by clicking it, and then, from the menu bar, select **View ▸ Simple Interface**, or use the **Simple** toggle switch at the lower-left corner of the JupyterLab window. The workspace should show only the active tab. If you toggle this on and off for the *globular* workspace, you might detect a drawback. When you return to the regular view, you *may* lose your preferred tab arrangement (compare Figure 6-18 to 6-11).

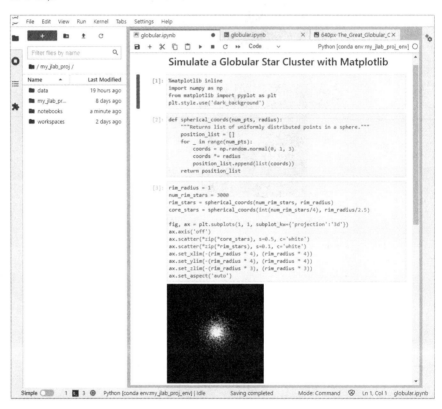

Figure 6-18: The workspace tab arrangement after toggling off Simple Interface mode for the globular *session*

If your view does change, you can restore the original layout either manually or by using a saved *.jupyterlab-workspace* file. Because this is a bit tedious, you'll only want to use simple interface options when you plan to spend a long time in a single document or activity.

Using the Text Editor

JupyterLab includes a text editor that you can use to write Python scripts. We didn't cover this in our *globular* session, so let's work a simple example here using the Pythagorean theorem. This is the famous $a^2 + b^2 = c^2$ formula used to find the hypotenuse of a right triangle.

If you've closed JupyterLab, start it from your *base* environment. Open Anaconda Prompt (Windows) or a terminal (macOS or Linux), and then enter the following:

```
jupyter lab
```

This should open the default layout shown in Figure 6-3.

If you already have JupyterLab up and running, return to the default workspace by editing the URL so that it ends in */lab*. For example:

> *http://localhost :8888/lab*

If for some reason your workspace doesn't look like the one in Figure 6-3, reset it by adding the *?reset* URL parameter, like this:

> *http://localhost:8888/lab?reset*

Now, from the Launcher pane, start a new text file or Python file. A new tab should open for the untitled file. Click in the file and enter this:

```python
def pythagoras(a, b):
    return (a**2 + b**2)**0.5

for i in range(9):
    a = i
    b = i + 1
    print(f"a = {a}, b = {b}, c = {pythagoras(a, b)}")
```

From the menu bar, select **File ▸ Save As** (or **File ▸ Save Python File As** if you chose the Python file option) and name the file *pythagoras.py*. When you click **Save**, the file should appear in your file browser. If you go back and click the File menu again, you'll note that the save options are now **Save Python File** and **Save Python File As**, even if you started with a text file. JupyterLab now recognizes that this as a Python file.

You'll need to save scripts before you run them. You can tell if a file has been saved by looking at the tab. Unsaved files will have a black dot adjacent to the filename, and saved files will have an X (Figure 6-19).

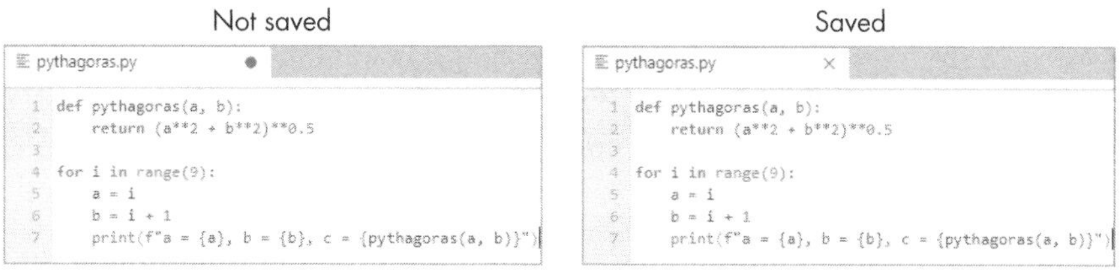

Figure 6-19: An unsaved text file with • in the tab versus a saved text file with X in the tab

Although the JupyterLab text editor is not as robust as the one in Spyder (Chapter 4), it's more sophisticated than a simple editor such as Notepad. If you click **Settings** on the menu bar, you'll see several submenus for altering its appearance and behavior, as illustrated in Figure 6-20.

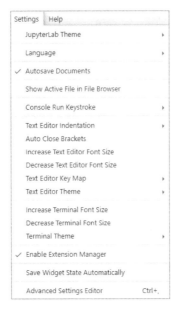

Figure 6-20: The Settings menu

There's a Key Map that lets you use the same keys as those used by the Sublime Text, vim, and emacs editors. There are multiple Editor Theme choices, options to change the font size, set tab indention levels, and automatically close brackets. Under the Advanced Settings Editor, you can change the Editor's configuration file. Keyboard shortcuts are also available and depend on which key mapping you chose. You can perform an online search for a list of each map's key bindings.

By default, the editor uses the *Plain Text* syntax highlighting style, but you can choose from an exhaustive list by selecting **View ▸ Text Editor Syntax Highlighting** (or **Text Editor Theme**) from the menu bar. Going forward, I will use the default Jupyter theme, key map, and syntax highlighting. For more on highlighting, see Chapter 3.

Back to our script. You have several options for running the code that you wrote in the editor. In the following sections, we'll look at options involving a terminal and a notebook.

Running a Script in a Terminal

To run the saved *pythagoras.py* file in a terminal emulator, on the menu bar, select **File ▸ New ▸ Terminal**. Next, click in the terminal pane and enter the following:

```
python pythagoras.py
```

Press ENTER, and the script should run (Figure 6-21).

Figure 6-21: Running a Python file in a terminal pane

Depending on your machine, you might need to use `python3` in place of `python`:

```
python3 pythagoras.py
```

If you edit the Python file and want to rerun it in the terminal, remember that you can use the arrow keys to select previous commands, saving you keystrokes.

Running a Script in a Notebook

To run the saved *pythagoras.py* file in a notebook, on the menu bar, select **File ▸ New ▸ Notebook**. If prompted for a kernel, accept **Python3** or choose the one in **my_jlab_proj_env** from the pull-down menu. Next, click in the notebook cell and enter the following:

```
%run pythagoras.py
```

Press CTRL-ENTER, and you should see the output in the notebook (Figure 6-22).

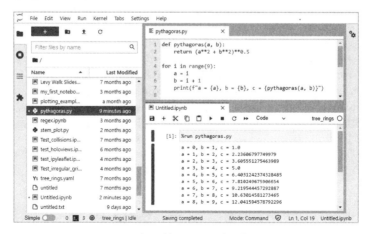

Figure 6-22: Running a Python file in a notebook

Note that you don't need to save or rename the notebook to use it to run scripts.

Simultaneously Writing and Documenting Code

JupyterLab lets you document your code, check that the code in the document runs, and preview the results, all in a single workspace. Let's look at an example.

In the file browser, navigate to your user directory. Open a new text editor from the menu bar using **File ▸ New ▸ Text File**. Rename it *doc.md* (*.md* files are plaintext format files that use Markdown language, like Notebook's Markdown Cells), and then enter the following:

```
## Example of Previewing Code Documentation in JupyterLab.
```

Now, in the Editor pane, open a context menu and then select **Show Markdown Preview**.

Back in the editor, enter the following:

```
### Let's run some code in a console.

import matplotlib.pyplot as plt

plt.plot([0, 1, 2, 3], [0, 1, 2, 3])
plt.savefig('doc_test.png')
```

In the editor pane, open a context menu and then select **Create Console for Editor**. If prompted for a kernel, choose one with Matplotlib installed, such as Python [conda env:my_jlab_proj_env]. Now, copy into the console the preceding code, starting with the import statement, and then run it using SHIFT-ENTER.

Next, enter the following code into the editor to show the plot in the Markdown preview:

```
![](doc_test.png)
```

Your layout should look similar to Figure 6-23.

You can also use the **Create Console for Editor** option to run code in the text editor. After opening the console, highlight the code inside the editor and then select **Run ▸ Run Code** from the menu.

JupyterLab's versatile layouts and sharable kernels support efficient workflows that boost productivity. If you find yourself constantly switching tabs and scrolling through panes while writing code, you might not be taking full advantage of JupyterLab's capabilities.

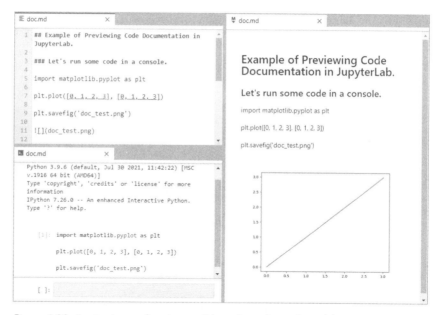

Figure 6-23: Previewing code using an Editor, Console, and Markdown pane

Using JupyterLab Extensions

JupyterLab extensions are plug-and-play add-ons to "extend" the functionality of JupyterLab. Each extension may contain one or more plug-ins (the basic unit of extensibility). Extensions can be created by anyone, including yourself. To quote the documentation, "[the] whole of JupyterLab itself is simply a collection of extensions that are no more powerful or privileged than any custom extension."

A small subset of popular JupyterLab extensions is listed in Table 6-1. Some previous popular extensions, such as the Table of Contents and Debugger extensions, are now built in to JupyterLab. There are also extensions for working with plotting and dashboarding libraries such as Plotly, Bokeh, and Dash. We look at those libraries in Chapter 16.

JupyterLab extensions contain JavaScript that's run in the browser. There are two types of extensions: *source* and *prebuilt*. Activating a source extension requires installation of Node.js and a rebuild of JupyterLab. Prebuilt extensions such as those published as Python packages do not require a rebuild of JupyterLab. Extensions can also include a server-side component necessary for the extension to function.

Table 6-1: Useful JupyterLab Extensions

Extension	Description	Website
nbdime	Tools for diffing and merging Jupyter notebooks	https://nbdime.readthedocs.io/en/latest/
jupyterlab-git	Version control using Git	https://github.com/jupyterlab/jupyterlab-git/
JupyterLab GitHub	Access notebooks from repositories	https://www.npmjs.com/package/@jupyterlab/github/
Jupyter-ML Workspace	IDE dedicated to machine learning	https://github.com/ml-tooling/ml-workspace/
JupyterLab System Monitor	Monitor memory and CPU usage	https://github.com/jtpio/jupyterlab-system-monitor/
jupyterlab_html	View rendered HTML files	https://github.com/mflevine/jupyterlab_html
jupyterlab matplotlib	Interactive inline Matplotlib	https://github.com/matplotlib/ipympl/
JupyterLab LaTeX	Live-edit LaTeX documents	https://github.com/jupyterlab/jupyterlab-latex/
JupyterLab Code Formatter	Use formatters like Black or Autopep8 to enforce style guidelines	https://github.com/ryantam626/jupyterlab_code_formatter/
jupyterlab-spellchecker	Spellchecker for Markdown cells and text files	https://github.com/ocordes/jupyterlab_spellchecker/
jupyterlab-google-drive	Cloud storage via Google Drive	https://github.com/jupyterlab/jupyterlab-google-drive

NOTE   *The classic Jupyter Notebook extensions we reviewed in Chapter 5 are not compatible with JupyterLab. Even though many useful extensions have been ported over to JupyterLab, others are still being updated. If an extension you want is unavailable, be patient and check the Extension Manager periodically for changes. The extension's website might also include news on updates.*

Installing and Managing Extensions with the Extension Manager

You can use the Extension Manager on the left sidebar (see Figure 6-3) to install and manage extensions that are distributed as single JavaScript packages on *npm*, the *node package manager* (*https://www.npmjs.com/*). The extension manager is off by default, but you can turn it on it by clicking the **Enable** button (Figure 6-24).

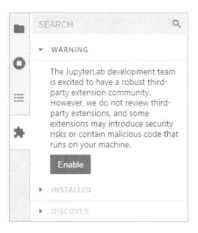

Figure 6-24: Turning on the Extension Manager from the left sidebar

Installing extensions allows them to execute arbitrary code on the server, kernel, and browser. Because third-party extensions are not reviewed and may introduce security risks or contain malicious code, you're asked to explicitly enable the action.

The extension manager pane has three sections: a search bar, a list of installed extensions, and a "Discover" section for all the JupyterLab extensions on the NPM registry. The results are listed according to the registry's sort order (see *https://docs.npmjs.com/searching-for-and-choosing-packages-to -download#package-search-rank-criteria/*). An exception to this order is extensions released by the Jupyter organization. These have a small Jupyter icon next to their names and will always appear at the top of the search results list (Figure 6-25).

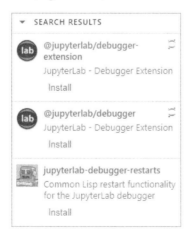

Figure 6-25: Extensions released by the Jupyter organization are clearly marked and appear at the top of the search results.

To find an available extension, you can scroll down the list or use the Extension Manager's search box. To learn more about an extension, click its name. This opens the extension's website (usually on GitHub) in a new browser window.

You can use the manager's **Install** button to install extensions. For source extensions, you'll need Node.js. To install it in your *base* environment *from the defaults channel*, open Anaconda Prompt (Windows) or a terminal (macOS or Linux) and enter the following:

```
conda install nodejs
```

To install it from the *conda-forge* channel, enter this:

```
conda install -c conda-forge nodejs
```

You're now ready to install extensions.

Because most extensions are *source* extensions, when you click the manager's **Install** button, a drop-down menu should appear under the search bar, indicating that the extension has been downloaded but that a rebuild is needed to complete the installation. You should click **Rebuild**, but if you ignore this for some reason, the next time you refresh your browser, change workspaces, or start JupyterLab, you'll be presented with a **Build** button. Click the button and you'll be asked to "Reload without Saving" or "Save and Reload."

If you want to manage additional extensions at the same time, you can ignore the rebuild notice until you have made all the changes you want. After that, click the **Rebuild** button to start a rebuild in the background. When it's complete, a dialog will open, indicating that a reload of the page is needed to load the latest build into the browser. At this point, the extension will appear in the Installed section of the manager, where you'll have options for uninstalling or disabling it (Figure 6-26). Disabling an extension will prevent it from being activated, but without rebuilding the application.

NOTE *Avoid installing extensions that you don't trust, and watch out for any extensions trying to masquerade as a trusted extension. Extensions released through the Jupyter organization will have a small Jupyter icon to the right of the extension name.*

During installation, JupyterLab will inspect the package metadata for any companion packages such as Notebook server extensions or kernel packages. If JupyterLab finds instructions for companion packages, it will present an informational dialog to notify you about these. It will be up to you to take these into account or not.

To read more about the Extension Manager, visit the documentation at *https://jupyterlab.readthedocs.io/en/stable/user/extensions.html.*

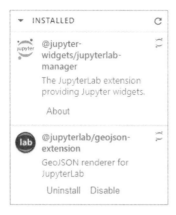

Figure 6-26: User-installed extensions can be uninstalled or disabled in the Extension Manager

Installing and Managing Extensions Using the CLI

There are other ways to install extensions besides the Extension Manager. Installing a source extension will still require that you install Node.js and rebuilding JupyterLab, however, and you'll need to be aware of the same security issues (see the previous section for details).

Clicking an extension name in the manager takes you to the extension's website. Here, you might find instructions for installing in the CLI. For example, to install the jupyterlab-git extension, which lets you use version control with Git, open Anaconda Prompt (Windows) or a terminal (macOS or Linux) and enter the following:

```
conda install -c conda-forge jupyterlab-git
```

To uninstall the extension, use this:

```
conda remove jupyterlab-git
```

Also in the CLI, you can use the jupyter labextension command to install or uninstall source extensions from NPM, list all installed extensions, or disable an extension.

To install an extension, use this format, where *<extension-name>* represents the extension's name:

```
jupyter labextension install <extension-name>
```

To install multiple extensions, enter this:

```
jupyter labextension install <extension-name> <another-extension-name>
```

To install a specific version of an extension, use the following:

```
jupyter labextension install <extension-name>@1.2.3
```

To uninstall extensions, use this:

```
jupyter labextension uninstall <extension-name> <another-extension-name>
```

If you are installing/uninstalling multiple extensions in several stages, you might want to defer rebuilding JupyterLab by including the flag --no-build in the install/uninstall step. When you are ready to rebuild, you can run the command:

```
jupyter lab build
```

You can list extensions using the following:

```
jupyter labextension list
```

NOTE *The* `jupyter labextension` *command uses the JavaScript package name for the extension, which can be different from the name of the conda package used to distribute the extension.*

To disable an extension without rebuilding JupyterLab, use this:

```
jupyter labextension disable <extension-name>
```

Disabling an extension leaves the code loaded but prevents the plug-ins from running.

You can enable a disabled extensions with the following:

```
jupyter labextension enable <extension-name>
```

Installed extensions are enabled by default unless there is a configuration explicitly disabling them.

For help with the `jupyter labextension` command, enter:

```
jupyter labextension --help
```

To read more about this command, visit the documentation at *https://jupyterlab.readthedocs.io/en/stable/user/extensions.html.*

Installing ipywidgets for JupyterLab

In Chapter 5, we worked with the ipywidgets extension to use widgets in classic Jupyter Notebook. Most of the time, installing ipywidgets automatically configures JupyterLab to use widgets, as it depends on the jupyterlab _widgets package, which configures JupyterLab to display and use widgets.

If you're using the modular approach, by which JupyterLab and the IPython kernel are installed in different environments, installing ipywidgets requires two steps:

1. Install the jupyterlab_widgets package in the environment containing JupyterLab.
2. Install ipywidgets in each kernel's environment that will use ipywidgets.

For example, with JupyterLab installed in your *base* environment and the kernel installed in your *my_jlab_proj_env* environment created earlier, use the following commands, substituting your path to the *my_jlab_proj_env* folder:

```
conda activate base
conda install -c conda-forge jupyterlab_widgets
conda activate C:\Users\hanna\my_jlab_proj\my_jlab_proj_env
conda install -c conda-forge ipywidgets
```

Creating Custom Extensions

A JupyterLab extension is a package that contains one or more JupyterLab plug-ins. You can write your own plug-ins and package them together into a JupyterLab extension. The details for this are beyond the scope of this book, but you can find what you need in the *Extension Developer Guide* at *https://jupyterlab.readthedocs.io/en/stable/extension/extension_dev.html*.

Sharing

When we talk about sharing in JupyterLab, we're mainly talking about sharing notebooks. Because we covered this subject in "Sharing Notebooks" on page 122, I won't repeat it here. To supplement that section, however, you can find more about using JupyterLab on JupyterHub at *https://jupyterlab .readthedocs.io/en/stable/user/jupyterhub.html*. For performing real-time collaboration with JupyterLab, see *https://jupyterlab.readthedocs.io/en/stable/user/ rtc.html*.

Summary

JupyterLab builds on Jupyter Notebook by providing an IDE-like environment for developing code, exploring datasets, and conducting experiments. With its extensible environment, JupyterLab takes us another step closer to true *literate programming*, wherein the exposition of logic is integrated into ordinary human language.

Although it's open for business, JupyterLab is still under development, and you'll want to consult the official documentation for the most recent additions, changes, and deprecations. In addition to work on the core program, development of third-party extensions will continue. New tools such as *nbdev* (*https://nbdev.fast.ai/*) and debuggers (*https://jupyterlab.readthedocs.io/en/stable/user/debugger.html*) are turning JupyterLab into a full-fledged IDE.

One development in late 2021 was the release of the cross-platform standalone *JupyterLab App* (*https://github.com/jupyterlab/jupyterlab-desktop/*). With the App, JupyterLab no longer "lives" in a web browser, but instead exists as a self-contained desktop application. For convenience, it bundles a Python environment with several popular libraries ready to use in scientific computing and data science workflows. These include pandas, NumPy, Matplotlib, SciPy, and more. A current drawback, however, is that the application provides only `pip` installations in place of conda installations. This means that it's not as easy to install some libraries compared to the web version.

This concludes Part I of the book. At the end of Chapter 4, readers new to Python were instructed to work through Part II, which is a Python primer. If you've done that—or don't need to—proceed to Part III, which provides an overview of important scientific and visualization packages available through Anaconda, including tips on how to choose the best ones for your needs.

PART II

A PYTHON PRIMER

If you've never used the Python programming language before, this primer will get you up to speed quickly. You'll learn the language's basics, as well as helpful hints and tips for solving real-world problems on your own. If you already know some Python, use this primer as reference material to jog your memory when needed.

When you learn a human language in school, you begin with the alphabet and parts of speech like nouns, verbs, and adverbs. Next, you might learn how to diagram sentences using these building blocks, stringing them together to form cohesive thoughts.

Learning a programming language works much the same way. Just as human language uses grammatical rules to join parts of speech into understandable expressions, Python uses syntactical rules to join objects into executable programs. But this isn't a linear process. Like a toddler learning to speak, a whole lot happens at the same time.

From the start, you'll acquire a lot of "nested" knowledge. You can't understand what a variable is without understanding what an object is, and you can't understand objects without understanding values, or values without data types. Therefore, if you browse the tables of contents in beginner programming books, you won't see a consistent approach to presenting the information.

In the chapters that follow, I'll attempt to progress through the language logically so that each step builds on the one that came before. There'll be times, however, when we'll need to run functions before we define them, or touch on a concept before fully developing it. That's okay. Humans learn by doing, and we're good at filling in knowledge gaps using context and interpolation.

Of course, this short introduction can't cover all the features of Python in detail, but it should give you a good foundation to begin programming on your own. If you want a more thorough introduction to Python, I suggest reading *Python Crash Course, 2nd edition: A Hands-On, Project-Based Introduction to Programming* (No Starch Press, 2019) by Eric Matthes. Alternatively, for a more technical and hard-core introduction, try *Learning Python,* 5th edition (O'Reilly Media, 2013) by Mark Lutz. To expand your knowledge beyond the beginner books, I suggest *Beyond the Basic Stuff with Python: Best Practices for Writing Clean Code* (No Starch Press, 2021) by Al Sweigart.

To find online tutorials, bootcamps, videos, and so on, visit *https:// wiki.python.org/moin/BeginnersGuide/Programmers/.* This Wiki page includes a section for nonprogrammers (*https://wiki.python.org/moin/BeginnersGuide/ NonProgrammers/*) as well as for those with a range of programming experience and will help guide you to additional resources. I've also found the *Real Python* site (*https://realpython.com/*) to be a great source of Python tutorials and information. It includes both free and paid content.

And to be a true Pythonista, you'll want to check out the *Zen of Python* (*https://www.python.org.dev/peps/pep-0020/*), a collection of 19 guiding principles that influence the design of the Python language. According to these principles, "There should be one—and preferably only one—obvious way to do something." In the spirit of providing a single obvious "right way" of doing things and building consensus around these practices, the Python community releases coding conventions known as *Python Enhancement Proposals,* or *PEPs.*

The most important PEP is *PEP 8* (*https://www.python.org/dev/peps/pep -0008/*), a set of standards for the style of your Python code. It includes naming conventions; rules about the use of blank lines, tabs, and spaces; maximum line lengths; the format of comments; and so on. The goal is to improve the readability of code and make it consistent across a wide spectrum of Python programs. Another useful style guide is PEP 257 (*https:// www.python.org/dev/peps/pep-0257/*), which covers code documentation. We look at both these guides in the chapters that follow.

Finally, if books and online searches fail to meet your needs, the next step is to ask someone for help. If no coworkers or classmates can help, you can do this online, either for a fee or at free forums like Stack Overflow (*https://stackoverflow.com/*). But be warned: the members of these sites don't suffer fools gladly. Be sure to read their "How do I ask a good question?" pages before posting. You can find advice and counsel for Stack Overflow at *http://stackoverflow.com/help/how-to-ask/.*

7

INTEGERS, FLOATS, AND STRINGS

In this chapter, you'll learn the difference between expressions and statements, discover how to assign values to variables, and become familiar with the most common types of data in Python: integers, floats, and strings. In the process, you'll likely be surprised by how much programming you can accomplish using simple mathematical operations.

While working through this and the following chapters, I recommend running the code examples rather than just reading them. Typing in the commands will help you to remember them and reduce any apprehension you might feel about coding. I'll be using the console and text editor in Spyder for the examples in this primer. I suggest you do the same so that you can follow along. If you need a refresher on these tools, see Chapter 3 for the Jupyter Qt console and Chapter 4 for the Spyder IDE.

Mathematical Expressions

In computer science, *expressions* are instructions that evaluate to a single value. The most familiar expressions are mathematical, such as 1 + 2, which evaluates to 3. Using Python, you can incorporate equations into your programs (and even use an interactive console as a calculator). To do so, you'll need to be familiar with the mathematical operators.

Mathematical Operators

Symbols used to represent an action or process are called *operators*. These perform a function or manipulate values in some way. Common operators are the plus (+) and minus (-) signs, used for addition and subtraction, respectively. Table 7-1 lists some of the available mathematical operators in Python. Most of these should be familiar to you, with a few exceptions that we'll expand on next.

Table 7-1: Mathematical Operators

Operator	Description	Example	Result
+	Addition	5 + 3	8
-	Subtraction	5 - 3	2
*	Multiplication	5 * 3	15
/	Division	5/3	1.6666666666666667
//	Division (floor or integer)	5 // 3	1
%	Modulus (remainder)	5 % 3	2
**	Power	5**3	125

Whereas the division operator (/) represents true division, *floor* division (//) returns an integer with any fractional part ignored. Note that floor division will not round up. If the result is 1.99999, you'll still get 1 as the answer.

If you want only the fraction, or *remainder*, of the division operation, use the *modulo* operator (%). The remainder may seem like a strange thing to separate out, but it can be useful. For example, you can use it to identify even and odd numbers. Enter the following code in a console:

```
In [1]: 4 % 2
Out[1]: 0
```

NOTE *The command for executing the code will depend on what tool you're using. For the Jupyter Qt console, this will be pressing ENTER (or SHIFT-ENTER if you're within indented code) on your keyboard.*

In the previous example, dividing 4 by 2 using the modulo operator returned 0, meaning the operation resulted in no remainder and thus 4 is even. Other uses for modulo include instructing your program to do something every *n*th time and converting seconds to hours, minutes, and seconds.

The power, or *exponent*, operator also has a non-intuitive feature. Not only can you raise numbers to a power, you can also calculate the root by using a decimal value after the ** operator. For example, to take the square root of 9, enter the following:

```
In [2]: 9**0.5
Out[2]: 3.0
```

To take the cube root of 27, enter:

```
In [3]: 27**(1/3)
Out[3]: 3.0
```

The Assignment Operator

Using Python as a hand calculator is a bit like calling in an air strike on an ant. For programs to be truly useful, you need to store the output of expressions in a reusable manner. That's where assignment statements, assignment operators, and variables come in.

Whereas expressions evaluate to a single value, *statements carry out* some action. The *assignment* statement, for example, creates a new *variable*. Variables are just *references* to data stored in memory. In an assignment statement, the equal sign (=) is an assignment *operator* that assigns a value or expression to a variable (Figure 7-1). A simple example is my_name = 'Lee'.

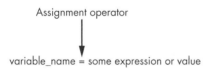

Figure 7-1: Rudiments of an assignment statement

In assignment statements, the item to the left of the equal sign is the variable's name. This acts as a label for accessing information in memory. The item on the right is the variable's value. These values don't have to be numeric. Text data, lists of items, even images and music can be stored as variables.

Now that you know about assignment statements, let's use them to make our math expressions more persistent and purposeful by assigning the results to a variable. Because this is such a common programming task, Python helps you out by providing special *augmented* assignment operators, which we'll talk about next.

Augmented Assignment Operators

For convenience, you can combine mathematical operators to form *augmented assignment operators* that let you perform two operations at the same time. Here's an example *without* an augmented operator:

```
In [4]: x = 5

In [5]: x = x + 5

In [6]: x
Out[6]: 10
```

Note that you can add a variable to itself, and entering the variable name in the console will display its value. In the text editor, you'll need to use print(x) to display the value to the screen.

With an augmented assignment operator (+=), you can add 5 to x without having to repeat x:

```
In [7]: x += 5

In [8]: x

Out[8]: 15
```

To make an augmented assignment operator, just add the mathematical operator (Table 7-1) before the equal sign (=). For example, to multiply x by 2, you could enter the following:

```
In [9]: x *= 2

In [10]: x
Out[10]: 30
```

Notice how, because you assign the result of each expression to the variable x, each expression can build on the one before.

Precedence

Mathematical expressions in Python use familiar rules of precedence (Table 7-2). Expressions bounded by parentheses are always performed first, and operations within the same precedence level are evaluated left to right.

Table 7-2: Mathematical Precedence

Level	Operator	Description
1 (highest)	()	Parentheses
2	**	Power
3	-n, +n	Negative and positive arguments

Level	Operator	Description
4	*, /, //, %	Multiplication, division, floor division, modulus
5	+, -	Addition and subtraction

Here's an example of precedence in action. Follow along in your head and see if you get the same answer as Python:

```
In [11]: 10**2 + (6 - 2) / 2 * 3
Out[11]: 106.0
```

The precedence level influences how you use *whitespace* within an expression. For example, the expression that follows will execute, but you might find it less readable than the previous version:

```
In [12]: 10 ** 2 + (6-2)/2*3
Out[12]: 106.0
```

You can find guidelines for improving the readability of expressions in PEP8 (*https://pep8.org/*). Although there are some set rules—such as never use more than one space, always have the same amount of whitespace on both sides of a mathematical operator, and surround assignment (=) and augmented assignment operators (like +=) with a single space—you're mostly free to use your own judgement. If you have poor eyesight, you might prefer to use more whitespace than is recommended.

The math Module

The Python standard library includes a math module that provides access to underlying C library functions. *Functions* are like mini-programs that perform some task or tasks. They hide the details of these programs from you so that you can write cleaner code.

To use a function, enter the function name followed by parentheses. Values or variables you enter in the parentheses will be input to the function. We look at functions in more detail in Chapter 11, including how to write your own custom versions.

Groups of related functions are often gathered into *modules*. The math module lets you efficiently perform common and useful mathematical calculations including working with factorials, quadratic equations, and trigonometric, exponential, and hyperbolic functions. It also includes constants including π and e. A subset of the available functions is listed in Table 7-3.

To use the math module, you first must *import* it using an import statement. Think of this as checking a book out of a library. As there are literally thousands of available modules, you don't want them all to load by default. This would be like emptying all the book shelves in a library onto your desk at once. Instead, you just take down books you need. Importing modules follows this principle with respect to your computer's memory.

Table 7-3: A Subset of Python Math Module Functions

Function	Description
`ceil(x)`	Returns the smallest integer greater than or equal to x
`fabs(x)`	Returns the absolute value of x as a floating-point number
`factorial(x)`	Returns the factorial of x
`floor(x)`	Returns the largest integer less than or equal to x
`frexp(x)`	Returns the mantissa and exponent of x as the pair (m, e)
`isnan(x)`	Returns True if x is a NaN (Not a Number)
`exp(x)`	Returns e**x
`log(x[, b])`	Returns the logarithm of x to the base b (defaults to e)
`log2(x)`	Returns the base-2 logarithm of x
`log10(x)`	Returns the base-10 logarithm of x
`pow(x, y)`	Returns x raised to the power y
`sqrt(x)`	Returns the square root of x
`acos(x)`	Returns the arc cosine of x
`asin(x)`	Returns the arc sine of x
`atan(x)`	Returns the arc tangent of x
`atan2(y, x)`	Returns the arc tangent of y / x
`cos(x)`	Returns the cosine of x
`hypot(x, y)`	Returns the Euclidean norm, sqrt(x**2 + y**2)
`sin(x)`	Returns the sine of x
`tan(x)`	Returns the tangent of x
`degrees(x)`	Converts x from radians to degrees
`radians(x)`	Converts x from degrees to radians

Let's use the math module to calculate the cosine of 45 degrees:

```
In [13]: import math

In [14]: x = math.radians(45)

In [15]: math.cos(x)
Out[15]: 0.7071067811865476
```

Start by importing the math module, converting 45 to radians (all trigo-nometric calculations in Python use radians), and assigning the result to the variable, x. Note that you enter the name of the module followed by a dot (.), and the radians() function with the angle you want to convert in the parentheses. Using a dot in this manner is called *dot notation*. It tells Python to use the math module's radians() function. You can think of it as an apostrophe indicating possession: "math's radians() function."

Finally, call the cos() function on x. You can also assign this value to a variable as follows:

```
In [16]: cos_x = math.cos(x)

In [17]: cos_x
Out[17]: 0.7071067811865476
```

Next, let's use math to access π and calculate the circumference of a circle with a diameter of 100 units:

```
In [18]: 100 * math.pi
Out[18]: 314.1592653589793
```

The math module handles basic math well, but for more advanced functionality, such as calculus, you'll want to use external libraries like *SymPy*, which we examine in later chapters. In the meantime, to learn more about math and see a complete list of the available functions and constants, along with detailed documentation, visit *https://docs.python.org/3/library/math.html*.

NOTE   *Recalling information that you've recently learned helps you retain the knowledge. Take a few minutes to complete this short quiz. You can find answers and suggestions in the appendix.*

TEST YOUR KNOWLEDGE

1. True or false: Statements are computational instructions that evaluate to a single value.

2. The expression 12%4 evaluates to:

   a. 3

   b. 48

   c. 0

   d. 12.4

3. The mathematical operator with the highest precedence is:

   a. Power (**)

   b. Floor division (//)

   c. Parentheses (())

   d. Negative and positive arguments (-n, +n)

4. Write a line of code that first takes the square root of 42 and then raises the result to the 4th power.

Error Messages

As soon as you start coding, you're going to make mistakes. One issue is that computers are much more literal than people. You and I can be very flexible when it comes to contextual meaning, grammar, and even spelling, but with computers, what you see is what you get (Figure 7-2).

Figure 7-2: Computers take everything literally.

You can't bend Python's syntactic rules like you can the grammatical rules that govern human speech. When you try to perform an illegal operation in Python, such as divide a number by zero, it halts execution and displays an error message, a process called *raising an exception*.

Let's look at an example that a human could handle but Python can't:

```
In [16]: 25 / 'five'
Traceback (most recent call last):

File "C:\Users\hanna\AppData\Local\Temp/ipykernel_8852/1797604750.py", line 1, in <module>
25 / 'five'

TypeError: unsupported operand type(s) for /: 'int' and 'str'
```

Python displayed an error message indicating a TypeError because you tried to divide an integer (int) by a string of text (str). Although you and I can easily guess the correct answer, Python won't even try, because you mixed *data types* (more on these in a moment). To Python, this is as silly as dividing 25 by "Steve."

Now, let's try to divide by zero:

```
In [20]: x = 42 / 0
Traceback (most recent call last):

File "C:\Users\hanna\AppData\Local\Temp/ipykernel_22688/3599633117.py", line 1, in <module>
42 / 0

ZeroDivisionError: division by zero
```

This raises the aptly named ZeroDivisionError and again provides a record, called a *traceback*, which describes where the interpreter encountered a problem in your code. In this case, the traceback includes the assignment statement that caused the exception and the type of error encountered. For some errors, it will also provide a pointer (^) to where the exception occurs in the line.

NOTE *In many cases, it's the line* before *the line referenced in the traceback that causes the problem. So always remember to look up!*

Knowing the type of error that the interpreter encountered will help you debug your code when you make mistakes. Table 7-4 lists some of the common error types that you'll encounter (you can find more at *https:// docs.python.org/3/library/exceptions.html*). Don't worry if you don't understand them all now. They should make more sense by the end of this primer.

Table 7-4: Some Common Python Error Types

Error type	Thrown when...
SyntaxError	a syntax error is encountered.
IndexError	trying to access an item at an invalid index.
ModuleNotFoundError	a module or package can't be found.
KeyError	a dictionary key can't be found.
ImportError	a problem occurs when loading a module or package.
StopIteration	the next() function goes beyond the iterator items.
TypeError	an operation or function is applied to data of an inappropriate type.
ValueError	a function's argument is of an inappropriate type.
NameError	an object (variable, function, and so on) can't be found.
RecursionError	the maximum recursion depth exceeded (long-running loop terminated).
ZeroDivisionError	the denominator in the division operation is zero.
MemoryError	an operation runs out of memory.
KeyboardInterrupt	the user presses the interrupt key (such as CTRL-C) during execution.

Errors are no big deal. The last line in a traceback includes the error type and a brief explanation (such as NameError: name 'load' is not defined). If you copy and paste this line into a search engine, you'll find lots of friendly explanations that are easier to understand than the overly technical ones provided in the traceback report and the official documentation.

Later, we'll look at ways to handle certain exceptions so that a program can keep running rather than crash when it encounters one. It's also possible to write custom exceptions for a specific program in the event that the supplied exceptions are insufficient.

Data Types

Just as errors have types, every value in Python is automatically assigned to a specific data type. This lets Python distinguish between the letters of the alphabet, like "abc," and numbers, like "123."

The same principles apply to humans. We wouldn't try to multiply letters together (unless we were doing algebra). Nor would we name our children using numbers (unless we were Elon Musk). Without conscious thought, our brains recognize different types of data, and after we've categorized that data, we know how to use it.

In computer science, a *data type* is a classification that dictates what values objects can hold (in other words, what input is acceptable) and how they can be used (what operations can be performed using them, such as converting text to lowercase). Whereas many programming languages use *static typing* that requires you to explicitly declare the data type for any variable you create, Python uses *dynamic typing*, wherein variables can be any data type and even change types during execution. This makes Python a friendlier language, though this comes at a cost. Languages using static typing are better at catching bugs because they can check that data is being used correctly before the program runs.

NOTE *Python permits optional static typing using* type hints. *We won't cover these here, but you can learn more at* https://www.python.org/dev/peps/pep-0484/.

Let's begin by looking at some of the built-in data types that you'll use with Python (Table 7-5). Because numbers and text occur in pretty much every computer program, here we'll focus on three data types: *strings, integers*, and *floating-point numbers (floats)*; we cover other data types in subsequent chapters. These three data types are highlighted in bold in Table 7-5.

Table 7-5: Some Common Data Types

Category	Data type	Examples
Numeric type	**Integer**	`-1, 0, 1, 4000`
Numeric type	**Float**	`-1.5, 0.0, 0.33, 4000.001`
Numeric type	Complex	`a = 4 + 3j`
Text type	**String**	`'a', "b", "Hello, world"`
Sequence type	Tuple	`(2, 5, 'Pluto', 4.56)`
Sequence type	List	`[2, 5, 'Pluto', 4.56]`
Sequence type	Range	`range(0, 10, 1)`
Set type	Set	`{2, 5, 'Pluto', 4.56}`
Set type	Frozenset	`frozenset({2, 5, 'Pluto', 4.56})`
Mapping type	Dictionary	`{'key': 'value'}`
Boolean type	Bool	`True, False`

Additional binary types, not listed in Table 7-5, include *Bytes*, *Bytearray*, and *Memoryview*. For more on all these built-in types, visit *https://docs.python .org/3/library/stdtypes.html*.

Accessing the Data Type

You can query for data type using the built-in type() function that ships with Python. Enter a value or variable in the parentheses, as in the following code:

```
In [21]: type(0.5)
Out[21]: float

In [22]: type(0)
Out[22]: int
```

You can also use the isinstance() function to check whether a variable is an instance of a particular data type. For example, to check whether the integer 42 is an integer or a string, enter 42 in the parentheses, along with the data type you're checking for, as follows:

```
In [23]: x = 42

In [24]: isinstance(x, int)
Out[24]: True

In [25]: isinstance(x, str)
Out[25]: False
```

Much like the human brain, Python can recognize data types based on context. Numbers without a decimal are considered integers. Numbers with a decimal point are floats, even if there are no values behind the decimal point (such as 5.). Strings are identified by enclosing characters in quotation marks (such as "Hello" or '123').

Integers

The *integer* type represents whole numbers, such as 0, 42, and 5,280. The length of an integer is limited only by your system's maximum available memory.

Python recognizes integers by the absence of a decimal point:

```
In [26]: whole_number = 42

In [27]: type(whole_number)
Out[27]: int
```

When working with large numbers, you can use an underscore (_) to separate thousands, such as 15_000_000 for 15000000. Python doesn't need this separator to understand these values, but you'll have an easier time reading them. It reduces key-in errors and saves you from having to count lots of zeros:

```
In [28]: 30_000_000 * 2
Out[28]: 60000000
```

Later in this chapter, we'll look at how to make the output more readable, as well.

Floats

Floats, or *floating-point numbers*, have decimal points. They include 0.0, 0.42, and 3.14159. With floats, you get 15 to 17 digits of precision. Small rounding errors, caused by the universal CPU need to store digits in the binary number system, mean that floats aren't always perfectly accurate. For example, notice that the following addition results in an extra 0.00000000000000004:

```
In [29]: 0.1 + 0.1 + 0.1
Out[29]: 0.30000000000000004
```

If you need more exact precision for scientific work, you can use the built-in decimal module (*https://docs.python.org/3/library/decimal.html*). For more on floating-point accuracy, see *https://docs.python.org/3/tutorial/floatingpoint.html*.

Converting Floats and Integers

Operations using integers sometimes return integers and sometimes return floats. Try the following in the console:

```
In [30]: x = 42 * 2

In [31]: x
Out[31]: 84

In [32]: type(x)
Out[32]: int

In [33]: y = 42 / 2

In [34]: y
Out[34]: 21.0

In [35]: type(y)
Out[35]: float
```

Even though most operations between integers always yield an integer, division may not (42 / 5, for example). Because dividing an integer by an integer can result in a float, Python automatically converts the quotient into a floating-point value, even if the result is still an integer.

The process of converting from one data type to another is known as *type casting*. This can occur *implicitly*, as in the previous example, or *explicitly*, in which you use predefined functions. Explicit type casting is commonly used on user input, to ensure that the input values are the proper type for subsequent operations.

With Python, you can convert integers to floats in several ways. One is to combine them in the same mathematical operation. Notice that adding a float value to an integer turns that integer into a float:

```
In [36]: x = 5

In [37]: type(x)
Out[37]: int

In [38]: x += 0.0

In [39]: type(x)
Out[39]: float
```

You can also use explicit type casting with the float() built-in function:

```
In [40]: x = float(5)
```

If x is an integer, the following would work as well:

```
In [41]: x = float(x)
```

To convert a float into an integer, use the int() built-in function:

```
In [42]: y = 5.8

In [43]: y = int(y)

In [44]: y
Out[44]: 5
```

Note that int() simply drops the decimal part and keeps the whole number to the left of the decimal point. If you want to account for any fractional remainder, you'll need to use rounding.

Rounding

To *round* a floating-point number up or down to the nearest integer rather than just remove the decimal part, you'll want to use the built-in round() function. In the following example, we use round() to convert the float 5.89 to the nearest integer, 6:

```
In [45]: y = 5.89

In [46]: y = round(5.89)

In [47]: y
Out[47]: 6
```

```
In [48]: type(y)
Out[48]: int
```

The round() function rounds to no decimal places by default and returns an integer. To specify the number of significant digits for rounding, include the number after the value to round. In the following example, we round the value of the y variable to one decimal place:

```
In [49]: y = 5.89
```

```
In [50]: y = round(y, 1)
```

```
In [51]: y
Out[51]: 5.9
```

Because you preserved a value after the decimal point, y is still a float.

When working in an interactive console, you can also round numbers directly, without the need for a variable:

```
In [52]: round(5.678, 2)
Out[52]: 5.68
```

If a float value is halfway between integer values, the rounding function rounds odd numbers up, and even numbers down, as follows:

```
In [53]: round(5.5)
Out[53]: 6
```

```
In [54]: round(4.5)
Out[54]: 4
```

As you can see from the previous examples, you should always be aware of data types when working with numbers. Integers can automatically convert to floats during the processing of your code, and vice versa. For example, doing any operation that *uses* a float (such as 5 *= 1.0), or *results* in a float (such as 5 /= 3), will produce a float.

TEST YOUR KNOWLEDGE

5. Write an expression that raises a SyntaxError.

6. What error would you expect from the expression, round('Alice'):

    a. a TypeError

    b. a ValueError

    c. a NameError

    d. a SyntaxError

7. Round π to five decimal places.

Strings

Strings, also known as *string literals*, are what we think of as text values. You can recognize them by their quotation marks. The most famous string value in programming, "Hello, World!", is generally the first thing you learn to print.

Strings should be surrounded by quotes, which mark the beginning and end of the string. In the console, enter:

```
In [55]: a_string = "Hello, World!"

In [56]: print(a_string)
Hello, World!

In [57]: type(a_string)
Out[57]: str

In [58]: type('1234')
Out[58]: str

In [59]: """Multiline strings can be encased in triple quotes \
    ...: so you don't have to type the marks over and over \
    ...: like a chump."""
Out[59]: "Multiline strings can be encased in triple quotes so you don't have
to type the marks over and over like a chump."
```

Generally, you should encase strings in single quotes, but if you need to include a single quote within your string, say, as an apostrophe, you can use double quotes, as in line In [55]. In line In [58], note how numbers are treated as strings if they're surrounded by quotes. You won't be able to use these numbers in mathematical expressions without converting them into a numeric type, like integers or floats.

Triple quotes (""") let you stretch strings across multiple lines. Although computers don't care how long a line of code is, humans do. For readability, PEP 8 recommends a maximum line length of 79 characters. If you want to write a very long string, say, for in-code documentation, you can use triple quotes at the start and end of the string, as in line In [59].

To honor the line length guideline, you can use the line continuation character (\) to break lines between triple quotes. But note that, if you're using strings with single or double quotes, you'll need to place it *outside* the strings, as shown here:

```
In [60]: 'Hello, ' \
    ...: 'World!'
Out[60]: 'Hello, World!'
```

Triple quotes also allow you to add simple drawings to your program, such as a grid to show board positions in a tic-tac-toe game:

```
"""

0 | 1 | 2
---------
3 | 4 | 5
---------
6 | 7 | 8

"""
```

Finally, you can convert other data types to strings using the str() function. The following example converts an integer into a string:

```
In [61]: x = 42

In [62]: type(x)
Out[62]: int

In [63]: x = str(x)

In [64]: type(x)
Out[64]: str
```

NOTE *Behind the scenes, strings are sequences of characters in* Unicode, *an international encoding standard in which each letter, digit, or symbol is assigned a unique numeric value. Unicode ensures that computers everywhere see an A as an A, and a ☺ as a happy face.*

Escape Sequences

Escape sequences are special characters that let you insert otherwise illegal text into a string. In the previous section, we were able to include a single quote apostrophe by first enclosing the string in triple quotes. With the backslash (\) escape character, used *within* the quotes, we could use single quotes exclusively:

```
In [65]: print('I don\'t have a banana.')
I don't have a banana.
```

Note that the backslash doesn't appear in the final string. To print a backslash character, you'll need to escape it with another backslash:

```
In [66]: print("I don't have an apple\\banana.")
I don't have an apple\banana.
```

Table 7-6 lists some useful escape sequences and their result.

Table 7-6: Useful Python Escape Sequences

Escape sequence	Result
\'	Single quote (')
\"	Double quote (")
\\	Backslash (\)
\a	Ring ASCII bell (such as print('\a') in Windows 10)
\n	ASCII linefeed (newline)
\r	ASCII carriage return
\t	ASCII tab

For the full list of escape sequences, visit the documentation at *https://docs.python.org/3/reference/lexical_analysis.html.*

Raw Strings

Raw strings don't recognize escape sequences. These are helpful when you need to deal with lots of backslashes, such as in a Windows path name. In a normal string, you must escape backslashes with the \\ escape sequence, which can become awkward:

```
In [67]: print('C:\\Users\\hanna\\anaconda3\\envs')
C:\Users\hanna\anaconda3\envs
```

With a raw string, what you see is what you get. To use one, just add an r prefix to the string, before the first quotation mark:

```
In [68]: print(r'C:\Users\hanna\anaconda3\envs')
C:\Users\hanna\anaconda3\envs
```

Operator Overloading

Python can apply context to certain operators depending on whether they're used with numbers or strings. An operator doing different things on different data types is known as *operator overloading*. That sounds bad, but it's not. To see an example, enter the following code:

```
In [69]: 'Hello, ' + 'world!'
Out[69]: 'Hello, world!'
```

When used with strings, the + addition operator becomes the *string concatenation* operator. Also note that spaces are legitimate characters, so I added a space before the single quote at the end of Hello. Alternatively, the space could have been added before world, or separated out entirely, as shown here:

```
In [70]: 'Hello,' + ' ' + 'world!'
Out[70]: 'Hello, world!'
```

Likewise, the * multiplication operator becomes the *string replication* operator when a string is multiplied by an integer:

```
In [71]: 'Ha' * 7
Out[71]: 'HaHaHaHaHaHaHa'
```

This can be useful for drawing in scripts, such as generating a dividing line within your code:

```
In [72]: '-' * 20
Out[72]: '--------------------'
```

Of course, you can't mix and match these easily among data types. You can't add a number to a string, for instance, or multiply two words together.

String Formatting

In many cases, you'll want to create a string that includes other strings. For example, you might want to reference a variable in the print() function. *Format strings*, also called *f-strings*, make this easy. You just need to prefix the string with an f and put the variable name in curly brackets, as follows:

```
In [73]: solute = 'salt'

In [74]: solvent = 'water'

In [75]: print(f'{solute} dissolves in {solvent}')
salt dissolves in water
```

NOTE *If you're working in the console, you can omit the print() function and apply the f-string directly (such as: f'{solute} dissolves in {solvent}').*

Within an f-string, expressions in curly brackets are evaluated at runtime:

```
In [76]: print(f"The circumference of a 10-inch circle is {10 * 3.14159}")
The circumference of a 10-inch circle is 31.4159
```

You can also specify the text's alignment with f-strings, letting you create tabular output. In the following example, 25 spaces are reserved and < justifies these spaces to the left, ^ centers the text, and > justifies to the right:

```
In [77]: print(f'{"output1" : <25}')
   ...: print(f'{"output2" : ^25}')
   ...: print(f'{"output3" : >25}')
output1
           output2
                        output3
```

You can use f-strings to format numeric values. To add commas to a long number, use this format:

```
In [78]: long_number = 93000000

In [79]: print(f'{long_number:,}')
93,000,000
```

To use exponential notation, use the e qualifier:

```
In [80]: speed_of_light = 299792458

In [81]: print(f'{speed_of_light:e}')
2.997925e+08
```

To format numbers to a specific precision point, use the f qualifier. For example, to print Euler's number, e, to three decimal places, enter:

```
In [82]: e = 2.718281828459045

In [83]: print(f'{e:.3f}')
2.718
```

To convert a number to a percent, use the % qualifier. Include a number to specify the number of decimal places to preserve:

```
In [84]: num = 0.456

In [85]: print(f'{num:.2%}')
45.60%
```

As you can see, f-strings make code very readable, so long as your variable names make sense.

10. If x = '30_000_000', what data type is x?
11. Which of the following is the result of running the code f'{3.14159:.2f}' in Jupyter Qt console?

    a. '3.14'

    b. '314,159'

    c. '3.141590e+00'

    d. '314.15%'

12. Draw an owl in the text editor and print it to the screen. Each line of code should not exceed 79 characters.

13. Write a program that converts 1,824 seconds into minutes and seconds and then print the results.

String Slicing

Each character in a string has a unique index that locates it within the string. Think of this as the character's address. Python starts counting with 0; thus, the index for the first character in a string is 0, not 1.

In the console, enter the following to retrieve the first and last characters in the string 'PYTHON':

```
In [86]: x = 'PYTHON'

In [87]: x[0]
Out[87]: 'P'

In [88]: x[5]
Out[88]: 'N'
```

To use an index, enter the variable name (such as x) with the index that you want in square brackets ([]). Note that, even though PYTHON has six characters, the last index is 5, again because Python starts counting at 0.

If you ask for an index beyond the end of the string, you will get the (very common) index out of range error:

```
In [89]: x[6]
Traceback (most recent call last):

File "<ipython-input-89-04aa5bc9ecce>", line 1, in <module>
x[6]

IndexError: string index out of range
```

You can also *slice* strings (and many other data types) using indexes. Slicing lets you chop a string into smaller pieces. For example, you might take the first three characters, the last two, the one in the middle, and so on.

To slice a string, enter endpoints that encompass the characters in which you're interested. For example, to get the first three characters in PYTHON, enter the following:

```
In [90]: x[0:3]
Out[90]: 'PYT'
```

Note that you get the characters at indexes 0, 1, and 2, but not at index 3. When slicing, Python takes everything *up to* but *not including* the ending index.

Because the starting and ending indexes are used so often, Python provides a shorthand technique in which you leave off those indexes. Rerun the preceding code, omitting the 0:

```
In [91]: x[:3]
Out[91]: 'PYT'
```

To return the whole string, you can use just the colon:

```
In [92]: x[:]
Out[92]: 'PYTHON'
```

You can also take larger strides through the string by specifying a step size. The default step size is 1. If you want to start at the beginning and take every other character, add another colon followed by a step size of 2:

```
In [93]: x[::2]
Out[93]: 'PTO'
```

The extra colon is needed because we are using the syntax x[start:end:step]. When no values are provided, Python defaults to taking the starting and ending indexes, for convenience.

You can slice going forward or backward through a string. To slice in reverse, you use *negative* indexes. For example, if you just want the end part of a string, you can use negative indexes when slicing. To get the last character and the last three characters, respectively, enter:

```
In [94]: x[-1]
Out[94]: 'N'

In [95]: x[-3:]
Out[95]: 'HON'
```

Note that the "first" index in reverse is -1, not 0 as you might expect.

To print the string in reverse, step backward one character at a time using -1:

```
In [96]: x[::-1]
Out[96]: 'NOHTYP'
```

Membership Operators

The in and not in operators tell you whether a character or substring exists within a string. For example:

```
In [97]: 'e' in 'scientist'
Out[97]: True

In [98]: 'engineer' not in 'I am a scientist'
Out[98]: True
```

This functionality is useful with conditional statements, which we'll cover in more detail later. For example, if you'd like to know whether 'Waldo' is part of the string contained in the variable x, you could enter the following:

```
In [99]: x = "Here's Waldo!"

In [100]: if 'Waldo' in x:
     ...:         print("I found Waldo!")
I found Waldo!
```

String Methods

A nice thing about data types is that they come with methods (a type of function) that helps you manipulate them. Methods represent actions that data types can perform. For example, although the in operator tells you that a character or substring is *present*, it doesn't tell you how many times it occurs. If you want to count number of occurrences, you can use the count() method.

To count the number of times lowercase "i" occurs in the string "I am a scientist," enter the string (or a variable representing the string), followed by a dot (.) and the count() method with the character or substring you're searching for in the parentheses:

```
In [101]: 'I am a scientist'.count('i')
Out[101]: 2
```

Python comes with a long list of string methods (see *https://docs.python .org/3/library/stdtypes.html#string-methods/*). Some of the more-used methods are listed in Table 7-7. You should replace text in bold with your specific string or substring. Text in italics is optional. For example, the *start* and *end* indexing options default to the starting and ending indexes of the string, respectively, if you choose to ignore them.

Table 7-7: Common String Methods

Method	Description
str.capitalize()	Capitalizes the first character with the rest lowercase
str.count(**sub**, *start*, *end*)	Counts occurrences of a character or substring
str.endswith(**suffix**, *start*, *end*)	Returns True if string ends with the suffix
str.find(**sub**, *start*, *end*)	Returns lowest index where substring is found within slice
str.isalnum()	Returns True if all string characters are alphanumeric
str.isalpha()	Returns True if all string characters are alphabetic
str.isdigit()	Returns True if all string characters are digits
str.islower()	Returns True if all string characters are lowercase
str.isupper()	Returns True if all string characters are uppercase
str.lower()	Changes all cased characters to lowercase
str.replace(**old, new,** *count*)	Replace old substring with new substring
str.split(sep=None, *maxsplit=-1*)	Return list of words with "sep" character as the delimiter
str.startswith(**prefix**, *start*, *end*)	Return True if string starts with the prefix
str.strip(**chars**)	Removes leading and trailing characters; removes whitespace if no character is specified
str.title()	Capitalizes the first character in each word
str.upper()	Changes all cased characters to uppercase

Because strings are *immutable* (unchangeable), these methods return *copies* of strings rather than alter the original object. Enter the following into the console:

```
In [102]: x = 'string'

In [103]: print(x.upper())
STRING

In [104]: x
Out[104]: 'string'
```

In this example, you assigned a lowercase string ('string') to the x variable. You then called the upper() string method on x as you printed it. Python was able to see and use the uppercase string, but when you use the x variable later, it's still in its original lowercase form.

To make x always refer to an uppercase string, you need to reassign it to itself, as follows:

```
In [105]: x = x.upper()

In [106]: x
Out[106]: 'STRING'
```

To see another example of string immutability, let's try to change the I in 'STRING' to A using its index:

```
In [107]: x[3] = 'A'
Traceback (most recent call last):

File "<ipython-input-106-124534701dc6>", line 1, in <module>
x[3] = 'A'

TypeError: 'str' object does not support item assignment
```

This raised a TypeError, as the string data type is immutable.

To (sort of) get around immutability, use the replace() method, which requires the creation of a new variable:

```
In [108]: old_string = "I'm the old string."

In [109]: new_string = old_string.replace('old', 'new')

In [110]: new_string
Out[110]: "I'm the new string."
```

The split() method breaks up a string and returns a *list* data type (covered in Chapter 9). For example:

```
In [111]: caesar_said = 'Tee-hee, Brutus.'

In [112]: words = caesar_said.split()

In [113]: print(words)
Out[113]: ['Tee-hee,', 'Brutus.']
```

If you look carefully at the results, you'll notice that punctuation marks get lumped in with words.

By default, split() uses a space as the *delimiter* character on which to split the string. You can also specify a delimiter such as a hyphen or, as in the following case, a comma with trailing whitespace:

```
In [114]: words = caesar_said.split(sep=', ')

In [115]: words
Out[115]: ['Tee-hee', 'Brutus.']
```

Note that, compared to line `Out[113]`, the comma is no longer attached to `Tee-hee`, but the period is still attached to the `Brutus` item.

To strip the punctuation marks, import the `string` module, which includes a `punctuation` string, and use the built-in `translate()` function to eliminate them:

```
In [116]: from string import punctuation

In [117]: print(punctuation)
!"#$%&'()*+,-./:;<=>?@[\]^_`{|}~

In [118]: no_punc = caesar_said.translate(str.maketrans('', '', punctuation))

In [119]: no_punc
Out[119]: 'Teehee Brutus'
```

The `maketrans()` method takes three arguments; the first two are empty strings (`''`), and the third is the list of punctuation marks to remove. This tells the function to replace all punctuation with `None`.

Now, you can split the string on white space and get a list of words only:

```
In [120]: no_punc.split()
Out[120]: ['Teehee', 'Brutus']
```

Note that the remaining comma in line `Out[120]` is part of the list and is used to separate items in the list, like `Teehee` and `Brutus`. It doesn't count as part of a string. Note also that the hyphen is missing from `Teehee`. This is because the `punctuation` string includes a hyphen.

There are a lot of ways to strip unwanted characters from strings with Python. For working with large bodies of text, you'll want to use either *regular expressions* (*regex*) or a *natural language processing* library. These tools are specially designed for working with text, and we explore some of them in Chapter 15.

TEST YOUR KNOWLEDGE

14. Which of the following is the result of running the code `'latchstring'[2:8]`?

    a. `'atchstr'`

    b. `'tchstr'`

    c. `'gnirts'`

    d. `'atchst'`

(continued)

15. To get a subset of a string value, you should use:

    a. negative indexes

    b. string iteration

    c. augmented operators

    d. string slicing

16. Running the code `'latchstring'[12]` results in:

    a. a `SyntaxError`

    b. a `StopIteration` error

    c. an `IndexError`

    d. a `ValueError`

17. Using the previous `caesar_said` example, use the `translate()` function to remove all the punctuation, *except for* the hyphen.

18. Convert the string `'impractical python projects'` to "title" format.

Summary

In this chapter, you learned that *expressions* are instructions that evaluate to a single value, like a mathematical formula. *Statements* express some action to be carried out but don't evaluate to a value. *Operators*, which are symbols used to represent an action or process, perform a function, or manipulate values in some way.

You also learned that a *variable* is a *label* for data stored in memory. Variables have *names* and *values*. Every value in Python is automatically assigned a *data type*, and you learned about the three most basic types: *integers*, *floats*, and *strings*.

In the next chapter, you'll learn more about variables, which are objects that let you connect to and manipulate data.

8

VARIABLES

Variables are one of the most important concepts for beginning programmers to understand, so this chapter explores these features in detail. Technically speaking, a variable is a reserved memory location used to store values. It's a reference, or *pointer*, to an object in memory, but it's not the object itself. Variables let you access and manipulate each object's associated metadata (*attributes*) and functionality (*methods*).

In the previous chapter, you learned how to assign variables. In this chapter, you'll learn more about assignment statements, discover how to name variables clearly, use a built-in function to get a user's input, and practice comparing one variable to another.

Variables Have Identities

Python encapsulates data and the functions that operate on that data into named entities known as *objects*. As the fundamental building block for the language, everything in Python is an object, and every object has an *identity* (memory address), a *type*, and a *value*. The number 42 is an object as well as the sentence "Hello, world!" An object's identity and type never change, but its values can sometimes be changed.

Variables can be thought of as labels for objects. Just as you can be referred to by multiple names and nicknames (Figure 8-1), an object in Python can be referenced by many variables.

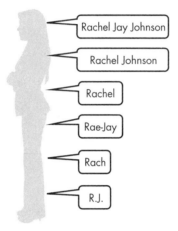

Rachel Jay Johnson

Rachel Johnson

Rachel

Rae-Jay

Rach

R.J.

Figure 8-1: We can go by multiple names; so can objects in Python.

When you use an assignment statement, such as x = 5, the variable x is initialized as a reference to the object on the right-hand side of the equal sign. It's also given an integer number as an identity. This number is unique for all existing objects. You can view this number with the built-in id() function that ships with Python. Note that the ID numbers you see on your computer might be different from mine.

```
In [1]: x = 5

In [2]: id(x)
Out[2]: 140718638636928
```

You can overwrite a variable by assigning it a new value:

```
In [3]: x = 15

In [4]: id(x)
Out[4]: 140718638637248
```

The reassigned x now has a new identity. In the first example, the variable x is a reference to an integer object with a value of 5. When you overwrite x with the new value of 15, the old object continues to exist, but if no variable is referencing it, its *reference count* goes to zero. At this point, it's subject to *garbage collection*, the process by which Python periodically reclaims blocks of memory that are no longer in use. In some other languages, you must manually designate and free memory allocations in your code. Python's ability to automatically manage memory and "clean up its own mess" makes it a very friendly language!

Because variables are just references, multiple variables can point to the same object:

```
In [5]: x = 42

In [6]: id(x)
Out[6]: 140718638638112

In [7]: y = x

In [8]: y
Out[8]: 42

In [9]: id(y)
Out[9]: 140718638638112
```

By assigning x to y in line In [7], both variables now reference the same object, as evidenced by their having the same identity. This type of memory-efficient behavior is called *aliasing*.

If you overwrite x, its identity will change, but y will still point to the "old" object:

```
In [10]: x = 50

In [11]: id(x)
Out[11]: 140718638638368

In [12]: y
Out[12]: 42

In [13]: id(y)
Out[13]: 140718638638112
```

As a result, the old object has a reference count greater than one, so it will not be deleted during garbage collection and will hang around for you to use in expressions, functions, and so on.

Assigning Variables

Assigning a value to a variable is known as *binding* in Python. In addition to the straightforward "x = y" method for assigning variables, you can also do so using expressions, operator overloading, functions, and more. Basically, both values and things that return values can be assigned to a variable.

Using Expressions

The result of expressions can be assigned to a variable:

```
In [14]: x = 6 * 7

In [15]: x
Out[15]: 42
```

Operator Overloading

Likewise, you can use operator overloading when assigning variables. As discussed in the previous chapter, operator overloading refers to the ability of an operator to work in a different manner with different data types. The classic example is using the + sign to add numbers *and* concatenate strings.

Operator overloading works directly with strings or with other variables. Here, we use it with strings:

```
In [16]: name = 'Hari ' + 'Seldon'

In [17]: print(name)
Hari Seldon
```

And here, with variables:

```
In [18]: first_name = 'Hari'

In [19]: surname = 'Seldon'

In [20]: full_name = first_name + ' ' + surname

In [21]: print(full_name)
Hari Seldon
```

In line In [20], notice how I added a space between names. Without it, the printout would be HariSeldon.

Here's another example of operator overloading:

```
In [22]: repeat_name = (full_name + ' ') * 5

In [23]: print(repeat_name)
Hari Seldon Hari Seldon Hari Seldon Hari Seldon Hari Seldon
```

Notice how you can use precedence with the operators. To see the impact, run line In [22] again *without* the parentheses and print the result.

Using Functions

Although we haven't covered them yet, you can use functions in assignment statements. Here, we'll use the built-in count() string method with an assignment statement:

```
In [24]: number_of_y_in_python = 'Python'.count('y')

In [25]: number_of_y_in_python
Out[25]: 1
```

In this case, the count() method returned the value 1, which was then stored in the variable.

Chained Assignment and Internment

You can simultaneously assign the same value to multiple variables using a *chained assignment*:

```
In [26]: answer_to_life = answer_to_the_universe = answer_to_everything = 42
```

Interestingly, Python doesn't create a new object for each of these variables; they all have the same identity:

```
In [27]: id(answer_to_life)
Out[27]: 140718641390624

In [28]: id(answer_to_the_universe)
Out[28]: 140718641390624

In [29]: id(answer_to_everything)
Out[29]: 140718641390624
```

To improve processing speed, Python creates a small cache of memory addresses at startup. It uses some of these for a list of small integer values (-5 to 256). The programming practice of using references in place of copies of equal objects is called *interning*. Larger values not in the cache will get new addresses. For example:

```
In [30]: big_var1 = 5**9

In [31]: big_var2 = 5**9

In [32]: id(big_var1)
Out[32]: 2642973757040

In [33]: id(big_var2)
Out[33]: 2642973756016
```

Some strings are also interned by Python as an optimization. As Python code is compiled, identifiers such as variable names, function names, and class names are interned. Other strings may be interned, as well. Python automatically makes this determination on a code-by-code basis.

An advantage of string interning is that when you're comparing two variables Python can compare *memory addresses*. This is faster than comparing each string character by character.

Generally, Python's default interning is more than sufficient, but you should never rely on it. If you ever need to *ensure* that a string is interned, import the system module (import sys) and use the sys.intern() method, with your string in the parentheses (see *https://docs.python.org/3/library/sys.html*).

Using f-Strings

You can use f-strings (see the section "String Formatting" on page 192) with variables. Just prefix the assignment with f, followed by a starting single or double quotation mark. Place the variables that you want to use within curly brackets and then add the ending quotation mark. Here's an example:

```
In [34]: first_component = 'hydrogen'

In [35]: second_component = 'sulfide'

In [36]: compound = f'{first_component} {second_component}'

In [37]: print(compound)
hydrogen sulfide
```

You can even format the strings within the assignment statement. For example:

```
In [38]: compound = f'{first_component.title()} {second_component.title()}'

In [39]: print(compound)
Hydrogen Sulfide
```

In this case, we called the title() string method on each variable in the curly brackets. This method capitalizes the first letter in a word and converts the remaining letters to lowercase.

Naming Variables

Programs are read far more often than they're written. Your code should be as readable as possible, and not just for other users, but for yourself. It's all too common to return to a program you wrote months ago and have no idea how it works.

A common saying is that "code should be self-documenting." This means that readers should be able to understand your code without relying on explanatory comments. To make your code "self-documenting," you'll want to pay a lot of attention to how you name variables. This involves ensuring that your names are legal and making them as logical and compact as possible.

There are three main rules for naming variables:

- Variables can contain only letters, numbers, or underscores (_)
- The first character cannot be a number
- The name cannot be a *reserved keyword*

Reserved Keywords

Python reserves a set of keywords for its own use (Table 8-1). You cannot use these as variable names, function names, or as any other identifiers.

Table 8-1: Python's Reserved Keywords

Keyword	Description
and	Logical operator
as	Used to alias an imported module or tool
assert	Used in debugging
async	Used to define an asynchronous function
await	Specifies the point in an asynchronous function at which control is given back to an event loop
break	Breaks out of a loop
class	For defining a class in object-oriented programming
continue	Continue to the next iteration of a loop
def	Define a function
del	Delete an object
elif	Else-if conditional statement
else	Conditional statement
except	Instructions on how to handle an exception
False	Boolean value
finally	Used for a block of code that executes in spite of exceptions
for	Creates a for loop
from	Imports specific parts of a module
global	Declares a global variable
if	Conditional statement
import	Loads modules
in	Checks whether a value is present
is	Tests whether two variables are equal
lambda	Creates an anonymous function on-the-fly
None	A null value
nonlocal	Declares a non-local variable
not	Logical operator
or	Logical operator
pass	Statement that will do nothing
raise	Raises an exception
return	Exits a function and returns a value or values
True	Boolean value

(continued)

Table 8-1: Python's Reserved Keywords *(continued)*

Keyword	Description
try	Makes a try/except statement
while	Creates a while loop
with	Simplifies exception handling; auto-closes files after loading
yield	Suspends a generator function and returns a value

You don't need to memorize all of these keywords; Python will raise a SyntaxError if you try to use one as a variable name. Here's what happens if you try to assign the value 5 to the pass keyword:

```
In [40]: pass = 5
File "<ipython-input-40-85539e45a032>", line 1
pass = 5
^
SyntaxError: invalid syntax
```

You can also view the list of keywords through Python. Just run import keyword followed by keyword.kwlist.

Besides keywords, you should avoid using the name of one of Python's built-in functions, like print() or id(), as a variable name. There's nothing to prevent this from happening, however. For example:

```
In [41]: print = 5

In [42]: print
Out[42]: 5
```

You'll regret doing this, as print now refers to the integer 5. If you try to use the print() function, Python will raise an error:

```
In [43]: print("Hello, World!")
Traceback (most recent call last):

File "<ipython-input-43-2223c92d0779>", line 1, in <module>
print(print)

TypeError: 'int' object is not callable
```

To fix this, you'll need to delete the print variable using del(print). This will restore the print() function and allow it to work again.

Many of these function names are ones that you'll be tempted to use, such as min, max, sorted, list, set, slice, and sum. We'll look at a list of these built-in functions and their purpose in Chapter 11. In the meantime, you just need to be observant when naming variables. The console and text editor in Spyder will highlight these special names with a unique color. I can't show this in a black-and-white book, but you can see for yourself by entering the following code in the console:

```
In [44]: spam = 42

In [45]: list
```

The spam variable should be colored differently than the list variable because list is a built-in function name. If list is colored purple in your console, avoid using purple names for variables.

NOTE *If you insist on using a reserved keyword or built-in function name, you can avoid conflicts by adding an underscore to the variable name, such as sum_, max_, or class_. Even better, add a descriptor after the underscore, such as max_pressure. Everybody wins!*

Variables Are Case Sensitive

Python is a case-sensitive programming language. Not only must you spell variable names correctly to access them, but you must also use the same arrangement of uppercase and lowercase characters. For example:

```
In [46]: declination = 80

In [47]: print(declination)
80

In [48]: print(Declination)
Traceback (most recent call last):

File "<ipython-input-48-d1839757958b>", line 1, in <module>
print(Declination)

NameError: name 'Declination' is not defined
```

Python doesn't recognize the Declination variable due to the capitalized first letter, and it raises a NameError as a result.

Best Practices for Naming Variables

Here are some suggestions for ensuring that your variable names are Pythonic. You can also find a section on naming conventions in Python's PEP8 style guide at *https://pep8.org/#naming-conventions/*.

You should use the underbar (_) to separate words within a variable name. For example:

```
In [49]: the_answer_to_life_the_universe_and_everything = 42
```

You should also use *lowercase* characters in most cases and reserve the use of capitals for special objects, such as *constants*. Constants are values that should not change during the program's execution, and you can let others know that a variable represents a constant by naming it with all caps. For example, to assign the speed of light to a constant, you could use the following:

```
In [50]: SPEED_OF_LIGHT = 299_792_458
```

Constants in Python have a *contextual* meaning that's not enforced by the Python interpreter. Using all caps for the name only alerts other programmers of your intent. Otherwise, constants can be overwritten like any other variable.

Variable names should be logical and descriptive but try to achieve this goal using as few characters as possible. Long variable names are not only difficult to type, but they can also cause lines to wrap, making the code challenging to read.

Here's an example of a name that's clear to a fault:

```
In [51]: distance_from_earth_to_the_sun_in_kilometers = 149_597_870
```

There are several ways in which you can shorten this name, such as:

```
In [52]: earth_sun_distance_km = 149_597_870
```

or:

```
In [53]: earth_to_sun_km = 149_597_870
```

Alternatively, you can include important information, such as units, in *nonexecutable comments*. These will help you to keep your variable names under control. We examine comments in Chapter 14, but here's an example of an inline comment:

```
In [54]: SPEED_OF_LIGHT = 299_792_458   # Meters per second in a vacuum.
```

Python will ignore all the text after the # mark, but humans reading your code will be able to glean more than what is conveyed in the variable name.

Naming variables is an exercise in optimization. You'll be surprised how often you can return to a program and improve your names. You should avoid taking this to extremes, however, as it's easy to over-optimize. If you're used to working in Imperial units, it might be obvious to you that mps means miles-per-second, but to most of the world, it's meters-per-second.

Likewise, eschew variables that end with a numerical suffix, like step1 or step2. These names aren't meaningful, and if you add or delete a step, you'll need to refactor the whole program. It's better to use descriptive names that reference the step, such as denoised_image, or kalman_filtered. Additionally, never use final in a name. This will anger the gods, and you'll surely need another variable after the "final" one.

You should also avoid the characters "l" (lowercase letter L), "O" (uppercase letter o), or "I" (uppercase letter i) as single character variable names. These characters can be confused for the numerals one and zero. In fact, you should avoid single character names altogether, except when the single letter is commonly understood, such as using x and y for Cartesian coordinate values. It's also acceptable to use short names when performing simple tutorial exercises, like the ones we've been using here.

Managing Dynamic Typing Issues

In the last chapter, we talked about how Python is a dynamically typed language, which means that Python can use context to assign a data type, and variables do not have a fixed type. This can lead to complex and difficult-to-debug code, as a variable named x might represent an integer, a string, or even a function at different places in a program.

One way to manage this issue is to change the name of the variable when it changes data types. Compare this code:

```
In [55]: x = '42'

In [56]: x = int(x)

In [57]: x = float(x)
In [58]: type(x)

Out[58]: float
```

with this code:

```
In [59]: x_string = '42'

In [60]: x_integer = int(x_string)

In [61]: x_float = float(x_integer)

In [62]: type(x_float)
Out[62]: float
```

In both examples, the x variable starts out as a string and ends up as a float. Careful and considerate naming practices in the second example, however, help you to track what's going on even if a program has many branches and loops and the assignment statements are many lines apart.

NOTE *Linters that you can get through Spyder, like Pylint (https://pylint.org/), will alert you if you reassign a variable to a different type.*

There's no reason to reuse variable names, because each assignment will create a new object. And you don't need to be as explicit as the previous example and include the data type in the name. The important thing is that you change names when you change data types.

Handling Insignificant Variables

Variables that serve as placeholders are often named using a single lower-case letter, typically "i". Here's an example using a for loop (which we cover in Chapter 10):

```
In [63]: for i in 'Python':
   ...:     print(i)
P
y
t
h
o
n
```

Remember to use SHIFT-ENTER after print(i) to execute the code in the Qt console.

Although there's technically nothing wrong with this strategy, linters that check your code's conformance to the PEP8 guidelines (like we discussed in Chapter 4) will flag the i as an "unused variable." Although you can ignore this, it gets annoying.

To keep from violating a coding standard, you can use an underscore as a throwaway variable name, such as in this example:

```
In [64]: for _ in 'Python':
   ...:     print(_)
P
y
t
h
o
n
```

This will keep the linters quiet and happy.

TEST YOUR KNOWLEDGE

1. Which variable names are valid?

    a. _steve

    b. br549

    c. light-speed

    d. 0579

---

2. Which naming style is recommended for a constant?

   a. GravConstant

   b. GRAV_CONSTANT

   c. GRAV_constant

   d. grav_constant

3. When should you use a single underscore as a variable name?

   a. When you want to use a reserved keyword

   b. When you want to keep the variable private

   c. When you need a placeholder for iteration

   d. When you can't think of a good name

4. When you change a variable's data type, you should _____.

5. Create a new variable and then delete it.

Getting User Input

So far, *we've* been assigning values to variables. In many cases, you'll want to get input directly from a user; for example, for a unit conversion program. Because this is such a common practice, Python provides the built-in input() function. Here's how it works:

```
In [65]: first_name = input('Enter your first name: ')

Enter your first name: Robert

In [66]: first_name
Out[66]: 'Robert'
```

The input() function takes a question, known as a *prompt*, and presents it to the user. This prompt should be as unambiguous as possible so that the user knows what to enter (and in what format). The function then pauses the program until the user enters a value. Because the function expects strings, you don't need to use quotation marks with the input.

The input() function returns a *string*, which means that numbers might need to be converted to integers or floats depending on what your program will do with the input. To make the conversion, call either the int() or float() function in the assignment statement. For example, to ensure that the age variable is an integer, enter the following:

```
In [67]: age = int(input('Enter your age in years: '))

Enter your age in years: 42

In [68]: type(age)
Out[68]: int
```

You might also want the input to be consistent. Because Python is case sensitive, it's common to convert input to lowercase to avoid running into any problems later. Enter the following in the console:

```
In [69]: name = input('Enter your full name: ').lower()

Enter your full name: Chesterfield Walkingstick

In [70]: print(name)
chesterfield walkingstick
```

In this case, we called the lower() string method to automatically convert the input to lowercase.

NOTE   *When your program interacts with actual human users, if anything can go wrong, it will. There's nothing to stop a user from entering their age as "forty-two" rather than "42." Fortunately, Python provides things like* while *loops,* try *statements, and conditional statements, that let you check input for errors and either fix the problem directly or request that the user reenter the value in the format that it expects. We explore these in later chapters.*

The input() function is a fairly primitive way to get user input. Later, in the book, we'll look at more sophisticated methods, such as using a GUI with menus, radio buttons, text boxes, and so on.

Using Comparison Operators

Python provides *comparison* operators, also called *relational* operators (see Table 8-2), that let you compare variables and determine the relationship between them. Each operator returns either True or False.

Table 8-2: Python Relational Operators

Operator	Description	Example
==	If values are equal, condition is True	(a == a) is True
!=	If values not equal, condition is True	(a != b) is True
<>	If values not equal, condition is True	(a <> b) is True

Operator	Description	Example
>	If left value is greater than right, then True	(2 > 6) is False
<	If left value is less than right, then True	(2 < 6) is True
>=	If left is greater than or equal to right, then True	(2 >= 6) is False
<=	If left is less than or equal to right, then True	(2 <= 6) is True
is	Object identity	(a is a) is True
is not	Negated object identity	(a is not b) is True

The operators evaluate from left to right. For example, to check that 10 is larger than 2:

```
In [71]: 10 > 2
Out[71]: True
```

Comparison operators evaluate to a *Boolean* data type that has two values: True and False. These values are always capitalized and unlike strings, they don't require quotes. You can check their type like any other value:

```
In [72]: type(False)
Out[72]: bool
```

Because computers work in binary, True represents 1 (or 1.0), and False represents 0 (or 0.0). You can test this for yourself in the console, as follows:

```
In [73]: a = True
In [74]: int(a)
Out[74]: 1

In [75]: b = False

In [76]: float(b)
Out[76]: 0.0
```

As you can see from this example, you can store Booleans in variables, use them in expressions, and convert them into integers and floats without raising an exception.

The first two operators, both of which evaluate equality, work with any data type:

```
In [77]: 42 == 42
Out[77]: True

In [78]: 'Steve' != 'Steve'
Out[78]: False
```

The other operators, such as > and <=, work only with floating-point values and integers. For convenience, you can chain these together in a single line, such as 2 < x < 5, which states that x is greater than 2 but less than 5.

You can use these relationships to control what your program does. Here's an example using conditional statements, which we investigate in Chapter 10. This example compares the pH values of three samples to determine which is the most acidic:

```
In [79]: sample1_pH = 1.6

In [80]: sample2_pH = 6.0

In [81]: sample3_pH = 7.8

In [82]: if sample1_pH <= sample2_pH:
   ...:         print('Sample 1 is more acidic.')
   ...: else:
   ...:         print('Sample 2 is more acidic.')
Sample 1 is more acidic.
```

In the preceding example, the relational operators determine which statement is printed. You can also chain together comparisons for multiple variables, as demonstrated here:

```
In [83]: sample1_pH < sample2_pH < sample3_pH
Out[83]: True
```

The is and is not operators check whether two objects share the same identity (in other words, point to the same object in memory). This is not the same as the equal to (==) and not equal to (!=) operators, that check *equality*. Let's look at an example:

```
In [84]: x = 1_000_000

In [85]: y = 1_000_000

In [86]: x == y
Out[86]: True

In [87]: x is y
Out[87]: False
```

Because we assigned a large number to x and y, they are equal in value but are not the same object. If we use small values, however, they will be the same object, due to Python's use of a startup memory cache, as discussed in "Chained Assignment and Internment" on page 205. For example:

```
In [88]: a = 256

In [89]: b = 256

In [90]: a is b
Out[90]: True
```

6. If x = 257 and y = 257, what does the code x is y evaluate to?

    a. True

    b. False

7. Write a code snippet that prompts a user for their name and then prints the name in reverse.

8. Write an assignment statement that generates a NameError.

9. Write an assignment statement that generates a TypeError.

10. In the console, what is the output of 'hydrogen sulfide'.title()?

    a. 'Hydrogen Sulfide'

    b. 'HYDROGEN SULFIDE'

    c. AttributeError: 'str' object has no attribute 'title'

    d. NameError: name 'title' is not defined

Summary

In this chapter, you got a closer look at how variables work and how to use them. You learned more about assigning variables as well as some rules and suggestions on how to name them, compare them, and solicit user input.

Because variables are just labels for objects in memory, they can sometimes behave unexpectedly, especially when working with mutable objects. This will become apparent in the next chapter, in which we explore the container data types such as lists, sets, and dictionaries.

9

THE CONTAINER DATA TYPES

According to *Merriam-Webster*, *data* is plural in form. If you're working with data, you're working with *collections* of things such as the names of students in a class or the luminosity of stars in the galaxy. You're going to need somewhere to hold all of these collections. That's where the container data types like tuples, lists, sets, and dictionaries come in handy. Each one serves a special purpose and comes with special abilities. Together, they'll help you to keep your rats, genes, soil samples, and temperature measurements organized and under control.

In this chapter, we'll explore the built-in data structures listed in Table 9-1, along with some of their main characteristics. Remember, *mutability* refers to whether something can be changed (mutated) after it's created. Immutable objects must be copied to a new object for their values to be altered, appended, or removed.

Table 9-1: Container Data Types

Category	Data type	Mutability	Features	Examples
Sequence type	Tuple	Immutable	Fast, efficient, and unchangeable	`(2, 5, 'Pluto', 4.56)`
Sequence type	List	Mutable	Flexible with many built-in functions	`[2, 5, 'Pluto', 4.56]`
Set type	Set	Mutable	No duplicate elements, fast searches	`{2, 5, 'Pluto', 4.56}`
Set type	Frozenset	Immutable	No duplicate elements, fast searches	`frozenset({2, 5, 'Pluto', 4.56})`
Mapping type	Dictionary	Mutable	Maps unique keys to values	`{'key': 'value'}`

Let's begin by looking at the simplest of these, the tuple.

Tuples

A *tuple* (pronounced *TOO-pul*) is a fixed-length, iterable, immutable, ordered sequence of values. These values are typically referred to as *items* or *elements*. Here is an example of a tuple where each name represents an item:

```
('K. L. Putney', 'M. B. Clark', 'S. B. Vaughan')
```

NOTE *An* iterable *is a collection of items that you can loop through, such as strings, tuples, lists, and sets. A* sequence *is a positionally ordered collection of items that acts as a single storage unit.*

Tuples are a lot like strings, but whereas strings can contain only characters, tuples are heterogeneous and can hold any type of value, including mixtures of different types. You can even make tuples of tuples.

Tuples are less commonly used than the *list* data type, and they have fewer methods that work on them. But like the old saying goes, "horses for courses." There are times when tuples are preferable to other container types.

For example, because they're immutable, tuples make good places to store objects like passwords. When you use a tuple, other programmers will understand that you don't want those values to change. Tuples are also more memory efficient than lists due to their immutability, and tuple operations are smaller in size, making them slightly faster when working with many elements.

Creating Tuples

Tuples consist of a sequence of comma-separated values enclosed by parentheses. But Python is smart, and just as it can recognize different data types such as floats and integers by context, you can break the rules when creating a tuple and leave off the parentheses:

```
In [1]: tup = 1, 2, 3

In [2]: tup
Out[2]: (1, 2, 3)
```

Most of the time, however, you'll want to include the parentheses, both for clarity and for when you use more complicated code such as a nested tuple (a tuple stored in a tuple):

```
In [3]: nested = (1, 2, 3), ('Alice', 'Bob')

In [4]: nested
Out[4]: ((1, 2, 3), ('Alice', 'Bob'))
```

Because tuples can hold a single value, separating values with commas is more important than using parentheses. To see why, enter the following in a console and check the object's type:

```
In [5]: what_am_I = (1)
In [6]: type(what_am_I)
Out[6]: int
```

In this case, Python thinks you've just typed an integer in parentheses! Now, add a trailing comma:

```
In [7]: what_am_I = (1,)

In [8]: type(what_am_I)
Out[8]: tuple
```

The takeaway is that single-item tuples require a trailing comma.

Converting Other Types to Tuples

You can also create tuples by using the built-in tuple() function to convert other data types. Enter the following in the console:

```
In [9]: x = tuple('Hello, World!')

In [10]: x
Out[10]: ('H', 'e', 'l', 'l', 'o', ',', ' ', 'W', 'o', 'r', 'l', 'd', '!')
```

This snippet turned the string Hello, World! into a tuple. Note that each character in the string is now a separate element in the tuple.

NOTE   *Because "tuple" is the name of a function, you should never use it as a variable name.*

You can also use tuple() to turn a list (indicated by square brackets []) into a tuple, as demonstrated here:

```
In [11]: planet_list = ['Venus', 'Earth', 'Mars']

In [12]: planet_tup = tuple(planet_list)

In [13]: planet_tup
Out[13]: ('Venus', 'Earth', 'Mars')
```

In this case, the strings are preserved as words because each string was a separate item in the list.

Working with Tuples

Tuples can be indexed and sliced just like strings (see Chapter 7). Because tuples are immutable and can't be modified, they don't come with many built-in methods. For added utility, however, you can take advantage of general-purpose built-in functions from the Python Standard Library when you work with them. Some of these functions and methods are listed in Table 9-2. You should substitute real variable and value names for those shown in italics.

Table 9-2: Useful Built-in Functions and Tuple Methods

Function	Description	Example
tuple()	Converts a sequence into a tuple	tuple(*seq_name*)
len()	Returns the length of a sequence	len(*tuple_name*)
min()	Returns the sequence item with the minimum value	min(*tuple_name*)
max()	Returns the sequence item with the maximum value	max(*tuple_name*)

Method	Description	Example
count()	Returns the count of a specified value	*tuple_name*.count(*value*)
index()	Returns position of a specified value	*tuple_name*.index(*value*)

Note that tuple *methods* are called *after* the tuple name using dot notation, such as *tuple_name*.count(value). You saw tuple() in action in the previous section, so let's start by looking at len().

Getting a Tuple's Length

After you make a tuple, it always has a fixed length. You can find this length by using the built-in len() function. Let's use the tuples that we created in the previous section to see how this works:

```
In [14]: len(tup)
Out[14]: 3

In [15]: len(nested)
Out[15]: 2
```

The first result, 3, is intuitive, as there are three items in tup. But the length of the nested tuple is only 2, when there are clearly five items present.

To understand what's happening, let's use indexing to look at the first item in each tuple:

```
In [16]: tup[0]
Out[16]: 1

In [17]: nested[0]
Out[17]: (1, 2, 3)
```

Values enclosed by parentheses are considered a *single item* within the overall sequence of items in the tuple. If you need to access a value *within* an item, you can add another index. For example, to see the first element within each nested item, enter the following:

```
In [18]: nested[0][0]
Out[18]: 1

In [19]: nested[1][0]
Out[19]: 'Alice'

In [20]: nested[1][0][0]
Out[20]: 'A'
```

This is shown diagrammatically in Figure 9-1. Because the items in the second nested tuple are strings ('Alice', 'Bob'), which are in turn made up of elements (the letters), you need to index to three levels to access all of the elements in nested[1]. The first index gets the nested tuple, the second index gets you a string in the nested tuple, and the third index gets you a character within the string.

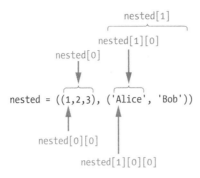

Figure 9-1: An example of indexing a nested tuple

Understanding this behavior is important in the event that you want to iterate over a tuple to get its values one by one (we haven't covered looping yet, so bear with me). This is straightforward with a non-nested tuple:

```
In [21]: for i in tup:
    ...:        print(i * 10)
10
20
30
```

For nested tuples, you need to loop through the elements in *each* nested tuple to get to all of the items:

```
In [22]: for item in nested:
    ...:        for element in item:
    ...:             print(element * 5)
5
10
15
AliceAliceAliceAliceAlice
BobBobBobBobBob
```

The terms "item" and "element" have no special meaning. You could just as easily call them "i" and "j" or "Fred" and "George."

Getting a Tuple's Minimum and Maximum Values

The min() and max() functions return the minimum and maximum values in a tuple, respectively. Here's an example:

```
In [23]: min(tup)
Out[23]: 1

In [24]: max(tup)
Out[24]: 3
```

That's easy, but what if we try it on the nested tuple? Let's see what happens:

```
In [25]: min(nested)
Traceback (most recent call last):

  File "C:\Users\hanna\AppData\Local\Temp/ipykernel_25576/1378168620.py", line 1, in <module>
    min(nested)

TypeError: '<' not supported between instances of 'str' and 'int'
```

Python raises a TypeError because it doesn't know how to distinguish a minimum string from a minimum integer. You can find a minimum value in a tuple of strings, however. Check this out:

```
In [26]: test = ('c', 'bob', 'z')

In [27]: min(test)
Out[27]: 'bob'

In [28]: test = ('a', 'A')

In [29]: min(test)
Out[29]: 'A'
```

The min() and max() functions use the ASCII sorting order to sort strings. In an ASCII table (see *https://www.asciitable.com/*), special characters, like punctuation marks, come before the alphabet, and uppercase letters come before lowercase letters.

Many other built-in functions work with multiple data types. The len() function, for example, works with strings, tuples, sets, lists, and dictionaries. The membership operators you learned about in Chapter 7 also work with multiple types. Here's an example:

```
In [30]: elements = 'carbon', 'calcium', 'oxygen'

In [31]: 'carbon' in elements
Out[31]: True
```

Unpacking Tuples

You can assign the values in a tuple to multiple variables at once using a process called *unpacking*. Let's pretend that you've written a function that returns a Cartesian coordinate (x, y) pair as a tuple. You want to use the individual x and y values later in the program. Here's how you can get at those values:

```
In [32]: coordinates = (45, 160)

In [33]: x, y = coordinates
```

```
In [34]: x
Out[34]: 45

In [35]: y
Out[35]: 160
```

To use nested tuples, assign variables using the same parenthesis structure. Let's revisit our nested tuple:

```
In [36]: nested = (1, 2, 3), ('Alice', 'Bob'), 549

In [37]: (a, b, c), (d, e), f = nested

In [38]: a
Out[38]: 1

In [39]: e
Out[39]: 'Bob'
```

You don't need to take all of the elements. Suppose that you want only the first three numbers in the nested tuple. If you try to take them directly, you'll raise an exception:

```
In [40]: (a, b, c) = nested
Traceback (most recent call last):

File "C:\Users\hanna\AppData\Local\Temp/ipykernel_25576/3799313898.py", line 1, in <module>
(a, b, c) = nested

ValueError: not enough values to unpack (expected 3, got 2)
```

Python expects you to unpack every item in the tuple. To get around this, you can unpack the tuple using the *splat* (or *star*) operator (*) with the insignificant variable symbol (_). Splat allows for an arbitrary number of items, so in this case, you're telling Python to "get the rest" and assign them to _:

```
In [41]: (a, b, c), *_ = nested

In [42]: a
Out[42]: 1

In [43]: b
Out[43]: 2

In [44]: c
Out[44]: 3

In [45]: _
Out[45]: [('Alice', 'Bob'), 549]
```

You don't need to use the values in `_`. These will be cleared from memory later when Python performs routine garbage collection.

Operator Overloading with Tuples

You can use operator overloading on tuples, just as with strings. For example, adding two tuples produces a new tuple containing the values of both tuples:

```
In [46]: tup1 = 1, 2, 3

In [47]: tup2 = 4, 5, 6

In [48]: tup3 = tup1 + tup2

In [49]: tup3
Out[49]: (1, 2, 3, 4, 5, 6)
```

Using the multiplication operator with an integer concatenates multiple copies of a tuple along with copies of the references to the objects they contain:

```
In [50]: tup1 * 3
Out[50]: (1, 2, 3, 1, 2, 3, 1, 2, 3)
```

Unexpected Tuple Behaviors

There are loopholes in the rule that tuples are immutable. For instance, if a tuple contains a *mutable* data type, you can change that item within a tuple. Let's try this using a mutable list (enclosed in square brackets; we'll look more at lists in the next section):

```
In [51]: tup_with_list = (1, 2, ['Alice', 'Bob'], 3)

In [52]: tup_with_list
Out[52]: (1, 2, ['Alice', 'Bob'], 3)

In [53]: tup_with_list[2][1] = 'Steve'

In [54]: tup_with_list
Out[54]: (1, 2, ['Alice', 'Steve'], 3)
```

In this example, we were able to change the second item in the list ([2][1]) from Bob to Steve, even though tuples are immutable. This falls under the category of things you *can* do, but *shouldn't*!

Printing Tuples

Running print() on a tuple can be frustrating, as the default display includes the commas and quotation marks:

```
In [55]: names = 'Harry', 'Ron', 'Hermione'

In [56]: print(names)
('Harry', 'Ron', 'Hermione')
```

To fix this, you can use the join() string method to print only the strings in a tuple:

```
In [57]: print(' '.join(names))
Harry Ron Hermione
```

In this example, we joined each item in the tuple using a space (' '). You can use other characters to join items, such as the newline escape sequence (\n):

```
In [58]: print('\n'.join(names))
Harry
Ron
Hermione
```

The join() method works only for sequences composed of strings. To handle mixed data types, include the map() built-in function:

```
In [59]: tup = 'Steve', 5, 'a', 5

In [60]: print(' '.join(map(str, tup)))
Steve 5 a 5
```

You can also use the splat operator (*) to print a tuple efficiently and attractively:

```
In [61]: print(*tup, sep='\n')
Steve
5
a
5
```

Splat takes the tuple as input and expands it into positional arguments in the function call. The last argument is the separator used between items for printing. The default separator is a space (sep=' ').

Lists

A list is a variable-length, iterable, mutable, ordered sequence of values. They look like tuples except they're enclosed in brackets rather than parentheses:

```
['K. L. Putney', 'M. B. Clark', 'S. B. Vaughan']
```

Because lists are mutable, you can change their values at will. You can add items, change items, and delete items. Otherwise, lists are like tuples. They can hold multiple data types, including mixtures of types. You can index them, slice them, concatenate them, nest them, use built-in functions, and more. Lists are true workhorses in Python, and you will use them all the time.

NOTE *Lists are objects that can be treated as values. That is, they can be stored in variables and passed to functions. If you hear the term* list value, *be aware that it refers to the whole* list, not some value inside it.

Creating Lists

To create a list, enclose a value or comma-delimited values in square brackets ([]):

```
In [62]: dna_bases = ['adenosine', 'guanine', 'cytosine', 'thymidine']

In [63]: dna_bases
Out[63]: ['adenosine', 'guanine', 'cytosine', 'thymidine']
```

Because lists are mutable, you can begin with an empty list. For example, you might set up an empty list to hold user input later in the program. Here's how:

```
In [64]: empty_list = []
```

You can also use the built-in list() function to convert other data types, like tuples and strings, into lists:

```
In [65]: my_tuple = 1, 2, 3

In [66]: my_tuple
Out[66]: (1, 2, 3)

In [67]: my_list = list(my_tuple)

In [68]: my_list
Out[68]: [1, 2, 3]
```

NOTE *Because "list" is the name of a function, never use it as a variable name.*

Working with Lists

Because lists are mutable, you can do a lot more with them than you can with tuples, and they come with more built-in functionality. Table 9-3 summarizes the list methods. You'll need to substitute your own names for those in italics. In addition, you can use the len(), min(), and max() built-in functions from Table 9-2 with lists.

Table 9-3: Built-in List Methods

Method	Description	Example
append()	Add a single item to the end of a list	*list_name*.append(*item*)
extend()	Add iterable items to the end of a list	*list_name*.extend(*iterable*)
insert()	Insert an item before a given index (i)	*list_name*.insert(*i, item*)
remove()	Remove first item from list with value = item	*list_name*.remove(*item*)
pop()	Remove and return item at given index	*list_name*.pop(*index*)
clear()	Remove all items from a list	*list_name*.clear()

Method	Description	Example
index()	Return index of first item with value = item	*list_name*.index(*item*)
count()	Return number of times an item appears in a list	*list_name*.count(*item*)
sort()	Sort list items in place	*list_name*.sort()
reverse()	Reverse list items in place	*list_name*.reverse()
copy()	Return a shallow copy of a list	*list_name*.copy()

List methods don't work like the string methods you learned about in Chapter 7. Whereas string methods perform their task and return a new string, list methods usually modify the list and return None. To sort a list, for example, you should use list_name.sort(), not list_name = list_name.sort().

 NOTE: *All of the methods for printing tuples also work with lists, so refer to "Printing Tuples" on page 228.*

Adding Items to Lists

The append item lets you add an item to the end of a list.

```
In [69]: patroni = ['stag', 'otter', 'dog']

In [70]: patroni.append('doe')

In [71]: patroni
Out[71]: ['stag', 'otter', 'dog', 'doe']
```

To add *multiple* items to the end of a list, the items need to be in the form of an iterable. Let's try adding a heron and a hare to the patroni list:

```
In [72]: patroni.extend('heron', 'hare')
Traceback (most recent call last):

File "C:\Users\hanna\AppData\Local\Temp/ipykernel_24452/4246633803.py", line 1, in <module>
patroni.extend('heron', 'hare')

TypeError: extend() takes exactly one argument (2 given)
```

You get a TypeError because the extend() method is looking for one argument (the thing between parentheses), not two. Now, try passing it a tuple of names, instead:

```
In [73]: extra_patroni = 'heron', 'hare'

In [74]: patroni.extend(extra_patroni)

In [75]: patroni
Out[75]: ['cat', 'stag', 'otter', 'dog', 'doe', 'heron', 'hare']
```

Success! Both append() and extend() are useful when either looping through values and adding some of them to a list, or when adding values returned from a function.

Inserting Values into Lists

If you need to insert an item at a specific location in a list, not just at the end, use the insert() method and *pass* it (add between the parentheses) the index *before which* you want to place the item, and then the item, separated by a comma. For example, to add an item to the start of the patroni list, use an index of 0:

```
In [76]: patroni.insert(0, 'cat')

In [77]: patroni
Out[77]: ['cat', 'stag', otter, dog, 'doe']
```

The insert() method shifts the index for each item to accommodate the new item. This is computationally expensive, however, and should be avoided when possible.

Removing Items from Lists

If you want to remove an item from anywhere in a list, use the pop() method. Let's remove the cat patronus. Because pop() returns the item as well as removing it, we can also use it in some way, such as by assigning it to a variable, though this is optional. Let's take a look:

```
In [78]: Umbridge_patronus = patroni.pop(0)

In [79]: Umbridge_patronus
Out[79]: 'cat'
In [80]: patroni
Out[80]: ['stag', 'otter', 'dog', 'doe', 'heron', 'hare']
```

We now have a new variable that holds the string cat, whereas the patroni list no longer contains that item.

If you don't specify an index, pop() removes the last item in the list.

Another way to remove items is to use the del operator, short for "delete." Just pass it the index:

```
In [81]: names = ['Harry', 'Ron', 'Hermione', 'Ginny']

In [82]: del names[1]

In [83]: names
Out[83]: ['Harry', 'Hermione', 'Ginny']
```

The del operator also permits slicing:

```
In [84]: del names[:2]

In [85]: names
Out[85]: ['Ginny']
```

You can also remove an item by naming it in the remove() method:

```
In [86]: my_list = ['a', 'b', 'c', 'a', 'b', 'c']

In [87]: my_list.remove('a')

In [88]: my_list
Out[88]: ['b', 'c', 'a', 'b', 'c']
```

Note that only the first occurrence of 'a' is removed. Also, if the specified item doesn't exist in the list, Python will raise a ValueError.

Changing the Value of Items in Lists

You can change the value of items within a list by using indexing. Let's change the hare patronus to a wolf:

```
In [89]: patroni[5] = 'wolf'

In [90]: patroni
Out[90]: ['stag', 'otter', 'dog', 'doe', 'heron', 'wolf']
```

Because hare came at the end of the list, we could have also used the built-in len() function to find the end of the list and use its return value:

```
In [91]: patroni[len(patroni) - 1] = 'wolf'
```

You need to subtract one from the length of the list because iterating and indexing in Python start at 0, so the final index is always one less than the length of the list.

Finding the Index of Items in Lists

Similar to the remove() method, the index() method will return the zero-based index of the first occurrence of a specified item in a list. It also raises a ValueError if the item doesn't exist. Let's fetch the index of dog in the patroni list:

```
In [92]: patroni

Out[92]: ['stag', 'otter', 'dog', 'doe', 'heron', 'wolf']

In [93]: patroni.index('dog')
Out[93]: 2
```

You can also use slice notation on the list to limit the search to a particular subsequence. Just add optional start and end arguments after the item name. The returned index is still computed relative to the beginning of the full sequence, however:

```
In [94]: patroni.index('dog', 2, 5)
Out[94]: 2
```

In this example, the index() method looked at the items between indexes 2 and 5 (dog *up to* wolf).

The count() method returns the number of times an item appears in a list:

```
In [95]: my_list.count('b')
Out[95]: 2
```

Sorting the Values in Lists

The sort() method sorts lists in place, either alphabetically or numerically. For example:

```
In [96]: letters = ['c', 'a', 'c', 'b', 'd']

In [97]: letters.sort()

In [98]: letters
Out[98]: ['a', 'b', 'c', 'c', 'd']
```

Computers are very literal, however, and things might not go as planned. Notice what happens if you try to alphabetize a list comprising letters with different cases:

```
In [99]: letters_mixed_case = ['C', 'a', 'c', 'B', 'd']

In [100]: letters_mixed_case.sort()

In [101]: letters_mixed_case
Out[101]: ['B', 'C', 'a', 'c', 'd']
```

The Python default is to place capitalized letters before lowercase ones. So, this mixed-case example is correct by Python's standards but probably isn't what you expected or wanted. To force Python to compare apples to apples, you can use the optional key argument to convert all strings to lowercase before sorting:

```
In [102]: letters_mixed_case.sort(key=str.lower)

In [103]: letters_mixed_case
Out[103]: ['a', 'B', 'C', 'c', 'd']
```

The sort() method also comes with a second optional argument for *reversing* the order of items in the list:

```
In [104]: letters_mixed_case.sort(reverse=True)

In [105]: letters_mixed_case
Out[105]: ['d', 'c', 'a', 'C', 'B']
```

You can pass sort() a *sort key* to let it know what parameter you want to sort by. In this example, we're sorting based on the length of strings using the len sort key:

```
In [106]: my_list = ['longest', 'long', 'longer']

In [107]: my_list.sort(key=len)

In [108]: my_list
Out[108]: ['long', 'longer', 'longest']
```

You can even write and pass sort() a custom function to do more complicated sorting. To find out more, visit the sorting tutorial at *https://docs .python.org/3/howto/sorting.html*.

The Curious Case of Copy

The act of copying a list reveals perhaps the greatest "gotcha" in the Python language. Pour yourself a cup of coffee, because this may be the most important thing you learn all day.

Remember that variable names are references to an object, but not the object itself? Likewise, when you copy an object using an assignment statement, you copy only the reference to that object. When this behavior is combined with mutable objects, mayhem can result.

Let's assign a list to another list, a seemingly straightforward thing to do:

```
In [109]: my_patroni = ['cat', 'hare', 'doe']

In [110]: your_patroni = my_patroni

In [111]: your_patroni
Out[111]: ['cat', 'hare', 'doe']
```

You might think my_patroni and your_patroni are separate lists containing identical values, but they're not. Each name points to the same object in memory. You can confirm this by checking the identity of each:

```
In [112]: id(my_patroni), id(your_patroni)
Out[112]: (2181240760640, 2181240760640)
```

They're the same object. So, if you alter one, you alter the other:

```
In [113]: my_patroni[0] = 'stag'

In [114]: my_patroni
Out[114]: ['stag', 'hare', 'doe']

In [115]: your_patroni
Out[115]: ['stag', 'hare', 'doe']
```

Changing the first item in my_patroni changed the same item in your _patroni. This kind of behavior can keep you up bug-hunting all night.

To properly copy a mutable object, like a list, use the copy() method:

```
In [116]: my_patroni = ['cat', 'hare', 'doe']

In [117]: your_patroni = my_patroni.copy()

In [118]: your_patroni
Out[118]: ['cat', 'hare', 'doe']
```

Alternatively, you can use slice notation to copy the whole list from start to finish:

```
In [119]: your_patroni = my_patroni[:]
```

Regardless of the method, each list object has a separate identity:

```
In [120]: id(my_patroni), id(your_patroni)
Out[120]: (2181240443968, 2181240620288)
```

That's great, but we're not through yet. The slice and copy() methods make a *shallow* copy. This means that if a list contains nested lists, copy() duplicates only *references* to the inner, nested lists. Let's look at an example:

```
In [121]: my_patroni = [['cat', 'hare'], ['doe', 'stag']]

In [122]: your_patroni = my_patroni.copy()

In [123]: id(my_patroni), id(your_patroni)
Out[123]: (2181240513024, 2181240710976)
```

As expected, these two lists have different identities, meaning that they're different objects. Now let's check the identity of the first nested list, at index 0:

```
In [124]: id(my_patroni[0]), id(your_patroni[0])
Out[124]: (2181240520640, 2181240520640)
```

This inner list is the same object in both lists. To prove it, change the first item in this list to `wolf`. Remember, the first index references the first nested list, and the second index references the first item in this list.

```
In [125]: my_patroni[0][0] = 'wolf'

In [126]: my_patroni
Out[126]: [['wolf', 'hare'], ['doe', 'stag']]

In [127]: your_patroni
Out[127]: [['wolf', 'hare'], ['doe', 'stag']]
```

Again, changing an item in one list has changed the same item in another. Note that this behavior extends only to the *nested* lists. If you append a new item to the my_patroni list, it won't affect your_patroni:

```
In [128]: my_patroni.append('Manx cat')

In [129]: my_patroni
Out[129]: [['wolf', 'hare'], ['doe', 'stag'], 'Manx cat']

In [130]: your_patroni
Out[130]: [['wolf', 'hare'], ['doe', 'stag']]
```

To avoid this type of behavior, you should import the built-in copy module and use its `deepcopy()` method:

```
In [131]: import copy

In [132]: their_patroni = copy.deepcopy(your_patroni)

In [133]: their_patroni
Out[133]: [['wolf', 'hare'], ['doe', 'stag']]

In [134]: id(your_patroni[0]), id(their_patroni[0])
Out[134]: (2181240520640, 2181240818368)
```

Now, the nested lists are separate objects, and you've created a true copy of the original. No more "quantum entanglement" with "spooky action at a distance."

For a slight slowdown in your code, `deepcopy()` will ensure that you're copying any internal object references. This includes all mutable objects within a list at every possible level, thereby avoiding bugs that can cost you much more time to find and correct.

Checking for Membership

You can check whether an item occurs in a list using the `in` and `not_in` keywords. These also work on other container data types, as shown here:

```
In [135]: my_patroni = ['cat', 'hare', 'doe']

In [136]: 'hare' in my_patroni
```

```
Out[137]: True

In [138]: 'wolf' in my_patroni
Out[138]: False
```

Doing this for large lists is not recommended, however. Checking for membership in a list is computationally expensive and thus slow. Python must check through every value in a list to perform this operation, whereas in other collection data types such as sets and dictionaries, it can use very fast hash tables for a noticeable improvement in performance. It's easy to convert a list into a set for this purpose, and we'll do it later in this chapter.

TEST YOUR KNOWLEDGE

6. Create an empty list named patroni and then add a tiger, shark, and weasel to it all at once.

7. Delete all of the items in the previous patroni list.

8. Which is the *wrong* way to add "shrew" to the patroni list?

   a. `patroni.append('shrew')`

   b. `patroni += ['shrew']`

   c. `patroni = patroni + 'shrew'`

   d. `patroni = patroni + ['shrew']`

9. Why shouldn't you use `patroni += 'shrew'`?

   a. You'll raise TypeError: can only concatenate list (not "str") to list

   b. Augmented operators work only with strings and mathematical expressions

   c. Each letter in "shrew" will become a separate item in the list

   d. No one uses a shrew as a patronus

10. J.K. Rowling's personal patronus is a:

    a. Manx cat

    b. Jack Russell terrier

    c. Heron

    d. Hummingbird

Sets

A *set* is a mutable, unordered, iterable collection of *unique* elements (no duplicates allowed). Sets are designed to work like their mathematical counterparts and thus support operations such as union, intersection, and difference. Sets look like tuples and lists except that they're enclosed in curly brackets:

{'K. L. Putney', 'M. B. Clark', 'S. B. Vaughan'}

Sets are based on a data structure known as a *hash table* that makes adding and searching for elements in them very fast (we looked at this data structure back in "The Debugger Pane" on page 90). A deep discussion of hash tables is beyond the scope of this book, but basically, hashing is a process whereby a key or string of characters is transformed into a shorter, fixed-length value that's easier to find, and these values are stored in a hash table for easy look-up.

Besides being significantly faster than tuples and lists for membership tests, sets let you efficiently remove duplicate values from those data types by converting them into a set. On the other hand, sets are slightly slower during iteration and take up more memory. And because sets are unordered, you can't access elements using indexes, as you can with the tuples and lists.

Creating Sets

Sets consist of a sequence of comma-separated values enclosed by curly brackets ({}):

```
In [139]: a_set = {1, 2, 3}

In [140]: a_set
Out[140]: {1, 2, 3}
```

You can also copy sets using the built-in set() function:

```
In [141]: new_set = set(a_set)

In [142]: new_set
Out[142]: {1, 2, 3}
```

The set() function also converts other data types to sets:

```
In [143]: a_string = ('Hello, World!')

In [144]: a_set = set(a_string)

In [145]: a_set
Out[145]: {' ', '!', ',', 'H', 'W', 'd', 'e', 'l', 'o', 'r'}
```

This snippet turned the string a_string into a set. Note that each character in the string is now a separate element in the set, duplicates have been removed, and the elements are unordered.

As a side note, to create an empty set you must use set(), not {}, as the latter creates an empty dictionary, a data structure that we'll cover in the next section.

NOTE *Because "set" is the name of a function, you should never use it as a variable name.*

The elements in sets must be *hashable*, which means they must be immutable. Because Python hashes each element in a set and stores the hash value, if you change an element in place, it gets hashed again, and the new hash value is stored in a different location in the hash table. This can cause conflicts and lost elements that prevent the set from working correctly.

Integers, floats, and strings are immutable, as are tuples composed of immutable items. You can check whether an object in Python is hashable using the hash() function:

```
In [146]: hash('astrolabe')
Out[146]: -4570350835965251752
```

Because the string 'astrolabe' is immutable, it was assigned a hash value. If you try this with a mutable list, however, you'll raise a TypeError.

To use a mutable sequence in a set, you first must convert it to a tuple. In this example, we convert a list to a tuple while assigning it to a set:

```
In [147]: my_set = {tuple(['a', 'list'])}

In [148]: my_set
Out[148]: {('a', 'list')}
```

This works only if the tuple doesn't contain mutable items, such as a list:

```
In [149]: a_tuple = (1, 2, 3, ['Hello, World!'])

In [150]: my_set = set(a_tuple)
Traceback (most recent call last):

File "C:\Users\hanna\AppData\Local\Temp/ipykernel_3856/1713377465.py", line 1, in <module>
my_set = set(a_tuple)

TypeError: unhashable type: 'list'
```

If you have lists in your tuple, you'll need to convert them into tuples, as well.

Because sets use hash tables, they take up more memory than tuples and lists. Enter the following code into the console to see the difference:

```
In [151]: import sys  # For system module.

In [152]: a_list = list(range(10_000))

In [153]: a_tuple = tuple(a_list)

In [154]: a_set = set(a_list)

In [155]: sys.getsizeof(a_list)
Out[155]: 87616

In [156]: sys.getsizeof(a_tuple)
Out[156]: 80040

In [157]: sys.getsizeof(a_set)
Out[157]: 524504
```

A set of 10,000 elements takes up roughly six times the memory of a list and 6.5 times the memory of a tuple of the same size.

Working with Sets

Table 9-4 lists some common methods for working with sets. You should substitute your own variable names for the ones shown in *italics*. In addition to these methods, you can use many of Python's built-in functions, such as min(), max(), and len(), with sets.

Table 9-4: Useful Built-in Set Methods

Methods	Operator syntax	Description
set1.add(*item*)		Add item to set
set1.clear()		Reset to empty
set1.copy()		Returns a shallow copy of a set
set1.difference(*set2*)	*set1* - *set2*	Return unshared items
set1.difference_update(*set2*)	*set1* -= *set2*	Set set1 to items not in set2
set1.discard(*item*)		Removes selected item from a set
set1.intersection(*set2*)	*set1* & *set2*	Return all items in both sets
set1.intersection_update(*set2*)	*set1* &= *set2*	Set set1 to intersecting items
set1.isdisjoint(*set2*)		Return True if no shared items

(continued)

Table 9-4: Useful Built-in Set Methods *(continued)*

Methods	Operator syntax	Description
`set1.issubset(set2)`	`set1 <= set2`	Return True if set2 contains set1
`set1.issuperset(set2)`	`set1 >= set2`	Return True if set1 contains set2
`set_name.pop()`		Remove arbitrary element from a set
`set_name.remove(item)`		Remove an item from a set
`set1.symmetric_difference(set2)`	`set1 ^ set2`	Return unshared set1 and set2 items
`set1.symmetric_difference_update(set2)`	`set1 ^= set2`	Set set1 to unshared items
`set1.union(set2)`	`set1	set2`	Return all unique items in both
`set1.update(set2)`	`set1 |= set2`	Set set1 to unique set1 and set2 items

The operator syntax column represents shorthand syntax that you can use with certain methods. These are obviously less readable than the full method names.

Sets are good choices for datasets that can't contain duplicates and that you need to compare to other datasets. Let's assume that you're studying the fauna in two farm ponds. We'll make the dataset short for convenience, but imagine it's a much more extensive list (and uses proper taxonomical naming):

```
In [158]: pond1 = {'catfish', 'bullfrog', 'snail', 'planaria', 'turtle'}
```

```
In [159]: pond2 = {'bullfrog', 'crayfish', 'snail', 'leech', 'planaria'}
```

Next, add a gar to pond1:

```
In [160]: pond1.add('gar')
```

```
In [161]: pond1
Out[161]: {'bullfrog', 'catfish', 'gar', 'planaria', 'turtle', 'snail'}
```

If you try to add another gar to pond1, everything will appear to work, but you'll still have only one gar entry in the set because duplicates aren't permitted.

Finding the Differences Between Two Sets

Let's assume that you've finished and have too many animals in each set to visually compare. That's okay; you can use the difference() method to look at animals that are unique to each pond. Figure 9-2 uses a Venn diagram to demonstrate what this method returns.

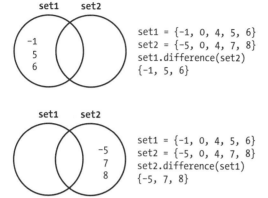

```
set1 = {-1, 0, 4, 5, 6}
set2 = {-5, 0, 4, 7, 8}
set1.difference(set2)
{-1, 5, 6}
```

```
set1 = {-1, 0, 4, 5, 6}
set2 = {-5, 0, 4, 7, 8}
set2.difference(set1)
{-5, 7, 8}
```

Figure 9-2: The difference() set method

Applying this to our pond sets yields the following:

```
In [162]: pond1_unique_animals = pond1.difference(pond2)

In [163]: pond1_unique_animals
Out[163]: {'catfish', 'gar', 'turtle'}

In [164]: pond2_unique_animals = pond2.difference(pond1)

In [165]: pond2_unique_animals
Out[165]: {'crayfish', 'leech'}
```

Because there's no such thing as a pond without a turtle, these results suggest that you need to make another field trip to pond2.

Finding the Duplicate Items in Two Sets

To see which animals are common to both ponds, use the intersection() method, described in Figure 9-3.

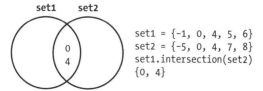

```
set1 = {-1, 0, 4, 5, 6}
set2 = {-5, 0, 4, 7, 8}
set1.intersection(set2)
{0, 4}
```

Figure 9-3: The intersection() set method

Using this method on our pond data yields the following:

```
In [166]: pond_common_animals = pond1.intersection(pond2)

In [167]: pond_common_animals
Out[167]: {'bullfrog', 'planaria', 'snail'}
```

You need to do this only for pond1 because you'll get the same result for pond2.intersection(pond1).

Combining Sets

Now, suppose that you sample a large lake and find more animal species than in the small ponds:

```
In [168]: lake1 = {'bream', 'planaria', 'mussel', 'catfish', 'gar', 'snail', 'crayfish',
'turtle', 'bullfrog', 'cottonmouth', 'leech', 'alligator'}
```

You're curious if the lake environment behaves like a big pond that includes the same animals plus a few more. To determine this, you first must combine the pond animals into a single set using the union() method, as demonstrated in Figure 9-4.

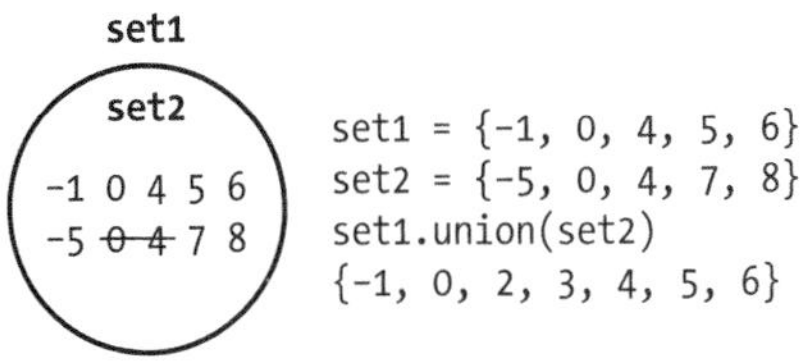

```
set1 = {-1, 0, 4, 5, 6}
set2 = {-5, 0, 4, 7, 8}
set1.union(set2)
{-1, 0, 2, 3, 4, 5, 6}
```

Figure 9-4: The union() set method

```
In [170]: pond_animals = pond1.union(pond2)

In [171]: print(pond_animals)
{'planaria', 'catfish', 'gar', 'snail', 'crayfish', 'turtle', 'bullfrog',
'leech'}
```

NOTE   *You can combine multiple sets at once using this syntax: set1.union(set2, set3, set4...).*

Determining Whether One Set Is a Superset of Another

If the lake1 set contains all the animals in the pond_animals set, it's considered to be a *superset*. Figure 9-5 shows how supersets and subsets work.

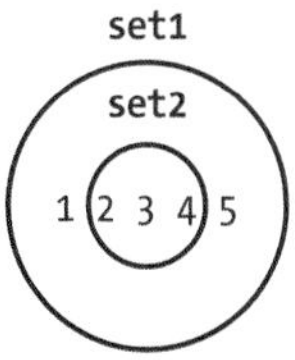

```
set1 = {1, 2, 3, 4, 5}
set2 = {2, 3, 4}
set1.issuperset(set2)
True
set2.issubset(set1)
True
```

Figure 9-5: The issuperset() and issubset() set methods

If you run issuperset() on the lake set, it will return True, indicating that all of the pond animals are present in the lake:

```
In [172]: lake1.issuperset(pond_animals)
Out[172]: True
```

Creating Frozensets

The built-in frozenset() function takes an iterable object as input and makes it immutable. The resulting "frozenset" is a set with elements that can't be added, removed, or altered.

Frozensets are mainly used as dictionary keys (which must be immutable) or as elements in other sets, because sets can't be inserted into sets. Frozensets are "safer" than sets, as there's no risk of accidentally changing elements in frozensets later in your code.

To make a frozenset, pass the function an iterable, such as another set:

```
In [173]: a_set = {1, 2, 3}

In [174]: a_frozenset = frozenset(a_set)

In [175]: a_frozenset
Out[175]: frozenset({1, 2, 3})
```

You can use the same functions and methods on frozensets that you use on sets, as long as they don't change the frozenset. Mathematical set operations such as intersection, difference, and union will all work on frozensets.

Dictionaries

Short for dictionary, the *dict* data type is an ordered, iterable, mutable collection of values indexed by *keys* rather than numbers. The keys can be almost any data type and are mapped to one or more values. Dictionaries are considered the most important Python structure for storing and accessing data.

A dictionary looks different than tuples, lists, and sets, as it has key-value pairs. But like a set, it's surrounded by curly brackets:

```
{'a_key': 'a_value', 'another_key': 'another_value'}
```

Keys have the same properties as elements in a set: they must be unique and immutable because they get hashed. In fact, sets are just collections of dictionary keys with no corresponding values. And like sets, dictionaries take up more memory due to the use of a hashing process.

The dictionary key is like a word in a language dictionary and the values represent the definition(s) for that word. Keys can be multiple objects (like a word pair), as long as the multiple objects are immutable (tuples can be used but not lists).

Values, on the other hand, can be mutable objects. And just as a word in a language dictionary can have more than one definition, it's fine to have multiple values that map to a single key.

Because dictionaries associate, or map, one thing with another, they tend to be used when items in a collection are labeled. You can use them as simple databases whose data has a key-value relationship, such as student names and student IDs.

Creating Dictionaries

A dictionary consists of comma-separated key-value pairs enclosed in curly brackets ({}). The key-value pairs are separated by colons, as shown here:

```
{'hello': 'hola', 'goodbye': 'adios'}
```

The use of the colon distinguishes dictionaries from sets, which also use curly brackets.

Now let's make a dictionary that maps some letters to their equivalent Morse code symbols:

```
In [176]: morse = {'e': '.', 'h': '....', 'l': '.-..', 'o': '---', 's': '...'}

In [177]: morse
Out[177]: {'e': '.', 'h': '....', 'l': '.-..', 'o': '---', 's': '...'}
```

As you can see, dictionaries preserve the *insertion* order of the key-value pairs. This mainly affects readability when viewing the dictionary, and you should take this into account when entering the data.

Now, using this morse dictionary, you can loop through a word and translate it to Morse code:

```
In [178]: for letter in 'hello':
     ...:     print(morse[letter])
....
.
.-..
.-..
---
```

Note that dictionary keys are case sensitive, so the following code will raise an KeyError:

```
In [179]: morse['S']
Traceback (most recent call last):

File "C:\Users\hanna\AppData\Local\Temp/ipykernel_6456/1793354668.py", line 1,
in <module>
morse['S']

KeyError: 'S'
```

If you're doing something like looping through the letters in the dictionary, you can avoid this error by converting the letters to lowercase as part of the key index, as follows:

```
In [180]: for letter in 'SOS':
    ...:         print(morse[letter.lower()])
...
---
...
```

You can also use the built-in dict() function to create a dictionary. An advantage here is that you can use keyword arguments for the keys and avoid typing as many single quotation marks:

```
In [181]: frank_sez = dict(bread='good', fire='bad')

In [182]: frank_sez
Out[182]: {'bread': 'good', 'fire': 'bad'}
```

NOTE   *Because "dict" is the name of a function, you should never use it as a variable name.*

Combining Two Sequences into a Dictionary

You can pair up two sequences such as a tuple or list into a dictionary. Of course, the sequences should contain the same number of items and they should be ordered appropriately, so that index 5 in one list pairs with whatever's at index 5 in the second list. Here's an example translating English words to Spanish words. The zip() built-in function pairs up the two lists item by item by mapping similar indexes:

```
In [183]: english = ['then', 'but', 'cold']

In [184]: spanish = ['entonces', 'pero', ['frio', 'fria']]

In [185]: translation = {}

In [186]: for key, value in zip(english, spanish):
    ...: translation[key] = value

In [187]: translation
Out[187]:
{'then': 'entonces',
 'but': 'pero',
 'cold': ['frio', 'fria']}
```

Because the Spanish word for cold is both masculine and feminine, the two forms are stored in a nested list (you could also use a tuple for memory efficiency). You can access items in the list using standard list indexing:

```
In [188]: translation['cold'][0]
Out[188]: 'frio'
```

Of course, you'll want either the masculine form or the feminine form to consistently come first in the nested lists, so you always use the same index to fetch it. Alternatively, you can nest a dictionary that specifies masculine (m) versus feminine (f):

```
In [189]: english = ['then', 'but', 'cold']

In [190]: spanish = ['entonces', 'pero', {'m': 'frio', 'f': 'fria'}]

In [191]: translation = {}

In [192]: for key, value in zip(english, spanish):
     ...:     translation[key] = value

In [193]: translation['cold']['f']
Out[193]: 'fria'

In [194]: translation['cold']['m']
Out[194]: 'frio'
```

By using a dictionary, you don't need to map translations to arbitrary indexes, and your code is much more readable and less prone to error. And if you want to get really pithy, you can build the translation dictionary using: translation = dict(zip(english, spanish)).

Creating Empty Dictionaries and Values

To make an empty dictionary, use curly brackets:

```
In [195]: empty_dict = {}
```

Values can also be empty. This is handy for setting-up placeholder keys to which you will assign values at a later time, like when you load a new list or a user provides some input. Here's an example:

```
In [196]: empty_dict['color'] = None

In [197]: empty_dict['weight'] = ''

In [198]: empty_dict
Out[198]: {'color': None, 'weight': ''}
```

Working with Dictionaries

Table 9-5 summarizes some dictionary methods. You'll need to substitute your own names for those in italics. In addition, you can use many built-in functions such as len(), min(), and max() with dictionaries.

Table 9-5: Built-in Dictionary Methods

Method	Description	Example
clear()	Remove all dictionary elements	*dict_name*.clear()
copy()	Return a copy of a dictionary	*dict_name*.copy()
fromkeys()	Return a dictionary with the specified keys and a value	*dict_name* = dict.fromkeys(*key _tuple*, *value*)
get()	Return the value of a specified key	*dict_name*.get(*key*)
items()	Return a tuple of all key-value pairs	*dict_name*.items()
keys()	Return a list of a dictionary's keys	*dict_name*.keys()
pop()	Remove the element with specified key	*dict_name*.pop(*key*)
popitem()	Remove the last inserted key-value pair	*dict_name*.popitem()
set default()	Insert specified key and value if no key, else return value if key exists	*dict_name*.setdefault(*key*, *value*)
update()	Update dictionary with specified key-value	*dict_name*.update({*key: value*})
values()	Return a list of the values in a dictionary	*dict_name*.values()

NOTE *Although you can change the value that's mapped to a key, there are no dictionary methods that let you* add *a value to an existing key. To do this, you'll need to import and use the* collections *third-party module. We'll look at* collections *and other helpful modules in Chapter 11.*

Getting the Contents of Dictionaries

The keys(), values(), and items() methods return the contents of dictionaries in list-like data types called dict_keys, dict_values, and dict_items(), respectively. You can iterate (loop) over these structures, but otherwise they don't behave like true lists. Here's how they work:

```
In [199]: chems = dict(HCl='acid', NaOH='base', HNO3='acid')

In [200]: chems.keys()
Out[200]: dict_keys(['HCl', 'NaOH', 'HNO3'])

In [201]: chems.values()
Out[201]: dict_values(['acid', 'base', 'acid'])

In [202]: chems.items()
Out[202]: dict_items([('HCl', 'acid'), ('NaOH', 'base'), ('HNO3', 'acid')])
```

If you want to use this output as a list, you can convert it using the
list() function:

```
In [203]: chems_keys = list(chems.keys())
In [204]: chems_keys
Out[204]: ['HCl', 'NaOH', 'HNO3']
```

This returns a list of key-value pair tuples when used with items():

```
In [205]: list(chems.items())
Out[205]: [('HCl', 'acid'), ('NaOH', 'base'), ('HNO3', 'acid')]
```

Getting the Value of a Dictionary Key

As you've seen, if you just want the value of a key in a dictionary, you can
use the key as you would an index with a list, as shown here:

```
In [206]: chems['HCl']
Out[206]: 'acid'
```

This works great until you ask for a key that doesn't exist, in which case
Python will raise a KeyError. To avoid this, use the get() method, which lets
you provide a default value for non-existent keys:

```
In [207]: chems.get('KOH', 'unknown')
Out[207]: 'unknown'
```

The second argument ('unknown') passed to the get() method is the
default return value. Now, when you ask for a missing key like potassium
hydroxide, the method returns 'unknown'.

You can also check whether a key is present using the in keyword:

```
In [208]: 'NaOH' in chems
Out[208]: True
```

Adding Key-Value Pairs to a Dictionary

To add a key-value pair to a dictionary, you can use the indexing approach
(see line In [196]:) or use the update() method:

```
In [209]: chems.update({'KOH': 'base'})

In [210]: chems
Out[210]: {'HCl': 'acid', 'NaOH': 'base', 'HNO3': 'acid', 'KOH': 'base'}
```

To add multiple key-value pairs, separate them with commas:

```
In [211]: chems.update({'KOH': 'base', 'Ca(OH)2': 'base'})
```

Combining Dictionaries

You can also add a dictionary to another dictionary with update(), but a more succinct method is to use the ** operator. Let's chain three dictionaries together to make a fourth:

```
In [212]: d1 = dict(Harry='good', Draco='bad')

In [213]: d2 = dict(Hermione='good', Tom='bad')

In [214]: d3 = dict(Ron='good', Dolores='bad')

In [215]: d4 = {**d1, **d2, **d3}

In [216]: d4
Out[216]:
{'Harry': 'good',
 'Draco': 'bad',
 'Hermione': 'good',
 'Tom': 'bad',
 'Ron': 'good',
 'Dolores': 'bad'}
```

Removing Key-Value Pairs from a Dictionary

To remove a key-value pair, pass the key to the pop() method:

```
In [217]: chems.pop('Ca(OH)2')
Out[217]: 'base'

In [218]: chems
Out[218]: {'HCl': 'acid', 'NaOH': 'base', 'HNO3': 'acid', 'KOH': 'base'}
```

Notice that this method returns the value, so you can assign it to a variable while popping if you want:

```
In [219]: val = chems.pop('KOH')

In [220]: val
Out[220]: 'base'
```

You can also use the del keyword to remove elements; for example, del chems['KHO'].

Creating Default Values for Keys

The setdefault() method lets you check whether a key exists and set a value for the key if it doesn't. Otherwise, it returns the value of the key. Here's an example:

```
In [221]: solar_system = {'Sol': 0, 'Mercury': 1, 'Venus': 2, 'Earth': 3}

In [222]: solar_system.setdefault('Mars', 4)
Out[222]: 4
```

```
In [223]: solar_system
Out[223]: {'Sol': 0, 'Mercury': 1, 'Venus': 2, 'Earth': 3, 'Mars': 4}
```

Because the key 'Mars' didn't exist, the method added it along with its order in the solar system, 4. But if you try to change an existing key, like 'Earth', the method will just return its value and make no changes:

```
In [224]: solar_system.setdefault('Earth', 42)
Out[224]: 3
```

```
In [225]: solar_system
Out[225]: {'Sol': 0, 'Mercury': 1, 'Venus': 2, 'Earth': 3, 'Mars': 4}
```

Suppose that you want to count the number of times the companies Pfizer, Moderna, and Johnson & Johnson were mentioned in articles related to the coronavirus in 2021. You plan to store the counts in a dictionary. The fromkeys() method will help you to set up this dictionary by populating it with keys with the same initial value. The default value is None, but in this case, use 0. You'll need to pass the keys in the form of a tuple, followed by a value:

```
In [226]: companies = ('Pfizer', 'Moderna', 'Johnson & Johnson')
```

```
In [227]: company_counts = dict.fromkeys(companies, 0)
```

```
In [228]: company_counts
Out[228]: {'Pfizer': 0, 'Moderna': 0, 'Johnson & Johnson': 0}
```

Performing a Reverse Lookup

Dictionaries are optimized to efficiently find the value or values for a given key. But sometimes you might want to find all the keys that correspond to a given value (such as looking up the name associated with a phone number). There's no built-in functionality for doing a "reverse lookup," so you need to define a function to perform the task. Let's take a look:

```
In [229]: def lookup_keys(d, v):
     ...:     keys = []
     ...:     for k in d:
     ...:         if d[k] == v:
     ...:             keys.append(k)
     ...:     return keys
```

We haven't covered functions yet, so let me explain. We used the def keyword to define a function named lookup_keys that has two parameters: d (for dictionary) and v (for value). Python automatically indents four spaces when you press ENTER, which designates that you're working within the function. Because the same value can be associated with multiple keys, we created an empty list named keys to hold the values. Next, we looped through the keys in the dictionary, and if the key's value matched the specified value (v), we appended it to the list. After the loop finished, we returned the list using the return keyword, which ended the function and made the list accessible to the rest of the program.

Let's test the function using the solar_system dictionary from the previous section. Pass it the name of the dictionary and 3, for the third planet, in parentheses:

```
In [230]: lookup_keys(solar_system, 3)
Out[230]: ['Earth']
```

As you might expect, reverse lookups run slower than forward lookups.

Printing Dictionaries

If you print a dictionary using the print() function, you'll get all the braces, quotation marks, and commas used to build the dictionary. To get around this, you can use "pretty printing" techniques.

If you do an online search for "pretty print a Python dictionary," you'll find numerous methods such as pprint(), that yield more readable output than the built-in print() function. Let's look at one of them here, the json .dumps() method.

The json.dumps() method converts a Python object into a JSON string. This, in turn, formats the dictionary into attractive JSON format. The method accepts three parameters used for pretty printing: the dictionary name, a Boolean value (True or False) for whether to sort the keys, and the number of spaces for indentation.

In the following example, we import json, create a dictionary (d), and then print it using the print() function followed by json.dumps(), for comparison:

```
In [231]: import json

In [232]: letter_order = dict(z=26, c=3, a=1, b=2, g=7, t=20)

In [233]: print(letter_order)
{'z': 26, 'c': 3, 'a': 1, 'b': 2, 'g': 7, 't': 20}

In [234]: print(json.dumps(letter_order, sort_keys=False, indent=4))
{
    "z": 26,
    "c": 3,
    "a": 1,
    "b": 2,
    "g": 7,
    "t": 20
}
```

The JSON output is easier to digest than the horizontal layout returned by print(). It works only with data types that JSON supports, however, which means that embedded sets and functions will fail.

Although we're getting ahead of ourselves, it's worth noting that you can traverse, sort, and print a dictionary using a for loop, the built-in sorted() function, and print(). You can even "pretty it up" with f-strings, as in this example:

```
In [235]: for k in sorted(letter_order):
     ...:         print(f'{k}: {letter_order[k]}')
a: 1
b: 2
c: 3
g: 7
t: 20
z: 26
```

You can indent the output similar to the json.dumps() method by adding spaces or tabs (\t) after the first single quote in the print command. Give it a try.

TEST YOUR KNOWLEDGE

15. What would you use to initialize a dictionary in which all the values are empty?

    a.   The setdefault() method

    b.   The fromkeys() method

    c.   The update() method

    d.   The built-in zip() function

16. Which statement about dictionaries is false?

    a.   Membership searches are very fast.

    b.   Dictionaries are optimized to find the key or keys for a given value.

    c.   You can pretty print dictionaries with the pprint() module.

    d.   Dictionaries are more memory intensive than lists.

17. Create a joke dictionary that maps setup lines to punchlines. Here's a few to get you started: "Did you hear about the kidnapping? He slept for three hours." "I started a band called '999 Megabytes.'" "You'll never get a gig." "I heard you had to shoot your dog. Was he mad?" "He wasn't too happy about it!"

18. What data type is held in the square brackets in contacts['Nix', 'Goaty'] = 'goatynix@gmail.com'?

    a.   String

    b.   List

    c.   Tuple

    d.   Set

Summary

In this chapter you learned about the four data types for working with collections in Python. A tuple is an immutable sequence type that holds a collection of objects, indexed by integers, in a defined order. A list is a mutable sequence type that holds a collection of objects, indexed by integers, in a defined order. A set is a mutable set type that holds an unordered collection of unique objects. A dictionary is a mutable mapping type that holds an ordered collection of unique objects (keys) that map to associated objects (values).

Each of these built-in data structures has its uses as well as its own peculiar behaviors. Tuples are memory efficient and good places to keep objects "safe" because they're unchangeable after they're created. Lists, being mutable, are flexible and useful for many jobs, though membership searches are slow. Sets can be used to efficiently remove duplicates from a dataset, provide very fast membership searches, and let you perform mathematical set operations, like union and intersection. Dictionaries also provide fast membership searches and let you easily set up associative databases for labeled data. If you are not sure how a data type is going to behave, take the time to test it in the interactive console before you incorporate it in your code.

There's a lot more left to learn. *Comprehensions*, which we'll cover in the next chapter, provide shorthand methods for creating lists, sets, and dictionaries. Importable modules, such as *collections* and *itertools*, provide useful tools for working with container data types. We'll look at these in Chapter 11. Then, in Chapter 12, we'll look at how to load external datasets into lists, sets, and so on, rather than typing them in item by item. For now, it's time to learn about *flow control*.

10

FLOW CONTROL

So far, we've been focusing on the components of programs, such as expressions, variables, and data types. We've strung a few of these together into simple executable instructions, but these have been mostly linear in nature; in other words, they were executed in the order in which they were written. More complex programs will include branching instructions that skip whole sections of code, jump back to the beginning, or decide among multiple options. To handle these situations, you'll need a way to control the flow of your code.

The *flow of execution* refers to the order in which statements are run in a program. Execution starts at the top of the code, with the first statement, after which point the statements are read in order. But this order doesn't have to be from top to bottom. In fact, the flow in most programs changes directions like cars in a busy intersection.

Flow control statements give Python the ability to make decisions about which instructions to execute next. You can think of these statements as the diamond shapes in flowcharts that indicate a decision is required to move forward (Figure 10-1).

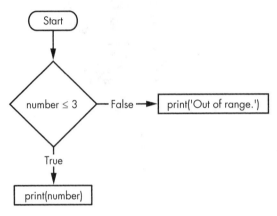

Figure 10-1: The diamond shape represents a decision in a flowchart.

This flowchart evaluates whether the number variable is greater than or equal to 3. The resulting decision causes the code to choose one path or another, a process called *branching*.

In this chapter, we'll discuss the if, else, elif, while, for, break, and continue flow control statements and clauses. We'll also look at ways to monitor the execution of flow and handle any exceptions that might occur.

The if Statement

The if statement is a conditional, or relational, statement. All control statements, including if statements, end with a colon (:) and are followed by an indented block of code. This indented clause executes only if the if statement's condition is True. Otherwise, the clause is skipped.

For example, this snippet checks whether 42 is less than 2. Only if the condition tests True will it print a message:

```
In [1]: if 42 < 2:
   ...:         print("That's crazy!")
In [2]:
```

All if statements must express a *condition,* which is an expression that is either true or false. This example used a comparison operator (<) to express a condition. Another option is to use Boolean values (covered in Chapter 8).

If you run this code, you should notice that nothing happens. This is because the statement evaluated to False. That's okay, but in most cases, you'll want to explicitly handle False outcomes, if only to make it clear that there's no missing code.

You can do that by adding an else clause, which executes if the if statement does not. The diamond in Figure 10-1 represents an if-else statement that works as follows:

```
In [3]: number = 2

In [4]: if number <= 3:
   ...:         print(number)
   ...: else:
   ...:         print('Out of range.')
2
```

The else clause represents the False branch in Figure 10-1. If the condition in the if statement is not met, it prints the string 'Out of range.'

Working with Code Blocks

Lines of code immediately underneath if statements and else clauses are indented. Indenting code tells the Python interpreter that a group of statements belongs to a specific *block* of code. These blocks execute as a single unit and end when the indentation level decreases back to zero or to the same level as a containing block.

Most programming languages make use of specific syntax to structure their code, such as braces ({}) for marking blocks, and semicolons (;) for ending lines. Python uses whitespace, because it is easier to understand visually, as demonstrated in Figure 10-2, which diagrams the previous if statement.

```
If number ≤ 3:
    ┌────────────────────┐
    │ print(number)      │   Block 1
    └────────────────────┘
else:
    ┌────────────────────┐
    │ print('Out of range.') │   Block 2
    └────────────────────┘
```

Figure 10-2: Example code blocks

The colon at the end of the first line lets Python know that a new code block is coming up. Each line in this block must be indented by the same amount. In the figure, Block 1 is the code that runs if the if statement's condition is True.

The following else clause returns to the previous indentation level. The colon after else denotes the start of another block (the block that runs if the if statement's condition is False, or Block 2), and this, too, must be indented.

We've been dealing with a single level of indentation, but blocks can contain more deeply indented, or *nested*, blocks. In the following example, each input line after line In [7] represents a new block of code:

```
In [5]: genus = 'rattus'

In [6]: species = 'norvegicus'

In [7]: if genus == 'rattus':
   ...:         if species == 'norvegicus':
   ...:             print('The common brown rat.')
The common brown rat.
```

If you make a mistake indenting code, don't worry, Python will let you know. Depending on where the mistake occurred (such as outside or inside a function), it will raise either a SyntaxError or an IndentationError.

NOTE *According to the PEP8 Style Guide (https://pep8.org/#indentation/), you should use four spaces per indentation level, and spaces are preferred to using tabs. By default, the Spyder text editor converts a tab into four spaces, so you can reduce your repetitive strain injury exposure. You can find this option under Tools ▸Preferences ▸Editor ▸Source Code ▸Indentation characters.*

Using the else and elif Clauses

The if statement comes with another optional clause, called elif (short for "else-if"), which tests another condition when the if statement evaluates to False. The elif clause lets you check multiple expressions for True and execute a block of code as soon as one of the conditions evaluates to True. You then can use the else clause as a final "catch all" that runs if none of the previous conditions are met.

Let's use elif and else to compare a single variable, representing the core temperature of a nuclear reactor in degrees Celsius, to several possible responses:

```
In [8]: core = 300

In [9]: if core < 200:
   ...:         print("Core is shut down")
   ...: elif 200 <= core < 300:
   ...:         print("Core is below optimum")
   ...: elif core == 300:
   ...:         print("Core is at optimum")
   ...: elif 300 < core < 1800:
   ...:         print("Core is above optimum")
```

```
...: else:
...:     print("Meltdown! Run for your life!")
Core is at optimum
```

The code starts by assigning an optimum operating temperature of $300°$ Celsius to the core variable. Next, an if statement tests whether the temperature is below $200°$. If so, the core should be shut down, so a message to that effect is printed. Next, a series of elif clauses look at other outcomes, such as a core temperature of exactly $300°$, and print appropriate responses. Finally, an else clause executes if all the proceeding conditions evaluate to False. This will catch a core value greater than or equal to 1800. Closing an if statement block with an else clause ensures that at least one clause is executed, and you won't be left with an empty response.

When using an else clause, you'll want to be very careful that your code properly handles the full range of possible values. For example, the following code prints the meltdown warning, even though the core temperature is only $200°$. See if you can figure out what went wrong:

```
In [10]: core = 200

In [11]: if core < 200:
    ...:     print("Core is shutdown")
    ...: elif 200 < core < 300:
    ...:     print("Core is below optimum")
    ...: elif core == 300:
    ...:     print("Core is at optimum")
    ...: elif 300 < core < 1800:
    ...:     print("Core is above optimum")
    ...: else:
    ...:     print("Meltdown! Run your life!")
Meltdown! Run your life!
```

Because this code failed to explicitly handle a core value of exactly 200, it was evaluated in the else clause, resulting in an incorrect message and a lot of unnecessary excitement.

Also make sure that *only one* condition evaluates to True. An advantage of using elif is that if a condition evaluates to True, the program will execute its corresponding block and exit the statement immediately. This is efficient, but if more than one elif condition evaluates to True, only the block associated with the *first* True condition will execute.

To illustrate, here's a poorly written piece of code that uses elif to increment multiple count variables whose conditions overlap:

```
In [12]: dogs = ('poodle', 'bulldog', 'husky')

In [13]: cats = ('persian', 'siamese', 'burmese')

In [14]: popular_breeds = ('poodle', 'persian', 'siamese')

In [15]: dog_count = 0
```

```
In [16]: cat_count = 0

In [17]: popular_breeds_count = 0

In [18]: animal = 'poodle'

In [19]: if animal in dogs:
   ...:         dog_count += 1
   ...: elif animal in cats:
   ...:         cat_count += 1
   ...: elif animal in popular_breeds:
   ...:         popular_breeds_count += 1

In [20]: dog_count
Out[20]: 1

In [21]: cat_count
Out[21]: 0

In [22]: popular_breeds_count
Out[22]: 0
```

This code starts by assigning tuples of dog breeds, cat breeds, and combined popular breeds. It then assigns count variables for each category, after which a poodle breed is assigned to the animal variable.

Next, a series of conditional statements evaluates the animal variable. If it's in the dogs tuple, the dog_count variable is incremented by 1. Otherwise, if it's in the cats tuple, the cat_count variable is incremented, and then, if it's only in the popular_breeds tuple, the popular_breeds_count is incremented.

When you run the code and check the counts, they're incorrect. Despite "poodle" being present in both the dogs and popular_breeds tuples, only the dog_count variable was updated. Because the first elif clause evaluated to True, the if statement terminated immediately, and the popular breeds evaluation was never performed.

Using Ternary Expressions

For convenience, Python lets you combine an if-else block into a single expression called a *ternary expression* whose syntax is as follows:

```
true expression if condition else false expression
```

Here's an example:

```
In [23]: core = 1801

In [24]: 'Run for your lives!' if core >= 1800 else 'So far so good!'
Out[24]: 'Run for your lives!'
```

Ternary expressions let you write pithy code at the cost of readability. They should be used with only simple and straightforward conditions and expressions.

Using Boolean Operators

To further aid you in making comparisons, Python provides the and, or, and not operators. These three operators compare Boolean values and evaluate to a Boolean value.

The possible outcomes for Boolean operators can be shown in a truth table, which we present in Table 10-1.

Table 10-1: Truth Table for and/or Operators

Expression	Evaluation
True and True	True
True and False	False
False and True	False
False and False	False
True or True	True
True or False	True
False or True	True
False or False	False

As you can see, the and operator evaluates an expression to True only if *both* Boolean values are True. The or operator evaluates to True if *either* of the Boolean values is True. For example, you could say you had "cereal" for breakfast if you had either corn flakes *or* raison bran, but you can't say you had "bacon and eggs" unless you had both bacon *and* eggs.

The not operator operates on only *one* expression or Boolean and evaluates to the *opposite* Boolean value. For example:

```
In [25]: not False
Out[25]: True
```

With and, or, and not, you can build more complex comparisons for directing your code's flow of execution. Try out a few examples in the console:

```
In [26]: 'a' == 'a' and 10 > 2
Out[26]: True

In [27]: (10 > 2) and (42 > 2) and ('a' == 'b')
Out[27]: False

In [28]: (10 < 2) or ('a' != 'b')
Out[28]: True
```

Python will evaluate each expression, from left to right, until it has a single Boolean value. It then evaluates these Booleans down to a single value, either True or False. The order of operations is as follows:

math operators → comparison operators → not operator → and operator → or operator

With Boolean operators, you can compare multiple variables within if statements. Here's an example in which you discriminate animals using the number of legs and the sound produced:

```
In [29]: legs = 4

In [30]: sound = 'bark'

In [31]: if legs == 4 and sound == 'bark':
   ...:         print('a dog')
   ...: elif legs == 4 and sound == 'meow':
   ...:         print('a cat')
a dog
```

In the previous example, both conditions had to be True for the if statement to execute. In the following example, only one of the conditions needs to be True:

```
In [32]: today = 'Sunday'

In [33]: if today in ('Saturday', 'Sunday'):
   ...:         print('Enjoy your weekend!')
Enjoy your weekend!
```

If today is either Saturday or Sunday, you're in the weekend, and the print() function is called.

Note that it's easy to slip-up when using if statement syntax. The following code looks logical, but it will evaluate to True no matter the value of the today variable:

```
In [34]: today = 'Saturday'

In [35]: if today == 'Saturday' or 'Sunday':
   ...:         print('Enjoy your weekend!')
Enjoy your weekend!
```

Loops

Loops permit the repetition of certain steps indented under a keyword. The repetition continues until some condition is met, making loops much like if statements, but they can run more than once.

Python uses the while and for keywords for loops. These correspond to *condition-controlled* loops and *collection-controlled* loops, respectively.

The while keyword, plus a condition, forms a while statement. These are used to execute a block of code repeatedly until the given condition evaluates to False. At this point, the line immediately after the loop in the program is executed. Here's the syntax:

```
while some condition is True:
    do something
```

The for keyword is used to repeat a block of code a fixed number of times or to iterate over a sequence of items. Here's the basic syntax:

```
for something in something:
    do something
```

When the for loop runs out of items, its underlying condition becomes False, and the loop ends and returns control to the first line of code under the for loop's block.

The while Statement

The while statement tests a condition and executes the block over and over until the condition is False (Figure 10-3) or you explicitly end the loop with a break statement (more on these later). In fact, a while loop can go on forever.

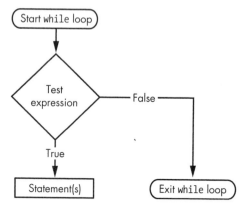

Figure 10-3: Flowchart for a generic while loop

A while loop is useful for performing some action until a target is reached. For instance, you can simulate the population growth of a herd of deer until the population reaches a target value, at which time the simulation loop could stop and log details, such as the time it took to reach the target or the average weight of an adult deer.

A much simpler example is testing for a password. In the code that follows, we give a user a set number of times to enter the correct value.

```
In [36]: password = ''

In [37]: count = 0
```

```
In [38]: ❶ while password != 'Python':
   ...:         password = input("Enter your password: ")
   ...:         count += 1
   ...:     ❷ if count > 3:
   ...:             print("No more tries.")
   ...:             break
```

In this example, we first create an empty password variable and set a count variable to 0. We then start a loop using the while keyword ❶. If pass word does not equal "Python," the indented while clause will prompt the user to enter a password. It then increments the count variable by 1 and uses an if statement to check if the number of allowable counts has been exceeded ❷. If this evaluates to True, the user is informed that they have exceeded the allowed number of attempts, and the break keyword ends the loop. If count is less than or equal to 3, the loop will continue prompting the user for a pass word. If the user enters the correct password, the loop ends without fanfare.

Figure 10-4 documents this loop in a flowchart. Note how both the while and if conditions are marked by diamonds. This is because they represent decision points.

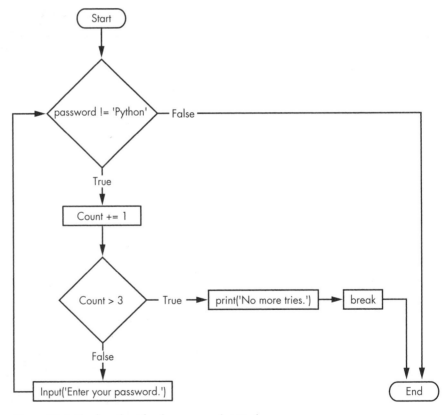

Figure 10-4: The flowchart for the password while loop

Each execution of a loop is called an *iteration*. The while loop can iterate *indefinitely* because the number of loops isn't explicitly specified at the start. A while loop can run a million times or end after the first iteration depending on whether its condition is met.

NOTE   *If your program is ever caught in an infinite loop, you can use CTRL-C to stop the program and escape.*

TEST YOUR KNOWLEDGE

1. Each new code block should be indented ____ spaces.

2. True or false: The or operator evaluates an expression to True only if *both* Boolean values are True.

3. Write a while loop that never ends, and then stop it using CTRL-C.

4. Which values can be used to represent False?

    a. 0

    b. 0.0

    c. F

    d. All the above

5. To form Pig Latin, one takes an English word that begins with a consonant, moves the consonant to the end, and adds "ay" after it. If the word begins with a vowel, one simply adds "way" to the end. Use the Spyder text editor to write a program that takes a word as input and uses indexing and slicing to return its Pig Latin equivalent. Keep the program running until the user decides to quit.

The for Statement

The for statement lets you execute a loop a set number of times. This number is often specified using the built-in range() function. This memory-efficient function returns a sequence of evenly spaced numbers starting from 0 (by default) and ending before a specified endpoint. Here's what the range() function's syntax looks like:

```
range(start, stop, step).
```

The start and step parameters are optional. If omitted, the start parameter defaults to 0, and the step defaults to 1.

Here's an example where a for loop uses range() to print five numbers:

```
In [39]: for i in range(5):
    ...:     print(i)
0
1
2
3
4
```

Note that 5 is not included in the output. This is because the function reads *up to* the stop value but doesn't include it. The block beneath the for statement is indented, just as in a while loop.

Here's an example using all three arguments (start, stop, step) in range() to print every other number from 1 to 6:

```
In [40]: for i in range(1, 6, 2):
    ...:     print(i)
1
3
5
```

Note that the i in the previous code is used by convention when iterating over a range of numbers. Any legal variable name (such as num or number) will work. An underscore (_) is preferred if you're using a linter to check your code; other names can sometimes trigger an unused variable warning.

You can use range()in conjunction with the len() function to obtain the endpoint for an arbitrarily sized sequence. Here's an example:

```
In [41]: my_list = ['a', 'b', 'c', 'd', 'e']

In [42]: for i in range(len(my_list)):
    ...:     print(my_list[i])
a
b
c
d
e
```

This code starts by assigning a list of letters to the my_list variable. In the following if statement, passing the range() function the length of this list, as determined by the len() function, sets the number of iterations. With each iteration, the current value of i is used as an index for the list, and the corresponding letter is printed.

Although it works, this code isn't very Pythonic. Fortunately, you can use iterables directly in the for statement. Remember, iterables are objects that can return their items one at a time. These include sequence types like range, list, tuple, string, set, and more. As a result, you don't need to keep track of the iterable's size or use a running index. Here's the previous snippet in this format:

```
In [43]: for item in my_list:
    ...:         print(item)
a
b
c
d
e
```

Notice how the code reads almost like English. You can't get more Pythonic than that!

NOTE *Never add or delete items to a list while looping over it. If you want to change a list during a loop, append the changes to a new list.*

You can even loop over a string and print its characters without using an intermediate variable to hold the string:

```
In [44]: for letter in "Python":
    ...:         print(letter)
P
y
t
h
o
n
```

In the event that you need an item's index during looping, the best solution is to use the built-in enumerate() function. This function adds a counter to each item of an iterable object and returns an *enumerate* object, allowing you to loop over an iterable and keep track of the number of iterations. The first index value starts from 0 by default, but you can override that by specifying a starting index. The following example produces a numbered list of items (starting at 1) from a list datatype:

```
In [45]: equipment_list = ['binoculars', 'rock hammer', 'hand lens']

In [46]: for index, item in enumerate(equipment_list, start=1):
    ...:         print(index, item)
1 binoculars
2 rock hammer
3 hand lens
```

Some other uses for enumerate() include selecting every *n* items from a list, ending a loop after *n* items, and using the indexes for line weights or symbol sizes when plotting.

Loop Control Statements

Loop control statements are used inside loops to change the normal sequence of execution. Earlier, you used a break statement to interrupt and end a while loop. Python also uses continue and pass statements to control loops.

The break Statement

The break keyword lets you exit a loop's code block at any time. A common usage is to set while to True (while True:) and then "manually" break out of the loop when a condition is met. Because True always evaluates to true, no condition can stop the loop, so you must force the issue using break.

In nested loops, break terminates only the block it's in, along with any inner blocks. Outer loops will continue to run. For example, the following example contains two nested loops:

```
In [47]: mythbusters = ['Kari', 'Grant', 'Tory']

In [48]: MYTHBUSTERS = ['Adam', 'Jamie']

In [49]: for star in mythbusters:
    ...:         for big_star in MYTHBUSTERS:
    ...:             if big_star == 'Adam':
    ...:                 break
    ...:         print(star, big_star)
Kari Adam
Grant Adam
Tory Adam
```

Even though the break statement interrupted the inner for loop and prevented "Jamie" from printing, the outer loop ran to completion.

The continue Statement

The continue keyword immediately returns control to the beginning of a loop, skipping any loop clauses beneath it. For instance, the following example uses a while loop to evaluate a user's name and password. Enter the code in the Jupyter Qt console and then run it using SHIFT-ENTER:

```
In [50]: while True:
    ...:         name = input('Enter your username: ')
    ...:         if name != 'Alice':
    ...:             ❶ continue
    ...:     ❷ pwd = input('Enter your password: ')
    ...:         if pwd == 'Star_Lord':
    ...:             ❸ break
    ...:         else:
    ...:             print('That password is incorrect')
```

The first step, of course, is to get the correct username. If the condition for the first if statement evaluates to True (the name is *not* "Alice"), a continue statement interrupts the loop and restarts the sequence ❶. If the name passes the test, the continue statement is skipped, and the user is prompted for a password ❷. If they enter the correct value, a break statement ends the loop ❸. Otherwise, they're alerted to the error, and the loop begins again.

The pass Statement

You can use the pass keyword to build "empty" loops, or blocks that take no action. These keywords serve as placeholders for future code or flag places where you intentionally omitted something. Take this snippet, for example, in which we choose to not print the h in Python:

```
In [51]: word = 'Python'

In [52]: count = 0

In [53]: while count < len(word):
   ...:        if count < 3:
   ...:            print(word[count])
   ...:    ❶ elif count == 3:
   ...:            pass
   ...:        else:
   ...:            print(word[count])
   ...:        count += 1
P
y
t
o
n
```

This code assigns "Python" to the word variable, and 0 to the count variable. While the count value is less than the length of word, the loop prints each character in word indexed at the current count value. If count is equal to 3, however, it uses the pass keyword to continue the loop with no action.

You could have accomplished the same thing by only printing count < 3 and count > 3, but this could be viewed in retrospect as an indexing mistake (see the earlier core temperature example at lines In [10]-[11]). By including the reference to h and using pass ❶, it's clear that you knew what you were doing. You basically said, "I know there's a letter at this index and I'm not printing it on purpose."

Replacing Loops with Comprehensions

In Python, a *comprehension* is a way to construct a new sequence such as a list, set, or dictionary from an existing sequence. For example, you might want to create a new list containing the logarithms of the numbers in another list. As a replacement for standard for loops, comprehensions are faster, more elegant, and more concise.

A downside to comprehensions is that you can't embed a print() function in the loop to help you track what's happening. They can also be quite difficult to read when using complicated expressions, but for simple expressions, they can't be beat. Python supports comprehensions for lists, dictionaries, sets, and something we'll talk about in the next chapter: generators.

List Comprehensions

To make a new list using *list comprehension*, encase the comprehension expression in square brackets. To use a condition to choose items from an existing iterable, use this syntax:

```
new_list = [item for item in iterable if item satisfies condition]
```

Use this syntax if you want to alter these items, or generate new items, before adding them to the new list:

```
new_list = [expression for item in iterable if condition]
```

For instance, here's an example that takes a string, loops through it, capitalizes each letter, and appends the capital letter to a new list:

```
In [54]: word = 'Python'

In [55]: letters = [letter.upper() for letter in word]

In [56]: letters
Out[56]: ['P', 'Y', 'T', 'H', 'O', 'N']
```

Here, we assigned a string to the word variable, and then we created the letters list using the list comprehension [letter.upper() for letter in word]. Notice how the syntax is a little backward, as the loop variable (letter) appears before it's defined.

The following example extracts the capital "P" from "Python." Notice how you can use an if statement within the comprehension:

```
In [57]: cap = [letter.upper() for letter in word if letter.isupper()]

In [58]: cap
Out[58]: ['P']
```

The single-line list comprehension in line In [57] is equivalent to the following code using a standard for loop:

```
In [59]: cap = []

In [60]: for letter in word:
   ...:         if letter.isupper():
   ...:             cap.append(letter)
```

In this case, list comprehension saved you three lines of code.

You also can create list comprehensions using nested for loops, as follows:

```
In [61]: first = ['Python is']

In [62]: last = ['fun', 'easy', 'neat']
```

```
In [63]: print([f + ' ' + l for f in first for l in last])
['Python is fun', 'Python is easy', 'Python is neat']
```

You can see more examples of list comprehensions at *https://docs.python.org/3/tutorial/datastructures.html*.

Dictionary Comprehensions

Dictionary comprehensions are like list comprehensions, except they use key-value (k, v) pairs in place of items, are enclosed in curly brackets, and return a dictionary rather than a list.

You can use dictionary comprehensions on existing dictionaries, in which case you either choose key-value (k, v) pairs from the existing dictionary based on some condition, or you use an expression on the keys and/or values based on a condition. The general syntax is:

```
new_dict = {k:v for (k, v) in dictionary if k, v satisfy condition}
```

or:

```
new_dict = {k-expr:v-expr for (k, v) in dictionary if condition}
```

When using dictionary comprehension on another type of iterable, like a list, you choose and/or change existing items in the iterable to serve as key-value pairs in the new dictionary. The syntax is varied, but looks something like this:

```
new_dict = {item:item-expr for item in iterable if condition}
```

Note that the expression can alter the item used for the key, for the value or for both.

Here's an example in which we extract even numbers from a list and map them to their square:

```
In [64]: inputs = [1, 2, 3, 4, 5, 6]

In [65]: new_dict = {item:item**2 for item in inputs if item % 2 == 0}

In [66]: new_dict
Out[66]: {2: 4, 4: 16, 6: 36}
```

And here we zip two tuples (the mineral and hardness variables) together to create part of Moh's famous mineral hardness scale:

```
In [67]: mineral = 'talc', 'gypsum', 'calcite'

In [68]: hardness = 1, 2, 3

In [69]: mohs = {m: h for (m, h) in zip(mineral, hardness)}

In [70]: mohs
Out[70]: {'talc': 1, 'gypsum': 2, 'calcite': 3}
```

Note that you also can generate the mohs dictionary by calling the built-in dict() function, as follows:

```
In [71]: mohs = dict(zip(mineral, hardness))
```

Set Comprehensions

Set comprehensions are enclosed in curly brackets and return unsorted sets. You can *choose* existing items from an iterable using the following syntax:

```
new_set = {item for item in iterable if item satisfies condition}
```

Alternatively, you can alter the items or derive new items by applying a condition-based expression before adding the item to the new set:

```
new_set = {expression for item in iterable if condition}
```

Here's an example that returns all the unique characters in a string, given that sets don't allow duplicate values (this is the same as the set() function):

```
In [72]: pond_animals = ['turtle', 'duck', 'frog', 'turtle', 'snail', 'duck']

In [73]: unique_animals = {animal for animal in pond_animals}

In [74]: unique_animals
Out[74]: {'duck', 'frog', 'snail', 'turtle'}
```

And here we use an expression to calculate word lengths:

```
In [75]: unique_word_lengths = {len(word) for word in pond_animals}

In [76]: unique_word_lengths
Out[76]: {4, 5, 6}
```

Handling Exceptions

An *exception* occurs when Python encounters an error during execution. This causes it to "raise an exception" and produce a Python object that represents an error. If not dealt with immediately, exceptions cause a program to terminate and display an error message.

Fortunately, exceptions can be "caught" and handled through flow control. This gives you the opportunity to fix the problem, try again, supply a more helpful error message, or suppress the error.

Using try and except

Python provides a try statement with an except clause to help you handle exceptions. The try statement lets you isolate code that could potentially contain an error and crash your program. If the code contains an error, the except clause will deal with it by providing code that executes only when an exception is raised.

The simplest exception handling merely prevents your program from crashing unceremoniously. It presents an alert and (hopefully) a helpful message. Here is an example that's designed to handle incorrect input from a user:

```
In [77]: try:
    ...:        age = int(input("Enter your age in years: "))
    ...: except:
    ...:        print("Please start again and enter a whole number.")

Enter your age in years: Harry
Please start again and enter a whole number.
```

In this example, the age value is converted to an integer, as we want age to be a whole number in years. But the user entered letters instead (Harry), which raised a ValueError exception because letters can't be converted into integers. You don't see this error message, however, as we caught the exception using the try statement, which does just what it says. It tries the code in isolation, giving us the opportunity to do something before the program crashes. Here, we printed a message to the user so that the program ends somewhat gracefully.

In most cases, you want to specify the type of exception to handle rather than catch all of the built-in exceptions, as we did in the previous snippet. You can find a list of exception types on Table 7-4 on page 183. You can also create the error you're looking for in the console and read off the exception name from the resulting Traceback message.

Let's rewrite the previous snippet to catch a ValueError. Just place the proper exception name after except, as follows:

```
In [78]: try:
    ...:        age = int(input("Enter your age in years: "))
    ...: except ValueError:
    ...:        print("Please start again and enter a whole number.")
```

You can also trap for multiple exception types by placing them after except as a tuple with parentheses:

```
In [79]: except (ValueError, TypeError):
```

If you want to use a different message or take a different action with each error type, simply use multiple stacked except clauses within a try statement, in this format:

```
try:
    something...
except ValueError:
    do something...
```

```
except TypeError:
    do something else...
```

You can also incorporate Python's error messages into your own customized versions. Exceptions have *arguments*, which are official messages from Python describing what happened. You can use an argument by specifying a variable after the exception type, preceded by the keyword as. Here's an example:

```
In [80]: try:
   ...:     age = int(input("Enter your age in years: "))
   ...: except ValueError as e:
   ...:     print(e)
   ...:     print("Please start again and enter a whole number.")

Enter your age in years: Steve
invalid literal for int() with base 10: "Steve"
Please start again and enter a whole number.
```

Now you have the best of both worlds: Python's precise but technical explanation combined with friendlier instructions for non-programmers.

Finally, you can add an else clause at the end of all the except clauses. This lets you do something, like confirm a successful operation, if no exceptions are raised elsewhere in the try block:

```
In [81]: try:
   ...:     age = int(input("Enter your age in years: "))
   ...: except ValueError as e:
   ...:     print(f'\n{e}')
   ...:     print("Please start again and enter a whole number.")
   ...: else:
   ...:     print(f"You entered an age of {age} years")

Enter your age in years: 42
You entered an age of 42 years
```

Forcing Exceptions with the raise Keyword

Python's raise keyword allows you to force a specified exception if a condition occurs. You can use it to raise built-in error types or your own custom errors. It's especially useful for validating inputs such as enforcing the use of maximum values, or handling the input of a negative number when you're working with positive numbers.

To see how to create your own custom error, enter the following in Spyder's text editor and save it with any name you please:

```
word = input("Enter Harry's last name: ")
if word.lower()!= 'potter':
    raise Exception('I was looking for Potter!')
```

Here, you accept a name from the user, convert it to lowercase, and compare it to "potter." If the name doesn't match, you use the raise keyword and call the Exception class, passing it a custom message.

Now, run the file and input "Houdini" when prompted. You should see this general output in the console (truncated for brevity):

```
Enter Harry's last name: Houdini
Traceback (most recent call last):

--snip--

  File "C:/Users/hanna/file_play/junk.py", line 1, in <module>
    raise Exception('I was looking for Potter!')

Exception: I was looking for Potter!
```

To force Python to raise one of its built-in exceptions, substitute the name of the built-in exception class (see Table 7-4) for the Exception class you used previously. In this example, we raise Python's built-in TypeError exception:

```
number = 'Steve'
if isinstance(number, int):
    pass
else:
    raise TypeError("Only integers are accepted.")
```

You can read more about raise at *https://docs.python.org/3/tutorial/errors.html.*

Ignoring Errors

What if you want to ignore errors while looping? For example, suppose that you've used Python's None keyword to define missing or null sample values in your dataset (data):

```
In [82]: data = [24, 42, 5, 26, None, 101]
```

You don't want to strip this placeholder out of the dataset, because it contains valuable information. It lets you know that the dataset is incomplete as well as the location of the missing data. But if you try to iterate over the data and do something to it such as divide each value by 2, the None value will raise a TypeError and crash the program:

```
In [83]: for sample in data:
    ...:     print(f'{sample / 2}')
12.0
21.0
2.5
13.0
Traceback (most recent call last):

File "C:\Users\hanna\AppData\Local\Temp/ipykernel_5140/163932511.py", line 2, in <module>
```

```
print(f'{sample / 2}')

TypeError: unsupported operand type(s) for /: 'NoneType' and 'int'
```

We can combine multiple flow control elements to handle this missing data. Here, we use a try statement with an except clause within the for loop:

```
In [84]: for sample in data:
    ...:     try:
    ...:         x = sample / 2
    ...:         print(x)
    ...:     except TypeError:
    ...:         print("missing data")
12.0
21.0
2.5
13.0
missing data
50.5
```

Now, the loop runs to completion and flags where data is missing.

However, what if you want to completely skip over the missing data? For example, you want to pass the output to some other mathematical operation where the "missing data" values would interfere? In this case, use the continue statement, as follows:

```
In [85]: for sample in data:
    ...:     try:
    ...:         x = sample / 2
    ...:         print(x)
    ...:     except TypeError:
    ...:         continue
12.0
21.0
2.5
13.0
50.5
```

The loop now treats the missing value as if it doesn't exist because it continues the loop when it encounters the value. Remember, continue immediately returns control to the beginning of a loop.

Tracing Execution with Logging

To control the flow of your program, you need to know what it's returning at key locations. One way to keep track of this is to use the print() function. This lets you print the output, the data type of a variable, or some other useful information about an important step.

The print() function works well for small programs, but if you're using it only to quality-control your code, it can come with a price. To unclutter your code and output, you might need to go back later and either delete all the lines containing print() or comment them out (by placing a # at the start of the line) so that they don't run.

For large programs, a better choice is to use the logging module. This module is part of the standard library that ships with Python, and it can provide a customized report on what your program is doing at any location you choose. Five *logging levels* let you categorize messages by importance. These are listed in Table 10-3.

Table 10-3: Python's Logging Levels

Level	Function	Description
DEBUG	logging.debug()	Detailed information for diagnosing problems
INFO	logging.info()	Confirmation that things are working as expected
WARNING	logging.warning()	Unexpected event or potential future problem in working code
ERROR	logging.error()	An error prevented the code from functioning as intended
CRITICAL	logging.critical()	A serious error that may halt the program

Large programs that require logged messages are difficult to write in a console, so enter the following example in Spyder's text editor and save it with a name like *logging.py*. This code uses logging to check that a vowel-counting program is working correctly:

```
import logging

❶ logging.basicConfig(level=logging.DEBUG,
                    format='%(asctime)s %(levelname)s - %(message)s')
word = 'scarecrow'
VOWELS = 'aeiouy'
num_vowels = 0

for letter in word:
    if letter in VOWELS:
        num_vowels += 1
❷   logging.debug('letter & vowel count = %s-%s', letter, num_vowels)
```

Save the file and press F5 (or click the **Run** button) to execute the code. You should see this general output:

```
In [86]: runfile('C:/Users/hanna/.spyder-py3/temp.py', wdir='C:/Users/hanna/.spyder-py3')
202x-09-27 14:37:30,578 DEBUG - letter & vowel count = s-0
202x-09-27 14:37:30,580 DEBUG - letter & vowel count = c-0
202x-09-27 14:37:30,581 DEBUG - letter & vowel count = a-1
202x-09-27 14:37:30,581 DEBUG - letter & vowel count = r-1
202x-09-27 14:37:30,582 DEBUG - letter & vowel count = e-2
202x-09-27 14:37:30,582 DEBUG - letter & vowel count = c-2
202x-09-27 14:37:30,582 DEBUG - letter & vowel count = r-2
202x-09-27 14:37:30,583 DEBUG - letter & vowel count = o-3
202x-09-27 14:37:30,583 DEBUG - letter & vowel count = w-3
```

Let's look at what you did. After importing the module, you used the basicConfig() method to set up and format the debugging information you wanted to see ❶. The DEBUG level is the lowest level of information and is used for diagnosing details. Adding a timestamp (%(asctime)s) is not necessary here, but it can become important when debugging long-running programs.

After setting up the word to count, a constant for vowels, and a count variable, you started a for loop through the letters in the word and compared each to the contents of VOWELS. If the letter matched, you incremented the num_vowels counter by 1.

For each letter evaluated, you used logging.debug() to enter the custom text message to display along with the current count ❷. The logging output displayed in the console. You can see the timestamp, the logging level, and the cumulative vowel count, along with which letters changed the count. In this case, only vowels change the count, so the program appears to be working as intended.

You can redirect the logged messages to a permanent *text file* rather than displaying them on the screen. Just use the filename keyword in the logging.basicConfig() function, as follows:

```
logging.basicConfig(filename='vowel_counter_log.txt',
                    level=logging.DEBUG,
                    format='%(asctime)s %(levelname)s - %(message)s')
```

As written, this code will save the log file to the same folder as your Python file. To save it elsewhere, you'll need to specify a path.

Both the print() function and logging can slow down your program. It's easier to disable the logging messages, however. With the logging.disable() function, you can turn off all the messages for a certain level with one line of code, as follows:

```
import logging
logging.disable(logging.CRITICAL)
```

Placing logging.disable() near the top of the program, just below the import statement, allows you to find it easily and toggle messages on and off by commenting them out with a hash mark, as follows:

```
import logging
#logging.disable(logging.CRITICAL)
```

The logging.disable() method will suppress all messages at the designated level and lower. Because CRITICAL is the highest level, you can use it to disable messages at every level. This is much easier than finding and deleting (or commenting-out) multiple calls to print().

For more details on the logging module, check out the documentation at *https://docs.python.org/3/library/logging.html*. For a basic tutorial, visit *https://docs.python.org/3/howto/logging.html*.

6. Write a code snippet that asks a user for a username and password. If the username is incorrect, keep asking until the user gets it right. If only the password is incorrect, keep asking for the correct password but don't repeat the username request.

7. The for loop is just a concise version of a while loop. Write a while loop that behaves like a for loop and prints the word "Python" five times.

8. Use list comprehension to make a list of all the *even* numbers between 1 and 10.

9. Use a for loop with the range() function to print a NASA-style countdown from 10 to 0.

10. A secret message is hidden at the center of each of the following words: "age," "moody," "knock," "adder," "project," "stoop," "blubber." Use a for loop to find and print the message.

11. Use the text editor to write a "guess my number" game that randomly chooses an integer between 1 and 100 (using random.randint()) and tells the player whether their answer is too high or too low until they guess correctly. Inform the player when they win and show them how many tries it took.

12. Use the text editor to write a "fortune cookie" program that presents the user with a menu of three options: Quit, Open a Fortune Cookie, or Open a Misfortune Cookie. Make a list of positive fortunes and a list of funny "misfortunes" and use the random module's choice() function to randomly choose from each list. Print the results to the screen.

Summary

The magic of programming lies in the ability of programs to make decisions during execution. These decisions are facilitated by conditions that evaluate to True or False. Using comparison and Boolean operators with *conditional statements* like if, elif, and else, you can control what your code does and when it does it.

Indentation (whitespace) is used to segregate code into functionally similar segments, called *blocks*. Indentation levels let Python know when blocks start and end. This, in turn, helps you control the flow of execution through a program.

A while loop causes code to execute over and over until a certain condition is met. The for loop, on the other hand, runs for a designated number of times or until it exhausts the items in a container data type, such as a list. Both types of loops can be manually interrupted with the break statement or forced to jump back to the start using the continue statement.

A for loop can be simplified into a single line of code using *comprehensions*. You can use these with lists, sets, and dictionaries. For simple expressions, comprehensions are not only more concise than for loops, they're faster, as well.

Because errors can affect the flow of your code, Python provides try statements with except clauses to help you handle errors by suppressing them, fixing them, creating customized error messages, or getting a user to try again. To help you find and debug errors and other issues, Python provides the logging module. Compared to the print() function, logging is a more sophisticated and manageable way to monitor the flow of execution of large programs.

Another way to control flow is to write functions. We'll look at these important "mini-programs" in the next chapter.

11

FUNCTIONS AND MODULES

A *function* is a reusable set of instructions that performs a specific task. When the function completes its task, the flow of execution returns to the proper place in the greater code structure. *Modules* are programs, usually comprising functions, that perform a task or group of related tasks. Whereas you can define functions in place, you must import modules into a Python program to use them.

Both functions and modules let you simplify code through the process of *abstraction*. Abstraction is the act of moving the details of some process into a seemingly simpler object away from the main routine. Later, you can perform the task by calling the object's name in a single line of code.

The best function and module names are short and descriptive. They allow you to skim the main routines of programs and get an idea of what's going on, as if you were reading a summary. A good analogy is the table of contents of this book. Although a great many details are hidden away in the actual chapter, the headings and subheadings give you a good idea of what each chapter entails.

In previous chapters, you imported modules like math and os, and you used built-in functions like print() and input(). Their code was abstracted to the point that you never saw it. You just called a function, and something happened. There'll be times, however, when a pre-built solution is either unavailable or insufficient, and you'll need to create a function yourself.

By writing your own functions to reuse units of code, you can create more readable, better organized, and less redundant programs. In this chapter, you'll write custom functions and modules and become familiar with additional built-in functions and third-party modules designed to make your life easier.

Defining Functions

To write a function in Python, you define it by using the def keyword followed by a name for the function, parentheses, and a colon. As always, code coming after the colon must be indented, and the indented lines represent executable code. Here's an example in the IPython console in the Spyder IDE:

```
In [1]: def warning():
   ...:     print("WARNING: Converting units from metric to Imperial!")
```

NOTE   *In the console, you can complete a function by pressing ENTER twice or by using SHIFT-ENTER. In the editor, a function's code block ends when you return to the same indentation level as the def keyword.*

You've now encapsulated a warning message within the warning() function. To use the message again, you need only *call* the function by entering its name and parentheses. This saves you from typing out the full message over and over:

```
In [2]: warning()
WARNING: Converting units from metric to Imperial!
```

In a function name, the parentheses (), sometimes referred to as the *call operator*, let Python know that an object can be *invoked*, which is a fancy way of saying "execute this command."

Like everything else in Python, functions are objects. They belong to the function data type:

```
In [3]: type(warning)
Out[3]: function
```

You can assign functions to variables, use them in other functions, define them in other functions, return them as values from other functions, and store them in data structures (for example, as an item in a list).

According to Python's PEP 8 Style Guide (*https://pep8.org/*), you should surround top-level functions (those defined at indentation level 0) with two blank lines. Within functions, you should use blank lines (sparingly) to indicate logical sections.

Using Parameters and Arguments

You can submit, or *pass*, input to a function; perform some operation on the input; and then output, or *return*, the result. To do so, you use parameters and arguments inside the parentheses and separate them by commas.

Parameters are special kinds of variables, defined by a function, that receive a value when the function is called. They refer to the pieces of data provided as input but are not the data itself. For example, the following code defines a function that calculates a force value, using the famous equation *F=MA*, when passed mass and acceleration parameters:

```
In [4]: def calc_force(mass, acceleration):
   ...:         return mass * acceleration
```

Arguments are the actual data values input when calling the function. For example, you could call the calc_force() function with these arguments:

```
In [5]: calc_force(15000, 9.78033)
Out[5]: 146704.94999999998
```

Figure 11-1 identifies the parameters in the calc_force() function definition and the arguments passed to it when called.

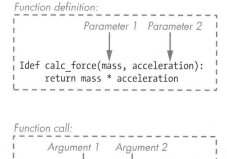

Figure 11-1: Function definitions *use*
parameters and function calls *use*
arguments

Functions like calc_force() that return a value are called *fruitful* functions. Functions that perform an action but don't return a value are called *void* functions. The warning() function in the previous section is an example of a void function.

With fruitful functions, the return statement causes execution to exit the function and resume at the point in the code immediately after the instruction that called the function, known as its *return address*. Values listed *after* the return keyword and on the same line are passed back to the code that called the function. For the calc_force() function, this would be the value in line Out[5].

The return keyword always ends a function and prevents execution of any subsequent code within the function.

NOTE *Technically, all functions need to evaluate to a return value. Void functions satisfy this requirement by automatically returning Python's null value, None, which belongs to the NoneType data type.*

Positional and Keyword Arguments

Function arguments can be of two types: *positional* and *keyword*. Positional arguments must be entered in the correct order, as defined by the order of the parameters in the function definition. As shown in Figure 11-1, the calc_force() function uses positional arguments, such that the first argument submitted corresponds to the mass, and the second to the acceleration.

Keyword (or *named*) arguments include a keyword and an equal sign before the submitted value. These are used to add clarity and make a function's intention clear. Here's how to call the calc_force() function using keywords:

```
In [6]: calc_force(mass=15000, acceleration=9.78)
Out[6]: 146700.0
```

NOTE *According to the Python Style Guide, no spaces should be used around the equal sign in keyword arguments.*

Another advantage to keyword arguments is that you don't need to remember the order in which the parameters were defined. Here, we enter arguments in reverse order:

```
In [7]: calc_force(acceleration=9.78, mass=15000)
Out[7]: 146700.0
```

You can enter both positional and keyword arguments when calling a function. However, after you use a keyword argument, you can't go back to using positional arguments in the same function call. So, this code works:

```
In [8]: calc_force(15000, acceleration=9.78)
Out[8]: 146700.0
```

But this code fails:

```
In [9]: calc_force(mass=15000, 9.78)

  File "C:\Users\hanna\AppData\Local\Temp/ipykernel_3212/3649549750.py", line 1
    calc_force(mass=15000, 9.78)
                              ^
SyntaxError: positional argument follows keyword argument
```

You can force the use of keyword arguments by including an asterisk (*) as the first parameter when defining a function:

```
In [10]: def calc_force(*, mass, acceleration):
   ...:         return(mass * acceleration)
```

Now, if you attempt to use positional arguments, Python will raise an exception and inform you that positional arguments are not accepted:

```
In [11]: calc_force(15000, 9.78)
Traceback (most recent call last):

  File "C:\Users\hanna\AppData\Local\Temp/ipykernel_3212/2133932729.py", line 1, in <module>
    calc_force(15000, 9.78)

TypeError: calc_force() takes 0 positional arguments but 2 were given
```

Using Default Values

You can specify a default value for one or more parameters. This lets you simplify the function call if a parameter generally uses a particular value. It also lets you guide users to an acceptable value if the user is not sure what to enter.

Default parameters should be placed *after* any non-default parameters. Here's an example of a function that uses a default value if the user presses ENTER without responding to the prompt question:

```
In [12]: def default_input(prompt, default=None):
   ...:     ❶ prompt = f'{prompt} [{default}]:'
   ...:       response = input(prompt)
   ...:     ❷ if not response and default:
   ...:             return default
   ...:       else:
   ...:             return response
```

This function takes a prompt and a default value as arguments. The prompt and default will be specified when the function is called, and the program will display the default in square brackets ❶. The response variable holds the user's input. If the user enters nothing and a default value exists, the default value is returned ❷. Otherwise, the user's response is returned.

Let's use this function to get a user's birth country. For users currently in the United States, we set the default value to "USA" so they can just press ENTER rather than type in the name. Note how this default lets you *control* the response when multiple choices are possible (such as "America," "United States," "United States of America," "US," and so on):

```
In [13]: birth_country = default_input('Enter your country of birth:', 'USA')
Enter your country of birth: [USA]:

In [14]: birth_country
Out[14]: 'USA'
```

A user can override the default by entering a response:

```
In [15]: birth_country = default_input('Enter your country of birth:', 'USA')

Enter your country of birth: [USA]: Scotland

In [16]: birth_country
Out[16]: 'Scotland'
```

In most cases, you'll want to avoid using mutable objects like dictionaries, sets, or lists as default argument values in Python. This is because the default mutable object is initialized only *once*, when the function is *defined* rather than each time the function is *called*. This can produce unexpected outputs. Here's an example:

```
In [17]: def dog_breeds(new, current=['bulldog', 'dachshund']):
    ...:     current.append(new)
    ...:     return current

In [18]: my_dogs = dog_breeds('pomeranian')

In [19]: my_dogs
Out[19]: ['bulldog', 'dachshund', 'pomeranian']

In [20]: your_dogs = dog_breeds('poodle')

In [21]: his_dogs = dog_breeds('mutt')

In [22]: his_dogs
Out[22]: ['bulldog', 'dachshund', 'pomeranian', 'poodle', 'mutt']
```

The naive expectation here is that everyone who calls the dog_breeds() function will start off fresh with a bulldog and dachshund and then add their dog breeds to this list. But because the current list was created once when the function was defined in line In [17], every subsequent call to the function appends items to this *same* list.

Returning Values

When functions return a value, you can store the result in a variable using an assignment statement. For example, the following code stores the value returned from running the calc_force() function in a variable called force:

```
In [23]: force = calc_force(15000, 9.78033)

In [24]: force
Out[24]: 146704.94999999998
```

You can even return multiple values, separated by commas. You'll need a variable to hold each value, as in this example, in which the function accepts a number as an argument and returns the square and cube of the number:

```
In [25]: def square_and_cube(a_number):
    ...:         return a_number**2, a_number**3

In [26]: squared, cubed = square_and_cube(2)

In [27]: squared, cubed
Out[27]: (4, 8)
```

Finally, functions can include multiple return statements. Each statement executes under a given condition, and as soon as one executes, the function ends. Try this in the console:

```
In [28]: def goldilocks(a_number):
    ...:         num = int(a_number)
    ...:         if num > 42:
    ...:             return "too high"
    ...:         elif num < 42:
    ...:             return "too low"
    ...:         else:
    ...:             return "just right!"

In [29]: goldilocks(43)
Out[29]: 'too high'

In [30]: goldilocks(41)
Out[30]: 'too low'

In [31]: goldilocks(42)
Out[31]: 'just right!'
```

In this example, the goldilocks() function accepts a number as an argument, converts it to an integer, and then compares it to 42. Each of the three possible outcomes (greater than, less than, or equal to) has its own return statement.

Naming Functions

The guidelines for naming functions are the same as those for naming variables (see "Naming Variables" on page 206). You can use letters, underscores, and numbers, as long as the first character isn't a number. All characters should be lowercase, and you should separate words with an underscore. You'll want to avoid reserved keywords and the names of built-in functions.

Because functions perform an action, a good naming strategy is to include a verb and a noun that describe that action. Some examples are reset_password(), register_image(), and plot_light_curve().

For more on naming and defining functions, visit the documentation at *https://docs.python.org/3/tutorial/controlflow.html#defining-functions/*.

Built-in Functions

Python comes with multiple built-in functions to make your coding life easier. You've already worked with many of these, including print(), len(), type(), list(), input(), round(), and more.

Table 11-1 lists some of the more frequently used built-in functions. To see the full list, along with detailed descriptions of each function, visit *https://docs.python.org/3/library/functions.html*.

Table 11-1: Frequently Used Built-in Functions

Function	Description
abs()	Return the absolute value of a number.
all()	Return True if all elements of an iterable are true or if the iterable is empty.
any()	Return True if any element of an iterable is true or False if iterable is empty.
chr()	Return a string representing an input Unicode code point (chr(97) returns 'a').
dict()	Create a new dictionary object.
dir()	Without argument, return names in the current local scope. If an object is passed as an argument, return list of attributes and methods for that object.
enumerate()	Adds a counter to each item of an iterable object and returns an enumerate object.
filter()	Return an iterator from those elements of an iterable for which function returns True.
float()	Return a floating-point number constructed from a number or string.
frozenset()	Return a frozenset object.
hash()	Return the hash value of an object if it has one.

Function	Description
help()	Invoke built-in help system (intended for interactive use).
hex()	Convert an integer to a lowercase hexadecimal string prefixed with "0x."
id()	Return the identity of an object.
input()	Get user input using a prompt and return it as a string.
int()	Return an integer number constructed from a number or string.
isinstance()	Return True if the specified object is of the specified type; otherwise, return False.
len()	Return the number of items in a sequence or collection (such as a string, list, or set).
list()	Create a new list object.
max()	Return largest item in an iterable or the largest of two or more arguments.
min()	Return smallest item in an iterable or the smallest of two or more arguments.
next()	Retrieve the next item from an iterator.
open()	Open a file and return a corresponding file object.
ord()	Return the Unicode code point of a character (ord('a') returns 97).
pow()	Return a number raised to the power specified.
print()	Print a specified message to the screen or other standard output device.
range()	Generate an immutable sequence of numbers for given start and stop integers.
repr()	Return a string containing a printable representation of an object.
reversed()	Return a reversed iterator.
round()	Return a number rounded to *n*-digits precision after the decimal point.
set()	Create a new set object.
sorted()	Return a new sorted list (forward or backward) from the items in an iterable.
str()	Return a string version of an object.
sum()	Return the sum of all items in an iterable.
tuple()	Create a new tuple object.
type()	Return the type of an object.
zip()	Iterate over several iterables in parallel, producing tuples with an item from each.

It's good practice to check whether a built-in function exists for a specific task before writing code on your own.

1. When you call a function that takes input, you pass it:

    a. parameters

    b. objects

    c. arguments

    d. the def keyword

2. Ideally, a function name should contain both a:

    a. noun and an underscore

    b. verb and an underscore

    c. verb and a noun

    d. number and an underscore

3. A function that returns no value is called a:

    a. fruitful function

    b. void function

    c. warning function

    d. module

4. Write a function that accepts a user's name and then returns their name stripped of vowels. You'll want to make a string of vowels, loop through the letters in the name, and compare each letter to the contents of the vowel string.

5. Write a function that calculates momentum (mass * velocity) using keyword arguments only.

Functions and the Flow of Execution

Like conditional statements and loops, functions can cause code to branch or jump around. In the following example, we define two functions and call the first function from within the second:

```
In [32]: def success():
    ...:         print("You found the number 3!")

In [33]: def find_3():
    ...:         for i in range(6):
    ...:             if i == 3:
    ...:                 success()

In [34]: find_3()
You found the number 3!
```

When you call the find_3() function, the flow of execution moves into the function. But rather than return a value—and control—back to the main routine, this function calls another function, which could theoretically call another function defined somewhere higher in the code.

The definition of these two functions doesn't need to be in order and can be separated by other code, as long as calls to the functions come *after* their definition. On the Spyder main menu, click **Consoles ▶ Restart kernel** and then enter the following code, which now defines find() before success(), with some other code in between:

```
In [35]: def find_3():
    ...:     for i in range(6):
    ...:         if i == 3:
    ...:             success()

In [36]: print("Here's some other code...")
Here's some other code...

In [37]: print("Here's some more code...")
Here's some more code...

In [38]: def success():
    ...:     print("You found the number 3!")

In [39]: find_3()
You found the number 3!
```

As you can see, the order in which you defined the two functions didn't matter; what's important is that you made the call to find_3() after they were defined.

Using Namespaces and Scopes

A *namespace* is a collection of names. Behind the scenes, Python uses namespaces to map names to corresponding objects in memory. This lets Python keep track of all the names currently in use and prevent *collisions*, wherein two different objects share the same name.

Different, isolated namespaces, called *scopes*, can exist at the same time within a single program. When you start typing a program in the console or the text editor, you are in the *global* scope, and all the object names share the same namespace. Every time you define a function, you enter the function's *local* scope, and all the names used within the function share a new namespace that's hidden from both the global scope and the local scope of other functions. Thus, it's possible to use the same object name within one function as you do within another function, or in the main program in the global scope (see Figure 11-2).

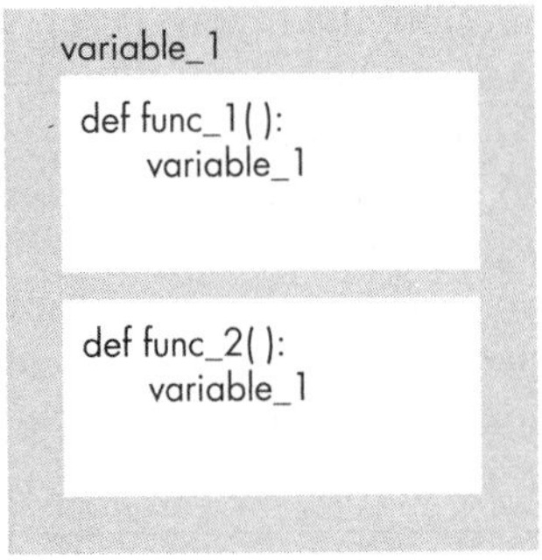

Figure 11-2: This program has a global scope (gray) and two isolated local scopes within functions (white).

Let's look at scope behavior in practice. Enter the following in the console:

```
In [40]: x = 42

In [41]: print(x)
42

In [42]: def local_scope():
    ...:     x = 5
    ...:     print(x)

In [43]: local_scope()
5

In [44]: print(x)
42
```

In the previous snippet, you used the same variable name (x) twice without a problem. This is because the first x is in the global scope, and the second x is safely tucked-away within the local scope of the function. As written, there's no way for the global scope to access the x in the local scope. So, when you print x in line In [44], you get the value in the global scope, despite the fact that x appears to have been reassigned to 5 in the function. After the function terminates, all its local variables are "forgotten" by Python, so no name conflicts occur.

Compartmentalizing the code using scopes also aids debugging. It's easier to track down the source of bad values because functions can only interact with the rest of the program through the arguments they're passed and the values they return.

Using Global Variables

Any variable assigned in the global scope is visible to both the global and local scopes. To indicate that you're accessing a global variable without passing it to the function as an argument, and to make it fully available to the function, you must use the *global statement* to specify it as a *global variable* within the function, as follows:

```
In [45]: x = 42

In [46]: def local_scope():
    ...:     global x
    ...:     x = 5
    ...:     print(x)

In [47]: local_scope()
5

In [48]: print(x)
5
```

By adding the line `global x` in the definition of the `local_scope()` function, you gave the function access to the x variable in the global scope. Now, when you change the value of x in the function, that change is reflected in the global scope, and printing x returns 5, not 42, as before.

NOTE *Variables in the global space can be changed from* within *a function, without the use of the* `global` *statement, if they are* mutable *objects.*

Because it's possible to use global variables in the local scope of functions, you should avoid using the same names for local and global variables. Likewise, you should avoid using the same name for variables in local scopes. Even though it's impossible to share a local variable either globally or with another function, this can become confusing. It's rarely a good idea to use the same name for two different things, even if they never interact.

Using global variables is generally discouraged, especially in large and complex programs. Imagine that you have hundreds of lines of code with dozens of functions. One of the functions changes a global variable to the wrong value, either due to a bug or to a failure in logic. To find and correct this problem, you must search through the *entire* program rather than focus on individual functions or function calls.

NOTE *An exception to the "don't use global variables" rule is the global constant. It's okay to assign constant values near the top of your program in the global scope. Because constants shouldn't change value, they shouldn't introduce complexity into your code.*

Using a main() Function

With the exception of short, simple programs, it's common practice to encapsulate the main code of a program into a function called `main()`. This code runs the rest of the program by executing expressions and statements and calling functions. Removing it from the global scope makes it easier to find and manage.

You can define the `main()` function anywhere, but generally it's near the start or end of a program. If your code and function names are very readable, placing `main()` at the start of a program can serve as a good summary of what the program does.

Here's a program that uses a `main()` function to calculate some statistics. Enter the following in the Spyder text editor and save it as *main_function _example.py*:

```python
from random import uniform

def main():
    data = generate_data()
    print(f"data = {data}")
    calc_mean(data)
    calc_max_value(data)
    calc_min_value(data)

def generate_data():
    samples = []
    for _ in range(10):
❶       sample = round(uniform(0.0, 50.0), 1)
        samples.append(sample)
    return samples

def calc_mean(data):
    print(f"\nMean = {round(sum(data) / len(data), 1)}")

def calc_max_value(data):
    print(f"Max = {max(data)}")

def calc_min_value(data):
    print(f"Min = {min(data)}")

main()
```

In this code, we first import the `uniform()` method from the `random` module so that we can generate random float values to use as data (in real life, you'd load or type some data into the program). Next, we define the `main()` function. All this function does, in this case, is call other functions. Note how it reads like a summary of what the program does.

The next function, `generate_data()`, returns a list of 10 random float values, rounded to one decimal place, from a uniform distribution. To use the `uniform` method, pass it the beginning and ending values of the range that you want to use, in this case, 0.0 and 50.0 ❶. The next three functions will take this list as input (an argument) and return the mean, maximum, and minimum values, respectively.

At this point, you've defined only functions. If you want the program to do something, you need to call the `main()` function before execution.

For code this simple, you could forgo use of a `main()` function and move its contents into the global scope, *below* the definitions of the functions being called. But as your code becomes longer and more complicated, a `main()` function will help you keep it clean and organized and make it easy to find and review what the program does.

Advanced Function Topics

At this point, you know enough about functions to handle most, if not all, of the coding problems you'll encounter. There's always more to learn, however. This section will give you a brief introduction to recursion, function design, lambda functions, and generators. Recursion is a particularly ambitious topic, and if you find it interesting or useful, I recommend reading more about it on your own.

Recursion

Recursion is a powerful programming technique in which a function calls itself. Although recursion can be accomplished using more efficient `for` and `while` loops, these loops can sometimes become complicated and messy.

For difficult problems, recursive functions can provide a simpler and more readable way to construct code. You'll commonly see recursion used for solving factorials, finding numbers in a Fibonacci sequence, and calculating compound interest for a loan using additional data, like regular payments.

Here's a simple example of a recursive function named `beer()`. Notice that the `elif` and `else` statements include calls to the `beer()` function.

```
In [49]: def beer(bottles):
   ...:     ❶ if bottles <= 0:
   ...:             print("No more bottles of beer on the wall!")
   ...:         elif bottles == 1:
   ...:             print(f"{bottles} bottle of beer on the wall!")
   ...:             beer(bottles - 1)
   ...:         else:
   ...:             print(f"{bottles} bottles of beer on the wall!")
   ...:             beer(bottles - 1)

In [50]: beer(3)
3 bottles of beer on the wall!
2 bottles of beer on the wall!
1 bottle of beer on the wall!
No more bottles of beer on the wall!
```

This function was inspired by the famous "99 Bottles of Beer" song. It accepts a number—representing bottles of beer—as an argument, and then it updates the number of bottles remaining and calls itself again until the number reaches zero. The `elif` clause in the middle is needed only to correct the grammar when one bottle remains.

The `if` statement ❶ does *not* include a recursive call to `beer()`, because this is the *base condition,* or *base case,* for the function. A base condition is one that will end the function if the condition is met.

You need a base case because recursions, like while loops, can go on forever. To see an example, enter the following in the console:

```
In [51]: def keep_on_keeping_on():
    ...:         print("Somebody stop me!")
    ...:         keep_on_keeping_on()

In [52]: keep_on_keeping_on()
```

This example will raise the following exception:

```
RecursionError: maximum recursion depth exceeded while calling a Python object
```

Because the keep_on_keeping_on() function kept calling itself, it created an infinite recursion, resulting in a *stack overflow*. This error occurs when you attempt to write more data to a memory block than it can hold. Inclusion of a reachable base case could have stopped this from happening, but not if it allows too many recursive calls.

To prevent infinite recursions, the Python interpreter limits the *depth of recursion*; that is, the number of recursive calls to a function, to a default value. To see this value, in the console, use the system module (sys), as follows:

```
In [53]: import sys

In [54]: print(sys.getrecursionlimit())
3000
```

Although you can increase this recursion limit by passing the sys.set recursionlimit() function an integer, you need to do this with care, as the highest possible limit is platform-dependent, and a high limit can still lead to a crash. A better option is to rewrite your code without recursion.

NOTE *The actual recursion limit is usually a bit less than the value returned by* sys.get recursionlimit(). *On my machine, a* RecursionError *is raised after 2,967 calls, despite the limit being set at 3,000.*

Designing Functions

When it comes to writing functions, there's a school of thought that believes a function "should do one thing and one thing only." Although keeping functions short and simple is a good guideline, there are many cases for which longer, more complex functions are the better choice.

Longer functions can merge related tasks under one umbrella while reducing the overall number of lines of code. Thus, adding a bit of complexity *locally* to a function can reduce the overall *global* complexity of a program.

Still, it's a good idea to keep the "one task only" guide in mind when writing functions. Here's a simple example involving an embedded print() function:

```
In [55]: def area_of_square(side_length):
    ...:     area = side_length**2
    ...:     print(f"Area is {area}")
    ...:     return area

In [56]: area_of_square(50)
Area is 2500
Out[56]: 2500
```

If this function is used as an intermediate step in a program—that is, if you're just calculating the area to pass it on to another function—do you really want it printing the answer to the screen? Unnecessary printing increases the runtime of programs and can clutter your screen with unneeded information.

On the other hand, suppose that you want to get a user's name, convert it to lowercase, and then sort the letters alphabetically to find anagrams for the name in a dictionary. It would be silly to break these tasks into multiple functions to honor the "one task only" guideline.

In his book *Beyond the Basic Stuff with Python* (No Starch Press, 2021), author Al Sweigart recommends that functions be as short as reasonably possible but no shorter. They should not exceed 200 lines of code and ideally contain fewer than 30 lines.

Lambda Functions

Remember how you can reduce for loops to a single line of code using comprehensions? Well, lambda functions let you do something similar with functions.

A *lambda function* is a single-use, unnamed function consisting of a single statement. They're sometimes called an *anonymous* function because they're defined with the lambda keyword rather than a name of their own. The syntax is as follows:

```
lambda parameter_1, parameter_2: expression
```

Words and characters that directly follow lambda are treated as parameters. Expressions come after the colon, and returns are automatic, with no need for the return keyword. Here's an example that multiplies two numbers together:

```
In [57]: multiply = lambda a, b: a * b

In [58]: multiply(6, 7)
Out[58]: 42
```

A nice thing about lambda functions is that you can create them on the fly, without the need for a variable assignment. Just put the function in parentheses and add the arguments, also in parentheses, to the end:

```
In [59]: (lambda a, b: a * b)(6, 7)
Out[59]: 42
```

Lambda functions are often used in conjunction with the built-in `filter()` function to select particular elements from a sequence. The lambda function defines the filtering constraint that the `filter()` function then applies to the sequence. Here's an example in which we return all the numbers with a value less than 10 from a list:

```
In [60]: numbers = [5, 42, 26, 55, 12, 0, 99]

In [61]: filtered = filter(lambda x: x < 10, numbers)

In [62]: print(list(filtered))
[5, 0]
```

Note that you need to type cast the `filtered` object to another data type such as a list or tuple before you can print it.

Lambda functions are useful in data analysis when you need to pass a function as an argument to a data transformation function. They'll also save you the effort of typing full function definitions while preserving the readability of your code.

Generators

A *generator* is a special routine for controlling the iteration behavior of a loop. It lets you generate a sequence *one value at a time* rather than all at once. Compare this to a regular function, which must create the entire sequence in memory before returning the result, regardless of the size of the sequence.

Generators use *lazy* evaluation, which means that they compute the value of an item only when invoked, without having to load everything in memory first. As a result, generator objects have a lower memory footprint than other iterables, such as lists.

Generators are useful when working with sequences large enough to occupy much (if not all) of your system's RAM. They're also a good choice when you need to use a sequence only once.

The most familiar generator is the built-in `range()` function, which you've used before. With `range()`, it doesn't matter for system memory if you set the upper limit to ten or a trillion, as each number is generated as it's needed, and then discarded.

Generator functions are defined like regular functions except that they use a yield statement in place of a return statement. Here's an example that yields the cube of each number in a sequence:

```
In [63]: def cubes(my_range):
    ...:     for i in range(1, my_range + 1):
    ...:         yield i**3
In [64]:
```

Whereas the return statement *ends* and *exits* a function, the yield statement *suspends* the function's execution and sends a value back to the caller. Later, the function can resume where it left off. When a generator reaches its end, it's "empty" and can't be called again.

If you try to call a generator function and pass it an argument in the same manner as a regular function, you might be surprised by the result:

```
In [65]: cubes(5)
Out[65]: <generator object cubes at 0x0000017FE06834A0>
```

The issue here is that the function returned a type of iterator called a *generator object*. This object won't begin executing its code until you *request* elements from it, for example, by using it in a for loop or by calling the built-in next() function.

Here's an example that creates a generator object (cube_gen) and then uses next() to get the next value from it. Behind the scenes, the generator pauses after each call to the next() function and resumes when the function is called again. This continues until the generator object is exhausted and raises a StopIteration exception:

```
In [66]: cube_gen = cubes(5)

In [67]: next(cube_gen)
Out[67]: 1

In [68]: next(cube_gen)
Out[68]: 8

In [69]: next(cube_gen)
Out[69]: 27

In [70]: next(cube_gen)
Out[70]: 64

In [71]: next(cube_gen)
Out[71]: 125

In [72]: next(cube_gen)
Traceback (most recent call last):
```

```
File "C:\Users\hanna\AppData\Local\Temp/ipykernel_23936/2492540236.py", line 1, in <module>
next(cube_gen)

StopIteration
```

At this point, the generator object is empty and can't be used again. If you try to iterate over it with a for loop, you get nothing:

```
In [73]: for i in cube_gen:
   ...:         print(i)

In [74]:
```

You must remake a generator to use it again:

```
In [75]: cube_gen = cubes(5)

In [76]: for i in cube_gen:
   ...:         print(i)
1
8
27
64
125
```

If your generator uses a simple expression, you can define it more concisely using a *generator expression*. A generator expression looks a lot like a list comprehension, but instead of square brackets, you surround the expression containing the for loop in *parentheses*:

```
In [77]: my_gen = (i for i in range(5))

In [78]: my_gen
Out[78]: <generator object <genexpr> at 0x000001C0DC3280B0>
```

Due to their efficiency, generator expressions are often used in place of list comprehensions in functions like min, max, and sum:

```
In [79]: sum(x**2 for x in range(500))
Out[79]: 41541750
```

Finally, you can convert generators to lists or tuples using type casting. In this example, we wrap the generator expression in the built-in list() function to convert the results to a list:

```
In [80]: my_list = list(range(5))

In [81]: my_list
Out[81]: [0, 1, 2, 3, 4]
```

You might perform this action when using a very large sequence to generate a smaller sequence with a memory footprint small enough to store in a list.

And here, we use the `tuple()` built-in function to convert the results to a tuple:

```
In [82]: my_tuple = tuple(i**2 for i in my_list)

In [83]: my_tuple
Out[83]: (0, 1, 4, 9, 16)
```

Again, you might undertake this action when you need to efficiently produce a relatively small tuple from a larger input sequence.

TEST YOUR KNOWLEDGE

6. A generator function always contains which keyword?

   a. return

   b. main

   c. yield

   d. range

7. Rewrite the generate_data() function in "Using a main() Function" on page 295 so that it uses list comprehension rather than a for loop.

8. Write a lambda expression that prints the multiples of 5 in this list: [3, 10, 16, 25, 88, 75].

9. True or false: The purpose of defining a main() function at the end of your code is to grant it access to any preceding functions.

10. To run a lambda function without assigning it to a variable, you must enclose it in:

    a. Curley brackets

    b. Square brackets

    c. Parentheses

    d. You don't need to enclose it at all

Modules

Modules are files—usually written in Python—that contain collections of related functions. Modules can be embedded in Python programs and used to perform both common and specialized tasks. Python's *standard library*, for example, includes the os module, which provides widespread utility related to operating systems. It also includes the more specialized math module, which provides basic mathematical functions.

Like functions, modules let you hide code that you don't want to see in all its gory detail. In fact, many modules, as well as built-in functions, aren't even *written* in Python. The standard library's familiar len() function, for example, is implemented in the C language. Here's some of its source code:

```
static PyObject *
builtin_len(PyObject *module, PyObject *obj)
/*[clinic end generated code: output=fa7a270d314dfb6c
input=bc55598da9e9c9b5]*/
{
    Py_ssize_t res;

    res = PyObject_Size(obj);
    if (res < 0) {
        assert(PyErr_Occurred());
        return NULL;
    }
    return PyLong_FromSsize_t(res);
}
```

Imagine having to include code like this in your programs every time you want to get the length of a list or a string!

Through encapsulation, modules reduce complex code to one-line function calls. This, in turn, helps you to write cleaner code that's easier to read. And the modules themselves let you break code into functional groups that are easier to access and maintain.

Modules save you time, effort, and even money in so much as most third-party modules are open source. Best of all, modules let you leverage the battle-tested efforts of experts in a field. The OpenCV computer vision module, for example, lets you identify faces, track objects, manipulate images and more, even if you know very little about the subject. You can also write your own modules if third-party versions aren't available.

Importing Modules

Except for some modules in the standard library, you need to import modules prior to use. By convention, you should place these imports at the top of Python programs and insert an empty line after the last import. Consequently, you can think of imports as the "headwaters" of the flow of execution.

Importing modules at the top makes it easy to see which modules are being applied. This is important given that many times users will need to install the modules before running the program, and they don't want to go on a "module scavenger hunt" through your code.

Let's look at the import process using the random module, which lets you work with pseudo-random numbers. The simplest way to import this module is to use the import keyword followed by the module name:

```
In [84]: import random
```

Now, to use the functions in the random module, you need to use dot notation and enter the module name, followed by a period, followed by the function name. Here's an example in which you use the choice() function to randomly choose from items in a list:

```
In [85]: planets = ['Mars', 'Venus', 'Jupiter']

In [86]: planet = random.choice(planets)

In [87]: planet
Out[87]: 'Venus'
```

You can import multiple functions at a time using comma-separated values, like this:

```
In [88]: from random import choice, randint, shuffle
```

To save yourself the effort of typing random each time and to make your code lines shorter, you can just import choice, using the from keyword in the import statement:

```
In [89]: from random import choice

In [90]: planet = choice(planets)

In [91]: planet
Out[91]: 'Mars'
```

This is more concise but somewhat less readable because you can forget where choice() comes from (though you can always scroll up to the top to check).

Another way to reduce typing is to use an alias for the module name:

```
In [92]: import random as ran

In [93]: planet = ran.choice(planets)

In [94]: planet
Out[94]: 'Jupiter'
```

In general, I would avoid this, except for modules for which the alias is widely used, such as sns for the seaborn plotting library and pd for the pandas data analytics library, among others.

Likewise, never use the * wildcard to import all the functions in a module, like this:

```
In [95]: from random import *
```

This basically says, "import all the functions available in the random module." You might encounter this in the literature or in other people's code, but it's considered bad practice. It imports all the functions and classes in a module into your namespace. As a result, names in the module might clash with the names of functions you define or functions of other libraries that you import. Although clashes rarely happen, it's a good habit to keep your namespace as uncluttered as possible, so avoid import *.

Finally, when importing multiple modules, the best practice is to import each module on a *separate line*. This is more readable and lets you group modules in the preferred order of Python standard library → third-party modules → user-defined modules. Each group should be separated by a blank line, and a blank line should follow the last import statement.

If you're concerned that multiple imported modules might use the same function name or names, import the modules by name—or with a short alias for the name—and call them by using dot notation. This way, the module name will be clearly linked to the function name, avoiding both confusion and collisions.

NOTE   *Python libraries are collections of packages, and packages are collections of modules. Consequently, all three are imported in the same way: using an import statement made up of the import keyword and the name of the library, package, or module being imported.*

Inspecting Modules

You can use the built-in dir() function to see the functions available in a module. Let's look at the random module, used for generating random numbers. The output is long, so I've truncated it here:

```
In [96]: import random

In [97]: dir(random)
Out[97]:
['BPF',
--snip--
'betavariate',
'choice',
'choices',
'expovariate',
'gammavariate',
'gauss',
'getrandbits',
'getstate',
'lognormvariate',
'normalvariate',
'paretovariate',
'randint',
'random',
'randrange',
'sample',
'seed',
'setstate',
'shuffle',
'triangular',
```

```
'uniform',
'vonmisesvariate',
'weibullvariate']
```

To view the source code for each function, you can use the getsource() method from the inspect module. Let's look at the random module's choice() function, used for randomly choosing an element from a sequence. Note that these modules are open source and subject to updates and revisions, so your output might look different:

```
In [98]: import inspect

In [99]: print(inspect.getsource(random.choice))
def choice(self, seq):
    """Choose a random element from a non-empty sequence."""
    try:
        i = self._randbelow(len(seq))
    except ValueError:
        raise IndexError('Cannot choose from an empty sequence') from None
    return seq[i]
```

You can see that choice() is just a function like you've defined before. There's really nothing magic about modules.

If you want to see only the module's documentation, use the getdoc() method:

```
In [100]: print(inspect.getdoc(random.choice))
Choose a random element from a non-empty sequence.
```

As mentioned earlier, the built-in functions in the Python standard library are written in C and thus can't be accessed by inspect. To view their source code, you'll need to download it from *https://www.python.org/downloads/source/*.

Besides checking what a module is doing, inspecting the source code can help you to learn how to write your own custom functions that expand on or modify an existing module's functionality.

Writing Your Own Modules

A Python (*.py*) file can serve as a module. After you import it, it becomes a special module object whose functions can be called with dot notation.

Let's assume that you're working on a project for which you need to repeatedly solve the quadratic equation and calculate the volume of a sphere. As these equations aren't part of the standard math module, you'll need to implement them on your own. Rather than define functions for these tasks in every program for which you'll need to perform them, you can instead define them once in a reusable module named mymath and import that where needed. The filename is used as the module name.

Next, we need to determine where to save the module. When a module is imported, the Python interpreter first searches for a *built-in* module with that name. If no built-in module is found, it then searches for the filename in a list of directories given by the sys module's built-in *sys.path* variable. According to the documentation, this path is initialized from these locations:

- The directory containing the input script (or the current directory when no file is specified).
- *PYTHONPATH* (a list of directory names, with the same syntax as the shell variable PATH).
- The installation-dependent default (by convention including a site-packages directory, handled by the site module).

Going forward, we'll use the first option and store your custom modules in your project's directory. This will be the simplest and most straightforward approach for beginners and non-developers such as scientists and engineers. However, the module will be available only to scripts run from the project directory. To use the module in other projects, you'll need to either copy the file to those directories or use one of the other options in the previous list. The easiest way is to add the path to the *PATH* variable, like so:

```
In [101]: import sys

In [102]: sys.path.append(r'/path/to/my_module')
```

The mymath module will contain functions for solving quadratic equations and for calculating the volume of a sphere. I'm going to save it in the *spyder_proj_w_env* project that we created in the "Creating a Project in an Existing Directory" on page 70. If you don't want to use this project, feel free to create your own project folder using the instructions in Chapter 4.

First, open the project by going to Spyder's top toolbar and then clicking **Projects ▸ Open Project ▸ spyder_proj_w_env**. You'll want to see Spyder's File Explorer, Text Editor, and IPython Console, such as it is presented in Figure 4-4.

Now, in the text editor, enter the following code:

```
import math

def quad(a, b, c):
    x1 = (-b - (b**2 - 4 * a * c)**0.5) / (2 * a)
    x2 = (-b + (b**2 - 4 * a * c)**0.5) / (2 * a)
    return x1, x2

def sphere_vol(r):
    vol = (4 / 3) * math.pi * r**3
    return round(vol, 2)
```

The quad() function accepts the standard a, b, and c coefficients for the quadratic equation as arguments. It then calculates and returns both solutions to the equation. The sphere_vol() function accepts a radius as an argument and returns the volume of a sphere with that radius rounded to two decimal places.

NOTE *The* mymath *module imports the built-in* math *module. This is fine, but be careful about writing and importing multiple modules that depend on one another. This results in* circular dependencies *that get messy and can cause an* ImportError.

Now, save the program as *mymath.py* in the *code* folder by clicking **File ▶ Save as** on the top toolbar. Alternatively, you could save it at the project folder level (Figure 11-3) and still access it from scripts in the *code* folder. Personally, I don't like to clutter the project folder with individual files, hence the decision to place it in *code*.

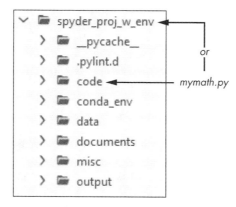

Figure 11-3: The mymath.py *module can be saved in either the* code *folder or the main project folder.*

If you're ever curious about what folder the Python interpreter is currently working in, import the operating system module (os) and use its getcwd() function to return the current working directory. Here's an example in the console:

```
In [103]: import os

In [104]: os.getcwd()
Out[104]: 'C:\\Users\\hanna\\spyder_proj_w_env\\code'
```

Because the current directory is the *code* folder, you don't need to specify a path to import or otherwise access other files in this folder.

Now, let's test the module in the console:

```
In [105]: import mymath

In [106]: mymath.quad(2, 5, -3)
Out[106]: (-3.0, 0.5)
```

```
In [107]: mymath.sphere_vol(100)
Out[107]: 4188790.2
```

If you want to assign the results of the quad() function to a variable, remember that the quadratic equation has two solutions, so you'll need to use two variables in the assignment statement:

```
In [108]: soln1, soln2 = mymath.quad(2, 5, -3)

In [109]: soln1, soln2
Out[109]: (-3.0, 0.5)
```

That's all there is to it! Now, any programs in the *code* folder can import and use the mymath module, just as they can use a built-in module.

NOTE *If you try to import a module that's already been imported, nothing will happen. So, if you change a module and want to reimport it, the best course of action is to restart the kernel and then import the module again. In fact, anytime Python is behaving strangely, you should consider restarting the kernel. As your IT support person likes to say, "Have you tried rebooting?"*

Naming Modules

When naming modules, the best practice is to use lowercase characters and separate words with underscores. Names should preferably be one word only, as names with underscores can be confused for variable names. You'll also want to avoid special symbols like the dot (.) and question mark (?). These symbols can cause problems due to the way Python looks for modules. A filename like *my.module.py*, for example, would indicate to Python that the *module.py* file should be found in a folder named *my*.

Writing Modules That Work in Stand-Alone Mode

The *mymath.py* program you wrote in "Writing Your Own Modules" on page 307 just defines two functions. It works great as a module, but it's not very usable on its own, because there's no call to the functions. So, let's turn *mymath.py* into a program that will run in stand-alone mode *and* work as a module.

In Spyder, open *mymath.py* in the text editor and make a copy of it using **File ▸ Save as** from the top toolbar. Name the new file *mymath2.py*.

Now, add the code blocks at ❶ and ❷ to define and call a main() function:

```
  import math

❶ def main():
      a = 2
      b = 5
```

```
        c = -3
        r = 100
        soln1, soln2 = quad(a, b, c)
        vol = sphere_vol(r)
        print(f'solution1 = {soln1}')
        print(f'solution2 = {soln2}')
        print(f'sphere volume = {vol}')

    def quad(a, b, c):
        x1 = (-b - (b**-2 - 4 * a * c)**0.5) / (2 * a)
        x2 = (-b + (b**-2 - 4 * a * c)**0.5) / (2 * a)
        return x1, x2

    def sphere_vol(r):
        vol = (4 / 3) * math.pi * r**3
        return round(vol, 2)

❷ if __name__ == '__main__':
        main()
```

At ❶, you define a `main()` function to run the program, assigning variables to serve as arguments to the module's functions, calling the two functions, and printing the results.

For Python to evaluate whether a program is being run in stand-alone mode or as an imported module, it's necessary for you to use the special built-in `__name__` variable ❷. If you run the program directly, `__name__` is set to `__main__`, and the `main()` function is called. If the program is imported, `__name__` is set to the module's filename, `main()` is not invoked, and the program won't execute until you call one of its functions, like `quad()` or `sphere_vol()`.

Save the program and run it using F5 or the "play" icon on the Run toolbar. You should see the following output in the console:

```
In [110]: runfile('C:/Users/hanna/spyder_proj_w_env/code/mymath2.py', wdir='C:/Users/hanna/
spyder_proj_w_env/code')
solution1 = -3.0
solution2 = 0.5
sphere volume = 4188790.2
```

The program ran as if you had simply called `main()` as the last line.

Built-in Modules

Python comes with multiple built-in modules. Covering all these is beyond the scope of this book, but Table 11-2 lists some commonly used ones, along with a brief description of each. You've already worked with several of these, including math, random, logging, and inspect. We'll look at some of the other ones in chapters to come.

Table 11-2: Frequently Used Built-in Python Modules

Module	Description
os	Operating system tasks like directory and file creation, deletion, identifying the current directory, and more.
sys	System operation and runtime environment tasks like exiting programs, getting paths, command line use, and more.
shutil	Shell utilities for high-level file operations like copying, moving, deleting directory trees, and more.
inspect	Functions to get information about live objects such as modules, classes, methods, functions, tracebacks, frame objects, and code objects.
logging	A flexible event logging system for monitoring a program's flow of execution.
math	Basic mathematical operations and constants.
random	Implements pseudo-random number generators for various distributions.
statistics	Functions for calculating mathematical statistics like mean, geometric mean, median, mode, covariance, and more.
collections	Provides specialized container datatypes providing alternatives to Python's general purpose built-in containers like dictionaries, lists, sets, and tuples. Useful tools include namedtuple(), deque, defaultdict, and Counter.
itertools	Creates iterators for efficient looping. Includes fast functions for zipping, computing cartesian products, generating permutations and combinations, cycling, and more.
datetime	Supplies tools for getting and manipulating dates and times.
re	Tools for working with regular expressions, that specify a set of matching strings. Used for searching and parsing text data.
http	Collects several modules for working with the HyperText Transfer Protocol
json	Methods for working with JSON-formatted data.
threading	Used for creating, controlling, and managing threads (smallest sequence of programmed instructions) that allow different parts of a program to run concurrently for speed and simplicity.
multiprocessing	Permits efficient use of multiple processors on a given machine.

It's a good idea to be aware of built-in modules so that you don't find yourself reinventing the wheel and duplicating modules that already exist. You can find the official documentation at *https://docs.python.org/3/tutorial/modules.html*. But don't think you need to memorize the modules or their contents. A simple online search for a particular task will generally return information on modules as well as actual code samples for accomplishing the task.

11. Write a function that calculates the force of gravity using the equation $F = (G * mass1 * mass2) / radius^2$, where G is the gravitational constant ($6.67 \times 10\text{-}11$ N-m^2/kg^2). Treat G as a *global* constant.

12. Import the math module and list all the functions it contains.

13. The preferred way to import *all* the functions available in a module is to use:

    a. `from module import *`

    b. `import module`

    c. `import module as *`

    d. `from module import func1, func2, func3...`

14. When you import a module, Python first searches for:

    a. a module with that name in the current working directory

    b. a module with that name in *PYTHONPATH*

    c. a module with that name in the site-packages directory

    d. a built-in module with that name

15. Write a function that accepts a variable in the global scope as an argument. Then, rewrite the function to use the same variable as a global variable.

Summary

Functions are callable collections of code that let you organize your program into modular, logical groups. If you find yourself repeating code, you should stop and write a function.

Recursion means to "run back," and recursive functions call themselves over and over. Recursive functions are used to solve complex problems that can be broken down into smaller problems of the same type and would be difficult to implement using a loop.

Lambda functions are one-off, unnamed functions consisting of single statements. For simple tasks, they save you the effort of defining a complete named function.

A generator is a function that returns an object that can be iterated over a single time. Rather than compute all of its values at once, a generator waits to be asked and then *yields* its values one at a time. As a result, a generator has a low memory footprint, making them useful for large data sets that you need to use only once.

A module is a Python file containing a collection of related functions. Modules must be imported into other Python files to be used. Modules let you take advantage of the expertise and efforts of others while keeping your code clean and uncluttered. You can also write customized modules for your own projects.

12

FILES AND FOLDERS

Files let you store data in a persistent and sharable manner. It's all but impossible to do any real work without them. Python comes with many modules and methods for working with files, folders, and directory paths. These let you read and write text files; preserve complex data after you exit your program; create, move, and delete folders; and perform other system-level tasks.

In this chapter, we'll use the built-in operating system (os), path library (pathlib), and shell utilities (shutil) modules to work with files, folders, and directory paths. We'll then use built-in functions to open, read, write, and close text files, and the built-in pickle, shelve, and json modules to preserve and store more complex data types such as Python lists and dictionaries. Finally, we'll look at ways to handle exceptions when opening files.

Creating a New Spyder Project

Let's make a new Spyder project to use in this chapter. If you need a refresher on Spyder projects, see "Using Project Files and Folders" on page 68.

To begin, start Spyder (either from the Start menu or from Anaconda Navigator) and then, in the top toolbar, click **Projects ▸ New Project**. In the Create New Project dialog that opens (Figure 12-1), make sure the Location box includes your home directory, set the Project name to **file_play**, and then click the **Create** button.

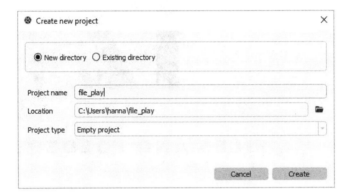

Figure 12-1: The Spyder Create New Project dialog

You should now see this new folder in Spyder's File Explorer pane.

Working with Directory Paths

Before you can work with files and folders (also called *directories*), you'll need to know how to find them and where to save them. And to do that, you'll need an address, otherwise known as a *directory path*.

A directory path is a string of characters used to uniquely identify a location in a directory structure. A path starts with a root directory designated by a letter (such as *C:*) in Windows and a forward slash (/) in Unix-based systems. Additional drives in Windows are assigned a different letter than *C*, those in macOS are placed under */volume*, and those in Unix under */mnt* (for "mount").

Pathnames appear differently depending on the operating system. Windows separates folders with a backslash (\), whereas macOS and Unix systems use a forward slash (/). Folders and filenames are also case sensitive in Unix.

These differences between operating systems can pose problems if you're trying to write code that can run on any system. If you're writing a program in Windows and enter pathnames with backslashes, other platforms won't recognize the paths. Fortunately, Python provides standard library modules such as os and pathlib to help you deal with this.

The Operating System Module

The operating system (os) module has been described as a "junk drawer for system-related stuff." Table 12-1 summarizes some of the most used methods in this module. For a complete list of the methods and details about their use, visit the documentation at *https://docs.python.org/3/library/os.html*.

Table 12-1: Useful os Module Methods

Method	Description
os.getcwd()	Return location of the current working directory (cwd)
os.chdir()	Change cwd to a specified path
os.getsize()	Return the size of a file in bytes
os.listdir()	Return list of files and folders inside specified directory (defaults to cwd)
os.mkdir()	Create a new directory based on a specified path
os.makedirs()	Create multiple nested directories based on a specified path
os.rename()	Rename a specified file or directory
os.rmdir()	Delete an empty directory
os.walk()	Generate filenames in a directory tree
os.path.join()	Join path components and return a string that contains a concatenated path
os.path.split()	Split a pathname into a head and tail (tail=last pathname component)
os.path.abspath()	Return a normalized absolute version of a specified path
os.path.normpath()	Correct path separators for the system in use
os.path.isdir()	Check whether a specified path corresponds to an existing directory
os.path.isfile()	Check whether a specified path corresponds to an existing file
os.path.isabs()	Check whether a specified path is absolute or not
os.path.exists()	Check whether a specified path exists or not

Several of these os methods are helpful for discovering pathnames you didn't already know. For example, to determine the name of the directory in which you're currently working (called the *current working directory*, or *cwd*), import the os module and enter the following in the console:

```
In [1]: import os

In [2]: os.getcwd()
Out[2]: 'C:\\Users\\hanna\\file_play'
```

In this example, you used the os.getcwd() method to get the path to your current working directory (your path will be different). This is a

Windows example, so backslashes separate directory names, and, because this is a string, the backslashes must be escaped with a backslash (see "Escape Sequences" on page 190 for a refresher on the escape sequence). The os.getcwd() method will insert these backslashes for you, but they will cause problems if you try to use this path with another operating system.

The current working directory is assigned to a *process* (a running instance of a program) when that process starts up. For a Python program, the current working directory is always the folder that contains the running program.

You can use os.chdir() to move from the current working directory to another directory, as follows:

```
In [3]: os.chdir('C:\\Users\\hanna')

In [4]: os.getcwd()
Out[4]: 'C:\\Users\\hanna'
```

As you can see, this new directory becomes the current working directory.

If you work in Windows and don't want to type the double backslash, you can enter an r before the pathname argument string to convert it to a raw string:

```
In [5]: os.chdir(r'C:\Users\hanna')

In [6]: os.getcwd()
Out[6]: 'C:\\Users\\hanna'
```

To make your program compatible with all operating systems, use the os.path.join() method and pass it the folder names and filenames without a separator character, as separate strings. The os.path methods are aware of the system you're using and return the proper separators. This allows for platform-independent manipulation of file and folder names. Here's an example:

```
In [13]: path = '/Users/'

In [14]: path2 = os.path.join(path, 'hanna', 'file_play')

In [15]: path2
Out[15]: '/Users/hanna\\file_play'

In [16]: os.chdir(path2)

In [17]: os.getcwd()
Out[17]: 'C:\\Users\\hanna\\file_play'
```

In this snippet, you assigned a pathname, as a string, to the path variable. Notice how you can safely use forward slashes in Windows. Next, you made a new path variable (path2) using the os.path.join() method. Even though the output in line Out[15] looks messy, the os.path.join() method knows which operating system you're using and corrects the separators as needed (lines In[16] - Out[17]).

You can also take an existing path with the wrong separators and *normalize* it to the system you're using with os.normpath(). Here's an example in which Unix forward slashes are changed to Windows backslashes:

```
In [18]: path = 'C//Users//hanna'

In [19]: os.path.normpath(path)
Out[19]: 'C\\Users\\hanna'
```

Absolute vs. Relative Paths

The full directory path, from the drive to the current file or folder, is called the *absolute path*. You can use shortcuts, called *relative paths*, to make working with directories easier.

Relative paths are interpreted from the perspective of the current working directory. Whereas absolute paths start with a forward slash or drive label, relative paths do not. In the following code snippet, you can change directories without entering an absolute path because Python is aware of folders within the current working directory:

```
In [20]: import os

In [21]: os.getcwd()
Out[21]: 'C:\\Users\\hanna'

In [22]: os.chdir('file_play')

In [23]: os.getcwd()
Out[23]: 'C:\\Users\\hanna\\file_play'
```

Behind the scenes, the relative path is joined to the path leading to the current working directory to make the complete absolute path shown in line Out[23].

In Windows, macOS, and Linux, you can identify folders and save yourself some typing by using dot (.) and dot-dot (..). For example, in Windows, .\ refers to the current working directory, and ..\ refers to the parent directory that holds the current working directory. You can also use a dot to get the absolute path to your current working directory:

```
In [24]: os.path.abspath('.')
Out[24]: 'C:\\Users\\hanna\\file_play'
```

If a file, folder, or user-defined module that you need to access is stored in the same folder as your code, you can simply refer to the item's name in your code, without the need for a path or a "dot" shortcut. Following is an example in which we create multiple nested folders within the *file_play* folder. Because *file_play* is the current working directory and these folders will exist within it, there's no need to include a file path:

```
In [25]: os.makedirs(r'test1/test2/test3')
```

In this example, the os.makedirs() method created three nested folders (*test1*, *test2*, and *test3*) using a raw string. You should now see three folders in your Spyder project in the File Explorer pane (Figure 12-2).

Figure 12-2: The three new folders in the Spyder project

The pathlib Module

The os module is widely used, and you should familiarize yourself with its methods and syntax. But it treats paths as strings, which can be cumbersome and requires you to use functionality from across the standard library (it takes three modules just to gather and move files between directories).

An alternative is to use the smaller and more focused pathlib module. This module treats paths as objects rather than strings, and gathers the necessary path functionality in one place. It's also agnostic to the operating system, making it useful for writing cross-platform programs.

The module's Path and PurePath classes not only help you work with directory paths, they also duplicate useful os module methods for tasks like the following:

- Getting the current working directory: Path.cwd()
- Making directories: Path.mkdir()
- Renaming directories: Path.rename()
- Removing directories: Path.rmdir()

NOTE   *Path classes in* pathlib *are divided into pure paths and concrete paths.* PurePath *objects act like strings and provide path-handling operations such as editing the path, joining paths, finding the parent path, and so on, but they don't access a filesystem. Concrete paths inherit from* PurePath *and provide both pure path operations and new methods to do system calls on path objects. Concrete paths let you access the filesystem to search directories, remove directories, write to files, and so on.*

Table 12-2 summarizes some of the more useful methods available through the `pathlib` module. For the full list, visit the documentation at *https://docs.python.org/3/library/pathlib.html*. This documentation also includes a complete mapping of various `os` methods to their corresponding `Path` and `PurePath` equivalents.

Table 12-2: Useful `Path` and `PurePath` Methods for Working with Paths

Method	Description
`Path.cwd()`	Return path object for the cwd
`Path.exists()`	Return Boolean that indicates whether path points to existing file or folder
`Path.home()`	Return path object representing the user's home directory
`PurePath.is_absolute()`	Return Boolean that indicates whether the path is absolute or not
`Path.is_dir()`	Return True if the given path points to a directory (or symbolic link)
`Path.iterdir()`	Yield contents of a given directory
`PurePath.joinpath()`	Combine a given path with each of the other arguments in turn
`Path.mkdir()`	Create a new directory at the given path
`Path.readlink()`	Return path for given symbolic link
`Path.resolve()`	Make path absolute, resolving any symbolic links; return new path
`Path.rmdir()`	Remove an empty directory
`Path.unlink()`	Remove a file or symbolic link

Here's how to make a path variable using `Path`. Start by importing the class from the module, as follows:

```
In [26]: from pathlib import Path

In [27]: a_path = Path('folder1', 'folder2', 'file1.txt')

In [28]: a_path
Out[28]: WindowsPath('folder1/folder2/file1.txt')
```

Note that `Path` returned a `WindowsPath` object. If you're using macOS or Linux, you should see a `PosixPath` object. Also note that although the `WindowsPath` object displays with forward slashes, it employs proper Windows backslashes behind the scenes:

```
In [29]: print(a_path)
folder1\folder2\file1.txt
```

`Path` includes methods that can make your code more readable and convenient to write. Suppose that you want to append a path to your home directory. Rather than type in the full path, you can just use the `home()` method to get the path:

```
In [30]: home = Path.home()

In [31]: another_path = Path(home, 'folder1', 'folder2', 'file1.txt')

In [32]: print(another_path)
C:\Users\hanna\folder1\folder2\file1.txt
```

Alternatively, you can do all this in one line and use forward slashes rather than commas to separate path components:

```
In [33]: another_path = Path.home() / 'folder1' / 'folder2' / 'file1.txt'

In [34]: another_path
Out[34]: WindowsPath('C:/Users/hanna/folder1/folder2/file1.txt')
```

Don't worry about those forward slashes if you're using Windows. As demonstrated previously, the path object is aware of the platform in use and will return the correct format for that system.

Each `Path` object includes handy attributes for working with files and folders. These let you get information like the `stem` of a path or a file's `name` or extension (`suffix`). The parent attribute, for example, returns the most immediate ancestor of a given file path. In the following example, we get the path *leading up to* the text file in the `another_path` variable:

```
In [35]: print(another_path.parent)
C:\Users\hanna\folder1\folder2
```

You can access this attribute multiple times to walk up the ancestry tree of a given file, like so:

```
In [36]: print(another_path.parent.parent.parent)
C:\Users\hanna
```

As mentioned previously, `pathlib` gives you access to basic filesystem operations like moving, renaming, and removing files and folders. These methods don't warn you or wait for confirmation before executing, so you'll want to be very careful when using them. Otherwise, you could easily delete or overwrite data that you want to keep.

The Shell Utilities Module

The built-in shell utilities module (`shutil`) provides high-level functions for working with files and folders, such as copying, moving, and deleting. Table 12-3 summarizes a few of the most popular methods. For a list of all of the available methods, along with detailed instructions for their use, visit the documentation at *https://docs.python.org/3/library/shutil.html*.

Table 12-3: Useful `shutil` Module Methods

Method	Description
`copy()`	Copy a file (if path is included, will copy to a new directory)
`copy2()`	Same as `copy()` but preserves all the metadata of the source file
`copytree()`	Recursively copy an entire directory tree rooted at a source directory to a new destination directory and return the destination directory path
`disk_usage()`	Return disk usage statistics about a file system as a named tuple with the attributes total, used, and free, in bytes
`move()`	Move a file or directory to another location and return the destination
`rmtree()`	Delete an entire directory tree (very dangerous)
`make_archive()`	Create an archive file (zip or tar) and return its name

Here's an example in which I get the current disk usage on my system using a dot to represent the absolute path:

```
In [37]: import shutil

In [38]: gb = 10**9

In [39]: total, used, free = shutil.disk_usage('.')

In [40]: print(f"Total memory (GB): {total / gb:.2f}")
Total memory (GB): 238.06

In [41]: print(f"Used memory (GB): {used / gb:.2f}")
Used memory (GB): 146.85

In [42]: print(f"Free memory (GB): {free / gb:.2f}")
Free memory (GB): 91.22
```

In the next example, we move the *test2* folder to a new location under the *file_play* folder. To accomplish this, we pass the current path (with the current working directory represented by a dot folder), followed by the target path, to the `move()` method (note that the path is configured for Windows):

```
In [43]: shutil.move('.\\test1\\test2', '.\\')
Out[43]: '.\\test2'
```

You should see this update reflected in Spyder's File Explorer (compare Figure 12-2 to Figure 12-3). Child folders move with parent folders, so the *test3* folder remains beneath *test2*.

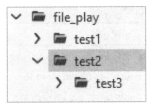

Figure 12-3: The test2 *folder moved beneath the* file_play *folder*

NOTE *Always be careful when using* shutil *methods; no warnings are provided, and unexpected behavior can result. The* rmtree() *method is especially dangerous because it* permanently *deletes folders and their contents. You can wipe much of your system, lose important documents unrelated to Python projects, and break your computer!*

Now that you have a feel for manipulating files and folders using Python, it's time to start writing and reading files. We'll begin with simple text files and then move on to more sophisticated data structures.

TEST YOUR KNOWLEDGE

1. The '.' folder represents:

    a. The current working directory

    b. The parent directory for the current working directory

    c. The absolute path

    d. The child directory for the current working directory

2. Which method should you be particularly careful about using?

    a. shutil.move()

    b. shutil.copytree()

    c. Path.resolve()

    d. shutil.rmtree()

3. True or false: a relative directory path is relative with respect to the root directory.

4. You can use the os.path.join() method to:

    a. Return a directory path as an object rather than a string

    b. Return a directory path as a list rather than a string

    c. Return the proper path separators for your operating system

    d. Correct existing path separators for your operating system

5. The pathlib module treats paths as ＿＿＿＿＿＿＿.

Working with Text Files

A *plaintext* file consists of human-readable characters encoded using some standard such as ASCII, with no formatting information other than space, tab, and newline characters. Some examples of plaintext files are text files (*.txt*), Python files (*.py*), and comma-separated values files (*.csv*). Plaintext files are cross-platform. You can open and read one using both Window's Notepad and macOS's TextEdit app.

Python's standard library includes built-in functions for reading and writing text files. The `pathlib` module also includes methods for working with text files. In the sections that follow, we'll first use the built-in functions, and then we'll look at the `pathlib` alternatives.

Reading a Text File

Using Python, you can read strings from a text file in multiple ways. For example, you can read individual characters, complete lines, the whole file, and so on. To demonstrate, open your system's text editor and enter the following. Be sure to press ENTER after the first two lines:

```
This is the first line.
This is the second line.
This is the third line.
```

Save the file in the *file_play* folder as *lines.txt*.

NOTE *You can double-click a text file in the Spyder File Explorer to edit and review its contents. You can also generate text files using File ▸ New file on the top toolbar. Use the Save as command to choose the .txt extension.*

Now, in the console, enter the following to open, read, and close the file:

```
In [44]: f = open('lines.txt', 'r')

In [45]: f
Out[45]: <_io.TextIOWrapper name='lines.txt' mode='r' encoding='cp1252'>
```

In the first line, we used the built-in `open()` function to open the file and assign its contents to the f variable (short for "file"). The `open()` function took two arguments. The first was the name of the text file. Because this file is in the current working directory, you didn't need to include a path. For files not in the current working directory, you would need to pass either an absolute or relative path.

The second argument was an *access mode*, which sets the type of operations possible in the opened file, such as read, write, append, and so on. The `'r'` informs Python that you want to open the file as *read-only*. This protects the file from modification. Although read-only is the default mode, explicitly including the `'r'` argument makes your intention clear. Table 12-4 includes some common file access modes in Python.

Table 12-4: Selected Text File Access Modes

Mode	Description
'r'	Read from a text file. Raise an exception if the file doesn't exist.
'w'	Write to a text file. Creates a new file, else overwrites existing files.
'x'	Write to a text file but return an error if the file already exists.
'a'	Append to a text file. Create a new file if one doesn't exist.
'r+'	Permit read and write mode.
'b'	Add to mode for binary files (such as 'rb').

The open() function returned a File object of type _io.TextIOWrapper. This is a type of object like a list or a tuple.

Now, let's look at some of the file object methods for reading files (Table 12-5). These are called on a file object using dot notation.

Table 12-5: Selected File Object Methods and Attributes

Method	Description
close()	Close a file.
closed	Attribute that returns True if a file is closed.
read()	Read the specified number of characters from a file and return a string.
readline()	Read the specified number of characters from a file and return a string. By default, return all characters from the current position to the end of a line.
readlines()	Read all the lines in a file and return them as items in a list.
seek()	Change the position of the file pointer to a specific position within the file.
tell()	Return the current position of the file read/write pointer within a file.
write()	Write the specified string to a file.
writelines()	Write the strings in a specified list to a file.

Among the most important methods is close(). Closing files before terminating the process is a good practice. If you don't close files, you could run out of file descriptors (numbers that uniquely identify open files in a computer's operating system), lock the files from further access in Windows, corrupt the files, or lose data if you are writing to the files.

To close a file, call close() using dot notation:

```
In [46]: f.close()
```

You can work with file objects only while they're open. After a file object is closed, you can no longer work with it.

Now, let's look at ways to get the file contents. In the following console snippet, open the file again and use the read() method to read the first character. This method returns a string data type (remember, in the console you can use the up and down arrow keys to retrieve previous commands):

```
In [47]: f = open('lines.txt', 'r')

In [48]: f.read(1)
Out[48]: 'T'

In [49]: f.read(10)
Out[49]: 'his is the'
```

Passing the read() method the value 1 returned the first character in the file. But passing it 10 did not return the first 10 characters in the file. That's because read() remembers where it left off. To find the current position in the file, use the tell() method:

```
In [50]: f.tell()
Out[50]: 11
```

To manually change the position of the pointer in the file, pass the seek() method a number, as follows:

```
In [51]: f.seek(12)
Out[51]: 12

In [52]: f.read(1)
Out[52]: 'f'

In [53]: f.close()
```

To restart at the beginning, you must either close and reopen the file or use seek() to return to the beginning.

If you don't specify the number of characters to read, Python returns the entire file. This is not a problem for small files, but it can become an issue with very large files. To demonstrate reading the entire file, reopen the file and call the read() method with no arguments:

```
In [54]: f = open('lines.txt', 'r')

In [55]: f.read()
Out[55]: 'This is the first line.\nThis is the second line.\nThis is the third line.'

In [56]: f.close()
```

Note that the file object includes the newline escape sequence (\n). This lets it know how to print the lines correctly:

```
In [57]: f = open('lines.txt', 'r')

In [58]: print(f.read())
```

```
This is the first line.
This is the second line.
This is the third line.

In[59]: f.close()
```

You can use the readline() method to read a line at a time, as follows:

```
In [60]: f = open('lines.txt', 'r')

In [61]: print(f.readline())
This is the first line.

In [62]: print(f.readline())
This is the second line.

In [63]: print(f.readline())
This is the third line.

In [64]: f.close()
```

In this case, "lines" are defined by the presence of the newline escape sequence (\n). Like the read() function, readline() remembers where it left off, so to start back at the beginning, you must close and reopen the file.

Be careful when you're using readline(). Don't assume that the value you pass it represents a line; it actually represents a character, just as with the read() method. In fact, you can duplicate the results from lines In[48]-In[49]:

```
In [65]: f = open('lines.txt', 'r')

In [66]: f.readline(1)
Out[66]: 'T'

In [67]: f.readline(10)
Out[67]: 'his is the'

In [68]: f.close()
```

To read in a whole file at once, you can use the readlines() method. Unlike the previous methods, which return strings, this method reads the file into a list. Each line in the file becomes a separate item in the list. Here's an example:

```
In [69]: f = open('lines.txt', 'r')

In [70]: lines = f.readlines()

In [71]: lines
Out[71]:
['This is the first line.\n',
```

```
'This is the second line.\n',
'This is the third line.']

In [72]: f.close()
```

Because the output is a list, you can get its length, iterate over it, and so on, as with any list:

```
In [73]: len(lines)
Out[73]: 3

In [74]: for line in lines:
    ...:     print(line)
This is the first line.

This is the second line.

This is the third line.
```

In the preceding methods, the end-of-line (EOL) markers are preserved. These are control characters used by character-encoding specifications such as ASCII to signify the end of a line of text. If you don't want these, you can strip them out using list comprehension:

```
In [75]: lines = [line.rstrip() for line in open('lines.txt', 'r')]

In [76]: lines
Out[76]:
['This is the first line.',
'This is the second line.',
'This is the third line.']
```

Compare the previous output to that in line Out[71.] The newline characters (\n) are gone. The rstrip() string method removes specified trailing characters from the right side of a string. If no character is specified, it removes any newline characters or whitespace at the end of a line.

Closing Files Using the with Statement

Because closing files is so important (and easily overlooked), Python provides the with statement, which automatically closes files after a nested block of code. In this example, we load the text file using the with statement and the open() function, and then we use the read() method to get the complete contents of the file and assign them to the lines variable:

```
In [77]: with open('lines.txt') as f:
    ...:     lines = f.read()
    ...:     print(lines)
This is the first line.
This is the second line.
This is the third line.
```

Whenever possible, try to use a with statement when opening files to ensure that the file is closed properly. To check that a file is closed, you can use its closed attribute, which returns True or False:

```
In [78]: f = open('lines.txt', 'r')

In [79]: f.closed
Out[79]: False

In [80]: f.close()

In [81]: f.closed
Out[82]: True
```

Writing to a Text File

You can write a string to a text file using the write() and writelines() file object methods (Table 12-5). Let's try this out using a haiku poem by yours truly.

To write to a file, you first must open it using the write ('w') file access mode (see Table 12-4). Enter the following in the console:

```
In [83]: f = open('haiku.txt', 'w')
```

Calling open() on a file in write mode either creates a new file with the specified name (if one doesn't exist) or completely overwrites an existing file with the same name, erasing its contents. In this case, we need only to enter a filename because we're writing to the current working directory. To write elsewhere, you need to either change directories using the chdir() method or include a directory path with the filename.

Now that we have a file object, we can write strings to it, using a newline character where we want carriage returns:

```
In [84]: f.write('Faraway cloudbanks\n')
Out[84]: 19

In [85]: f.write('That I let myself pretend\n')
Out[85]: 26

In [86]: f.write('Are distant mountains')
Out[86]: 21

In [87]: f.close()
```

The output represents the number of characters in each string, including the newline character. Closing the file at the end frees up system resources and prevents you from accidently writing more data to the file.

Let's check that it worked by using the read() method:

```
In [88]: with open('haiku.txt', 'r') as f:
   ...:     print(f.read())
```

```
Faraway cloudbanks
That I let myself pretend
Are distant mountains
```

Remember, when you open a file using the with statement, it closes automatically.

Entering lines one by one is tedious. The writelines() method lets you *write* a list of strings into a file, much like the readlines() method offers the ability to *read* a text file into a list. The following example creates a new haiku as a list, overwrites the existing *haiku.txt* file, writes the list to the file and then reads the file:

```
In [89]: poem = ['In city fields\n',
    ...:          'Contemplating cherry trees\n',
    ...:          'Strangers are like friends\n']

In [90]: with open('haiku.txt', 'w') as f:
    ...:     f.writelines(poem)

In [91]: with open('haiku.txt', 'r') as f:
    ...:     print(f.read())
In city fields
Contemplating cherry trees
Strangers are like friends
```

Oops, we forgot to attribute the haiku to the master Issa. No problem. With the append ('a') file access mode, you can add strings to an existing text file without overwriting the original contents:

```
In [92]: with open('haiku.txt', 'a') as f:
    ...:     f.write('                    --Issa')

In [93]: with open('haiku.txt', 'r') as f:
    ...:     print(f.read())
In city fields
Contemplating cherry trees
Strangers are like friends
                    --Issa
```

You can also use writelines() to generate new file contents on the fly, as follows:

```
In [94]: with open('a_random_thought.txt', 'w') as f:
    ...:     f.writelines(line for line in poem if line.startswith('C'))

In [95]: with open('a_random_thought.txt', 'r') as f:
    ...:     print(f.read())
Contemplating cherry trees
```

In this example, we filtered the poem list so that only lines beginning with C were written to the new file.

Reading and Writing Text Files Using pathlib

The Path class of the pathlib module also provides methods for working with files and folders (Table 12-6). These methods incorporate built-in functions like open() and can make simple reading and writing exercises more convenient (assuming that you like working with path objects).

Table 12-6: Some Useful Path Methods for Working with Files and Folders

Method	Description
Path.glob()	Yield all matchings files for a given pattern (such as *.py)
Path.is_file()	Return True if given path points to a regular file (or symbolic link)
Path.open()	Open a file based on name or path + name
Path.read_bytes()	Return the contents of a given file as a bytes object
Path.read_text()	Return the contents of a given file as a string and close the file
Path.rename()	Rename a file or directory and return new path
Path.replace()	Rename a file or directory unconditionally and return new path
Path.touch()	Create a file at the given path
Path.write_text()	Open a specified file in text mode, write to it, and then close the file

The Path.read_text() method calls open() behind the scenes and returns a file's contents as a string. It also closes the file automatically, like the with statement. Here's an example in the console using the *lines.txt* file from earlier in the chapter:

```
In [96]: from pathlib import Path

In [97]: p = Path('lines.txt')

In [98]: p.read_text()
Out[98]: 'This is the first line.\nThis is the second line.\nThis is the third
line.'
```

Note that you first must create a path object (p). For users unfamiliar with pathlib, this can be confusing compared to the more tradition file-opening techniques reviewed in the previous section.

Now, let's create a file in the test1 folder and write to it using Path. In the console, enter the following:

```
In [99]: path = Path(Path.cwd() / 'test1' / 'another_haiku.txt')

In [100]: lines2 = 'Desolate moors fray\nBlack cloudbank, broken, scatters\nIn the pines, the
graves'
```

```
In [101]: path.write_text(lines2)
Out[101]: 78

In [102]: print(path.read_text())
Desolate moors fray
Black cloudbank, broken, scatters
In the pines, the graves
```

The Path.write_text() method takes a string as an argument. Like open(), it will overwrite an existing file with the same name. Unlike open(), it doesn't permit use of an append mode. It will, however, close the file automatically.

You can read more about pathlib at *https://docs.python.org/3/library/pathlib.html*.

TEST YOUR KNOWLEDGE

6. Which statements or methods close a text file?

   a. The with statement

   b. The Path.read_text() method

   c. The Path.write_text() method

   d. The close() method

   e. All of the above

7. Rename the *another_haiku.txt* file created in the previous section to *haiku_2.txt*. Use either the os or pathlib modules.

8. Print the *haiku.txt* file starting at the 15th character.

9. Which file-access mode is used to add text to an existing text file?

   a. w

   b. r

   c. a

   d. b

10. True or false: The os.writelines() method writes a list to a file; the Path.write_text() method writes a string to a file.

Working with Complex Data

Text files are convenient and popular, but they're hardly the only game in town. The various file-writing methods we've reviewed so far accept only strings, or lists of strings, as input. But Python includes many different data types, such as dictionaries, that you'll use in your everyday work, and you'll need a way to save these, as well.

To save these other data types, you need to use *data serialization*. This process converts structured data such as a Python dictionary into a storable and sharable format. This format retains the information needed to reconstruct the object in memory when it's read from storage or transmitted. This process is called *de-serialization*.

In this section, we'll look at modules, like `pickle` and `json`, that serialize and deserialize data. The `pickle` module is Python's native serialization module. It converts objects into an ordered sequence of bytes (0s and 1s) known as a *byte stream*. Pickling and unpickling allow us to easily transfer data from one server or system to another and then store it in a file or database.

The `json` module converts Python objects to a serialized representation known as *JavaScript Object Notation*, or *JSON* for short, and deserializes them on demand. We used `json` for pretty-printing dictionaries back in Chapter 9. It works with just about every language.

These two modules have their strengths and weaknesses (Table 12-7). Pickling works on most Python objects and data types, whereas JSON is limited to certain objects and data types.

Table 12-7: Pickle vs. JSON for Serialization

Characteristic	Pickle	JSON
Storage format	Byte stream	Human-readable string object
Python objects	All objects	Limited to certain objects
Python data types	Almost every data type	Only lists, dictionaries, nulls, Booleans, numbers, strings, arrays, and JSON objects
Compatibility	Python only	Language independent
Speed	Relatively Slow	Relatively fast
Security	Has security issues	Safe and secure

NOTE *Pickling is less secure than using JSON. You should be very careful about unpickling data from an unknown source, as it may contain malicious data. Pickling is also intended for relatively short-term data storage because revisions to the module might not always be backward compatible.*

Pickling Data

To pickle something means to preserve it. The `pickle` module (*https://docs .python.org/3/library/pickle.html*) pickles Python data objects in binary files. Unlike text files, humans cannot read binary files.

Pickling is a lot like writing strings to a file, only you write pickled objects. The access modes are the same except for the addition of a 'b' for "binary" (Table 12-8).

Table 12-8: Selected Binary File Access Modes

Mode	Description
'rb'	Read from a binary file.
'wb'	Write to a binary file. Create or overwrite file, as required.
'ab'	Append to a binary file. Create or modify file, as required.

Let's pickle some lists. In the console, enter the following:

```
In [103]: import pickle

In [104]: dragon_prefix = ['Hungarian', 'Chinese', 'Peruvian']

In [105]: dragon_suffix = ['Horntail', 'Fireball', 'Vipertooth']

In [106]: f = open('dragons.dat', 'wb')

In [107]: pickle.dump(dragon_prefix, f)

In [108]: pickle.dump(dragon_suffix, f)

In [109]: f.close()
```

After importing the pickle module and creating two dragon lists, we opened a new binary file called *dragons.dat*. Next, we stored the two lists in this file using the pickle.dump() function, passing it the name of the list and the name of the file object as arguments. Finally, we closed the file (you should see it in your *file_play* folder).

The pickle.dump() function wrote each list to the file as a separate object. To retrieve these objects, we open the file again, in binary mode, and call the pickle.load() function, as follows:

```
In [110]: f = open('dragons.dat', 'rb')

In [111]: dragon_prefix = pickle.load(f)

In [112]: dragon_suffix = pickle.load(f)

In [113]: print(dragon_prefix)
['Hungarian', 'Chinese', 'Peruvian']

In [114]: print(dragon_suffix)
['Horntail', 'Fireball', 'Vipertooth']

In [115]: f.close()
```

The pickle.load() function accepts the file object as an argument and returns (or unpickles) the first pickled object, assigning it to the variable dragon_prefix. The next call to pickle.load() returns the next pickled object. One thing to note here is that you don't need to know the original names of the lists (like "dragon_prefix") to extract the data. You could have called these "poodledoodle" and "snickerdoodle," and you would have retrieved the same lists in the same order.

But what if you want to retrieve the pickled objects in some other order, such as retrieving only the dragon suffixes? For that, you'll need the shelve module, which takes pickling a step further.

Shelving Pickled Data

A *database* is a special file for storing data. Most databases resemble Python dictionaries, in that they map keys to values. Unlike a dictionary, however, databases persist after a program ends.

Python comes with the dbm module for creating and updating database files. This module has a limitation, though, as its keys and values must be either strings or bytes. The pickle module helps overcome this limitation by transforming multiple data types into strings suitable for use in a database.

Because the need to store non-string objects in a database is so common, the functionality has been incorporated into a module called shelve that helps you store and access pickled objects in a file. It builds on pickle and implements a serialization dictionary in which objects are pickled with an associated key, composed of strings. The keys let you load your shelved data file and randomly access the values, composed of pickled objects.

The shelve module produces a *shelf*, which is a persistent, dictionary-like object. Although it's possible to directly pickle a dictionary, using the shelve module is more memory efficient.

NOTE *Because this process involves the pickle module, loading a shelf can execute unexpected code, so it's unsafe to load a shelf from an untrusted source.*

Let's look at how shelving works using the dragon data from the previous section:

```
In [116]: import shelve

In [117]: s = shelve.open('dragon_shelf', 'c')

In [118]: type(s)
Out[118]: shelve.DbfilenameShelf
```

After importing the module, we used the shelve.open() method to create a new shelf named dragon_shelf in the current working directory, assigned it to the variable s, and then got the data type of s. To create the shelf, we used the 'c' access mode. Other shelve access modes are listed in Table 12-9.

Table 12-9: Shelve Access Modes

Mode	Description
'c'	Open a shelf for reading and writing, creating it if necessary
'n'	Create a new, empty shelf open for reading and writing, overwriting if needed
'r'	Open an existing shelf for reading only
'w'	Open an existing shelf for reading and writing

Now, let's add the dragon data to the shelf using a key-value combination. This will pickle the data behind the scenes. Although we create the list here, we could just as easily use a variable name assigned to a list, as we did in the earlier pickle.dump() example.

```
In [119]: s['prefix'] = ['Hungarian', 'Chinese', 'Peruvian']

In [120]: s['suffix'] = ['Horntail', 'Fireball', 'Vipertooth']

In [121]: s.close()
```

Closing the shelf *synchronizes* the data by ensuring that any data in the memory cache, or buffer, is written to the disk. It then releases system resources by clearing the cache.

Two things to note here are that shelve will automatically add a *.dat* extension to the filename, and it will create additional files that support the shelf (highlighted in gray in Figure 12-4). These additional files are operating system specific. On macOS, for example, you might see only a file named *dragon_shelf.db*.

Figure 12-4: Files related to dragon_shelf
in Windows

Binary files in Spyder's File Explorer include "01" on the file icon. The text file icon uses two straight lines.

Now, let's reopen the shelf and retrieve some data:

```
In [122]: s = shelve.open('dragon_shelf', 'r')

In [123]: type(s['prefix'])
Out[123]: list

In [124]: print(f"Dragon suffixes: {s['suffix']}")
Dragon suffixes: ['Horntail', 'Fireball', 'Vipertooth']

In [125]: s.close()
```

After opening the *dragon_shelf* file in read-only mode, you can see that the prefix key refers to a list object. You can also print the suffix list first, despite it being the second list loaded into the shelf. Compare this to the pickle.load() method from the previous section, which returns pickled objects in order.

Closing Shelves Using the with Statement

Shelving a large volume of data can use a lot of memory, so it's important to close a shelf when you're finished. Because this easily can be overlooked, Python lets you use the with statement when opening shelves so that the files automatically close after some action. Here's an example:

```
In [126]: with shelve.open('dragon_shelf', 'r') as s:
     ...:     print(type(s['prefix']))
<class 'list'>
```

Because the with statement closed the shelf after its block executed, subsequent actions on s will raise a ValueError:

```
ValueError: invalid operation on closed shelf
```

Using Shelve Methods

Shelf objects support most of methods and operations supported by dictionaries (Table 12-10). This is by design and is intended to ease the transition from dictionary-based scripts to those requiring persistent storage.

If you forget the key names in a shelf or if you're using a shelf that you didn't create, you can use the keys() method to retrieve the names. Note that you need to convert the output into a list with the list() function:

```
In [127]: with shelve.open('dragon_shelf', 'r') as s:
     ...:     print(list(s.keys()))
['prefix', 'suffix']
```

Table 12-10: Shelve Module Methods

Method	Description
close()	Synchronize and close the shelf object
get()	Return shelf values associated with a key
items()	Return shelf key-value pairs as tuples
keys()	Return list of shelf keys
pop()	Remove specified shelf key and return associated shelf value
sync()	Write back all entries in the cache if shelf was opened with writeback set to True
update()	Update shelf from another dict or iterable
values()	Return list of shelf values

Some other methods return an iterable that you can loop over. Here's an example using the items() method, which returns the key-value pairs as tuples:

```
In [128]: with shelve.open('dragon_shelf', 'r') as s:
   ...:        print(s.items())
ItemsView(<shelve.DbfilenameShelf object at 0x000001D3956BAF70>)
```

Printing the output yielded an object name, not the key-value pairs that you probably expected. To get the key-value tuples, loop over the output, as follows:

```
In [129]: with shelve.open('dragon_shelf', 'r') as s:
   ...:        for item in s.items():
   ...:            print(item)
('prefix', ['Hungarian', 'Chinese', 'Peruvian'])
('suffix', ['Horntail', 'Fireball', 'Vipertooth'])
```

You can read more about shelve and its methods at *https://docs.python .org/3/library/shelve.html*.

Storing Data with JSON

Using the json module (*https://docs.python.org/3/library/json.html*), you can store data as a single human-readable string. Here's an example of a Python dictionary stored in JSON format:

```
'{"key1": "value1", "key2": "value2", "key3": "value3"}'
```

It looks just like a regular Python dictionary except for one thing: it's enclosed in single quotes, making the whole thing a string.

Compared to pickle and shelve, the json module offers a faster and more secure way to store and retrieve complex Python data types. It supports fewer data types than pickle, however, because it's limited to dictionaries, lists, nulls, Boolean values, numbers (integers and floats), strings, and JSON objects.

JSON will also help you access information on the worldwide web. As a lightweight data-interchange format that's easy for humans to read and for machines to parse, the application programming interfaces (APIs) for many websites pass data using JSON format.

Saving Data in JSON Format

To see how `json` works, let's create a Python dictionary for the crew capacity of three famous spacecraft and save it in JSON format. Enter the following in the console:

```
In [130]: import json

In [131]: crew = dict(Mercury=1, Gemini=2, Apollo=3)

In [132]: crew
Out[132]: {'Mercury': 1, 'Gemini': 2, 'Apollo': 3}

In [133]: capsules_data = json.dumps(crew)

In [134]: capsules_data
Out[134]: '{"Mercury": 1, "Gemini": 2, "Apollo": 3}'

In [135]: with open('capsules_data.json', 'w') as f:
     ...:        f.write(capsules_data)
```

The `json.dumps()` method turns the dictionary into a JSON string. You can write JSON strings to persistent files using the `open()` function in write mode, as you've done before. The new *capsules_data.json* file should show up in the Spyder File Explorer pane (Figure 12-5).

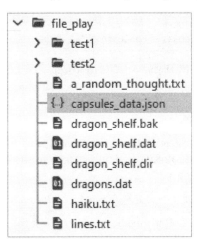

Figure 12-5: The capsules_data.json *file in the File Explorer pane*

Note how Spyder uses a special icon to denote the file. Because it's human-readable, you can open it and read its contents just like a text file.

Loading Data in JSON Format

Now, let's open, load, and use the JSON file. We'll continue to work in the console, but this following example could easily be done in a saved program written in a text editor or Jupyter Notebook:

```
In [136]: with open('capsules_data.json', 'r') as f:
     ...:     crew = json.load(f)

In [137]: print(f"The Mercury capsule had {crew['Mercury']} seat.")
The Mercury capsule had 1 seat.

In [138]: print(f"The Apollo capsule had {crew['Apollo']} seats.")
The Apollo capsule had 3 seats.
```

Saving Tuples in JSON Format

There is no concept of a tuple in the JSON format. If you save a tuple in JSON format, you'll receive a list. In the console, enter the following to see an example:

```
In [139]: import json

In [140]: t = (1, 2, 3)

In [141]: type(t)
Out[141]: tuple

In [142]: t_json = json.dumps(t)

In [143]: t_json
Out[143]: '[1, 2, 3]'

In [144]: t2 = json.loads(t_json)

In [145]: t2
Out[145]: [1, 2, 3]

In [146]: type(t2)
Out[146]: list
```

In simple cases, you can handle this by converting the output back into a tuple:

```
In [147]: t2 = tuple(t2)

In [148]: t2
Out[148]: (1, 2, 3)

In [149]: type(t2)
Out[149]: tuple
```

For more sophisticated cases, you'll want to do an online search for using tuples with JSON.

Catching Exceptions When Opening Files

Reading and writing files falls under the category of user interactions, and as we saw in Chapter 10, a lot of things can go wrong when users get involved. For working with files, these include trying to open files or use paths that don't exist, trying to open files or folders without the proper permissions, trying to open a folder instead of a file, and so on.

These problems can't be fixed within your code, but you can catch these exceptions and provide the user with some helpful advice, rather than allow the program to crash and spew gobbledygook all over the screen.

Most of the common file-loading errors fall under the operating system exception class called OSError. These include the errors shown in Table 12-11.

Table 12-11: Common Errors Associated with File Loading

Class	Subclass
BlockingIOError	
ChildProcessError	
ConnectionError	BrokenPipeError
ConnectionError	ConnectionAbortedError
ConnectionError	ConnectionRefusedError
ConnectionError	ConnectionResetError
FileExistsError	
FileNotFoundError	
InterruptedError	
IsADirectoryError	
NotADirectoryError	
PermissionError	
ProcessLookupError	
TimeOutError	

Here's an example in which we use OSError to catch the exception thrown by a nonexistent file (*fluffybunnyfeet.lol*). For a refresher on using try and except, see Chapter 10.

```
In [150]: try:
     ...:     with open('fluffybunnyfeet.lol', 'r') as f:
     ...:         data = f.read()
```

```
...: except OSError as e:
...:     print(e)
...: else:
...:     print('File successfully loaded.')
...: finally:
...:     print('File load process complete.')
[Errno 2] No such file or directory: 'fluffybunnyfeet.lol'
File load process complete.
```

The except clause printed a useful message informing the user that the file doesn't exist (as this was a FileNotFoundError). The finally clause let the user know that the file loading process has terminated. Note that the finally block executes regardless of the outcome, whereas the else code block executes only for a successful outcome.

Here's an example of a successful outcome using the *haiku.txt* file that we created earlier:

```
In [151]: try:
...:         with open('haiku.txt', 'r') as f:
...:             data = f.read()
...: except OSError as e:
...:     print(e)
...: else:
...:     print('File successfully loaded.')
...: finally:
...:     print('File load process complete.')
File successfully loaded.
File load process complete.
```

For more on Python's built-in exceptions, visit the documentation at *https://docs.python.org/3/library/exceptions.html.*

Other Storage Solutions

If your data is sufficiently complex, it might require more powerful storage solutions. The *eXtensible Markup Language (XML)* is designed to store and transport small to medium amounts of data and is widely used for sharing structured information. *YAML* is another human-readable data-serialization language used for configuration files and in applications where data is being stored or transmitted. It has a minimal syntax compared to XML. *SQLite* is a lightweight database that can provide a relational database management system with zero-configuration. *Hierarchical Data Format (HDF5)* is for storing large volumes of scientific array data. Covering these storage systems is beyond the scope of this book, but you can find copious information for each online.

11. True or false: The shelve module helps you to store and access pickled objects in a file.

12. Of the methods for saving and loading complex data discussed in this chapter, the most secure is:

    a. Pickling

    b. Syncing

    c. JSON format

    d. Shelving

13. Rewrite the crew capacity program from the "Storing Data with JSON" section of this chapter so that it automatically prints the name of the capsule and the grammatically correct version of "seat" (*seat* or *seats*) depending on the number of crew members.

14. Use the console to investigate how JSON handles quotation marks. Use the lists ["don't", "do"] and ['don\'t', 'do'].

15. Built-in Python exceptions for opening and closing files fall under which exception class?

    a. `IOError`

    b. `FileNotFoundError`

    c. `PermissionError`

    d. `OSError`

16. Use Python modules to move a copy of the *lines.txt* file to the *test1* folder and then archive it as a ZIP file.

Summary

Files let you save your work—including variables you assign in a program—in a persistent and sharable manner. To work with files, you need a base understanding of how your computer's filesystem works, how to manipulate directory paths, and how to open, read, and write files.

The *absolute* directory path refers to the full directory path, starting with the root directory (such as *C:* on windows). The *relative* directory is defined relative to the current working directory. You can use shortcuts, such as "." for the absolute directory and ".\\" for the current working directory, to make working with directories easier.

Python's built-in os, pathlib, and shutil modules include useful high-level methods for working with files and folders. These methods execute without warning, however, so you'll need to be careful when moving, renaming, or deleting data.

Python has other built-in tools for working with human-readable text files. To read a file, you first must open it as a file object using the open() function. Methods such as read() and readlines() can then be called on this object. To write to a file, you must open it in write mode and then call methods like write() and writelines(). To add data to an existing file without overwriting its contents, you must open it in append mode.

You should always close files when you're through with them to release system resources and protect the file from being accidently overwritten. You can manually close files using the close() method, or automatically by opening the file using a with statement.

More complex data, such as Python dictionaries and lists, can be saved in binary format using the pickle module, or as human-readable strings using the json module. The shelve module helps you to store and access pickled objects in a *shelf* file, which is a persistent, dictionary-like object that assigns each pickled object a unique *key* name. Using the JSON format is faster and more secure than pickling, but not all Python objects and data types can be stored with JSON.

Although it's important to understand the basic tools for file and folder management with Python, if much of your work involves data stored on disk, you'll want to read about the *Python Data Analysis Library*, otherwise known as *pandas*. This library contains high-level tools for moving data from disk into Python data structures and back again. Many file formats are accommodated, including Excel, CSV, TXT, SQL, HTML, JSON, Pickle, and HDF5. We look at pandas in Chapters 15 and 20.

13

OBJECT-ORIENTED PROGRAMMING

So far in this primer, you've been writing code using procedural programming techniques built around performing actions and evaluating logic. You've learned how to organize code using functions and modules, and you've used built-in data types to organize data. In this chapter, you'll learn how to use object-oriented programming to define your own types to organize both code and data.

Object-oriented programming (OOP) is a language model that lets you bundle together related data with functionality that acts on that data. The data consists of *attributes* (akin to variables) that are manipulated by *methods* (akin to functions). These "bundles" form custom data types called *classes*. Classes help you split your program into different sections that deal with different pieces of information rather than letting it all blend into an unstructured mess.

Classes let you create individual *objects* with specific properties and behaviors. Using a class template, you can efficiently "stamp out" multiple objects, just as a set of blueprints lets you build multiple versions of the same car. Each car will start off with the same attributes, such as color and mileage, and have the same methods, such as for accelerating and braking, but after they leave the factory, these can change. Some cars may be repainted, others can lose wheel alignment and pull to the left, their mileage will vary, and so on.

In this chapter, you'll learn how to define classes that create objects, write attributes and methods for the objects, and then instantiate those objects. You'll also write classes that inherit attributes and methods from other classes and use dataclasses to reduce code redundancy. This introduction to the topic should give you an understanding of the basics of OOP and an appreciation for how you can benefit from it as a programmer.

When to Use OOP

OOP is easier to appreciate when you're writing large, complex programs because it helps you to structure your code into smaller parts that are easier to understand. It also reduces code duplication and makes code easier to maintain, update, and reuse. As a result, most commercial software is now built using OOP.

Because Python is an object-oriented programming language, you've already been using objects and methods defined by other people. But unlike languages such as Java, Python doesn't force you to use OOP for your own programs. It provides ways to encapsulate and separate abstraction layers using other approaches such as procedural or functional programming.

Having this choice is important. If you implement OOP in small programs, most of them will feel overengineered. To quote computer scientist Joe Armstrong, "The problem with object-oriented languages is they've got all this implicit environment that they carry around with them. You wanted a banana, but what you got was a gorilla holding the banana and the entire jungle!"

As a scientist or engineer, you can get a lot done without OOP, but that doesn't mean you should ignore it. OOP makes it easy to simulate many objects at a time, such as a flock of birds or a cluster of galaxies. It's also important when things that are manipulated, like a GUI button or window, must persist for a long time in the computer's memory. And because most of the scientific packages you'll encounter are built using OOP, you'll want more than a passing familiarity with the paradigm.

Creating a New Spyder Project

Let's make a new Spyder project to use in this chapter. If you need a refresher on Spyder projects, see "Using Project Files and Folders" on page 68.

Start by launching Spyder from your *base* (root) environment (either from the Start menu or from Anaconda Navigator). In the Start window, this may show as Spyder (anaconda3). For a refresher on conda environments see Chapter 2.

Next, on the top toolbar, click **Projects ▸ New Project**. In the Create New Project dialog that opens (Figure 13-1), make sure the Location box includes your home directory, set the Project name to **oop**, and then click the **Create** button.

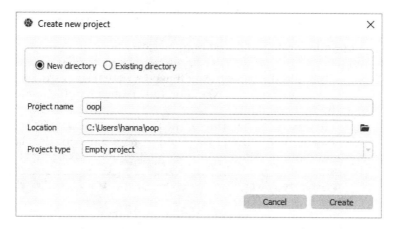

Figure 13-1: The Spyder Create New Project dialog

You should now see this new folder in Spyder's File Explorer pane.

For convenience, we'll use the default conda environments folder that's located within the *anaconda3* folder to hold third-party libraries. If you want to use an environments folder tied to this project, see the instructions in "Specifying an Environment's Location" on page 37.

Defining the Frigate Class

It's a lot easier to demonstrate OOP than it is to talk about it, so let's build some ship objects that might be used in a (very) simple war game simulator. Each unique type of ship will need its own class that can generate multiple versions of that ship type. We then can track and manipulate each of these objects independently. With OOP, the relationships among the ship class, the ship objects, and the methods that act on those objects will be clear, logical, and compact.

Let's begin by defining a class to build the most common type of warship, known as a "frigate" (Figure 13-2). Designed to be fast, maneuverable, and versatile, frigates escort and protect larger vessels from air, surface, and underwater threats.

Figure 13-2: Brazilian Tamandaré-class frigate

To build the frigates, you'll need a virtual shipyard, so, in your *oop* project, open up Spyder's text editor and create a new file named *ships.py*. Save it wherever you'd like.

To serve as a blueprint for the frigates, define a `Frigate` class using a `class` statement. After the `class` keyword, enter a name for the class followed by parentheses:

```
class Frigate(object):
❶ designation = 'USS'

❷ def __init__(self, name):
        self.name = name
        self.crew = 200
        self.length_ft = 450
        self.tonnage = 5_000
        self.fuel_gals = 500_000
        self.guns = 2
    ❸ self.ammo = self.guns * 300
        self.heading = 0
        self.max_speed = 24
        self.speed = 0
```

According to the PEP 8 Style Guide, class names should be capitalized. If you need to use multiple words, use the *CapWords* convention, where each new word is capitalized without spaces between them (also called *CamelCase*).

The `Frigate` class uses a single parameter, `object`. This `object` parameter represents the base class of all types in Python. Because `object` is the default parameter, you can omit stating it explicitly when defining a class.

Next, assign the string `USS`, for "United States Ship," to an attribute named `designation` ❶. This is the ship's name prefix. You can also use HMS ("Her Majesty's Ship"), INS ("Indian Naval Ship"), or whatever you'd prefer. In Python, an attribute is any variable associated with an object. Like in a function, a class makes a fresh local namespace for the attributes.

Classes are objects, too, so they can have their own attributes. *Class attributes* are common to all objects made from the class and behave sort of like global variables. In this case, all the frigates you build will have the "USS" designation, such as the "USS Saratoga." Class attributes are efficient, as they let you store a shared attribute in a single location.

Next, you define an *initialization* method ❷ that sets up the initial attribute values for an object. Methods are just functions that are defined within classes. The __init__() method is a special built-in method that Python automatically invokes as soon as a new object is created. In this case, it takes two parameters, self and the name of the object, which will be whatever you want to call the ship.

The __init__() method is a dunder *(double underscore) method, meaning its name is preceded and followed by double underscores. Also called* magic *or* special *methods, they let you create classes that behave like native Python data structures such as lists, tuples, and sets. They're also the magic behind operator overloading and behavior customization of other functions. When you call the built-in* len() *function, for example, a* __len__ *method is called behind the scenes.*

The first argument of every class method, including __init__(), is always a reference to the current instance of the class, called self by convention. (A new object is known as an *instance* of a class, and the process of setting the initial values and behaviors of the instance is called *instantiation*.)

You can think of self as a placeholder for the actual name you'll give an object. If you create a ship object and name it "Intrepid," self will become Intrepid. The self.speed attribute will become a reference to "Intrepid's speed." If you instantiate another ship object named "Indefatigable," self for that object will become Indefatigable. This way, the scope of the Intrepid object's speed attribute is kept separate from that of the Indefatigable object.

Now it's time to list some attributes for a frigate. You'll want to give each ship a name so that you can distinguish one ship from another. You'll also want to specify the value of key operational and combat characteristics, like fuel, heading, and speed. Because these are associated with an instance of the class, they're called *instance attributes*, and are assigned inside the __init__() method (with code such as self.name = name).

Some of these attributes, like the number of guns and the length of the ship, represent values common to each ship that shouldn't change over time. It's best not to make these *class* attributes, however, because they *could* change. For example, an individual ship could be outfitted with an extra experimental gun, or its helicopter pad could be extended off the stern. Other attributes such as the heading and speed represent placeholders that are *expected* to change. In general, you should set attributes to good default values, such as filling the fuel tank to capacity.

While it's possible to use methods to assign new attributes later, it's best to initialize them all within the __init__ method. This way, all the available attributes are conveniently listed in an easy-to-find location.

Take a moment to look through the list of attributes in the initialization method. For brevity, I've left off some that you might need in a true simulation, like the ship's current location, it's current "health," and a maximum limit for reverse speed. You might even want "cost" attributes for building and operating ships, forcing you to stay in budget.

Note also that you can use expressions when assigning attributes, just as you can with variables. For example, we've assumed the ship carries 300 rounds of ammunition for each of its big guns ❸.

Now let's define some methods for piloting the ship and firing its guns.

Defining Instance Methods

An *instance method* accesses or modifies the state of an object. They must have a self parameter to refer to the current object.

Let's define a helm() method to set the ship's heading and speed, and clip the speed to a maximum value. In the text editor, enter the following (the def statements should be indented four spaces relative to the class definition):

```
    def helm(self, heading, speed):
  ❶ self.heading = heading
        self.speed = speed
        if self.speed > self.max_speed:
          ❷ self.speed = self.max_speed
        print(f"\n{self.name} heading = {self.heading} degrees")
        print(f"{self.name} speed = {self.speed} knots")
```

In addition to the self parameter, you'll also pass the method a heading (between 0 and 359 degrees) and a speed (in knots).

The code within the method definition updates the object's existing attributes using the values passed to it as arguments. To access and change an attribute, use dot notation. You've used this syntax before to call methods in modules such as os and random. To override the heading and speed attributes initially assigned in the __init__() method, simply set them as equal to the values passed to the method ❶.

At this point, you'll want to validate the user inputs. To make sure the speed value doesn't exceed the ship's maximum speed. Compare the self .speed attribute to the self.max_speed attribute. If it's greater, set it equal to self.max_speed ❷. Complete the method by printing the heading and speed to the screen.

Now, let's define a method called fire_guns() that fires all the big guns at once. You won't need to pass this method any arguments other than self:

```
    def fire_guns(self):
      ❸ if self.ammo >= self.guns:
            print("\nBOOM!")
          ❹ self.ammo -= self.guns
            print(f"\n{self.name} ammo remaining = {self.ammo} shells")
        else:
            print("\nInsufficient ammunition!")
```

First, check that you're not out of ammunition ❸. If you're not, print "BOOM!" Then, decrement the self.ammo attribute by the number of guns (self.guns) ❹ and display the number of rounds left. Otherwise, print a message that you're out of ammunition.

One thing to note here is that the methods we've defined are not returning anything. Instead, they're changing attribute values in place. This behavior is very similar to the ill-advised technique of altering global variables within functions (see Chapter 12). What makes these methods acceptable, however, is that the attributes exist under the class umbrella rather than in the global namespace. Because changes are confined to the local namespace of the class, it's easier to track and debug issues than if you used global variables.

Instantiating Objects and Calling Instance Methods

We've defined a Frigate class and some methods for working with a Frigate object. Now, let's instantiate a ship and start using it. Add the following code, unindented, to *ships.py* and save the file:

```
garcia = Frigate('garcia')
print(f"\n----------{Frigate.designation} {garcia.name.upper()}----------")
print(f"\nCrew complement = {garcia.crew}")
garcia.fire_guns()
garcia.fire_guns()
garcia.helm(heading=180, speed=30)
```

This code first instantiates a new Frigate object named "Garcia" and assigns it to the garcia variable. It then prints the name of the ship, calling the built-in upper() method to print in uppercase characters.

Note that this code prints the "USS" designation by accessing the designation class attribute using the class name (Frigate.designation). You might have noticed that it would be easier to type "USS" here rather than access the designation attribute. We're using this attribute to demonstrate how class attributes work, but also to highlight a common problem with class attributes: in many cases, you can find an equally good alternative to avoid using them.

Next, the code prints the crew complement, using dot notation to access the crew attribute. Finally, it fires the guns twice and then changes the ship's direction and speed.

If you run the file, you should get this output:

```
----------USS GARCIA----------

Crew complement = 200

BOOM!

garcia ammo remaining = 598 shells

BOOM!
```

```
garcia ammo remaining = 596 shells

garcia heading = 180 degrees
garcia speed = 24 knots
```

Using our Frigate class template, we can create as many ships as we want. Let's make another one named "Boone." In the text editor, enter the following and save it:

```
boone = Frigate('Boone')
print(f"\n----------{Frigate.designation} {boone.name.upper()}----------")
boone.fire_guns()
boone.fire_guns()
boone.helm(heading=270, speed=-1)
```

Now, run it to see this output:

```
----------USS BOONE----------

BOOM!

Boone ammo remaining = 598 shells

BOOM!

Boone ammo remaining = 596 shells

Boone heading = 270 degrees
Boone speed = -1 knots
```

You now have two ships that use similar code but have different speeds and headings. With the Frigate class and OOP, you can easily create and track hundreds of ships with dozens of attributes.

TEST YOUR KNOWLEDGE

1. OOP makes code easier to read, maintain, and update by:

    a. Removing the need for functions

    b. Reducing code duplication

    c. Using methods instead of functions

    d. Providing bananas to gorillas

2. What is the name of an object created from a class:

    a. Child

    b. Attribute

    c. Instance

    d. Method

Defining a Guided-Missile Frigate Class Using Inheritance

Today, guns on warships have largely been replaced by missile systems (Figure 13-3). We can easily build new guided-missile frigates by simply refitting the existing Frigate class using the technique of inheritance.

Figure 13-3: The frigate HMS Iron Duke, firing her Harpoon anti-ship missile system

A key concept in OOP, *inheritance* lets you define a new child class based on an existing parent or ancestor class. (Technically, the original class is called a *base class* or *superclass.* The new class is called a *derived class* or *subclass.*) The new subclass inherits all of the attributes and methods of the existing superclass. This makes it easy to copy and extend an existing base class by adding new attributes and methods specific to the subclass.

Let's make a new guided-missile class called `GMFrigate` that inherits from and modifies our current `Frigate` class. Enter the following at the bottom of your *ships.py* program:

```
class GMFrigate(Frigate):
❶ designation = Frigate.designation

    def __init__(self, name):
    ❷ Frigate.__init__(Frigate, name)
    ❸ self.missiles = 100
        self.ammo = self.guns * 100

    def fire_missile(self):
        if self.missiles > 0:
            print("\nSSSSSSSSSSSStttttt!")
            self.missiles -= 1
            print(f"\n{self.name} missiles remaining = {self.missiles}")
        else:
            print("\nMissiles depleted")
        self.missiles -= 1
```

To create a child class, pass the `class` statement the name of the parent, or superclass, which in this case is `Frigate`. Remember that, when you first defined `Frigate`, you passed it `object`. This means `Frigate` inherited from the `object` class, which is the root of all Python objects. The `object` class provides the default implementation of common methods that all derived classes might need. By now passing `Frigate` instead of `object`, you get the attributes and methods under `object` as well as the new ones you added to the `Frigate` class.

Guided-missile frigates will have the same "USS" designation as frigates, so assign a `designation` class attribute to the same `Frigate` class attribute, referenced using dot notation ❶. You could skip doing this and just use the `Frigate.designation` attribute when you need it, but by explicitly reassigning the class attribute, you add clarity to the code.

Next, we define the `__init__()` initialization method for the `GMFrigate` class, which, like the `Frigate` class, has a `self` and `name` parameter. Immediately beneath it, we call the initialization method from the `Frigate` class ❷ and pass it `Frigate` instead of `self`, along with a `name` parameter. Passing in the `Frigate` class gives you access to all the attributes in the `Frigate.__init__()`method, so you don't need to duplicate any code, such as for crew, tonnage, guns, and so on.

If you don't define an `__init__()` method for a child class, it will use the `__init__()` method from the parent class. If you want to override some of the attribute values in the parent class, or add new attributes, you'll need to include an `__init__()` method for the child class, as we did in this example.

Our original frigate class did not allow for missiles, so add a new `self.missile` attribute ❸. Set the complement of missiles to 100. Because these missiles take up space, you have less room for other ammunition, so *override* the self.ammo attribute by setting it to 100x the number of guns rather than 300x, as we used before. Note that this won't affect the ammunition count for ships instantiated directly from the original Frigate class; they will use the superclass's ammunition setting.

Your ship will need a way to fire the missiles, so define a new method called fire_missile(), which will behave much like the fire_guns() that you defined earlier, but fires only one missile at a time.

Instantiating a New Guided-Missile Frigate Object

You can now instantiate a new guided-missile frigate. Let's name it "Ticonderoga":

```
ticonderoga = GMFrigate('Ticonderoga')
print(f"\n------{ticonderoga.designation} {ticonderoga.name.upper()}------")
for _ in range(3):
    ticonderoga.fire_guns()
ticonderoga.fire_missile()
ticonderoga.helm(95, 22)
```

This code generates the following output:

```
------USS TICONDEROGA------

BOOM!

Ticonderoga ammo remaining = 198 shells

BOOM!

Ticonderoga ammo remaining = 196 shells

BOOM!

Ticonderoga ammo remaining = 194 shells

SSSSSSSSSSSttttt!

Ticonderoga missiles remaining = 99

Ticonderoga heading = 95 degrees
Ticonderoga speed = 22 knots
```

By having your new class inherit attributes and methods from the Frigate class, you were able to follow the *DRY* ("don't repeat yourself") principle of software development, aimed at reducing the repetition of software patterns. You'll need to be careful not to make any changes to the Frigate class, however, unless you want those changes to be reflected in the GMFrigate class, as well.

Python permits the use of multiple inheritance, by which a child class inherits from more than one parent class. This is accomplished by passing the names of the parent classes, separated by commas, to the class definition. Using multiple parents is straightforward if none of the method names in the parent classes overlap. When they do, Python uses a process called Method Resolution Order (MRO) *to sort them out. This can be tricky, so in most cases, you'll want to stick to single inheritance, no inheritance, or cases where all parent classes contain distinct attribute and method names.*

Using the super() Function for Inheritance

The super() built-in function removes the need for an explicit call to a base class name when invoking base class methods. It works with both single and multiple inheritance. For example, in the GMFrigate class definition, you called the Frigate class's __init__() method within the GMFrigate class's __init__() method, as follows:

```
def __init__(self, name):
    Frigate.__init__(Frigate, name)
```

This lets the GMFrigate class inherit from Frigate. Alternatively, you could have used the super() function, which returns a proxy object that allows access to methods of the base class:

```
def __init__(self, name):
    super().__init__(name)
```

In this case, super() removes the need for an explicit call to the Frigate class. When using single inheritance, super() is just a fancier way to refer to the base type. It makes the code a bit more maintainable. For example, if you are using super() everywhere and want to change the name of the base class (such as from Frigate to Type26Frigate) you need to change the name only once, when defining the base class.

Another use for super() is for accessing inherited methods that have been overridden in a new class. Let's look at an example in which we define a Destroyer class that includes the guns found on a smaller corvette (another class of warship), plus some larger guns. Start a new *super_destroyer.py* file in the text editor and then enter the following:

```
class Corvette:
    def fire_guns(self):
        print('boom!')

❶ class Destroyer(Corvette):
    def fire_guns(self):
      ❷ super().fire_guns()
        print('BOOM!')
```

First, we define a Corvette class with a method for firing its guns. Because these guns are relatively small, they make a lowercase "boom." Next, we define a Destroyer class that inherits from Corvette class ❶. It has its own fire_guns() method that prints "BOOM!" for its large guns.

To fire the small guns available on the destroyer, use the super() function ❷. Because "super" refers to the *base* class, it calls the Corvette class's fire_guns() method.

Now, let's instantiate a corvette and destroyer object and fire their guns:

```
print('-----A Corvette-----')
corvette = Corvette()
corvette.fire_guns()

print('\n-----A Destroyer-----')
destroyer = Destroyer()
destroyer.fire_guns()
```

Here's the output. Notice that both versions of "boom" are printed by the destroyer object:

```
-----A Corvette-----
boom!

-----A Destroyer-----
boom!
BOOM!
```

NOTE   *The use of super() is somewhat controversial. On one hand, it makes code more maintainable. On the other, it makes it less explicit, which violates the Zen of Python edict "Explicit is better than implicit."*

Objects Within Objects: Defining the Fleet Class

Returning to our wargame simulation, let's create a Fleet class for manipulating all the ship objects we've been instantiating. That's right: using OOP, objects can control other objects.

In the editor, add the following code to the bottom of your *ships.py* file:

```
class Fleet():

    def __init__(self, name, list_of_ships):
        self.fleet_name = name
        self.ships = list_of_ships
        self.fleet_heading = 0
        self.fleet_max_speed = 0
        self.fleet_speed = 0
```

The initialization method for this class looks a lot like the one for the Frigate class, except now it has a parameter for a list of ships. This will be a list data type whose items are previously instantiated ship objects, like garcia and boone.

Now define some methods for the class:

```
def find_fleet_max_speed(self):
  ❶ max_speeds = [ship.max_speed for ship in self.ships]
    print(f'\nMaximum ship speeds = {max_speeds} knots')
  ❷ self.fleet_max_speed = min(max_speeds)
    print(f'Fleet maximum speed = {self.fleet_max_speed} knots')

def fleet_helm(self, heading, speed):
    self.fleet_heading = heading
    self.fleet_speed = speed
  ❸ if self.fleet_speed > self.fleet_max_speed:
        self.fleet_speed = self.fleet_max_speed
    print(f"\n{self.fleet_name} heading = {self.fleet_heading} degrees")
    print(f"{self.fleet_name} speed = {self.fleet_speed} knots")
    for ship in self.ships:
        ship.heading = self.fleet_heading
        ship.speed = self.fleet_speed
```

A fleet can travel no faster than its slowest ship, so define a method for setting the fleet's maximum speed, just as we did earlier for individual ships. The first step is to use list comprehension to loop through the ships in the self.ships list and append their maximum speeds, as found in the ship.max_speed attribute, to a new list named max_speeds ❶.

When the list is complete, you print it and then set the self.fleet_max _speed attribute to the maximum speed of the slowest ship, found by calling the built-in min() function on the list ❷. End the method by printing the fleet's maximum speed attribute.

Next, define a method for setting the heading and speed of the fleet. Again, this is similar to the technique we used for setting these values on an individual ship. As before, we clip the speed to the maximum speed limit, in the event that the user inputs an invalid speed ❸. We then print the information and loop through each ship in the self.ships list, setting its heading and speed.

Let's test the Fleet class by instantiating a "Seventh" fleet comprising the Garcia, Boone, and Ticonderoga ship objects created earlier. Enter the following and then save and run the program:

```
ships = [garcia, boone, ticonderoga]
seventh = Fleet("Seventh", ships)
print(f"\nShips in {seventh.fleet_name} fleet:")
for ship in seventh.ships:
    print(f"\t{ship.name.capitalize()}")

seventh.find_fleet_max_speed()
seventh.fleet_helm(42, 28)
print(f"\ngarcia helm = {garcia.heading, garcia.speed}")
```

```
print(f"boone helm = {boone.heading, boone.speed}")
print(f"ticonderoga helm = {ticonderoga.heading, ticonderoga.speed}")
```

This produces the following output:

```
Ships in Seventh fleet:
    Garcia
    Boone
    Ticonderoga

Maximum ship speeds = [24, 24, 24] knots
Fleet maximum speed = 24 knots

Seventh heading = 42 degrees
Seventh speed = 24 knots

garcia helm = (42, 24)
boone helm = (42, 24)
ticonderoga helm = (42, 24)
```

As all the ships are frigates, there's no difference in their maximum speeds, but if you had destroyers, aircraft carriers, and so on, you would see a mix of values in the max_speeds list.

With the Fleet class and its fleet_helm() method, you can simultaneously assign your ships the same heading and speed. You can also override these settings if you want by calling the self.helm() method of individual ships, like this:

```
garcia.helm(heading=50, speed=24)
print(f"\ngarcia helm = {garcia.heading, garcia.speed}")
print(f"boone helm = {boone.heading, boone.speed}")
print(f"ticonderoga helm = {ticonderoga.heading, ticonderoga.speed}")
```

Now, the Garcia's heading is different than those of the rest of the fleet:

```
garcia heading = 50 degrees
garcia speed = 24 knots

garcia helm = (50, 24)
boone helm = (42, 24)
ticonderoga helm = (42, 24)
```

Reducing Code Redundancy with Dataclasses

The built-in dataclass module introduced in Python 3.7 provides a convenient way to make classes less verbose. Although primarily designed for classes that store data, data classes work just like regular classes and can include methods that interact with the data. Some use cases include classes for bank accounts, the content of scientific articles, and employee information.

A dataclass comes with basic "boilerplate" functionality already implemented. You can instantiate, print, and compare dataclass instances straight out of the box, and many of the common things you do in a class, like instantiating properties based on the arguments passed to the class, can be reduced to a few basic instructions.

NOTE *Code linters will typically complain if you use more than seven or so instance attributes in a class. This seems to contradict the purpose of a dataclass, which is to store data. In addition, this limit can be difficult to honor in the scientific domain, where many attributes are often needed. Although the linter recommendations can be ignored, you should still strive to limit the number of instance attributes per class to reduce complexity. You might be able to treat some as class attributes, move others into parent classes, merge some into a single attribute, and so on.*

Using Decorators

Dataclasses are implemented using a helpful and powerful Python tool called a *decorator*. A decorator is a function designed to wrap around (encapsulate) another function or class to alter or enhance the wrapped object's behavior. It lets you modify the behavior without permanently changing the object. Decorators also let you avoid duplicating code when you're running the same process on multiple functions, such as checking memory use, adding logging, or testing performance.

Decorator Basics

To see how decorators work, let's define a function that squares a number. Then, we'll define a decorator function that squares that result. Enter the following in the console:

```
In [1]: def square_it(x):
   ...:     return x**2

In [2]: def square_it_again(func):
   ...:     def wrapper(*args, **kwargs):
   ...:         result = (func(*args, **kwargs))**2
   ...:         return result
   ...:     return wrapper
```

The first function, square_it(), takes a number, represented by x, and returns its square. The second function, square_it_again(), will serve as a decorator to the first function and is a little more complicated.

The decorator function has a func parameter, representing a function. Because functions are objects, you can pass a function to another function as an argument and even define a function within a function. When we call this decorator function, we'll pass it the square_it() function as an argument.

Next, we define an inner function, which we'll call wrapper(). Because square_it() takes an argument, we need to set up the inner function to handle arguments by using the special positional and keyword arguments *args and **kwargs.

Within the `wrapper()` function, we call the function we passed to the decorator (`func`), square its output, assign the resulting number to the `result` variable, and return `result`. Finally, we return the `wrapper()` function.

To use the `square_it_again()` decorator, call it, pass it the function that you want to decorate (`square_it()`), and assign the result to a variable (`square`), which also represents a function:

```
In [3]: square = square_it_again(square_it)
```

```
In [4]: type(square)
Out[4]: function
```

You can now call the new function and pass it an appropriate argument:

```
In [5]: print(square(3))
81
```

In this example, we manually called the decorator function. This demonstrated how decorators work, but it's a bit verbose and contorted. In the next section, we'll look at a more convenient method for using a decorator.

Decorator Syntactic Sugar

In computer science, *syntactic sugar* is clear, concise syntax that simplifies the language and makes it "sweeter" for human use. The syntactic sugar for a decorator is the @ symbol, which must be immediately followed by the name of the decorator function. The next line must be the definition statement for the function or class being wrapped, as follows:

```
@decorator_func_name
def new_func():
    do something
```

In this case, `decorator_func_name` represents the decorator function, and `new_func()` is the function being wrapped. A class definition can be substituted for the `def` statement.

To see how it works, let's re-create our number-squaring example. Use the arrow key to bring up the previously defined `square_it_again()` function in the console. Because we must invoke the decorator *before* defining the function to be wrapped, we must rewrite the code in the reverse order compared to the previous example.

Now, add the decorator and define the `square_it()` function in the next line. Note that, when using the @ symbol, you use the decorator function name with no parentheses:

```
In [7]: @square_it_again
   ...: def square_it(x):
   ...:     return x**2
```

To use the decorated function, simply call it and pass it a number:

```
In [8]: square_it(3)
Out[8]: 81
```

Notice that with the @ decorator we didn't need to use the square function, as in line In [3].

If decorators make your head spin a little, don't worry. If you can type @ dataclass, you can use dataclasses. This decorator modifies regular Python classes so that you can define them using shorter and sweeter syntax.

Defining the Ship Class

To see the benefits of dataclasses, let's define a regular class and then repeat the exercise using a dataclass. Our goal will be to make generic ship objects that we can track on a simulation grid. For each ship, we'll need to supply a name, a classification (like "frigate"), a country of registry, and a location.

Defining Ship as a Regular Class

To define a regular class called Ship, in the text editor, enter the following and then save it as *ship_tracker.py*:

```
class Ship:
  ❶ def __init__(self, name, classification, registry, location):
        self.name = name
        self.classification = classification
        self.registry = registry
        self.location = location
  ❷     self.obj_type = 'ship'
        self.obj_color = 'black'
```

This code looks a lot like the Frigate class we defined earlier. This time, however, the __init__() method includes more parameters ❶. All this data will need to be passed as arguments when instantiating an object based on this class.

Note how we're forced to duplicate code by repeating each parameter name, like classification, three times: once as a parameter and twice when assigning the instance attribute. The more data you need to pass to the method, the greater this redundancy.

In addition to the parameters passed to the initialization method, the Ship class includes two "fixed" attributes representing the object type ❷ and color. These are assigned using an equal sign, as with a regular class. Because these attributes are always the same for a given object, there's no need to pass them as arguments.

Now, let's instantiate a new ship object. Enter the following, save the file, and run it:

```
garcia = Ship('Garcia', 'frigate', 'USA', (20, 15))
print(garcia)
```

This created a US frigate named garcia at grid location (20, 15). But when you print the object, the output isn't very helpful:

```
<__main__.Ship object at 0x0000021F5FF501F0>
```

The issue here is that printing information on an object requires you to define additional dunder methods, like __str__ and __repr__, that return string representations of objects for informational and debugging purposes. Another useful method is __eq__, which lets you compare instances of a class. The list of special methods in Python is long, but a few basic examples are listed in Table 13-1.

Table 13-1: Basic Special Methods

Special Method	Description
__init__(self)	Called when initializing an object from a class.
__del__(self)	Called to destroy an object.
__repr__(self)	Returns a printable string for the object to use in debugging.
__str__(self)	Returns a string for pretty-printing useful information about an object. If not implemented, __repr__ is used instead.
__eq__(self, other)	Performs an equal to (==) comparison of two objects.

Defining these methods for each class you write can become a burden, which is where dataclasses come in. Dataclasses automatically handle the redundancy issues around attributes and dunder methods.

Defining Ship as a Dataclass

Now, let's define the Ship class again as a dataclass. Do this in a new file named *ship_tracker_dc.py* (for "ship tracker dataclass"):

```
from math import dist
from dataclasses import dataclass

@dataclass
❶ class Ship:
    ❷ name: str
    classification: str
    registry: str
    location: tuple
    ❸ obj_type = 'ship'
    obj_color = 'black'
```

Start by importing the math and dataclass modules. We'll use the dist method from math to calculate the distance between ships, and dataclass to decorate our Ship class. To use dist, you'll need Python 3.8 or higher.

Next, prefix dataclass with the @ symbol to make it a decorator. Define the Ship class on the following line to let the decorator know it's wrapping this class ❶.

Normally, the next step would be to define the __init__() method with self and other parameters, but dataclasses don't need this. The initialization is handled behind the scenes, removing the need for this code. You'll still need to list the attributes, however, but with a lot less redundancy than before.

For each attribute that must be passed as an argument, enter the attribute name, followed by a colon, followed by a *type hint* ❷. A type hint, or type *annotation*, tells people reading your code what types of data to expect. Static analysis tools can use type hints to check your code for errors. Type hints were introduced in PEP 484 (*https://www.python.org/dev/peps/pep-0484/*).

A class variable with a type hint is called a *field*. The @dataclass decorator examines classes to find fields. Without a type hint, the attribute won't become a field in the dataclass. In this example, all the fields in the Ship class use the string data type (str), except for location, which uses a tuple (for a pair of x, y coordinates). For a reminder of some common data types see Table 7-5 on page 184.

NOTE *You can use default values with the type annotations. For example, location: tuple = (0, 0) will place new Ship objects at coordinates x = 0, y = 0 if none are specified when the object is created. When you use a default parameter, however, all subsequent parameters must have default values.*

Because we don't need to pass the obj_type and obj_color attributes as arguments when creating a new object, we define them using an equal sign rather than a colon, and with no type hints ❸. By assigning them as we would in a regular class, every Ship object will, by default, be designated a "ship" and have a consistent color attribute for plotting.

Dataclasses can have methods, just like regular classes. Let's define a method that calculates the Euclidian distance between two ships. The def statement should be indented four spaces relative to the class definition:

```
def distance_to(self, other):
    distance = round(dist(self.location, other.location), 2)
    return str(distance) + ' ' + 'km'
```

The distance_to() method takes the current ship object and another ship object as arguments. It then uses the built-in dist method to get the distance between them. This method returns the Euclidean distance between two points (x and y), where x and y are the coordinates of that point. The distance is returned as a string, so we can include a reference to kilometers.

Now, in the global scope with no indentation, create three ship objects, passing them the following information:

```
garcia = Ship('Garcia', 'frigate', 'USA', (20, 15))
ticonderoga = Ship('Ticonderoga', 'destroyer', 'USA', (5, 10))
kobayashi = Ship('Kobayashi', 'maru', 'Federation', (10, 22))
```

As soon as you began entering the `Ship()` class arguments, a window should have appeared in the Spyder text editor, prompting you on the proper inputs (Figure 13-4).

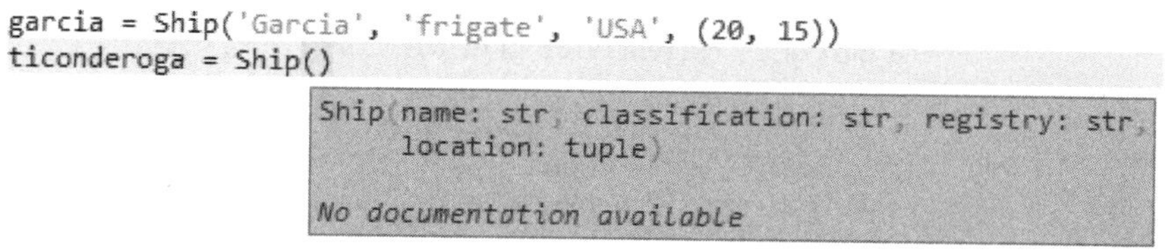

```
garcia = Ship('Garcia', 'frigate', 'USA', (20, 15))
ticonderoga = Ship()
```
```
Ship(name: str, classification: str, registry: str,
     location: tuple)

No documentation available
```

Figure 13-4: The Spyder text editor pop-up, showing the names and data types for the parameters in the Ship class.

Because classes you create are legitimate datatypes in Python, they behave like built-in datatypes. As a result, the Spyder editor will use the type hints to guide you when creating the ship objects. In the next chapter, we'll look at how to properly document classes so that the "No documentation available" message in Figure 13-4 is replaced with a one-line summary of the class, such as "Object for tracking a ship on a grid."

It's also worth noting that you don't need to use the correct data type for a parameter. Because Python is a *dynamically* typed language (see page 184 in Chapter 7), you can assign an integer as the `classification` argument, and the program will still run. Here's an example with the incorrect parameter highlighted in gray (don't add this to your code):

```
test = Ship('Test', 42, 'HMS', (15, 15))
```

NOTE *Even though the Python interpreter ignores type hints, you can use third-party static type-checking tools, like Mypy (https://mypy.readthedocs.io/), to analyze your code and check for errors before the program runs.*

The `@dataclass` decorator is a code generator that automatically adds methods under the hood. This includes the __repr__ method. This means that you now get useful information when you call `print(garcia)`:

```
print(garcia)
Ship(name='Garcia', classification='frigate', registry='USA', location=(20, 15))
```

Now, let's check that our data is there and the method works. Add the following lines and rerun the script:

```
ships = [garcia, ticonderoga, kobayashi]
for ship in ships:
    print(f"The {ship.classification} {ship.name} is visible.")
    print(f"{ship.name} is a {ship.registry} {ship.obj_type}.")
    print(f"The {ship.name} is currently at grid position {ship.location}\n")

print(f"Garcia is {garcia.distance_to(kobayashi)} from the Kobayashi")
```

By putting the ship objects in a list, we can loop through the list, access attributes using dot notation, and print the results:

```
The frigate Garcia is visible.
Garcia is a USA ship.
The Garcia is currently at grid position (20, 15)

The destroyer Ticonderoga is visible.
Ticonderoga is a USA ship.
The Ticonderoga is currently at grid position (5, 10)

The maru Kobayashi is visible.
Kobayashi is a Federation ship.
The Kobayashi is currently at grid position (10, 22)

Garcia is 12.21 km from the Kobayashi
```

The Ship dataclass lets you instantiate a ship object and store data such as the ship's name and location in type-annotated fields. By reducing redundancy and automatically generating required class methods such as __init__() and __repr__(), the @dataclass decorator lets you produce code that's easier to read and write.

NOTE   *The @classmethod and @staticmethod decorators let you define methods inside a class namespace that are not connected to a particular instance of that class. Neither of these are commonly used and can often be replaced with regular functions. You should be aware of their existence, however, as they're commonly mentioned in OOP tutorials and can be useful in some cases.*

Plotting with the Ship Dataclass

To get a better feel for how you might use OOP, let's take this project a step further and plot our ship objects on a grid. To plot the ships, we'll use the Matplotlib plotting library. (We look at Matplotlib in more detail later in the book.) To install the library in your base environment, open Anaconda Prompt (in Windows) or a terminal (in macOS or Linux) and enter the following:

```
conda activate base
conda install matplotlib
```

Enter **y** if prompted, and don't worry if you already have Matplotlib installed because Anaconda will just update the package, if needed.

NOTE   *If you're working in Spyder and aren't sure which conda environment is currently active, enter conda info in the console. This will display the active environment and its path.*

In the text editor, save or copy your *ship_tracker_dc.py* file to a new file called *ship_display.py* and edit it as follows:

```python
from math import dist
from dataclasses import dataclass
❶ import matplotlib.pyplot as plt


@dataclass
class Ship:
    name: str
    classification: str
    registry: str
    location: tuple
    obj_type = 'ship'
    obj_color = 'black'

    def distance_to(self, other):
        distance = round(dist(self.location, other.location), 2)
        return str(distance) + ' ' + 'km'


garcia = Ship('Garcia', 'frigate', 'USA', (20, 15))
ticonderoga = Ship('Ticonderoga', 'destroyer', 'USA', (5, 10))
kobayashi = Ship('Kobayashi', 'maru', 'Federation', (10, 22))

❷ VISIBLE_SHIPS = [garcia, ticonderoga, kobayashi]

❸ def plot_ship_dist(ship1, ship2):
    sep = ship1.distance_to(ship2)
    for ship in VISIBLE_SHIPS:
    ❹    plt.scatter(ship.location[0], ship.location[1],
                    marker='d',
                    color=ship.obj_color)
        plt.text(ship.location[0], ship.location[1], ship.name)
    ❺ plt.plot([ship1.location[0], ship2.location[0]],
             [ship1.location[1], ship2.location[1]],
             color='gray',
             linestyle="--")
    plt.text((ship2.location[0]), (ship2.location[1] - 2), sep, c='gray')
    plt.xlim(0, 30)
    plt.ylim([0, 30])
    plt.show()

❻ plot_ship_dist(kobayashi, garcia)
```

Start by adding a line to import Matplotlib ❶. After instantiating the three ship objects, replace the remaining code starting at line ❷. This line assigns a list of the three ship objects to the variable VISIBLE_SHIPS, which represents the ships you can see on the simulation grid. We'll treat this as a constant, hence the all-caps format.

Next, define a function for calculating the distance between two ships (ship1 and ship2) and for plotting all the visible ships ❸. Call the Ship class's distance_to() method on the two ships, assign the result to a variable named sep (for *separation*), and then loop through the VISIBLE_LIST, plotting each ship in a scatterplot ❹. For this, Matplotlib needs the ship's x and y locations, a marker style ('d' represents a diamond shape), and a color (the ship.obj_color attribute). Note how you can enter a return after the comma associated with each argument, for more readable "stacked" input.

Now, use Matplotlib's plt.plot() method to draw a dashed line between the ships used for the distance measurement ❺. This method takes the x–y locations of each ship, a color, and a line style. Follow this with the plt.text() method, for adding text to the plot. Pass it a location, the sep variable, and a color as arguments.

Complete the function by setting x and y limits to the plot and then calling the plt.show() method to display the plot. Back in the global scope, call the function and pass it the kobayashi and garcia ship objects ❻.

Save and run the file. You should see the plot shown in Figure 13-5.

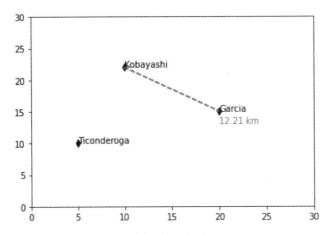

Figure 13-5: The output of the ship_display.py program

Bundling data and methods into classes produces compact, intuitive objects that you can manipulate en masse. Thanks to OOP, we could easily generate and track thousands of ship objects on our grid.

Identifying Friend or Foe with Fields and Post-Init Processing

Sometimes you'll want to initialize an attribute that depends on the value of another attribute. Because this other attribute must already exist, you'll need to initialize the second attribute outside the __init__ function. Fortunately, Python comes with the built-in __post_init__ function that's expressly designed for this purpose.

Let's look at an example based on our war game simulation. Because alliances can change through time, a ship registered to a certain country might switch from ally to enemy. Although the `registry` attribute is fixed, its allegiance is uncertain, and you might want to evaluate its friend-or-foe status post-initialization.

To create a version of the `Ship` dataclass that accommodates this need, in the text editor, enter the following and then save it as *ship_allegiance_post _init.py*:

```
from dataclasses import dataclass, field

@dataclass
class Ship:
    name: str
    classification: str
    registry: str
    location: tuple
    obj_type = 'ship'
    obj_color = 'black'
❶ friendly: bool = field(init=False)

❷ def __post_init__(self):
        unfriendlies = ('IKS')
        self.friendly = self.registry not in unfriendlies
```

In this case, we start by importing both `dataclass` and `field` from the dataclasses module. The `field` function helps you change various properties of attributes in the dataclass, such as by providing them with default values.

Next, we initialize the `Ship` class like we did in the *ship_tracker_dc.py* program, except that we add a new attribute, `friendly`, that's set to a Boolean data type with a default value of `False` ❶. Note that we set this default value by calling the `field` function and using the keyword argument init.

Now we define the `__post_init__()` method with `self` as a parameter ❷. We then assign a tuple of unfriendly registry designations to a variable named `unfriendlies`. Finally, we assign `True` or `False` to the `self.friendly` attribute by checking whether the current object's `self.registry` attribute is present in the `unfriendlies` tuple.

Let's test it out by making two ships, one friendly and one unfriendly. Note that you don't pass the `Ship` class an argument for the `friendly` attribute; this is because it uses a default value and is ultimately determined by the `__post_init__()` method:

```
homer = Ship('Homer', 'tug', 'USA', (20, 9))
bortas = Ship('Bortas', 'D5', 'IKS', (15, 25))

print(homer)
print(bortas)
```

This produces the following result:

```
Ship(name='Homer', classification='tug', registry='USA', location=(20, 9), friendly=True)
Ship(name='Bortas', classification='D5', registry='IKS', location=(15, 25), friendly=False)
```

You may have noticed that you didn't need to explicitly call the __post_init__() method. This is because the dataclass-generated __init__() code calls the method automatically if it's defined in the class.

NOTE   *Inheritance mostly works the same with dataclasses as with regular classes. One thing to be careful of is that dataclasses combine attributes in a way that prevents the use of* attributes *with defaults in a parent class when a child contains attributes without defaults. So, you'll want to avoid setting field defaults on classes that are to be used as base classes.*

Optimizing Dataclasses with __slots__

If you're using a dataclass for storing lots of data, or if you expect to instantiate thousands to millions of objects from a single class, you should consider using the class variable __slots__. This special attribute optimizes the performance of a class by decreasing both memory consumption and the time it takes to access attributes.

A regular class stores instance attributes in an internally managed dictionary named __dict__. The __slots__ variable stores them using highly efficient, array-related data structures implemented in the C programming language.

Here's an example using a standard dataclass called Ship, followed by a ShipSlots dataclass that uses __slots__. Enter this code in the text editor and save it as *ship_slots.py*:

```
from dataclasses import dataclass


@dataclass
class Ship:
    name: str
    classification: str
    registry: str
    location: tuple


@dataclass
class ShipSlots:
 ❶ __slots__ = 'name', 'classification', 'registry', 'location'

    name: str
    classification: str
    registry: str
    location: tuple
```

The only difference between the two class definitions is the assignment of a tuple of attribute names to the __slots__ variable ❶. This variable lets you explicitly state which instance attributes you expect your objects to have. Now, instead of having a *dynamic* dictionary (__dict__) that permits you to add attributes to objects after the creation of an object, you have a *static* structure that saves the overhead of one dictionary for every object that uses __slots__. Because it's considered good practice to initialize all of an object's attributes at once, the inability to dynamically add attributes with __slots__ is not necessarily a bad thing.

Using __slots__ with multiple inheritance can become problematic, however. Likewise, you'll want to avoid using it when providing default values via class attributes for instance variables. You can find more caveats in the official docs at *https://docs.python.org/3/reference/datamodel.html#slots/* and in this *Stack Overflow* answer at *https://stackoverflow.com/questions/472000/usage-of-slots/*.

Making a Class Module

In the previous chapter, we used modules to abstract away code. A program with one or more class statements can serve as a module, too, letting you use the classes without having to define them in your current code.

Let's walk through an example using the *ship_slots.py* program you made in the previous section. In the console, begin by importing the *ship_slots.py* program:

```
In [9]: import ship_slots as slots
```

Now you can use its classes as if you'd defined them in the console:

```
In [10]: garcia = slots.Ship('Garcia', 'frigate', 'USS', (10, 20))

In 11]: garcia
Out[11]: Ship(name='Garcia', classification='frigate', registry='USS',
location=(10, 20))
```

Creating a class module would be useful if you were building a complete war games simulation. You could turn the various ship, fleet, and display classes into modular class libraries and then import these modules when building individual simulations. This would let you focus on the code for the current simulation without encountering the "clutter" of the class statements.

If you forget class names or the arguments each class takes, just start instantiating a new object. Spyder will launch a pop-up window to prompt you on these values (Figures 13-6 and 13-7).

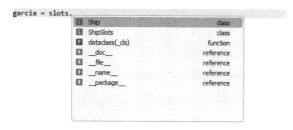

Figure 13-6: The Spyder pop-up window listing the classes in the ship_slots *module*

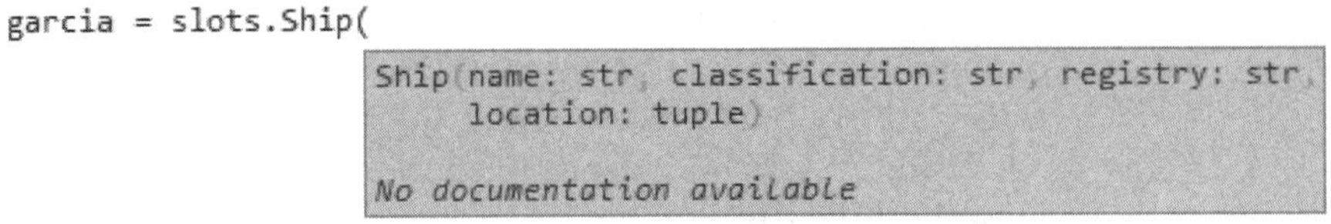

```
garcia = slots.Ship(
```

```
Ship(name: str, classification: str, registry: str,
     location: tuple)

No documentation available
```

Figure 13-7: The Spyder pop-up window listing parameters and data types used by the Ship class

These prompts will be less detailed if you're using regular classes instead of dataclasses. In the next chapter, you'll learn about documenting classes, and the documentation comment in Figure 13-7 will be useful.

TEST YOUR KNOWLEDGE

6. The super() built-in function removes the need for:

    a. Specifying the appropriate data type for each instance attribute

    b. Class attributes common to every class instance

    c. An explicit call to a base class name when invoking methods

    d. An initialization (__init__()) method

7. The dataclass was designed for when you have:

    a. Lots of methods that take lots of arguments

    b. A class with lots of attributes but few methods

    c. An initialization method that takes lots of arguments

    d. a. and b.

    e. b. and c.

8. The "syntactic sugar" symbol for a decorator is:

    a. #

    b. @

    c. **

    d. //

9. True or False: A type hint specifies the type of data you should use with an attribute.

10. Post-initialization processing is used to:

    a. Replace the __init__() method in a class definition

    b. Initialize attributes outside the __init__() method

    c. Optimize memory usage for large datasets

    d. Optimize processing speed when creating many objects

11. The class variable __slots__ reduces memory footprint by:

    a. Replacing the __dict__ dictionary normally used to store instance attributes

    b. Implementing C behind the scenes

    c. Using a dynamic dictionary versus a static data structure

    d. a. & b.

    e. b. & c.

12. Edit the *ship_display.py* program so that it moves the Garcia across the grid while continuously updating the distance to the Kobayashi (hint: you'll need a for loop).

Summary

Object-oriented programming helps you to organize code while reducing its redundancy. *Classes* let you combine related data, and functions that act on that data, into new custom data types.

Functions in OOP are called *methods*. When you define a class using a class statement, you couple related elements together so that the relationship between the data and the methods is clear, and so the proper methods are used with the appropriate data. Consequently, you'll want to consider using classes when you have multiple kinds of data, multiple functions that go with each kind of data, and a growing codebase that's becoming increasingly complex.

A class serves as a template or factory for making *objects*, also called *instances* of a class. You create objects by calling the class's name using function notation. As with regular functions, this practice introduces a new local name scope, and all names assigned in the class statement generate object *attributes* shared by all instances of the class. Attributes store data, and each object's attributes might change over time to reflect changes in the object's state.

Classes can *inherit* attributes and methods from other classes, letting you reuse code. In this case, the new class is a child or *subclass*, and the preexisting class is the parent or *base* class. Inherited attributes and methods can be overwritten in the subclass to modify or enhance the inherited behaviors. With the super() function, you can call original methods from a base class in the event that they've been modified in the subclass. Because Python lets classes inherit from multiple parents, this can result in complex code that's difficult to understand.

Decorators are functions that modify the behavior of another function without permanently changing the modified function. They also help you to avoid duplicating code. The @dataclass decorator decorates class statements and makes them more concise. Although *dataclasses* were designed for classes that mainly store data, they can still be used as regular classes. A downside, however, is that the use of multiple inheritance can be more difficult with dataclasses than with regular classes.

The __slots__ class variable optimizes both memory usage and attribute access speeds. It comes with some limitations, however, such as, but not limited to, the inability to dynamically create attributes after initialization and increased complexity when using multiple inheritance.

You can combine related class statements and save them as Python files. These *class libraries* then can be imported in other programs as modules. IDEs like Spyder will prompt users with the proper class names, arguments, methods, and documentation, removing the need to see all of the class definition code.

There's a lot more to OOP than what we've covered here; it *is* the entire jungle, after all. If you think your projects would benefit from OOP and want to explore the topic further, you can find the official Python tutorial on classes at *https://docs.python.org/3/tutorial/classes.html*, the official dataclass documentation at *https://docs.python.org/3/library/dataclasses.html*, and the PEP 557 dataclass enhancement proposal at *https://www.python.org/dev/peps/pep-0557/*.

14

DOCUMENTING YOUR WORK

Python is famous for the readability of its code, but this readability can take you only so far. To collaborate with others, and to remind yourself why you did what you did, you'll need to rely on natural human language to convey information, make your meaning as clear as possible, or explain the purpose of your program. Python enables this through comments and docstrings.

A *comment* is a non-executable annotation within a computer program. A *docstring*, short for *documentation string*, is a multiline string, unassigned to any variable, used to add documentation to Python modules, classes, methods, and functions. Together, comments and docstrings comprise code documentation.

Good documentation makes your intentions clear and saves future users (including yourself) both time and effort. There should be no reason to reverse engineer parts of the code or waste time trying to understand arcane arguments or numbers applied without context.

Proper documentation might also include lessons learned during programming and can flag potential problems such as those encountered when working across operating systems. These will let you pass on valuable knowledge and save others from discovering and dealing with these problems on their own.

Given that code generated in a console is usually temporary, you'll need only to document *persistent* programs, such as those generated in a text editor or Jupyter Notebook. These types of files are saved to disk and reused, sometimes months later, so it's important to record any intentions and assumptions that aren't clearly self-evident.

Comments

Comments are notes that you add to code to remind you of what you were doing, explain the purpose of a new block of code, flag a to-do item, and temporarily "turn off" code that you don't want to run. They're especially helpful when other people need to understand and modify your work.

Comments start with the hash (#) symbol, which tells Python to ignore (not execute) any remaining code on the same line. Here's an example:

```
# Step 1: Crop image to 50x50 pixels.
```

In consoles and text editors, comments display with a different color than regular code. If you're using the "Spyder" syntax highlighting theme (see "Configuring the Spyder Interface" on page 64), comments will be colored gray, and docstrings will be green.

Comments can occur on a single line, extend over multiple lines, or be embedded in a line of code. The latter are called *inline* comments.

Like variable names, comments should be as concise as possible, and it will take multiple iterations to get them right. If comments are too long or if there are too many, they'll become distracting, and users might ignore them. If they're too short and cryptic, their purpose will be wasted. If they're lacking, users might end up squandering time deciphering the code. And that user could be you!

Of course, you'll always want to avoid rude comments:

```
# Added this to fix Steve's stupid mistake.
```

Comments like this offend people, adversely affect teamwork, and make you look unprofessional.

Another commenting error is to violate the DRY (Don't Repeat Yourself) maxim and elaborate on code that's already readable and explicit. Here's an example of a redundant comment that adds no value and creates visual noise:

```
force = mass * acceleration  # multiply mass variable by acceleration variable.
```

The following comments state the obvious and clutter the code without adding much value, as the code itself is easy to understand:

```
# As Step 1, enter the mass of the object.
mass = 549
# As Step 2, enter the acceleration of the object.
acceleration = 42
# As Step 3, calculate Force.
force = mass * acceleration
```

The cryptic inline comment that follows was probably meant as a temporary reminder, but the coder forgot to remove it so now it adds confusion rather than clarity:

```
acceleration = 42  # Intermediate for now.
```

Along these lines, comments that contradict the code are worse than no comments at all. Consequently, you should keep comments up to date and address any code changes. This is difficult to do in practice and is a good argument for limiting the number of comments to those that are strictly necessary.

You can find the official Python guidelines for comments in the PEP8 *Style Guide for Python Code* at *https://pep8.org/*. Most of this content will be summarized in the sections that follow.

Single-Line Comments

A comment will often occupy a single line and summarize some code that follows, like this:

```
# Use Cartesian product to generate permutations with repetition.
for perm in product([0, 1, 2, 3, 4, 5, 6, 7, 8, 9], repeat=len(combo)):
```

Because users might not be familiar with the product function from the built-in itertools module, the comment saves them the effort of looking it up.

When writing single-line comments, you should insert a single space after the hash mark and use complete sentences with periods. If the comment contains multiple sentences, each period should be followed by two spaces. Comments should start with capital letters unless the first word is an identifier that begins in lowercase.

Additionally, all comments should be indented to the same level as the code they address. For example, because matter can't reach or exceed the speed of light (C), the following comment explains the purpose of reassigning the velocity variable to the speed of light minus 0.000001:

```
if velocity >= C:
    # Don't let the ship reach light speed.
    velocity = C - 0.000001
```

Because the referenced variable assignment occurs *within* the if statement block, the comment is indented four spaces.

Multiline Comments

Comments that span multiple lines are known as *multiline* comments or *block* comments. Python does not have an official syntax for multiline comments. One way to handle them is to treat them as a series of single-line comments beginning with hash marks, as follows:

```
# This is a really long-winded comment that probably should be
# shortened or left off or broken up and inserted before various
# bits of code or in a docstring somewhere.
```

The drawback to this method is that it's somewhat unreadable. An alternative is to use a multiline string with triple quotes. This works because Python ignores strings that aren't assigned to a variable. It's also more readable:

```
"""
This is a really long-winded comment that probably should be
shortened or left off or broken up and inserted before various
bits of code or in a docstring somewhere.
"""
```

You can also place the triple quotes on the same lines as the comments, as follows:

```
"""This is a really long-winded comment that probably should be
shortened or left off or broken up and inserted before various
bits of code or in a docstring somewhere."""
```

If a block comment contains more than one paragraph, separate the paragraphs with a blank line.

Block comments break up the continuity of code and should be used only in special circumstances. These include documenting important lessons learned, adding license and copyright information, and inserting temporary reminders such as TODO lists, FIXME flags, and warnings.

Inline Comments

An inline comment occurs at the end of a statement. A common use is to specify measurement units, as follows:

```
C = 299_792_458  # Speed of light in a vacuum in meters per second.
```

By including a comment to specify the value's unit instead of including that unit in the name of the variable, we were able to use a more concise variable name.

Inline comments should be separated from the code by at least two spaces, and the # should be followed by a single space. If the comment won't fit on the same line as the code, use a single line or multiline comment above the statement instead.

Inline comments are distracting and should be used sparingly. They should never state the obvious and should add clarity. For example, some functions and methods come with non-intuitive argument values, like the built-in turtle module's screen() method, which sets up a drawing window. Normally, you pass it the size of the window that you want, in pixels, such as width=800, height=900, but to use the whole screen, you just pass it 1. An inline comment can make this clear:

```
screen.setup(width=1.0, height=1.0)  # For fullscreen view.
```

An inline comment can also provide context to a variable assignment:

```
apogee = 25_500  # Highest point in the orbit.
```

Alternatively, the inline comment can provide a formatting tip:

```
url = https://www.python.org/  # Cut and paste from website address.
```

And here, the comment adds clarity to an argument in a user-defined function:

```
trajectory = rocket(dx=25, dy=-100)  # Negative y moves down the screen.
```

You'll be tempted to use inline comments far more than they're really needed. In most cases, they can be avoided or minimized by using clear object names.

Commenting-Out Code

Because Python ignores comments, you can use the # symbol to block the execution of some code. This can help you test and debug code by turning parts of it off and on.

For example, you might want your program to print out a good deal of information, but during development, these printouts can slow down the code and obscure other outputs that you want to see. While working on the code, you can *comment-out* these lines by turning them into comments, as follows:

```
# print(key_used)
# print(ciphertext)
# print(plaintext)
# print('Program complete.')
```

For convenience, you can highlight and comment-out blocks of code using keyboard shortcuts. In Spyder, you can see the shortcuts for your system by clicking **File ▸ Edit** on the top toolbar. For example, in Windows, you can toggle code on and off with CTRL-1. To comment out a block of code that includes an explanatory comment, use CTRL-4 to comment-out the block and CTRL-1 to restore it.

Docstrings

A docstring is a triple-quoted string literal that occurs as the first statement in a module, function, class, or method definition. Because of this positioning and the use of triple quotes, various types of help tools can discover and display docstrings.

Docstrings usually consist of a summary line followed by a more elaborate description:

```
"""
A one-line summary.

More info such as:
    function summaries
    method summaries
    attribute summaries
    exceptions raised
    and so on
"""
```

Because the summary line can be used by automatic indexing tools, it should fit on one line and be separated from the rest of the docstring by a blank line. The summary line can be on the same line as the opening quotes or on the next line. Unless the entire docstring fits on a single line, you should place the closing quotes on a line by themselves. The docstring should be indented to the same level as the quotes at its first line.

When docstrings are properly set up, you can access them with the special __doc__ attribute. To see an example using the pickle module that we used in Chapter 12, enter the following in the console:

```
In [1]: import pickle

In [2]: print(pickle.__doc__)
```

This will display the module's docstring:

```
Create portable serialized representations of Python objects.

See module copyreg for a mechanism for registering custom picklers.
See module pickletools source for extensive comments.

Classes:

    Pickler
    Unpickler

Functions:

    dump(object, file)
    dumps(object) -> string
    load(file) -> object
    loads(string) -> object

Misc variables:

    __version__
    format_version
    compatible_formats
```

You can also see this in Spyder's Help pane, by typing **pickle** in the Object box (Figure 14-1).

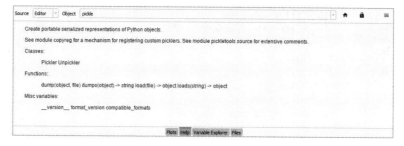

Figure 14-1: The pickle module docstring displayed in Spyder's Help pane

For simple functions or methods, the docstring can consist entirely of the one-line summary. Even though this summary doesn't span multiple lines, you should still use triple quotes, as follows:

```
"""Accept number as n and return cube of n."""
```

This is about as terse as a docstring can get, but it's sufficient for simple functions and functions you define for your own use. However, if you plan to work on enterprise-scale code or contribute to open source projects, you'll want to follow the instructions in PEP 257, which covers docstring conventions (*https://www.python.org/dev/peps/pep-0257/*). Some of these cases can be quite elaborate, with docstrings several screens long.

In the sections that follow, we'll look at docstring conventions appropriate for scientists and engineers working alone or in close groups. In these cases, users will be applying the code more often than modifying it, and simple docstrings should address their needs.

Documenting Modules

The docstring of a module should be placed at the top of the module above any import statements. The first line should describe the module's purpose. The rest of the docstring should generally list the classes, exceptions, functions, and any other objects that are exported by the module, with a one-line summary of each. It's okay if these summaries provide less detail than the summary line in the object's own docstring.

Here's how the pickle module's docstring looks in the actual code:

```
"""Create portable serialized representations of Python objects.

See module copyreg for a mechanism for registering custom picklers.
See module pickletools source for extensive comments.

Classes:

    Pickler
    Unpickler

Functions:

    dump(object, file)
    dumps(object) -> string
    load(file) -> object
    loads(string) -> object

Misc variables:

    __version__
    format_version
    compatible_formats
"""
```

As modules become larger and more complex, their docstrings can become quite technical. This makes them difficult for beginners and non-developers to both write and read. For programs written for your own use or for that of your immediate team, simpler summaries might be appropriate. Here's a friendly module docstring (in bold) for the *mymath.py* module we wrote in Chapter 11:

```
"""
Functions to solve the quadratic equation and get the volume of a sphere.

Functions:
quad(a, b, c) -> soln1, soln2
sphere_vol(radius) -> volume rounded to 2 decimal places
"""
import math

def quad(a, b, c):
    x1 = (-b - (b**2 - 4 * a * c)**0.5) / (2 * a)
    x2 = (-b + (b**2 - 4 * a * c)**0.5) / (2 * a)
    return x1, x2

def sphere_vol(r):
    vol = (4 / 3) * math.pi * r**3
    return round(vol, 2)
```

You can get this documentation using __doc__:

```
In [3]: import my_math

In [4]: print(my_math.__doc__)

Functions to solve the quadratic equation and get the volume of a sphere.

Functions:
quad(a, b, c) -> soln1, soln2
sphere_vol(radius) -> volume rounded to 2 decimal places
```

Likewise, the built-in help() function can retrieve this docstring with more information, including the location of the file:

```
In [5]: help(my_math)
Help on module my_math:

NAME
my_math - Functions to solve the quadratic equation and get the volume of a sphere.

DESCRIPTION
    Functions:
    quad(a, b, c) -> soln1, soln2
    sphere_vol(radius) -> volume rounded to 2 decimal places
```

```
FUNCTIONS
    quad(a, b, c)
    sphere_vol(r)

FILE
    C:\Users\hanna\spyder_proj_w_env\code\my_math.py
```

This docstring gives the user a nice overview of the my_math module. Don't worry that the description of the functions is a little sparse. As you'll see in a later section, functions get their own docstrings, in which you can expand on the function's purpose, parameters, outputs, and so on.

Documenting Classes

The docstring for a class should follow the same pattern as a module-level docstring. It should summarize the class behavior and list the public methods and instance variables. Any subclasses, constructors, and methods should have their own docstrings. You should insert a blank line after all docstrings that document a class.

Here's an example of a docstring for a Starship class:

```
class Starship:
    """
    A class to represent a starship.

    Attributes
    ----------
    name : str
        name of the ship
    torpedoes : int
        number of photon torpedoes
    phasers: int
        number of phaser banks
    crew: int
        number of crew members

    Methods
    -------
    info():
        Print the ship's attributes.

    fire_all():
        Return the sum of the weapon attributes as an integer.
    """
```

This docstring is simple, but that's okay because one of the main uses of docstrings is to provide dynamic hints when using the class (see Figure 13-4). Consequently, you'll want to present the information as concisely as possible.

In this case, we start the docstring with a one-line summary followed by a list of attributes. This listing includes the attribute name, its data type, and a brief description. Next, we list the class methods along with a single-line summary of each.

Documenting Functions and Methods

The docstring for a function or method should summarize its behavior and document its arguments, return values, and side effects as well as any exceptions raised and restrictions on when it can be called (if applicable). You should indicate optional and keyword arguments as such.

In general, if your function or method takes no arguments and returns a single value, a one-line summary should provide enough documentation. This summary should use *imperative mood*; in other words, use "Return" not "Returns":

```
def warning():
    """Print structural integrity warning message."""
    print("She canna take it Capt'n! She's gonna blow!")
```

Here's a longer docstring for a function that accepts two words and returns True if the words are anagrams (composed of the same letters in different orders) or False if they're not. It provides information on the function's arguments and return value:

```
def is_anagram(word1, word2):
    """
    Check if two strings are anagrams and return a Boolean.

    Arguments:
        word1: a string
        word2: a string

    Returns:
        Boolean
    """
    return sorted(word1.lower()) == sorted(word2.lower())

print(is_anagram('forest', 'softer'))
```

Here's the output of this code. Because "softer" is an anagram of "forest," the comparison returns True:

```
In [6]: runfile('C:/Users/hanna/oop/junk.py', wdir='C:/Users/hanna/oop')
True
```

If a function's arguments have default values, you should mention them. Here's an example using the tax_rate parameter:

```
def calc_taxes(taxable_income, tax_rate=0.24):
    """
```

```
        Calculate Federal taxes based on taxable income and rate.

        Args:
            taxable_income: int
                            Income after qualified deductions.
            tax_rate: float
                      Federal tax rate as decimal value.
                      Defaults to 24% tax bracket.

        Returns: int
                 Federal taxes owed.
        """
```

Keeping Docstrings Up to Date with doctest

It's easy to update a program and forget to edit the associated docstring. With the doctest built-in module, you can embed usage examples in docstrings to check whether there's a divergence between the code and its documentation.

The doctest module searches for pieces of text that look like interactive Python sessions and then executes those sessions to verify that they work exactly as shown. Let's look at a simple function that takes a starship's warp factor value and adjusts it so that it falls within acceptable operating limits. The code highlighted in bold represents the embedded test cases:

```
def warp(factor):
    """Return input warp factor adjusted to allowable values.

    Args:
        factor: int
                warp factor

    Returns: int
             warp factor adjusted to operating limits

    Raise: ValueError
           factor value must be float or integer

    >>> warp(5)
    5
    >>> warp(3.5)
    3
    >>> warp(12)
    10
    >>> warp(-4)
    0
    >>> warp(0)
    0
    >>> warp('ten')
    Traceback (most recent call last):
        ...
```

```
    ValueError: factor must be a number
    """
    if isinstance(factor, (int, float)):
        speed = int(factor)
        if speed < 0:
            speed = 0
        elif speed > 10:
            speed = 10
        return speed
    else:
        raise ValueError("factor must be a number")

if __name__ == "__main__":
    import doctest
    doctest.testmod()
```

The test cases check both acceptable and unacceptable values. Unacceptable values are those that will fail the comparison statements, such as 12 and -4.

You can run doctest in several ways. One way is by running the script from the text editor by pressing F5. Another is to open the console, import the doctest module and your custom module (without the *.py* extension), and call the testmod() method, as follows:

```
In [7]: import doctest
In [8]: import set_warp

In [9]: doctest.testmod(set_warp)
Out[9]: TestResults(failed=0, attempted=6)
```

Because none of the tests failed, you got an abbreviated summary of the test result. If you go back to the docstring and edit the expected result for warp(-4) from 0 to 4, you'll see this output when you rerun the method (remember to save your script first):

```
In [10]: doctest.testmod(set_warp)
**********************************************************************
File "C:\Users/hanna/file_play\set_warp.py", line 21, in set_warp.warp
Failed example:
warp(-4)
Expected:
4
Got:
0
**********************************************************************
1 items had failures:
1 of 6 in set_warp.warp
***Test Failed*** 1 failures.
Out[10]: TestResults(failed=1, attempted=6)
```

To print a detailed log of what the doctest module is trying, what it's expecting, and what it found, pass verbose=True to testmod(). Here's the result for a no-failure case:

```
In [11]: doctest.testmod(set_warp, verbose=True)
Trying:
    warp(5)
Expecting:
    5
ok
Trying:
    warp(3.5)
Expecting:
    3
ok
Trying:
    warp(12)
Expecting:
    10
ok
Trying:
    warp(-4)
Expecting:
    0
ok
Trying:
    warp(0)
Expecting:
    0
ok
Trying:
    warp('ten')
Expecting:
    Traceback (most recent call last):
        ...
    ValueError: factor must be a number
ok
1 items had no tests:
    set_warp
1 items passed all tests:
    6 tests in set_warp.warp
6 tests in 2 items.
6 passed and 0 failed.
Test passed.
```

You can also run doctest from Anaconda Prompt or a terminal. Just navigate to the directory that holds your Python file and run the following using the -v switch (for verbose mode):

```
python <your_filename.py> -v
```

Leave off the -v switch for a simple summary.

Besides checking that a module's docstrings are up to date, you can use doctest to verify that interactive examples from a test file or a test object work as expected. This is known as *regression testing*, and it ensures that previously developed and tested software still performs after a change.

You can also use doctest to write tutorial documentation for a package, liberally illustrated with input-output examples. To learn more, visit *https://docs.python.org/3/library/doctest.html*.

Checking Docstrings in the Spyder Code Analysis Pane

You can use the Spyder IDE to check how your docstrings conform to established guidelines. The results are presented in the code analysis pane, which was introduced on page 85 in Chapter 4, and within the text editor itself.

Setting Spyder Preferences

To set up Spyder to check docstrings, on the top toolbar, click **Tools ▸ Preferences**. In the Preferences window, click **Completion and linting**. Then, choose the **Docstring style** tab. You should see the window shown in Figure 14-2. Ensure that the **Enable docstring style linting** checkbox is selected.

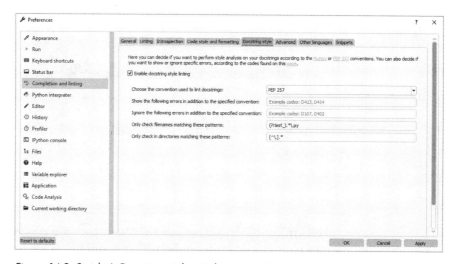

Figure 14-2: Spyder's Docstring style window

The **Choose the convention used to lint docstrings** drop-down menu offers you three choices: PEP 257, NumPy, and Custom. As previously discussed, PEP 257 is Python's official docstring guide, so we'll use it here.

In addition to PEP 257, some members of the scientific community use a NumPy docstring standard (*https://numpydoc.readthedocs.io/en/latest/install.html*). You can find examples of this style at *https://sphinxcontrib-napoleon.readthedocs.io/en/latest/example_numpy.html*).

You can also choose to show or ignore certain errors, based on the codes found at *http://www.pydocstyle.org/en/stable/error_codes.html*.

In addition to PEP 257 and NumPy, there are other docstring formats that you can follow. Google has its own format and an excellent style guide (https://google .github.io/styleguide/pyguide.html). You can see examples of this style at https://sphinxcontrib-napoleon.readthedocs.io/en/latest/example_google. html. In addition, reStructuredText is a popular format used mainly in conjunction with a tool called Sphinx. Sphinx uses docstrings to generate documentation for Python projects in formats such as HTML and PDF. If you've ever read the docs (https://readthedocs.org/) for a Python module, you've seen Sphinx in action.

Running the Analysis

To see how checking docstrings with Spyder works in practice, let's write some docstring-challenged code. Open the text editor, enter the following, and save it as *test_docs.py* (you can do this in the *oop* Spyder project from the previous chapter or somewhere else):

```python
class Volcano():
    'A volcano object'
    def __init__(self, name, classification, active):
        """sfsds"""
        self.name = name
        self.classification = classification
        self.active = active

    def erupt(self):
        'lsjljl'
        if self.classification == 'stratovolcano' and self.active is True:
            print("\nRUMBLE!\n")

    def pyroclastic_cloud(self):
        if self.classification == 'stratovolcano' and self.active is True:
            print("\nWHOOSH!\n")

mountain = Volcano('Krakatoa', 'stratovolcano', True)
mountain.erupt()
mountain.pyroclastic_cloud()
```

Hopefully, you've noticed several documentation errors here, but if you haven't, don't despair, Spyder will find and flag these for you. To start, on the top toolbar, click **Source**. This will produce the menu shown in Figure 14-3.

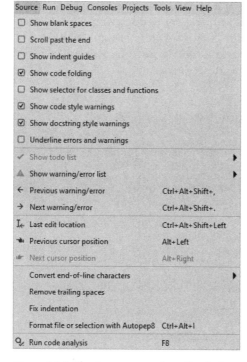

Figure 14-3: The Source menu from Spyder's
top toolbar

Be sure that the **Show docstring style warnings** checkbox is selected, and then, at the bottom of the menu, click the **Run code analysis** option (or press the F8 shortcut). The code analysis pane should appear (Figure 14-4).

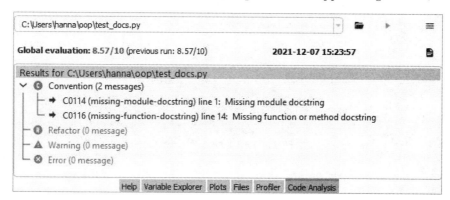

Figure 14-4: The code analysis pane with messages related to the docstrings

Click the right-facing arrow (>) symbol next to the **Convention** heading to expand the style messages for code and docstrings. In the *test_docs.py* example, we're missing two recommended docstrings: one for the entire program, called the module docstring, and one for the `pyroclastic_cloud` method.

The messages in the code analysis pane are from the code linting tool and lack granularity with respect to docstrings. To see specific documentation errors, hover your cursor over the orange triangles to the left of the line numbers in the text editor (Figure 14-5). You'll see multiple error codes along with their descriptions.

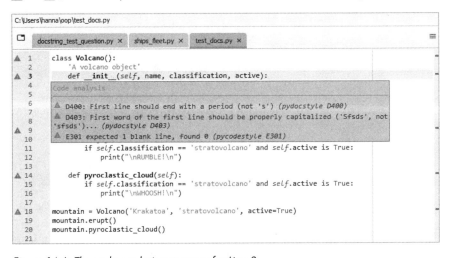

```
C:\Users\hanna\oop\test_docs.py

    docstring_test_question.py ×    ships_fleet.py ×    test_docs.py ×

▲ 1    class Volcano():
  2       Code analysis
▲ 3
  4         ▲ D100: Missing docstring in public module (pydocstyle D100)
  5         ▲ D204: 1 blank line required after class docstring (found 0) (pydocstyle D204)
  6         ▲ D300: Use """triple double quotes""" (found '-quotes) (pydocstyle D300)
  7         ▲ D400: First line should end with a period (not 't') (pydocstyle D400)
  8
▲ 9         def erupt(self):
  10            'lsjljl'
  11            if self.classification == 'stratovolcano' and self.active is True:
  12                print("\nRUMBLE!\n")
  13
▲ 14        def pyroclastic_cloud(self):
  15            if self.classification == 'stratovolcano' and self.active is True:
  16                print("\nWHOOSH!\n")
  17
▲ 18   mountain = Volcano('Krakatoa', 'stratovolcano', active=True)
  19   mountain.erupt()
  20   mountain.pyroclastic_cloud()
  21
```

Figure 14-5: The code analysis messages for Line 1

The message about the missing module docstring is repeated, but with a different error code than in the code analysis pane. Then, for the Volcano class definition, there are messages flagging a missing blank line, the incorrect use of quotes, and the need for a period at the end of the docstring.

If you hover your cursor over the Line 3 triangle, which refers to the __init__() method, you'll see a window with similar errors (Figure 14-6).

```
C:\Users\hanna\oop\test_docs.py

    docstring_test_question.py ×    ships_fleet.py ×    test_docs.py ×

▲ 1    class Volcano():
  2         'A volcano object'
▲ 3         def __init__(self, name, classification, active):
  4       Code analysis
  5
  6         ▲ D400: First line should end with a period (not 's') (pydocstyle D400)
  7         ▲ D403: First word of the first line should be properly capitalized ('Sfsds', not
  8         'sfsds')... (pydocstyle D403)
▲ 9         ▲ E301 expected 1 blank line, found 0 (pycodestyle E301)
  10
  11            if self.classification == 'stratovolcano' and self.active is True:
  12                print("\nRUMBLE!\n")
  13
▲ 14        def pyroclastic_cloud(self):
  15            if self.classification == 'stratovolcano' and self.active is True:
  16                print("\nWHOOSH!\n")
  17
▲ 18   mountain = Volcano('Krakatoa', 'stratovolcano', active=True)
  19   mountain.erupt()
  20   mountain.pyroclastic_cloud()
  21
```

Figure 14-6: The code analysis messages for Line 3

Note that the tool checks for the *presence* of a summary description, but it doesn't evaluate the *content* of the description. A nonsensical summary such as """sfsds""" will pass the test.

Spyder's code analysis tool is a great way to ensure that your code, and its documentation, conform to Python community standards.

TEST YOUR KNOWLEDGE

5. Which of the following help you access docstrings?

   a. The __doc__ special attribute

   b. help()

   c. The Spyder Help pane

   d. All the above

6. Import the built-in itertools module and get help on its product() method.

7. Which Python Enhancement Proposals provide guidance on code documentation?

   a. PEP 248

   b. PEP 8

   c. PEP 549

   d. PEP 257

8. Which of the following make docstrings accessible to automatic help tools?

   a. Use of triple quotes

   b. Description of input and output data types

   c. That it immediately follows a def statement

   d. Final triple quote followed by a space

9. Spyder's code analysis tool can check your docstrings for conformity to:

   a. PEP 8 conventions

   b. PEP 257 and Google conventions

   c. PEP 8 and Google conventions

   d. PEP 257 and NumPy conventions

10. Write a docstring for the Frigate class defined in the *ships.py* program from Chapter 13.

Summary

Good documentation maximizes the usability of your code as well as its maintainability over time. In the Python community, well-written code is synonymous with well-documented code. Comments and docstrings let you add human language to programs to fill in any explanatory gaps about their purpose, meaning, and usability.

Comments represent non-executable notes for annotating code or for temporarily commenting-out lines so that they don't run. You should use comments sparingly to explain your intent, capture important programming lessons learned, provide warnings, include legal information such as for license and copyright data, specify units, and so on. Most comments occupy a single line, or are placed inline, and start with a # symbol. Multiline comments can use triple-quotes for readability.

Docstrings are special triple-quoted strings that occur at the top of a module or immediately after a class or def statement. They provide a user with an overview of what the code does and how to use it, and you can access them using automatic help tools. You should use docstrings with every module, class, method, and function, and these should be kept up to date as the code changes.

Various tools can help you check that your docstrings are up to date and well formatted. With the built-in doctest module, you can embed testable cases in docstrings. These let you check that code updates don't change expected behaviors. They can also provide example use cases for new users. The Spyder IDE includes a tool for checking your docstrings against the PEP 257 and NumPy guidelines. The results are displayed in the code analysis pane and along the margin of the text editor, adjacent to the problem lines.

PART III

THE ANACONDA ECOSYSTEM

The Python scientific ecosystem is *robust*, which means it attempts to satisfy the same needs in multiple ways. As discussed in this book's introduction, this can lead to a bewildering maze of packages and tools for users to negotiate. It's easy to become overwhelmed by the literally thousands of Python packages available to you through conda and conda-forge.

Fortunately, only a few packages are considered essential to scientific work. These libraries form the basic Python ecosystem for scientific research and are shown, with a few overlapping, ancillary, and competing libraries, in Figure III-1 (PyCharm and Spyder, for example, are similar programming tools available through Anaconda).

At the core of this ecosystem is Python. In Figure III-1, at the outer edge, Anaconda coils around Python and the other libraries and tools to help you use them efficiently. Between Python and Anaconda are several rings, or layers, meant to convey that some libraries are built atop others.

The outer two rings hold tools that help you to write code, run code, and review the output. These include the (by-now-familiar) *Jupyter Qt console*, *Jupyter Notebook*, *Spyder*, and *JupyterLab*.

The innermost three rings hold some of the many scientific and plotting libraries available through Anaconda. We'll take a quick look at many of these in Part III so that you can become conversant on what they do, why you need them, and how you might choose among overlapping or competing versions. In Part IV, we'll take a deeper—though not comprehensive—dive into some of the more important ones.

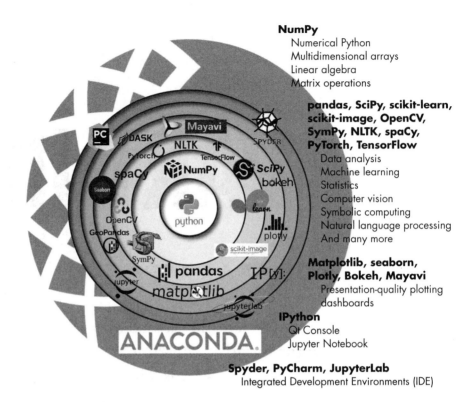

NumPy
Numerical Python
Multidimensional arrays
Linear algebra
Matrix operations

pandas, SciPy, scikit-learn, scikit-image, OpenCV, SymPy, NLTK, spaCy, PyTorch, TensorFlow
Data analysis
Machine learning
Statistics
Computer vision
Symbolic computing
Natural language processing
And many more

Matplotlib, seaborn, Plotly, Bokeh, Mayavi
Presentation-quality plotting
dashboards

IPython
Qt Console
Jupyter Notebook

Spyder, PyCharm, JupyterLab
Integrated Development Environments (IDE)

Figure III-1: Basic Python ecosystem for scientific research (after https://indranilsinharoy .com/, 2013)

15

THE SCIENTIFIC LIBRARIES

In this chapter, we'll look at high-level summaries of the core Python libraries for mathematics, data analysis, machine learning, deep learning, computer vision, language processing, web scraping, and parallel processing (Table 15-1). We'll also look at some guidelines for choosing among competing products. In subsequent chapters, we'll dive deeper into the functionality of several of these libraries and then apply them in real-world applications.

Table 15-1 organizes these libraries into subcategories, lists their websites, and provides a brief description of each. As these are popular and, in many cases, mature libraries, you should have no problem finding additional guidance for each online and in bookstores.

Table 15-1: Essential Python Scientific Libraries for Python

Task	Library	Description	Website
Math and data analysis	NumPy	Numerical computing tools for arrays	https://numpy.org/
	SciPy Library	Friendly and efficient numerical routines	https://www.scipy.org/
	SymPy	Symbolic math/computer algebra tools	https://www.sympy.org/
	Pandas	Data manipulation, analysis, and visualization tools	http://pandas.pydata.org/
Machine and deep learning	Scikit-learn	General-purpose machine learning toolkit	https://scikit-learn.org/
	TensorFlow	Symbolic math library for deep learning neural nets	https://www.tensorflow.org/
	Keras	Friendlier wrapper for TensorFlow	https://keras.io/
	PyTorch	Fast and efficient artificial neural networks	https://pytorch.org/
Image processing	OpenCV	Real-time computer vision library	https://opencv.org/
	Scikit-image	Scientific image processing and analysis tools	https://scikit-image.org/
	Pillow	Basic image processing tools	https://python-pillow.org/
Language processing	NLTK	Symbolic and statistical language processing library	http://www.nltk.org/
	spaCy	Fast production grade language processing library	https://spacy.io/
Helper libraries	requests	Webscraper for HTTP requests	https://pypi.org/project/requests/
	BeautifulSoup	Tools to extract text from HTML and XML files	https://www.crummy.com/software/
	re	Library for working with regular expressions	https://docs.python.org/3/library/re.html
	Dask	Library for parallel computing with Python	https://dask.org/
	Spark	"Heavier" alternative to Dask for Big Data	https://spark.apache.org/

The SciPy Stack

The SciPy stack of open source libraries comes preinstalled on Anaconda and includes NumPy, the SciPy library, Matplotlib, IPython, SymPy, and pandas (Figure 15-1). These have been called "the bedrock of number-crunching and visualization in Python" and are among the most used scientific libraries.

 **NumPy**
Base *N*-dimensional
array package

 **SciPy Library**
Fundamental
library for scientific
computing

 **Matplotlib**
Comprehensive
2D plotting

IP[y]:
IPython
iPython
Enhanced
interactive console

 SymPy
Symboloic
mathematics

 pandas
Data structure
and analysis

Figure 15-1: The core components of the SciPy ecosystem (courtesy of https://SciPy.org)

In the following sections, we take a high-level look at these libraries. Then, in later chapters, we take deeper dives into NumPy, Matplotlib, and pandas.

NumPy

Short for *Numerical Python*, NumPy is Python's dedicated library for performing numerical calculations. It supports the creation of large, multidimensional arrays and matrices and provides a large collection of high-level mathematical functions to operate on these arrays. NumPy is considered a basic package for scientific computing with Python, but I would also call it *foundational* because many other important libraries such as pandas, Matplotlib, SymPy, and OpenCV are built on top of it.

NumPy includes data structures, algorithms, and "glue" needed for most scientific applications involving numerical data. Operations in NumPy are faster and more efficient than competing functionality in the Standard Library that ships with Python. Having knowledge of NumPy is important to being able to use most, if not all, scientific Python packages, so we'll take a closer look at it in Chapter 18.

SciPy

The scientific library *SciPy* is designed for mathematics, science, and engineering, and addresses many standard problem domains in scientific computing. It's built on and supplements NumPy and provides many user-friendly and efficient numerical routines, such as routines for numerical integration, interpolation, optimization, linear algebra, statistics, fast Fourier transforms, signal and image processing, and the solving of differential equations. It extends the linear algebra routines and matrix decompositions provided in NumPy and provides access to many physical constants and conversion factors.

SymPy

SymPy is an open source library for symbolic mathematics. Its goal is to be a full-featured *computer algebra system (CAS)*.

Whereas most computer algebra systems invent their own language, SymPy is written and executed in Python. This makes it easier for those familiar with Python to use. It also allows you to use it as a library. So, in addition to using SymPy in an interactive environment, you can import it into your own Python application, where you can automate or extend it.

SymPy gives you the ability to do all sorts of computations symbolically. It can simplify expressions; compute derivatives, integrals, and limits; solve equations; work with matrices, and more. It includes packages for plotting, printing (including pretty printed output of math formulas or LaTeX), code generation, physics, statistics, combinatorics, number theory, geometry, logic, and more.

A simple way to appreciate SymPy is to consider the irrational number $\sqrt{8}$, calculated with Python's basic `math` library:

```
In [1]: import math

In [2]: math.sqrt(8)
Out[2]: 2.8284271247461903
```

The output is a truncated numeric answer, as $\sqrt{8}$ can't be represented by a finite number. With SymPy, the square roots of numbers that are not perfect squares are left unevaluated by default; thus, the symbolic results are symbolically simplified by default (as $2 \times \sqrt{2}$):

```
In [2]: import sympy
   ...: sympy.pprint(sympy.sqrt(8))
Out[2]: 2·√2
```

As stated previously, SymPy includes lots of useful methods, such as for solving equations. For example, to solve $x^2 - 2 = 0$

```
In [3]: import sympy

In [4]: x = sympy.symbols('x')

In [5]: sympy.pprint(sympy.solve(x**2 - 2, x))
[-√2, √2]
```

SymPy conveniently comes with its own plotting modules:

```
In [6]: from sympy import symbols, cos
   ...: from sympy.plotting import plot3d

In [7]: x, y = symbols('x y')

In [8]: plot3d(cos(x * 2) * cos(y * 4) - (y / 4), (x, -1, 1), (y, -1, 1))
```

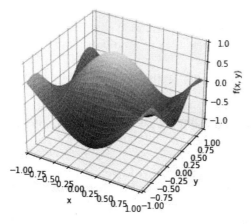

To see more of SymPy's capability, go to *https://docs.sympy.org/latest/tutorial/index.html.*

You might be wondering, why use SymPy when there's NumPy and the SciPy Library? The short answer is that SymPy is for working with algebra and doing theoretical math or physics; NumPy and SciPy are for performing analyses on actual data.

pandas

The *Python Data Analysis* library is the most popular open source library for data science. Called *pandas* for short, it contains data structures and manipulation tools designed to facilitate data extraction, cleaning, analysis, and visualization. It adopts significant parts of NumPy and works well with other libraries like SciPy, statsmodels, scikit-learn, and Matplotlib.

In addition, pandas is very useful for working with tabular data and common data sources, such as SQL relational databases and Excel spreadsheets. It's especially well suited for handling time-indexed data and incorporates plotting functionality—based on Python's core visualization library, Matplotlib—that makes it easy to visualize your data.

The most common data structure in pandas is the *DataFrame*, a tabular format similar to a spreadsheet, with columns, rows, and data. You can construct DataFrames from many types of input, in the case shown in the following example, from a list of lists using Jupyter Notebook:

```
import pandas as pd
data = [['Carbon', 'C', 6], ['Nitrogen', 'N', 7], ['Oxygen', 'O', 8]]
df = pd.DataFrame(data, columns=['Element', 'Symbol', 'Atomic #'])
df
```

	Element	Symbol	Atomic #
0	Carbon	C	6
1	Nitrogen	N	7
2	Oxygen	O	8

With DataFrames, you have the equivalent of an Excel spreadsheet or SQL table in Python. DataFrames, however, tend to be faster, easier to use, and more powerful because they're an integral part of the Python and NumPy ecosystems.

The pandas library is one of the most important for scientists, who, as the old joke goes, spend 80 percent of their time finding and preparing data and the other 20 percent complaining about it! Mastering pandas is therefore essential, and you'll get a good start in Chapter 20, which covers some of the basics.

NOTE   *Other libraries are beginning to challenge pandas by preserving its simplicity while addressing some of its efficiency issues such as the inability to scale projects through use of multicore processing, GPU processing, or cluster computing. Modin provides full drop-in replacement for pandas, letting you use pandas with more access to optimizations. Vaex (https://vaex.io/) helps you to explore and visualize large datasets on normal hardware by using efficient lazy evaluation and clever memory mapping. Dask (https://dask.org/) implements many of the same methods of pandas and offers more functionality than Modin or Vaex. Dask is more complex to use but helps you handle huge datasets and use computer clusters for improved processing speeds.*

A General Machine Learning Library: scikit-learn

Part of data analysis is the construction and validation of *predictive models* that use known results to forecast future outcomes or explain past behaviors. This falls under the category of *machine learning,* itself a category of artificial intelligence (Figure 15-2). Machine learning deals with methods for pattern recognition in datasets, making it possible for machine learning algorithms to improve automatically through experience. These algorithms build *supervised* models based on training data, and *unsupervised* models, in which the model "discovers" patterns on its own. The algorithms can use these models to make decisions without being explicitly programmed to do so.

The open source scikit-learn library is built on NumPy, SciPy, and Matplotlib. Considered the premier general-purpose machine learning toolkit for Python programmers, scikit-learn has been critical for enabling Python to be a productive data science tool. Preinstalled on Anaconda, scikit-learn is also easy to use, making it a great entry point to machine learning.

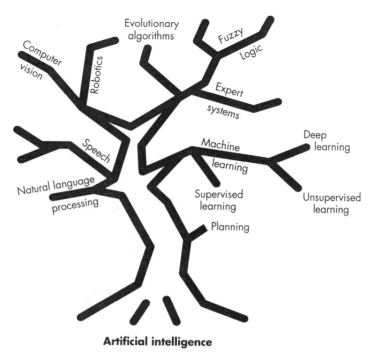

Artificial intelligence

Figure 15-2: Some of the branches of artificial intelligence

As shown in Figure 15-3, the scikit-learn library includes packages for predictive data analysis, including classification (support for vector machines, random forests, nearest neighbors, and so on), regression, clustering (*k*-means, spectral, and so forth), dimensionality reduction (principal component analysis, matrix factorization, feature selection, and more), preprocessing (feature extraction and normalization), and model selection (metrics, cross-validation, and grid search). Both supervised and unsupervised methods are addressed. You can get a feel for how scikit-learn works in Chapter 20.

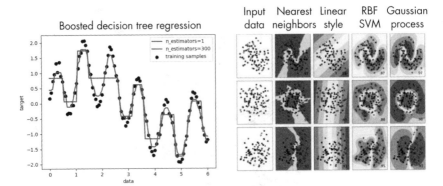

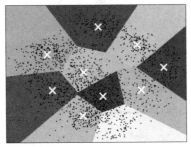

k-means clustering on the digits dataset (PCA-reduced data).
Centroids are marked with white cross.

Figure 15-3: Examples of regression, classification, and clustering analysis using scikit-learn (courtesy of https://scikit-learn.org/)

NOTE   *Complementing scikit-learn is a library called* statsmodels *(*https://www.statsmodels .org/*), which contains algorithms for classical statistics and economics. Whereas scikit-learn is more concerned with predictions, statsmodels is more about statistical inference,* p-*values, and uncertainty estimates.*

The Deep Learning Frameworks

Deep learning is a branch of machine learning that goes beyond the methods incorporated in scikit-learn. Rather than modify parameters of a fixed model through structured training sets *intended* for learning, deep learning networks are capable of learning *unsupervised* from data that is unstructured or unlabeled. Thus, deep learning systems imitate the nonlinear workings of the human brain in processing data and creating patterns for use in decision making.

The best-known of these systems are referred to as *artificial neural networks (ANNs)*. These networks generally require very complex mathematical operations with millions to billions of parameters and are only possible thanks to the speed and efficiency of graphics processing units (GPUs) developed for videogames. Example applications include self-driving cars and Google Translate.

Python libraries designed for deep learning are referred to as deep learning *frameworks*. These interfaces abstract away the details of the underlying algorithms, allowing you to define models quickly and easily using collections of prebuilt and optimized components. Of the many frameworks available, three dominate: TensorFlow, Keras, and PyTorch.

Although these three systems are still evolving, they already have good documentation, training sets, tutorials, and support, and you can count on all three to provide robust deep learning solutions.

TensorFlow

The oldest, most popular deep learning framework for Python is an open source library called *TensorFlow*. Created by Google to support its large-scale applications, TensorFlow is an end-to-end platform for multiple machine learning tasks. Thanks to its large user base, good documentation, and ability to run on all major operating systems, TensorFlow is popular in industry and academia, across a variety of domains. You can find many articles on the web to help you implement solutions to complex problems. You can also complete a certification program online.

TensorFlow is very powerful and can handle large datasets efficiently by distributing the computations across hundreds of mutli-GPU servers. It provides lots of functionality, including a tool called *TensorBoard* that helps you to create beautiful visualizations that are easy to understand and from which you can derive useful analytics.

Keras

Although TensorFlow is well documented and comes with walkthroughs to guide you, it's still considered one of the most challenging deep learning frameworks, with an intricate interface and a steep learning curve. Fortunately, François Challet, a Google engineer, has written another library, *Keras*, that acts as an interface for TensorFlow. Although it's now part of TensorFlow's core application programming interface (API), you can also use Keras in a stand-alone manner.

Like TensorFlow, Keras is open source and works on all platforms. Unlike TensorFlow, Keras is written in Python, which makes it more user friendly. Designed for quick prototyping and fast experimentation on smaller datasets, its lightweight, simplistic interface takes a minimalist approach, making it easy to construct neural networks that can work with TensorFlow. And because Keras can act as a wrapper, you can always "drop down" into TensorFlow when you need to use a feature not included in Keras's simpler interface.

Keras runs seamlessly on both CPUs and GPUs. It is primarily used for classification, speech recognition, and text generation, summarization, and tagging. By minimizing actions and making models easy to understand, Keras is a great deep learning tool for beginners.

PyTorch

PyTorch, developed by Facebook's AI Research Lab, is a direct competitor to TensorFlow. PyTorch works on all platforms and has recently incorporated *Caffe*, a popular deep learning framework developed at Berkeley and geared toward the image processing field. PyTorch comes preinstalled on Anaconda.

PyTorch is becoming the preferred framework for academic research, though it's still widely used among industry giants such as Facebook, Microsoft, and Wells Fargo. It excels at prototyping and is great for projects that have more of a non-production implementation. Among its strengths are flexibility, debugging capabilities, and short training durations.

Unlike TensorFlow, PyTorch is described as feeling more "native" to Python, making it easy to develop and implement machine learning models. Its syntax and application are very Pythonic, and it integrates seamlessly with essential libraries like NumPy. Although Keras seems to hold the upper hand with respect to ease of use, especially for those new to deep learning, PyTorch is faster, more flexible, and has better memory optimization.

Another strength of PyTorch is debugging. Keras hides a lot of the nitty-gritty details of building a neural network through encapsulation into various functions. This means that you can build an artificial neural network with only a few lines of code. With PyTorch, you need to specify a lot more details explicitly in your code; thus, finding an error becomes a much simpler task. It's also simpler to change weights, biases, and network layers and rerun the model.

Overall, PyTorch is considered easier to use than TensorFlow, but harder to use than Keras. TensorFlow's visualization capabilities are also held in higher esteem.

CHOOSING A DEEP LEARNING FRAMEWORK

According to Mark Twain, "All generalizations are wrong including this one." Many personal and project-related issues can affect which deep learning framework you choose. Still, if you search the internet enough, you can find some general guidelines for the selection of a deep learning framework:

- If you're brand new to deep learning, consider Keras, followed by PyTorch.
- If you're new and part of a research community, consider PyTorch.
- If you're an experienced researcher, you'll probably prefer PyTorch.
- Developers wanting a quick plug-and-play framework will prefer Keras.
- If you're experienced and want an industry job, consider TensorFlow.
- If you're working with large datasets and need speed and performance, choose either PyTorch or TensorFlow.

- If debugging is a concern, use PyTorch, as standard Python debuggers can be used (though with Keras, debugging is seldom needed due to the simple interface).
- If you need multiple backend support, choose Keras or TensorFlow.
- Keras and TensorFlow provide more deployment options and simplify model export to the web; with PyTorch you must use Flask or Django as the backend server.
- For fast prototyping, use Keras, followed by PyTorch.
- If visualization is a priority, choose Keras or TensorFlow.
- If you're already working with Keras or TensorFlow, use Keras for deep neural networks and TensorFlow for machine learning applications.

The Computer Vision Libraries

Computer vision is a branch of artificial intelligence focused on training computers to see and process digital images and videos in much the same way as human vision. The goal is for computers to gain a high-level understanding of the state of the world from images and return appropriate outputs. For example, a self-driving car should detect when you've drifted out of your lane and either warn you or automatically steer the car back. This requires detecting, tracking, and classifying features in images. In addition to autonomous cars, common applications include face detection and recognition, skin cancer diagnoses, event detection, and camera autofocusing.

There are quite a few Python libraries dedicated to computer vision and image manipulation, but three, OpenCV, scikit-image, and Pillow, should easily cover most of your needs. Let's take a quick look at these in the following sections.

OpenCV

OpenCV, short for *Open-Source Computer Vision*, is the world's most popular open source computer vision library. Its key focus is on real-time applications, like identifying faces in streaming video, but it can do everything from simple image editing to machine learning applications. OpenCV is written in C++ for speed but has a Python wrapper that works on Windows, Linux, Android, and macOS.

OpenCV has a modular structure that includes thousands of optimized algorithms, including ones for simple image processing, video analysis, 2D feature framework, object detection, object tracking, camera calibration, 3D reconstruction and more. OpenCV converts images into efficient NumPy arrays and, because it's written in optimized C/C++, it can take advantage of fast multicore processing.

OpenCV has been around for more than 20 years and has a large and supportive user base. Many major companies such as Google, Yahoo, Microsoft, Intel, IBM, Sony, and Honda actively use OpenCV. Thanks to its maturity and popularity, you can find many books and online tutorials to help you use the library.

scikit-image

The open-source *scikit-image* library is the image processing toolbox for SciPy. Its mission is to be *the* reference library for scientific image analysis in Python. It comes preinstalled with Anaconda.

The scikit-image library includes lots of algorithms and utilities for use in industry, research, and education. It's written in Python and, like OpenCV, uses NumPy arrays as image objects by transforming the original pictures. Although it lacks some of the sophisticated OpenCV algorithms for working with images in real time, it still has a lot of algorithms useful for scientists, including feature and blob detection. It also contains a few algorithm implementations that OpenCV does not.

The library is fairly easy to use and well documented with lots of examples and use cases. All of the code is peer reviewed and of high quality. It provides a consistent interface to many machine-learning models, making it relatively easy to learn a new model. It also provides many options—with sensible defaults—for tuning the models for optimal performance. You can find a gallery of examples at *https://scikit-image.org/docs/stable/auto_examples/*.

PIL/Pillow

Pillow is the "friendly" fork of the *Python Image Library (PIL)*, one of the oldest core libraries for image manipulation in Python. Pillow runs on all major operating systems, comes preinstalled on Anaconda, and is primarily designed for basic image processing.

If you don't need functionality from OpenCV or scikit-image, Pillow is widely used for image transformations in web projects given that it is more lightweight and usable. It supports a large selection of image file types and predefined image enhancement filters for sharpening, blurring, contouring, smoothing, finding edges, resizing, manipulating pixels, and more. It's especially useful for automatically processing large numbers of images.

CHOOSING AN IMAGE MANIPULATION LIBRARY

Here are some tips for choosing a library for manipulating images. Only open source libraries are considered.

- If your job or research involves computer-vision applications in *real time*, you'll want to learn OpenCV.

- If your datasets include a mixture of static images and streaming video, you should consider both OpenCV and scikit-image. Some of the latter's methods and utilities can complement OpenCV. For a short example of how these two can work together, visit Adrian Rosebrock's tutorial on detecting low-contrast images (*https://www.pyimagesearch.com/2021/01/25/detecting-low-contrast-images-with-opencv-scikit-image-and-python/*).

- If you mainly work with static images, scikit-image or Pillow should suffice and save you all the "overhead" of OpenCV. Between the two, scikit-image will be more appropriate if you regularly work with images and perform fairly sophisticated analyses and manipulations.

- For basic image manipulation, such as loading images, cropping images, or simple filtering, Pillow should be sufficient. Likewise, you can realize a large number of simple operations directly within NumPy and SciPy's ndimage module.

The Natural Language Processing Libraries

Natural language processing (NLP) is a branch of linguistics and artificial intelligence concerned with giving computers the ability to derive meaning from written and spoken words. Some familiar applications of NLP include speech recognition; text-to-speech conversion; machine translations; chatbots; spam detection; word segmentation (called tokenization); sentiment analysis, optical character recognition (OCR), in which an image of handwriting or printed text is converted into digital text; and, of course, Amazon's Alexa.

Among the more popular NLP libraries are NLTK, spaCy, Gensim, Pattern, and TextBlob. NLTK and spaCy are all-purpose NLP libraries and are discussed in more detail in the sections that follow. Others, like Gensim, are more specialized and focus on subdisciplines such as semantic analysis (detecting the meaning of words), topic modeling (determining a document's meaning based on word statistics), and text mining.

NLTK

The *Natural Language Tool Kit*, or *NLTK* for short, is one of the oldest, most powerful, and most popular NLP libraries for Python. NLTK is open source and works on Windows, macOS, and Linux. Created in 2001 as part of a computational linguistics course at the University of Pennsylvania, it has continued to develop and expand with the help of dozens of contributors. NLTK comes preinstalled on Anaconda.

Because it's designed by and for an academic research audience, NLTK is versatile but can be somewhat slow for quick-paced production usage. It's also considered a bit difficult to learn, though this is mitigated to a fair degree by the free and useful online textbook, *Natural Language Processing with Python* (*http://www.nltk.org/book/*), written by its developers.

A strength of NLTK is that it comes packaged with lots of corpora (bodies of text) and pretrained models. As a result, it can be considered the de facto standard library for academic researchers in NLP.

spaCy

The *spaCy* library is younger than NLTK and designed to work well with machine learning frameworks like scikit-learn, TensorFlow, PyTorch, and other NLP libraries like Gensim. It's advertised as being "industrial strength," meaning that it's scalable, optimized, and very fast for production applications. Like NLTK, it has great documentation and comes prepackaged with useful language models. Its support community is not as large as that for NLTK but it's growing rapidly and may someday overtake NLTK in popularity.

CHOOSING AN NLP LIBRARY

Although there are dozens of libraries in the NLP stack, you need to know only a few to be proficient in the field. Some guidelines for choosing an NLP library are presented here:

- If you're in academia or otherwise doing research, you'll probably want to take the time to learn NLTK.
- The spaCy library will be useful for mixing NLP with machine learning models.
- If you need highly optimized performance, consider spaCy.
- If all you plan to do is scrape websites and analyze the results, consider *Pattern* (*https://github.com/clips/pattern/*), a specialized web miner with basic NLP capabilities.
- If you're a beginner or plan to use NLP lightly in your work, consider *TextBlob* (*https://textblob.readthedocs.io/en/dev/*). TextBlob is a user-friendly frontend to the NLTK and Pattern libraries, wrapping both in high-level, easy-to-use interfaces. It's good for learning and for quick prototyping, and as you become more confident, you can add functionality to refine your prototypes.
- If you're into topic modeling and statistical semantics (analyzing and scoring documents on their similarity), you may want to consider *Gensim* (*https://radimrehurek.com/gensim/*). Gensim can handle very large file sizes by streaming documents to its analysis engine and performing unsupervised learning on them incrementally. Its memory optimization and fast processing speed are achieved through the use of the NumPy library. Gensim is a specialized tool, and not for general-purpose NLP.
- If you want to perform NLP on multiple languages at once, consider *Polyglot* (*https://polyglot.readthedocs.io/en/latest/index.html*).

The Helper Libraries

Helper libraries assist you in using the scientific libraries discussed in this chapter. The ones discussed here help you download data, prepare it for use, and analyze it as quickly as possible.

Requests

Data wrangling (or *munging*) refers to the process of transforming data from its "raw" form into a more usable format for analysis. This involves processes such as checking, correcting, remapping, and so on. You can do a lot of this with the pandas library, discussed previously, but first you need to get your hands on the data.

Given that the lion's share of human knowledge is available online, you're probably going to need a way to pull data off the World Wide Web. Note that I'm not talking about simply downloading an Excel spreadsheet from an online database, which is easy enough, or about manually copying and pasting text from a web page. I'm referring to automatically extracting and processing content, a process called *web scraping*. Let's look at two open source libraries to help with this, requests and Beautiful Soup, and a third, re, that helps you clean and correct the data.

The popular and trusted requests library is designed to make *HyperText Transfer Protocol (HTTP)* requests simpler and more human friendly. HTTP is the foundation of data communication for the World Wide Web, where hypertext documents include hyperlinks to other resources that users can easily access with, for example, a mouse click or by tapping the screen in a web browser. The requests library comes preinstalled with Anaconda.

Let's look at an example where you scrape Dr. Martin Luthor King, Jr.'s "I Have a Dream" speech from a website (*http://www.analytictech.com/mb021/ mlk.htm*) using Jupyter Notebook:

```
import requests

url = 'http://www.analytictech.com/mb021/mlk.htm'
page = requests.get(url)
```

After importing requests, you provide the url address as a string. You can copy and paste this from the website from which you want to extract text. The requests library abstracts the complexities of making HTTP requests in Python. The get() method retrieves the url and assigns the output to a page variable, which references the Response object the web page returned for the request. This object's text attribute holds the web page, including the speech, as a readable text string.

At this point, the data is in *HyperText Markup Language (HTML)*, the standard format used to create web pages:

```
<!DOCTYPE HTML PUBLIC "-//IETF//DTD HTML//EN">
<html>

<head>
```

```
<meta http-equiv="Content-Type"
content="text/html; charset=iso-8859-1">
<meta name="GENERATOR" content="Microsoft FrontPage 4.0">
<title>Martin Luther King Jr.'s 1962 Speech</title>
</head>
--snip--
<p>I am happy to join with you today in what will go down in
history as the greatest demonstration for freedom in the history
of our nation. </p>
--snip--
```

As you can see, HTML has a lot of *tags* such as <head> and <p> that let your browser know how to format the web page. The text between starting and closing tags is called an *element*. For example, the text "Martin Luther King Jr.'s 1962 Speech" is a title element sandwiched between the starting tag <title> and the closing tag </title>. Paragraphs are formatted using <p> and </p> tags.

Because these tags are not part of the original text, they should be removed prior to any further analysis, such as natural language processing. To remove the tags, you'll need the Beautiful Soup library.

Beautiful Soup

Beautiful Soup is an open source Python library for extracting readable data from HTML and XML files. It comes preinstalled with Anaconda.

Let's use Beautiful Soup (simplified as *bs4*) on the HTML file returned by requests in the previous section:

```
import bs4

soup = bs4.BeautifulSoup(page.text, 'html.parser')
p_elems = [element.text for element in soup.find_all('p')]
speech = ' '.join(p_elems)
print(speech)
```

After importing bs4, we call the bs4.BeautifulSoup() method and pass it the string containing the HTML. The soup variable now references a BeautifulSoup object, which means that you can use the find_all() method to locate the speech buried in the HTML document between paragraph tags (<p>). This makes a list, which you can turn into a continuous text string by joining the paragraph elements on a space (' '). The (truncated) printed results follow:

```
I am happy to join with you today in what will go down in
history as the greatest demonstration for freedom in the history
of our nation.  Five score years ago a great American in whose
symbolic shadow we stand today signed the Emancipation Proclamation.
This momentous decree came as a great beckoning light of hope to
--snip--
```

You now have text that you can easily read as well as analyze with Python's many language processing tools.

Regex

No matter where you get your raw data, it will probably contain spelling errors, formatting issues, missing values, and other problems that will keep you from using it immediately. You'll need to process it in some way, such as reformatting, replacing, or removing certain parts, and you'll want to do it in *bulk*. Fortunately, *regular expressions* provide you with a wide variety of tools for parsing raw text and performing these tasks.

A regular (or *rational*) expression, usually shortened to *regex*, is a sequence of characters that specifies a *search pattern*. You're probably familiar with these patterns if you've ever used "find" or "find and replace" operations in a text editor. Using pattern matching, regex helps you to isolate and extract text you want from text you don't want.

Regex can do tedious but important things that you'd normally assign to an assistant or technician. For example, it can scan text for information related to your field of study. If you're a seismologist interested in earthquakes, you can write programs that scan news feeds for reports on these events, grab the data, format it, and store it in a database.

Python has a built-in module called re which you can use to work with regular expressions. Let's look at an example in which you're searching texts for 10-digit phone numbers. Within this database, people have entered phone numbers in multiple ways, such as with the area code in parentheses, using dashes, using spaces, and so on, but you want to use 10 consecutive numbers with no spaces. Here's how re and Python can help you to extract and format the numbers:

```
import re

data = 'My phone number: (601)437-4455, also my number: (601) 437-4455, \
        again my number: 601-437-4455, still my number: 601.437.4455'

nums = re.findall(r'[\(]?[1-9][0-9\ \.\-\(\)]{10,}[0-9]', data)

print(nums)
```

After importing re and entering the data, you assign a variable, named nums, and call the re.findall() method. This diabolical syntax looks like some kind of code, pun intended, and like any code you must know the key. Without going into the gory details, you're basically telling the findall() method the following:

- The matched text string might start with the (symbol or the number one to nine [\(]?[1-9].
- There can be a number, space, period, dash, or parentheses in between [0-9\ \.\-\(\)].
- The matched string must contain at least 10 characters {10,}.
- Finally, it must end with a number between zero and nine [0-9].

This first attempt finds all the input numbers:

```
['(601)437-4455', '(601) 437-4455', '601-437-4455', '601.437.4455']
```

Next, you need to remove the non-number characters using the `re.sub()` method, which substitutes a character you provide for the targeted characters. The `^` tells the method to find everything but digits between zero and nine and replace them with nothing, signified by `''`:

```
nums_nospace = re.sub('[^0-9]', '', str(nums))
print(nums_nospace)
```

This produces a continuous string of numbers:

```
6014374455601437445560143744556014374455
```

You can now use list comprehension to loop through this string and extract the 10-digit groupings you desire:

```
phone_list = [nums_nospace[x:x+10] for x in range(0, len(nums_nospace), 10)]
print(phone_list)
```

This yields a list of numbers (as strings) in the format you desire:

```
['6014374455', '6014374455', '6014374455', '6014374455']
```

This simple example demonstrates both the power of regex and the daunting nature of its syntax. In fact, regex is probably the most unPythonic thing in Python. Fortunately, because just about everyone struggles with the syntax, there are a lot of tools, tutorials, books, and cheat sheets available to help you use it.

You can find a nice "how to" tutorial at *https://docs.python.org/3/howto/regex.html* and at *https://realpython.com/regex-python/*. Chapter 7 of Al Sweigart's book, *Automate the Boring Stuff with Python*, 2nd edition (No Starch Press, 2019), provides a high-level overview of pattern matching with regular expressions, and Jeffrey Friedl's *Mastering Regular Expressions* (O'Reilly, 2006) covers them in-depth. You can find cheat sheets with examples in many places online, including *https://learnbyexample.github.io/python-regex-cheatsheet/*. Other websites such as *https://regexr.com/* and *https://www.regexpal.com/* let you play with regular expressions to learn how they work.

If you have to wrangle a lot of text, regular expressions will dramatically reduce the amount of code that you need to write, saving you both time and frustration. With a little effort, you'll achieve complete mastery over your data, solve problems, and automate things you probably never realized could be automated.

Dask

Dask is an open source library for parallel computing written in Python. It was developed to scale Python ecosystem libraries such as pandas, NumPy, scikit-learn, Matplotlib, Jupyter Notebook, and so on, from a single computer to multicore machines and distributed clusters. Dask comes preinstalled on Anaconda.

To understand the benefits Dask provides, let's talk terminology for a moment. A *thread* is the smallest sequence of programmed instructions that can be managed independently by a scheduler. *Parallel processing* refers to dividing different parts of a computing task—the threads—among two or more processors for the purpose of accelerating program execution.

In the old days, the central processing unit (CPU) of computers had a single microprocessor, or *core*, that executed code one step at a time, like an army marching in single file. Nowadays, computers come with at least a dual-core CPU consisting of a chip with two complete microprocessors that share a single path to memory and peripherals. High-end workstations can have eight or more cores. So, theoretically, your programs no longer need to walk single file; they can run abreast. That is, if there are non-dependent threads, they can run simultaneously, saving lots of time.

But Python has limitations when it comes to parallel computing. Even though computers now have more than one CPU, Python uses *Global Interpreter Lock (GIL)* to boost the performance of single threads by encouraging only a single thread to execute at a time. This hinders the use of multiple CPU cores for speedier computing.

With Dask, you can use Python and perform parallel computations locally on a multicored machine or remotely across thousands of machines. And Dask does it efficiently, as well, by managing memory at the same time. To maintain a low-memory footprint, it stores large datasets on disk and copies off chunks of data for processing. It also discards intermediate values as soon as possible after they are generated. As a result, Dask permits manipulation of datasets of 100GB and larger on laptops, and larger than 1TB on workstations.

Dask is composed of two parts: distributed data structures, with APIs similar to pandas DataFrames and NumPy arrays, and a task grapher/scheduler (Figure 15-4). It implements many of the same methods as pandas, which means that it can fully replace it in many cases. Dask also offers NumPy and scikit-learn replacements and has the capability to scale *any* Python code.

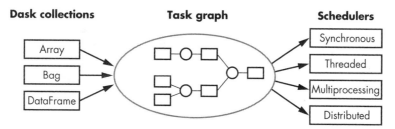

Figure 15-4: Dask collections generate graphs that are executed by schedulers (courtesy of https://dask.org/).

Dask will add extra complexity to your projects, thus you should use it mainly when you have huge datasets and need to use cluster computing. Documentation for Dask is excellent, and there are many tutorials available online to help you use the library.

NOTE   *You might hear about* Apache Spark, *a more mature and "heavier" alternative to Dask that has become a dominant and well-trusted tool in the Big Data enterprise world. It's an all-in-one project that has its own ecosystem and is written in Scala with some support for Python. You can find a comparison of the two libraries at* https://docs.dask.org/en/latest/spark.html. *In general, if you're already using Python and associated libraries like NumPy and pandas, you'll probably prefer working with Dask.*

Summary

Python supports scientific work through its easy-to-use core language and the many libraries built upon it. Not only are these packages free, but they're also robust, reliable, and well documented, thanks to the enormous and active user community.

You learned about some of the most important and popular libraries for science, including NumPy and SciPy, for numerical and array calculations; pandas, for data analysis; scikit-learn, for machine learning; Tensorflow, Keras, and PyTorch, for neural networks; OpenCV, for computer vision; and NLTK, for language processing. In Chapter 16 and Chapter 17, we cover libraries for working with geographic data and creating visualizations. In Chapters 18, 19, 20, and 21, we dive deeper into NumPy, Matplotlib, pandas, and scikit-learn.

16

THE INFOVIS, SCIVIS, AND DASHBOARDING LIBRARIES

Visualizing data is an integral part of science. Humans are visual creatures by nature, and viewing data graphically is more efficient and intuitive than reading through lists of strings or numbers. Effective plots help you to clean, prepare, and explore data. You can use them to reveal outliers and spurious samples, identify patterns, and compare datasets. Perhaps most important, they help you to communicate clearly with others and convey your ideas in an easily consumed manner. It's little wonder that graphics have been called the "pinnacle of communication."

Data visualization is a very broad category that includes everything from simple charts used for data exploration and reporting, to complex interactive web applications that operate in real time. With Python, you can easily cover this range. In fact, when it comes to creating graphics, Python suffers from an embarrassment of riches. With more than 40 different plotting libraries, there's something for everyone. But that's part of the problem.

Wading through Python's plotting APIs is exhausting. Users can be overwhelmed by all the choices, which cover a wide range of functionality, both unique and overlapping. As a result, they usually focus more on learning APIs than on their real job: exploring their data. In fact, this book was inspired by conversations with other scientists who were frustrated by this very problem.

Another issue with Python's plotting libraries is that the vast majority force you to write code to create even the simplest of visualizations. Compare this to software like Tableau or Excel, in which sensible, attractive graphs require just a few clicks of a mouse with little cognitive burden on the user.

Fortunately, many users share similar needs, and with a little forethought you can avoid going down suboptimal paths. In general, this involves selecting a high-level tool that covers the most common tasks succinctly and conveniently, typically by providing a simpler API on top of an existing tool.

In the sections that follow, we'll take a broad look at some of Python's most popular and useful plotting and dashboarding libraries. Then, we'll review some logical questions that should help guide you to the best plotting library, or libraries, for your needs.

NOTE *The plotting examples in this chapter are intended to demonstrate the complexity of the code and the types of plots produced. You're not expected to run the code snippets, as many of the libraries discussed do not come preinstalled with Anaconda. But if you do want to test them for yourself, you can find installation instructions in the product web pages. I recommend that you install them all in a dedicated conda environment (see Chapter 2), rather than dump them in your base environment.*

InfoVis and SciVis Libraries

We can divide visualizations into three main categories: *InfoVis, SciVis,* and *GeoVis* (Figure 16-1). InfoVis, short for *Information Visualization*, refers to 2D or simple 3D static or interactive representations of data. Common examples are statistical plots such as pie charts and histograms. SciVis, short for *Scientific Visualization*, refers to graphical representations of physically situated data. These visualizations are designed to provide insight into the data, especially when it's studied by novel and unconventional means. Examples are magnetic resonance imaging (MRI) and simulations of turbulent fluid flow. GeoVis, short for *Geovisualization*, refers to the analysis of geospatial (geographically located) data through static and interactive visualization. Examples include satellite imagery and map creation.

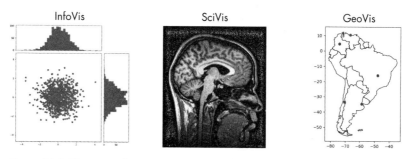

| InfoVis | SciVis | GeoVis |

Figure 16-1: Three visualization categories with examples

Tables 16-1 lists some of Python's more important InfoVis and SciVis plotting libraries. We'll take a closer look at some of these in the sections that follow before turning to the dashboard libraries. Then, in Chapter 17, we'll repeat this exercise for the GeoVis libraries.

Table 16-1: Python's Major InfoVis and SciVis Libraries

Type	Library	Description	Website
InfoVis	Matplotlib	Publication-quality 2D and simple 3D plots	https://matplotlib.org/
	seaborn	Matplotlib wrapper for easier, prettier plots	https://seaborn.pydata.org/
	pandas	Matplotlib wrapper for easy DataFrame plotting	http://pandas.pydata.org/
	Altair	Easy and simple 2D plots for small datasets	https://altair-viz.github.io/
	ggplot	Simple "grammar of graphics" plots with pandas	https://yhat.github.io/ggpy/
	Bokeh	Web interactivity tool with large or streaming datasets	https://bokeh.org/
	Chartify	Bokeh wrapper for easier charting	https://github.com/spotify/chartify/
	Plotly	Dynamic, interactive graphics for web apps	https://plotly.com/python/
	HoloViews	Viz data structures usable by many libraries	http://holoviews.org/
	hvPlot	Easy interactive plotting library built on HoloViews/Bokeh	https://hvplot.holoviz.org/
	Datashader	Tools for rasterizing giant datasets for easy visualization	https://datashader.org/
SciVis	VTK	Visualization toolkit for 3D computer graphics	https://vtk.org/
	Mayavi	3D scientific visualization tool with interactivity	https://docs.enthought.com/mayavi/
	ParaView	3D scientific visualization tool with interactivity	https://www.paraview.org/

If you're curious about how we got into this mess, take a few minutes to look at James Bednar's blog post "Python Data Visualization 2018: Why So Many Libraries?" (https://www.anaconda.com/blog/python-data-visualization-2018-why -so-many-libraries/). You should also check out his ebook, Python Data Visualization, *and* PyViz *site (https://pyviz.org/), which are designed to help users decide on the best open source Python data visualization tools for their purposes, with links, overviews, comparisons, examples, and exhaustive tool lists.*

Matplotlib

The *Matplotlib* library is an open source, comprehensive library for creating manuscript-quality static, animated, and interactive visualizations in Python. These are mainly 2D plots, such as bar charts, pie charts, scatterplots, and so on, though some 3D plotting is possible (Figure 16-2). Matplotlib is almost 20 years old and was designed to provide early versions of Python with a familiar MATLAB-type interface. MATLAB is a proprietary scientific programming language that has been displaced in popularity by Python.

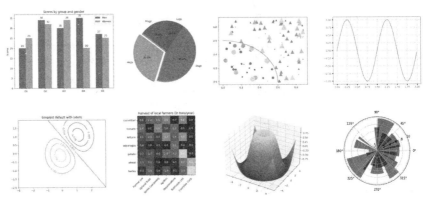

Figure 16-2: A small sampling of Matplotlib plot types (courtesy of https://matplotlib.org/)

Matplotlib's focus is on creating static images for use in publications and interactive figures for data exploration and analysis. These interactive figures use GUI toolkits like Qt, rather than web applications. The library comes preinstalled with Anaconda.

Matplotlib is the King, Grandaddy, and Big Kahuna of Python visualization. It's a massive, exhaustive library, and many alternative products are built on top of it, just as others are built on NumPy (including Matplotlib). Likewise, the internal visualization tools of libraries like pandas leverage Matplotlib methods.

The Matplotlib motto is that it "makes easy things easy and hard things possible." It works on all operating systems and handles all the common image formats. It has broad functionality, allowing you to build just about any kind of chart you can imagine, and it's very compatible with other popular science libraries like pandas, NumPy, and scikit-learn, thanks to collaborations between the Matplotlib and IPython communities.

Matplotlib is a powerful but low-level plotting engine. This means that you have lots of flexibility and options for precisely controlling plots by assembling them component by component. But this freedom comes with complexity. When creating anything beyond a simple plot, your code can become ugly, dense, and tedious.

The unfriendliness of Matplotlib's API is offset somewhat by its popularity and maturity. A simple online search will yield example code for just about any plot that you want to make. Its greatest resource is undoubtably the Matplotlib *gallery* (*https://matplotlib.org/gallery/index.html/*), a "cookbook" of code recipes for making a huge variety of plots.

Other issues with Matplotlib are the appearance and "explorable nature" of its plots. Although Matplotlib plots come with interactive features like zooming, panning, saving, and posting the cursor's *location* (Figure 16-3), they are somewhat antiquated compared to what's directly available in more modern libraries.

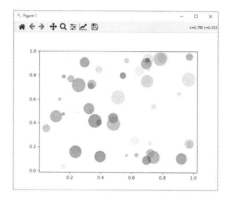
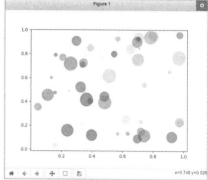

Figure 16-3: Matplotlib plot in an external Qt window (left) versus inline in a Jupyter notebook (right)

By default, Matplotlib's interactivity is designed to work in *external* windows rather than *inline* on the same screen as your code. You can force inline interactivity in Jupyter Notebook and JupyterLab, but the results can be buggy. For example, the Save button might simply open a blank web page rather than downloading the plot. Other libraries also provide more intelligent cursor hovering capabilities that can display custom information about posted data.

As a testament to Matplotlib's dominance and usefulness, a number of external packages extend or build on Matplotlib functionality (see *https:// matplotlib.org/3.2.1/thirdpartypackages/*). Two of these, *mpldatacursor* and *mplcursors*, let you add *some* interactive data cursor functionality to plots using only a few lines of code.

Likewise, there are add-on visualization toolkits that rely on Matplotlib under the hood. One of the most important is *seaborn*, which is designed to simplify plotting and to generate more attractive plots than those produced by Matplotlib's defaults. Both seaborn and pandas are wrappers over Matplotlib, which lets you access some of Matplotlib's methods with less code.

seaborn

The *seaborn* library is a free, open source visualization library built on Matplotlib. It provides a higher-level (that is, easier-to-use) interface for drawing attractive and informative *statistical* graphics such as bar charts, scatterplots, histograms, and so on. It also comes with built-in functions for density estimators, confidence bounds, and regression functions. Not surprisingly, it's well integrated with data structures in pandas and NumPy. Seaborn comes preinstalled with Anaconda.

A goal of seaborn is to make visualization a central part of exploring and understanding data through the use of dataset-oriented plotting functions. It makes default plots more attractive and supports the building of complex visualizations. It helps reveal data patterns through the use of high-level multiplot grids and different color pallets (visit *https://seaborn .pydata.org/examples/index.html* for some examples).

Seaborn is designed to work well with the popular DataFrame objects in pandas, and you can easily assign column names to the plot axes. It's also considered preferrable to Matplotlib for making multidimensional plots.

In the example that follows, the last line of code generated an attractive scatterplot including a linear regression line with 95 percent confidence interval, marginal histograms, and distributions:

```
import seaborn as sns
tips = sns.load_dataset('tips')
sns.jointplot(data=tips, x='total_bill', y='tip', kind='reg');
```

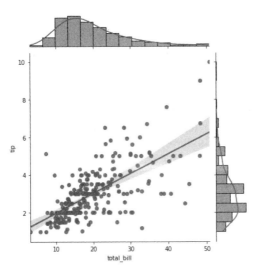

One of the best features of seaborn is the *pairplot*. This built-in plot type lets you explore the pairwise relationships in an entire dataset in one figure, with the option of viewing histograms, layered kernel density estimates, scatterplots, and more. Following is an example of a pairplot created using the Palmer Archipelago dataset for identifying penguin species. The data is loaded as a pandas DataFrame (see the pandas section in Chapter 15 for an overview of the pandas library).

```
import seaborn as sns
penguins = sns.load_dataset('penguins')
sns.pairplot(data=penguins, hue='species', markers=['o', 'X', 's']);
```

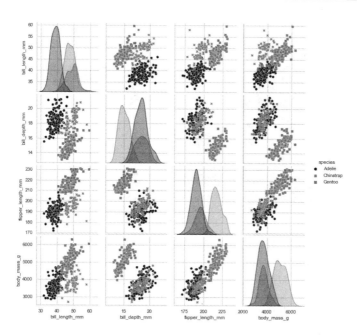

Another built-in plot type, *stripplot*, is a scatterplot in which one variable is categorical. It's perfect for comparing the lengths of bills among penguin species:

```
sns.set_theme(style='whitegrid')
strip = sns.stripplot(x='bill_length_mm', y='species', data=penguins);
```

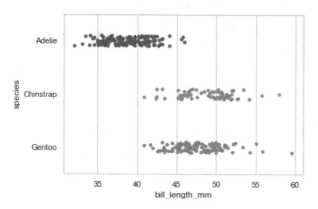

Unlike Matplotlib, seaborn lets you manipulate data *during* the plotting operation. For example, you can calculate the number of body mass samples in the penguins dataset by calling the built-in length function (len) from within the barplot() method:

```
bar = sns.barplot(data=penguins, x='species', y='body_mass_g', estimator=len)
bar.set(xlabel='Penguin Species', ylabel='Number of Samples');
```

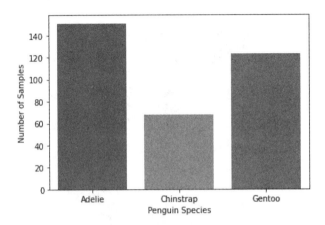

Let's take a look at how easy it is to customize a plot using seaborn. Table 16-2 lists the top 10 countries most affected by COVID-19 (based on number of cases) in roughly the first year of the virus's spread. The Fatality Rate column lists the number of deaths per 100 confirmed cases. The Deaths per 100,000 column calculates deaths based on a country's general population.

Table 16-2: COVID-19 Statistics

Country	Region	Cases	Deaths	Deaths/100K popl	Fatality rate
United States	North America	31,197,873	562,066	171.80	0.018
India	Asia	13,527,717	170,179	12.58	0.013
Brazil	Latin America	13,482,023	353,137	168.59	0.026
France	Europe	5,119,585	98,909	147.65	0.019
Russia	Asia	4,589,209	101,282	70.10	0.022
UK	Europe	4,384,610	127,331	191.51	0.029
Turkey	Middle East	3,849,011	33,939	41.23	0.009
Italy	Europe	3,769,814	114,254	189.06	0.030
Spain	Europe	3,347,512	76,328	163.36	0.023
Germany	Europe	3,012,158	78,500	94.66	0.026

Source: *https://coronavirus.jhu.edu/data/mortality*

Let's save Table 16-2 as a comma-separated value (*.csv*) file and use it with seaborn to look at the relationship among deaths, the death rate per 100,000 people, and the fatality rate:

```
import pandas as pd
import seaborn as sns

sns.set_style('whitegrid')
df = pd.read_csv('johns_hopkins_covid_stats_apr_2021.csv')
scatter = sns.scatterplot(data=df,
                          x='Deaths',
                          y='Deaths/100K Popl',
                          hue='Country',
                          style='Country',
                          size='Fatality Rate',
                          sizes=(50, 200))
scatter.legend(loc='center right', bbox_to_anchor=(1.4, 0.5), ncol=1);
```

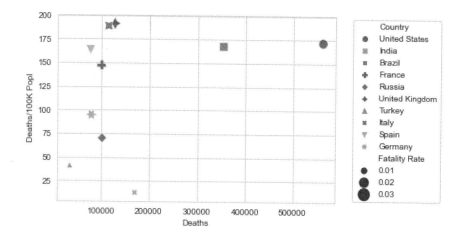

After importing pandas and seaborn, you set the style of the plot to give it a white background with gridlines. The data, in *.csv* format, is then loaded as a pandas DataFrame named df. Creating a scatterplot (scatter) takes a single command. The marker color (hue) and shape (style) are based on the country and their size reflects the fatality rate, with a size range of 50 to 200. You finish by creating a legend and calling the plot. Note how, by using the DataFrame column names from Table 16-2, the code is easy to read and understand.

Despite being an abstraction layer on top of Matplotlib, seaborn provides access to underlying Matplotlib objects, so you can still achieve precise control over your plots. Of course, you'll need to know Matplotlib to some degree to tweak the seaborn defaults in this manner.

Seaborn plots are considered more attractive, and thus better for publications and presentations, than those produced by Matplotlib. It's a good choice if all you want are static images made with simpler code and better defaults.

Even if you choose to use Matplotlib instead of the seaborn wrapper, you can still import seaborn and use its themes to improve the visual appearance of your plots. For examples, see https://www.python-graph-gallery.com/106-seaborn-style -on-matplotlib-plot *and* https://seaborn.pydata.org/generated/seaborn .set_theme.html?highlight=themes.

The pandas Plotting API

The pandas library discussed in the previous chapter has its own plotting API, Pandas.plot() (*https://pandas.pydata.org/pandas-docs/stable/user_guide/ visualization.html*). This API has emerged as a de facto standard for creating 2D charts because it can use Matplotlib and many other libraries as its plotting backend. This makes it possible to learn one set of plotting commands using pandas and then apply them with a wide range of libraries for static or interactive plots.

Plotting in pandas is arguably the easiest way to create visualizations using Python. It's especially good at quick "throwaway" plots for data exploration. Let's take a look:

```
import pandas as pd

female_ht_vs_wt = {'height': [137, 152, 168, 183, 198, 213],
                   'weight': [31.2, 45.2, 58.8, 72.3, 85.5, 108.3]}
df = pd.DataFrame(female_ht_vs_wt)
df.plot(kind='scatter', x='weight', y='height')
df.plot.bar('weight');
```

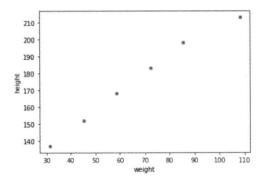

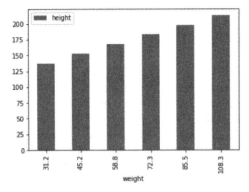

After importing pandas and making a Python dictionary of some measurements of female height verses weight, we turn the dictionary into a pandas DataFrame. The last two lines of code can then immediately build two plots! What could be easier?

The plots are very plain and lack any kind of interactivity, but never fear, pandas plays well with the other plotting libraries. With little effort, you can switch to an alternative plotting tool for additional functionality. By changing the plotting backend for pandas to HoloViews, a library we'll discuss shortly, you can produce an interactive plot that lets you zoom, pan, save, and hover the cursor over points to see their values. Here's an example of the code and its results:

```
import pandas as pd

pd.options.plotting.backend = 'holoviews'
female_ht_vs_wt = {'height': [137, 152, 168, 183, 196, 213],
                   'weight': [31.2, 45.2, 58.8, 72.3, 84.5, 108.3]}
df = pd.DataFrame(female_ht_vs_wt)
df.plot(kind='scatter', x='weight', y='height')
```

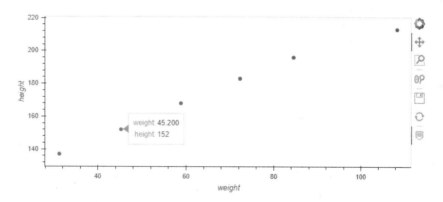

Note that, despite changing the plotting library, you didn't need to change a single line of the original plotting code. To see some other drop-in replacements for the Pandas .plot() API, see *https://pyviz.org/high-level/index.html#pandas-plot-api/*.

Altair

Altair is an open source statistical visualization library in Python that's closely aligned with pandas DataFrames. It's popular with people looking for a quick way to visualize small datasets.

Altair handles a lot of plotting details automatically, letting you focus on what you want to do rather than the button-pushing "how to do it" part. Much like the female height-verses-weight example in the previous section, you only need to link your data columns to encoding channels, such as the x- and y-axes, to make a plot. But this ease of use comes with a few down-sides. The plots are not as customizable as those made in Matplotlib, and there's no 3D plotting capability.

On the other hand, all Altair plots can be made interactive, meaning that you can zoom, pan, highlight plot regions, update linked charts with the selected data, enable *tooltips* that let you hover the cursor over points for detailed information, and so on. Altair visualizations require a JavaScript frontend to display charts and so should be used with Jupyter notebooks or an integrated development environment (IDE) with notebook support.

Unlike Matplotlib and other *imperative* plotting libraries that build plots step by step with no intermediate stages, Altair is *declarative* by nature, and generates a plot object, in JSON format, from which the plot can be recon-stituted. JSON, short for JavaScript Object Notation, is a file and data inter-change format that uses human-readable text to store and transmit data objects. Thus, Altair does not produce plots consisting of pixels, but plots consisting of data plus a visualization specification.

Because declarative plotting objects store your data and associated metadata, it's easy to manipulate the data during the plot render command or visualize it alongside or overlaid with other data. It can also result in very large visualization file sizes or entire datasets stored in your Jupyter note-book. Although there are some workarounds to help you manage memory and performance issues, the library's documentation recommends plotting no more than 5,000 rows of data (see *https://altair-viz.github.io/user_guide/faq .html#altair-faq-large-notebook/*).

Another drawback of using JSON is that it can be hacked if used with untrusted services or untrusted browsers. This can make the hosting web application vulnerable to a variety of attacks.

Bokeh

Bokeh is an open source visualization library that supports the creation of interactive, web-ready plots from very large or streaming datasets. Bokeh (pronounced "BO-kay") takes plots defined using Python and automatically renders them in a web browser using HTML and JavaScript, the dominant programming languages used for interactive web pages. It's one of the better-maintained and supported libraries and comes preinstalled with Anaconda.

Bokeh can output JSON objects, HTML documents, or interactive web applications. It has a three-level interface that provides increasing control over plots, from the simple and quick to the painstakingly detailed. However, unlike Matplotlib, Bokeh does not have high-level methods for some common diagrams such as pie charts, donut charts, or histograms. This requires extra work and the use of additional libraries such as NumPy. Support for 3D plotting is also limited. Thus, from a practical standpoint,

Bokeh's native API is mainly useful for publishing plots as part of a web app or HTML/JavaScript-based report, or for when you need to generate highly interactive plots or dashboards.

Bokeh works well in Jupyter notebooks and lets you use *themes*, for which you stipulate up front how you want your plots to look, such as font sizes, axis ticks, legends, and so on. Plots also come with a toolbar (Figure 16-4) for interactivity, including zooming, panning, and saving.

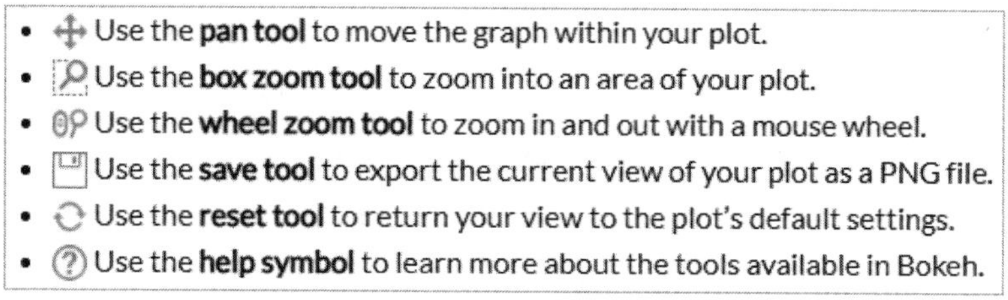

- ✛ Use the **pan tool** to move the graph within your plot.
- 🔎 Use the **box zoom tool** to zoom into an area of your plot.
- ⊕🔎 Use the **wheel zoom tool** to zoom in and out with a mouse wheel.
- 💾 Use the **save tool** to export the current view of your plot as a PNG file.
- ↻ Use the **reset tool** to return your view to the plot's default settings.
- ⑦ Use the **help symbol** to learn more about the tools available in Bokeh.

Figure 16-4: The Bokeh plot toolbar (courtesy of https://bokeh.org/*)*

Finally, if you keep your data in pandas, you can use a library called Pandas-Bokeh (*https://github.com/PatrikHlobil/Pandas-Bokeh/*), which consumes pandas data objects directly and renders them using Bokeh. This results in a higher-level, easier-to-use interface than Bokeh alone. Other high-level APIs built on Bokeh include HoloViews, hvPlot, and Chartify for plotting, and Panel for creating dashboards. We'll look at most of these later in the chapter.

Plotly

Plotly is an open source web-based toolkit for making interactive, publication-quality graphics. It's similar to Bokeh in that it builds interactive plots, generating the required JavaScript from Python. And like Bokeh and Matplotlib, Plotly is a core Python library on which multiple higher-level libraries are built.

Plotly graphs are stored in the JSON data format. This makes them portable and readable using scripts of other programming languages such as R, Julia, MATLAB, and more. Its web-based visualizations can be displayed in Jupyter notebooks, saved as standalone HTML files, or incorporated into web applications. Because Plotly uses JSON, it suffers similar memory and security issues as Altair (see "Altair" on page 429).

Unlike Matplotlib and seaborn, Plotly is focused on creating dynamic, interactive graphics in Python for embedding in web apps. You can create basic plots as well as more unique contour plots, dendrograms, and 3D charts (Figure 16-5).

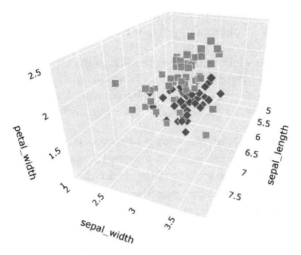

Figure 16-5: A 3D scatterplot made with Plotly Express

Figure 16-6 shows an example of a 3D mesh. You can even display LaTeX equations in legends and titles.

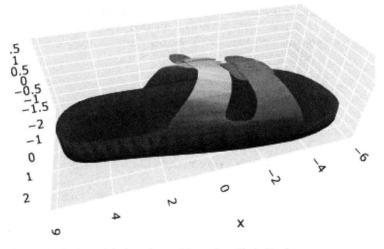

Figure 16-6: A sandal plotted as a 3D mesh in Plotly/Dash

Plotly also recognizes sliders, filters, and mouseover and cursor-click events. With only a few lines of code, you can create attractive interactive plots that save you time when exploring datasets and can be easily modified and exported. The toolkit also permits complex visualizations of multiple sources, in contrast to products like Tableau, which accept only one data table as input per chart.

Plotly is written in JavaScript and powers *Dash* (*https://dash.plotly.com/introduction*), an open source Python framework for building web analytic applications (called dashboards). Dash is written on top of Plotly.js and greatly simplifies the building of highly customized dashboards in Python. These apps are rendered in a web browser and can be deployed to servers and shared through URLs. Dash is cross-platform and mobile ready. We'll look at Dash a little more in "Dashboards" on page 445.

Plotly also comes with a high-level, more intuitive API called *Plotly Express* (*https://plotly.com/python/plotly-express/*) that provides shorthand syntax for creating entire figures at once. It has more than 30 functions for creating different types of graphics, each carefully designed to be as consistent and easy to learn as possible, allowing you to effortlessly switch from a scatterplot to a bar chart to a sunburst chart, and so on throughout a data exploration session. As such, Plotly Express is the recommended starting point for creating common figures with Plotly.

Plotly Express charts are easy to style so that they do really useful things. Suppose that you want to look at monthly rainfall totals over a two-decade period and see how the months of August and October compare to the rest. With Plotly Express, you can easily highlight the lines for these months so that they stand out. And with the interactive toolbar, you can toggle spike lines and the hover feature to query values (Figure 16-7).

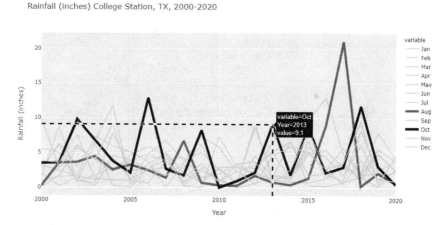

Figure 16-7: A Plotly Express line chart with highlighted lines, spike lines, and hover box

Another useful feature of Plotly Express is that legends are "alive." Click a category in a legend once and you temporarily remove it from the plot. Click it twice and all other lines will vanish, leaving that category isolated. This was done for the August (Aug) category in Figure 16-8. You can even animate the plot to see how things change over time. What a great way to untangle confusing "spaghetti" plots!

Rainfall (inches) College Station, TX, 2000-2020

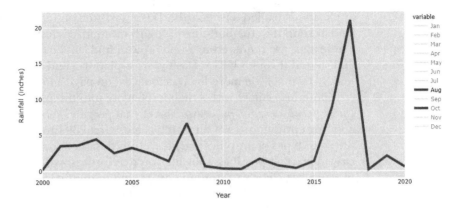

Figure 16-8: Double-clicking a legend category isolates that category by removing the other data.

Let's revisit the COVID-19 dataset that captures fatality statistics from the first year of the virus's spread. You'll want to compare the code and results that follow to the seaborn example on page 427.

```
import pandas as pd
import plotly.express as px

df = pd.read_csv('johns_hopkins_covid_stats_apr_2021.csv')
fig = px.scatter(data_frame=df,
                 x='Deaths',
                 y='Deaths/100K Popl',
                 color='Country',
                 size='Fatality Rate',
                 text='Country')
fig.update_layout(showlegend=False)
fig.show()
```

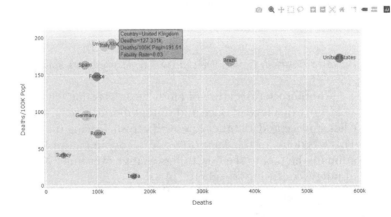

Like the previous seaborn code, it's very readable and easy to understand. Also note that Plotly Express has a specific parameter called data
_frame that lets you know without a doubt that it's built for working with
pandas.

A nice feature here is that you can easily post the country name over
the markers, letting you use a consistent marker shape for easy size comparisons. You don't get the automatic "size" legend that you get with seaborn,
but Plotly Express makes up for this by automatically permitting mouseover
events, as shown in the plot for the United Kingdom.

Another useful Plotly Express feature is the *facet plot*, which lets you
view the previous scatterplot by geographical region:

```
--snip--
fig = px.scatter(data_frame=df,
                 x='Deaths',
                 y='Deaths/100K Popl',
                 color='Country',
                 size='Fatality Rate',
                 text='Country',
               ❶ facet_col='Region')
fig.update_layout(showlegend=False)
fig.show()
```

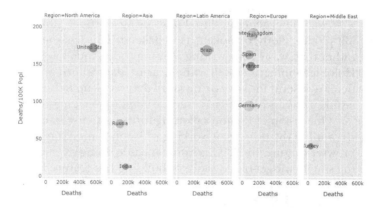

We did this by adding a single argument ❶ to the px.scatter() method.

Plotly Express is designed mainly for exploratory data analysis. Your
data must be in very specific formats (it's targeted at pandas DataFrames),
your overall ability to customize plots is limited, and you might have trouble
putting the visualizations into a presentation. To be able to do everything
you'll probably want to do, you'll need to occasionally drop down into the
full Plotly API or use Plotly Express in conjunction with other libraries like
Matplotlib or seaborn.

There also exists an independent third-party wrapper library around
Plotly called *cufflinks* (*https://github.com/santosjorge/cufflinks/*) that provides
bindings between Plotly and pandas. This helps you create plots from pandas DataFrames using the Pandas.plot() interface but with Plotly output.

Both Plotly and Plotly Express facilitate building charts for the web directly from pandas DataFrames. And plots you create in Jupyter notebooks can essentially be copied and pasted into a Dash app for quick implementation of a dashboard. You can see an example of some scientific charts built with Plotly at *https://plotly.com/python/scientific-charts/*.

HoloViews

HoloViews is an open source library (note that I didn't say *plotting* library) designed to make visualization simple by abstracting away the process of plotting. HoloViews makes it easier to visualize data interactively by providing a set of declarative plotting objects that store your data with associated metadata. The goal is to support the entire life cycle of scientific research, from initial exploration to publication to reproduction of the work and new extensions.

HoloViews lets you combine various container types into data structures for visually exploring data. Some example container types are *Layout*, for displaying elements side by side as separate subplots; *Overlay*, for displaying elements on top of one another; and *DynamicMap*, for dynamic plots that automatically update and respond to user interactions. To appreciate the DynamicMap container, check out *https://holoviews.org/user_guide/Streaming_Data.html* and *https://holoviews.org/user_guide/Responding_to_Events.html* to view animated examples.

HoloViews generates final plots using a proper plotting library such as Matplotlib, Plotly, or Bokeh, as a backend. This lets you focus on your data rather than waste time writing plotting code. And as a plotting "middleman," HoloViews integrates well with libraries like seaborn and pandas and is particularly useful for visualizing large datasets—up to billions—using libraries like *Dask* and *Datashader* (such as *https://holoviz.org/tutorial/Plotting.html*).

One vision of Python's plotting future is to use a set of libraries to streamline the process of working with small and large datasets in a web browser (Figure 16-9). This would include doing exploratory analysis, making simple widget-based tools, or building full-featured dashboards.

Panel   hvPlot   HoloViews   GeoViews   Datashader   Param   Colorcet

Figure 16-9: The HoloViz-maintained libraries (courtesy of holoviz.org)

In this coordinated effort, HoloViews and GeoViews provide a single, concise, and high-level API for libraries like Matplotlib, Bokeh, Datashader, Cartopy, and Plotly. Panel provides a unified approach to dashboarding, and Datashader allows for the plotting of very large datasets. Param supports declaring user-relevant parameters for working with widgets inside or outside of a notebook context. This arrangement permits you to easily switch between backends without having to learn commands for each new plotting library.

Recognizing that a typical figure is an object composed of many visual representations combined together, HoloViews makes it trivial to compose elements in the two most common ways: concatenating multiple representations into a single figure or overlaying visual elements within the same set of axes. When making multiplot figures, HoloViews helps by automatically linking axes and selections across each figure. It's also useful for creating charts that update dynamically, especially those using sliders. With the Bokeh backend, you can combine various widgets with zooming and panning tools to aid data exploration.

Let's take a look at a Jupyter Notebook example, adapted from the Holo-Views gallery (*https://holoviews.org/gallery/index.html*), that uses both HoloViews and Panel to generate a plot. For data, we'll again use the Palmer Archipelago dataset that quantifies the morphologic variations among three penguin species. Thanks to Panel, you'll be able to use drop-down menus to switch out and decorate the displayed data inside the single plot.

```
import seaborn as sns  # For access to penguins dataset.
import holoviews as hv
import panel as pn, panel.widgets as pnw
hv.extension('bokeh')

❶ hv.opts.defaults(hv.opts.Points(height=400, width=500,
                                   legend_position='right',
                                   show_grid=True))

penguins = sns.load_dataset('penguins')
columns = penguins.columns
discrete = [x for x in columns if penguins[x].dtype == object]
continuous = [x for x in columns if x not in discrete]
❷ x = pnw.Select(name='X-Axis', value='bill_length_mm', options=continuous)
y = pnw.Select(name='Y-Axis', value='bill_depth_mm', options=continuous)
size = pnw.Select(name='Size', value='None', options=['None'] + continuous)
color = pnw.Select(name='Color', value='None',
                   options=['None'] + ['species'] + ['island'] + ['sex'])
@pn.depends(x.param.value, y.param.value,
            color.param.value, size.param.value)

❸ def create_figure(x, y, color, size):
    opts = dict(cmap='Category10', line_color='black')
    if color != 'None':
        opts['color'] = color
    if size != 'None':
        opts['size'] = hv.dim(size).norm() * 20
    return hv.Points(penguins, [x, y], label="{} vs {}".
                     format(x.title(), y.title())).opts(**opts)

widgets = pn.WidgetBox(x, y, color, size, width=200)
pn.Row(widgets, create_figure).servable('Cross-selector')
```

After importing seaborn (for the data), HoloViews, and Panel, you tell HoloViews which plotting library to use. Bokeh is the default, but you can easily change this to Matplotlib or Plotly by changing the line to `hv.extension` `('matplotlib')` or `hv.extension('plotly')`. Most of the time, changing the backend doesn't require any change to the rest of the code.

The next line ❶ is optional but demonstrates a nice feature of Holo-Views: the ability to set your own defaults for how you want your plots to look. In this case, you set the size of the figure, position of the legend, and background grid to be used for all scatterplots.

Next, you load the penguins dataset, which conveniently ships with the seaborn library as a pandas DataFrame. To provide the user with menu choices, go through the columns in the `penguins` DataFrame and assign the contents to either a list called `discrete` or a list called `continuous`. The `discrete` list holds objects, such as species name, island name, or the penguin's sex. The `continuous` list is for numerical data, like the bill lengths and bill depths.

Starting at ❷, you must specify what choices the Panel widget will show for the x- and y-axes and the marker size and color, including the default options for what's initially shown. After this, you define a function to create the figure ❸ and return a HoloViews *Points* element. The final two lines create the figure with the menu widgets.

The output from this program is shown in Figure 16-10. Note the pull-down menus along the left side of the plot and the interactive toolbar along the right. Because we set the `size` and `color` default values to `'None'`, the points all look the same.

You can now use the menu widgets to color the points by species (Figure 16-11), which generates a legend at the lower-right corner of the plot. Setting the size option to body mass allows you to qualitatively incorporate a third measurement into the 2D scatterplot. Now you can see that the Gentoo species is clearly larger than the other two.

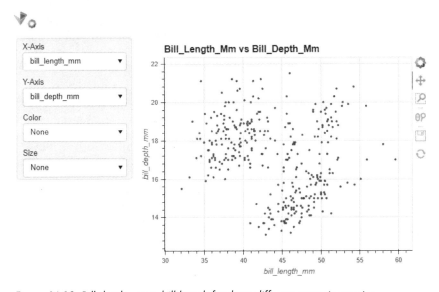

Figure 16-10: Bill depth versus bill length for three different penguin species

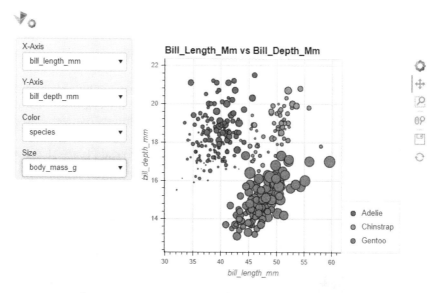

Figure 16-11: Bill depth versus bill length, colored by species and sized by body mass

In Figure 16-12, we've used the drop-down menus to change out both the data and size parameters. As you can see, this is a great way to interactively explore and familiarize yourself with a dataset without generating lots of plots.

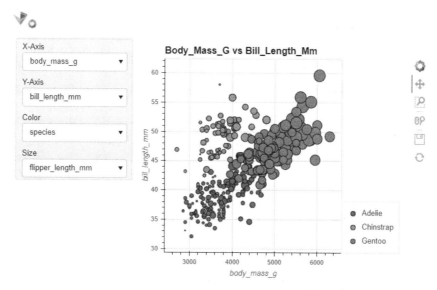

Figure 16-12: Bill length versus body mass, colored by species and sized by flipper length

A key point here is that the code references the DataFrame to make a HoloViews *Points element*. This object is basically the DataFrame plus knowledge of what goes on the x- and y-axes. This makes the DataFrame plottable. But unlike plot objects in other libraries, the hv.Points element holds onto your raw data. This makes it usable later in a processing pipeline (for a dynamic demonstration, see the HoloViews Showcase at *http://holoviews.org/Tutorials/Showcase.html*).

Just as Plotly has Plotly Express, the HoloViz libraries have *hvPlot*, a simpler plotting alternative built on top of HoloViews. This fully interactive high-level API complements the primarily static plots available from libraries built on Matplotlib, such as pandas and GeoPandas, that require support from additional libraries for interactive web-based plotting. It's designed for the PyData ecosystem and its core data containers, which allow users to work with a wide array of data types (Figure 16-13).

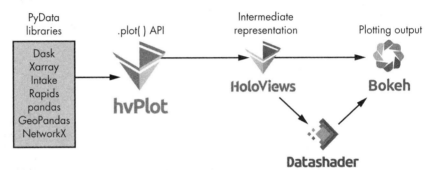

Figure 16-13: The hvPlot library provides a high-level plotting API for HoloViews.

The hvPlot library's interactive Bokeh-based API supports panning, zooming, hovering, and clickable/selectable legends. In the following example, hvPlot is used in conjunction with pandas to produce an interactive plot:

```
import hvplot.pandas
from bokeh.sampledata.degrees import data as degrees

degrees.hvplot.line(x='Year', y=['Art and Performance',
                                 'Business', 'Biology',
                                 'Education', 'Computer Science'],
                    value_label='% of Degrees Earned by Women',
                    legend='top')
```

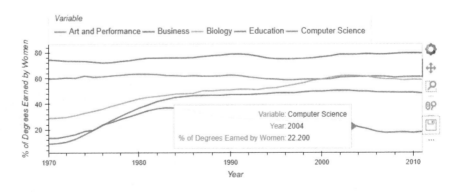

This is just as simple as plotting in pandas, but note the toolbar along the right side of the chart with icons for panning, zooming, saving, and hovering. The latter lets you query the graph details using the cursor, as shown by the pop-up window for the computer science variable. These options aren't available when plotting from native pandas.

For more on these libraries, check out *HoloViz* (*https://holoviz.org/*), the coordinated effort to make browser-based data visualization in Python easier to use, easier to learn, and more powerful.

Datashader

Datashader is an open source library designed for visualizing very large datasets. Rather than passing the entire dataset from the Python server to a browser for rendering, Datashader rasterizes (pixelates) it to a much smaller heatmap or image, which is then transferred for rendering. Whereas popular libraries like Matplotlib can suffer from performance issues with only 100,000 points, Datashader can handle hundreds of millions, even billions, of them. For example, Figure 16-14 plots 300 million data points.

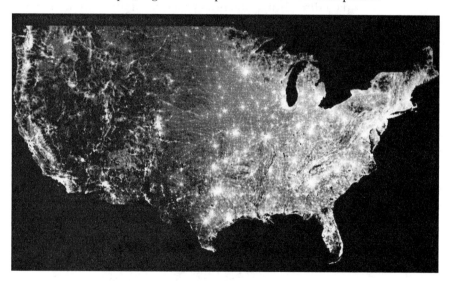

Figure 16-14: A Datashader-created plot of 300 million data points from the 2010 census (courtesy of Datashader)

Datashader makes it possible to work with very large datasets on standard hardware such as your laptop. Although the computationally intensive steps are written in Python, they're transparently compiled to machine code using a tool called *Numba* (*https://numba.pydata.org/*) and distributed across multiple processors using Dask.

The Datashader documentation highlights the tool's function in a preprocessing stage for plotting. What this means is that Datashader is often used with other plotting libraries to perform the heavy lifting associated with large datasets. Thus, although it's more focused on performance and efficiency than on directly generating basic statistical plots, it can work with other tools to help you plot large datasets—say, in a scattergram—by handling the common over-posting of points problem, where the density of the distributed points is obscured (Figure 16-15).

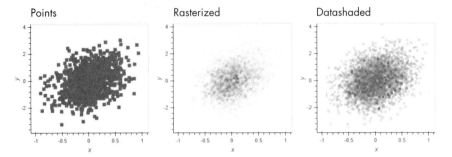

Figure 16-15: Datashader (right) handles over-posted points well (courtesy https:// holoviews.org/).

In another example, imagine that you're using Bokeh to copy your data directly into the browser so that a user can interact with the data even without a live Python process running. If the dataset contains millions or billions of samples, you'll run up against the limitations of the web browser. But with Datashader, you can prerender this huge dataset into a fixed-size raster image that captures the data's distribution. Bokeh's interactive plot can then dynamically re-render these images when zooming and panning, making it easier to work with the huge dataset in the web browser (Figure 16-16).

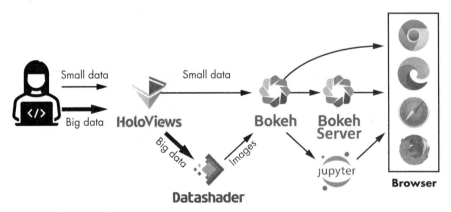

Figure 16-6: Generating interactive Datashader-based plots using HoloViews + Bokeh (courtesy of https://datashader.org/)

You can see a fantastic instance of Datashader in action in the "gerrymandering" example at *https://examples.pyviz.org/*. Working in concert with HoloViews and multiple plotting libraries, Datashader produces a map of

Houston's population, color-coded by ethnicity, that turns plotting into fine art, with a gorgeous watercolor-like rendering that has to be seen in color to be appreciated.

For a nice example of using Datashader with statistical plots, see *https:// holoviews.org/user_guide/Large_Data.html*. Peter Wang, co-creator of Datashader, gives an easily digestible video overview of the library at *https://www .youtube.com/watch?v=fB3cUrwxMVY/*.

In all of these examples, be aware that you'll lose some interactivity with Datashader. You'll still be able to zoom and pan, but mouseover events and the like will no longer work without special support, because the browser doesn't hold all of your datapoints ready for inspection. In return, you'll be able to visualize millions of datapoints without watching your computer grind to a halt.

Mayavi and ParaView

A common scientific practice is to visualize point clouds, such as those you might find in a Light Detection and Ranging (LIDAR) scan. General-purpose workhorse libraries like Matplotlib are capable of performing this task to a certain degree, but performance deteriorates quickly when interactively visualizing point clouds and other 3D plots. Matplotlib, for example, will be slow and might even crash your computer if you try to interact with a large number of samples. Even if the 3D representations successfully render, they won't look very nice, and you'll probably have trouble understanding what you see.

Datashader can help with this, but for graphics-intensive 3D and 4D visualizations such as those used for physical processes, you need a dedicated library like Mayavi (pronounced MA-ya-vee) that can handle *physically situated* regular and irregularly gridded data. This discriminates Mayavi from Datashader somewhat, as the latter is focused more on visualizations of information in *arbitrary spaces*, not necessarily the three-dimensional physical world.

Mayavi2 is an open source, general-purpose, cross-platform tool for 3D scientific data visualization. It's been designed with scripting and extensibility in mind from the ground up. You can import Mayavi2 into a Python script and use it as a simple plotting library like Matplotlib. It also provides an application (Figure 16-17) that is usable by itself.

Mayavi2 is written in Python, uses powerful Visualization Toolkit (VTK) libraries, and provides a GUI via Tkinter. It's cross-platform and runs on any platform where both Python and VTK are available (almost any Unix, macOS, or Windows systems). To a limited extent, you can use Mayavi in Jupyter notebooks. To see some examples of Mayavi2 plots, visit the gallery at *https://docs.enthought.com/mayavi/mayavi/auto/examples.html*.

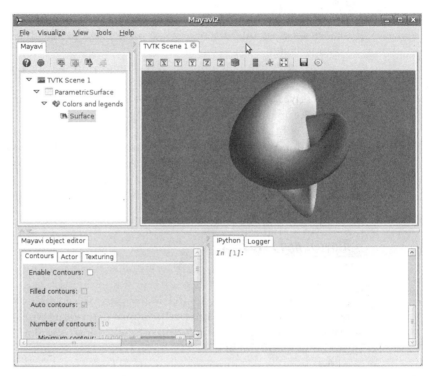

Figure 16-7: Mayavi2 application for 3D visualization. Note the Python console in the lower-right corner.

An alternative to Mayavi2 is *ParaView* (Figure 2-18). Although designed for 3D, it does 2D as well, is very interactive, and has a Python scripting interface.

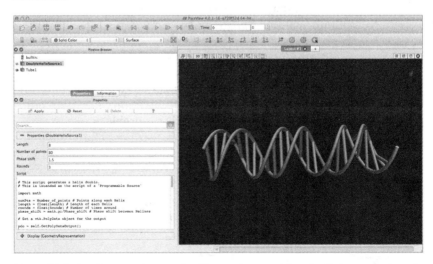

Figure 16-8: ParaView application for 3D visualization. Note the Python console in the lower-left corner.

ParaView was developed by Sandia National Laboratories, whereas Mayavi is a product of Enthought, whose Canopy distribution is a direct competitor of Anaconda.

Dashboards

A *dashboard* is a type of easy-to-read interactive GUI, often presented in real time. Dashboards are usually displayed on a single web page linked to a database, which allows the displayed information to be constantly updated. Example scientific dashboards include weather stations, earthquake monitoring, and spacecraft tracking (Figure 16-19).

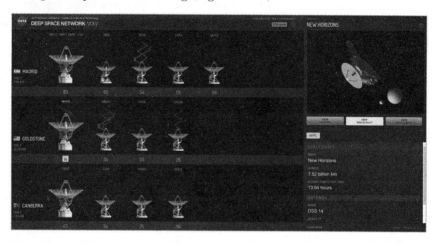

Figure 16-9: NASA spacecraft tracking dashboard (courtesy of https://www.nasa.gov)

Dashboards can really open up the usability and interactivity of your data, especially for nontechnical users. They also make the data accessible from anywhere, as long as you have an internet connection. This can be important when collaborating with external parties or providing results to scattered stakeholders.

Dashboards need to perform multiple tasks like analyzing and visualizing data, listening for and accepting user requests, and returning web pages via a web server. You can cobble together different libraries to handle these, or you can just use a dedicated dashboarding library.

Python supports higher-level web-based dashboarding with five main libraries: Dash, Streamlit, Voilà, Panel, and Bokeh (Table 16-4). These libraries let you create dashboards with pure Python, so you don't have to learn the underlying enabling languages like JavaScript and HTML. We looked at Bokeh earlier, so here we'll focus on the other four.

Table 16-4: Python's Most Important Dashboarding Libraries

Library	Description	Website
Plotly Dash	Advanced production-grade/enterprise dashboards	*https://plotly.com/dash/*
Streamlit	Fast and easy web apps from multiple plotting libraries	*https://streamlit.io/*
Voilà	Jupyter notebook rendering as stand-alone web apps	*https://voila.readthedocs.io/*
Panel	Interactive web apps with nearly any library	*https://panel.holoviz.org/*
Bokeh	Web interactivity with large or streaming datasets	*https://bokeh.org/*

Before we take a quick look at these four tools, note that it's possible to do some aspects of dashboarding in other libraries. The plotting stalwart Matplotlib supports several GUI toolkit interfaces, such as Qt, that can generate native applications you can use as an alternative to a web-based dashboard. Whereas several libraries make use of JavaScript to help build dashboards, *Bowtie* (*https://bowtie-py.readthedocs.io/*) lets you build them using pure Python. You can use *ipywidgets* with Jupyter Notebook to build a dashboard, but you need to use a separate deployable server, like Voilà, to share it.

For more insight, PyViz hosts a page on dashboarding that includes blog posts, links to comparison articles, and lists of alternative or supporting tools. You can find it at *https://pyviz.org/dashboarding/*.

NOTE *Bokeh, which we looked at previously, includes a widget and app library and a server for both plots and dashboards. It also supports live streaming of large datasets. However, if you intend to develop complex data visuals with Bokeh, you'll need some knowledge of JavaScript. Panel is built on Bokeh, just as seaborn is built on Matplotlib, and in the same way provides a higher-level toolkit to make dashboarding easier. It also supports multiple plotting libraries in addition to Bokeh.*

Dash

Dash is an open source Python framework developed by Plotly as a complete solution for deploying web analytic applications. Dash is built on Plotly. js, React.js, and *Flask* (a lower-level framework for building web apps from the ground up). Dash apps are rendered in a web browser deployed to servers and shared through a URL. This makes Dash platform agnostic and mobile ready. In 2020, Plotly released *JupyterDash* (*https://github.com/plotly/jupyter-dash/*), a new library designed for building Dash apps from Jupyter environments.

With Dash, it's possible to build a responsive, custom interface with pure Python in just a few hours. *Responsive*, by the way, means that the web page will render well on a variety of devices and screen sizes. Dash uses simple patterns to abstract away much of the dashboard-building process, such as generating the required JavaScript, React components, HTML, and server API. In fact, you can basically copy and paste Plotly graphs straight from a Jupyter notebook into a Dash app.

As far as how your dashboard looks, Dash provides an attractive out-of-the-box default stylesheet but also allows you to easily add third-party styling. *Dash-bootstrap-components* (*https://dash-bootstrap-components.opensource .faculty.ai/*) is an open source library that makes it easier to build consistently styled apps with complex, responsive layouts. You can also use any of the themes from Bootswatch themes (*https://www.bootstrapcdn.com/ bootswatch/*). These time-saving add-ons will let you build professional-looking dashboards with little effort.

Because of its relative maturity, expanding user community, and adoption by large enterprise organizations, Dash now has a large library of specialized modules, a host of repositories, and great documentation and tutorials to aid with the construction of customized dashboards. Whereas most scientists might aim to produce simple single-page dashboards, Dash can also build multipage, scalable, high-performance dashboards capable of incorporating organization style guides in the final layouts. This is a distinguishing feature of Dash versus friendlier tools like Streamlit and Voilà.

On the flip side, Dash is primarily designed for Plotly, though it's possible to use other third-party plotting libraries (see *https://github.com/plotly/ dash-alternative-viz-demo/*). Dash also requires you to work with HTML and Cascading Style Sheets (CSS) syntax, which isn't something Python users generally want to do. This has led to the development of simpler tools, like Streamlit, which we'll look at next.

Streamlit

Streamlit is a relatively new open source library for quickly building attractive dashboard web applications. As an all-in-one tool, it addresses web serving as well as data analysis.

Streamlit's simple API lets you concentrate on your data analysis and visualization rather than on frontend and backend technology issues. Sharing and deploying is fast and easy, and the learning curve is arguably the shortest of any of Python's dashboarding tools. As a result, Streamlit's popularity has risen rapidly, and new features are constantly being added.

Whereas Dash focuses on production and enterprise settings, Streamlit is designed for rapid prototyping. It lets you do more with less code, and unlike Dash, which is designed to work primarily with Plotly, Streamlit lets you easily mix and match plots from multiple libraries, including Plotly, Altair, Bokeh, seaborn, and Matplotlib. This gives you the option to choose the best tool for the particular plotting job and allows contributing team members to use their preferred plotting library.

For existing Python scripts, Streamlit is arguably the best way to quickly and easily turn them into interactive dashboards. However, it provides no support for Jupyter Notebook, and you'll encounter some friction moving your code into Streamlit. On the other hand, it's very compatible with major libraries like scikit-learn, TensorFlow/Keras, NumPy, OpenCV, PyTorch, pandas, and more. If you're happy with Streamlit's design defaults and don't need to do a lot of customization, it's a great choice for getting a dashboard up and running quickly.

Voilà

Voilà is an open source library that lets you quickly convert a Jupyter notebook into a stand-alone interactive dashboard sharable with others. As a thin layer built over Jupyter, it represents a very specific use case rather than a complete dashboarding solution.

Voilà allows nontechnical people associated with your project to use your Jupyter notebooks without having to know Python or Jupyter or have them installed on their computer. And if you already have a notebook with all the interactivity you need, it's the shortest path to turning your work into a dashboard.

Voilà is mostly about rendering. A common approach is to add interactivity (widgets) to a Jupyter notebook using a Python library like bqplot, Plotly, or ipywidgets, all of which are supported by Voilà. (We looked at ipywidgets in Chapter 5 on Jupyter Notebook.) You might then need to format the notebook to suppress and hide unused code and markdowns.

Voilà runs the code in the notebook, collects the outputs, and converts them to HTML. By default, the notebook code cells are hidden from view. The outputs are displayed vertically in the order in which they appear in the notebook (Figure 16-20), but you can use *widget layout templates* to change the position of the cell outputs, for example, by dragging them into a horizontal configuration. The page is then saved as a web application where the widgets on the page have access to the underlying Jupyter kernel.

At this point, the dashboard is only on your computer. For others to have access, you need to deploy your dashboard on the cloud using a public cloud computing platform such as Binder, Heroku, Amazon Web Services (AWS), Google Cloud Platform (GCP), IBM Cloud, or Microsoft Azure.

Binder, a free open source web application for managing digital repositories, is one of the most accessible ways to deploy Voilà applications. Use cases involve workshops, scientific workflows, and streamlined sharing among teams. Heroku (*https://www.heroku.com/*) is also a good choice for the less tech-savvy and those with limited budgets. It manages the supporting hardware and server infrastructure allowing you to focus on perfecting your app. On the downside, the app might run slowly due to low network performance. You can see more deployment options at *https://voila .readthedocs.io/en/stable/deploy.html*.

Voilà produces dashboards broadly similar to Streamlit and can be simpler to use, assuming that you already have a Jupyter notebook ready to go. Jupyter aficionados will also appreciate that Voilà shares Jupyter's widget library, whereas Streamlit requires you to learn its own set of custom widgets. You can see some example dashboards at *https://voila-gallery.org/*.

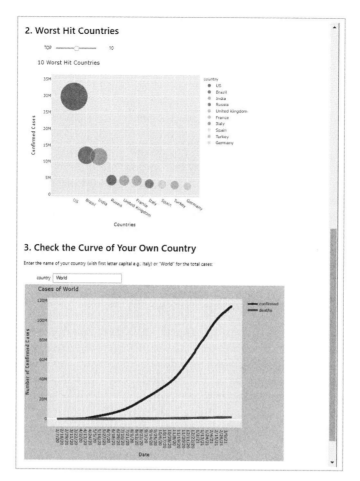

Figure 16-20: Dashboard elements retain Jupyter Notebook arrangement (courtesy of https://voila-gallery.org*).*

Panel

Panel is an open source Python library that lets you create custom interactive web apps and dashboards by connecting user-defined widgets to plots, images, tables, or text. Created and supported by Anaconda, Panel is part of the HoloViz family of unified plotting tools (see Figure 16-9) and uses the Bokeh server.

Panel helps support your entire workflow so that you never need to commit to only one way of using your data and your analyses, and you don't need to rewrite your code just to make it usable in a different way. You can move seamlessly from exploring data, creating reproducible steps, and telling a story in a notebook to creating a dashboard for a target audience, or even creating a notebook from a dashboard.

Panel automatically creates frontends based on Python syntax without requiring you to write in HTML or create style sheets with CSS. It integrates better with Jupyter Notebook than Dash or Streamlit. It's arguably the next choice if you're already using Jupyter Notebook, and Voilà is not flexible enough for your needs.

Like Streamlit, Panel works with visualizations from multiple libraries, including Bokeh, Matplotlib, HoloViews, and more, making them instantly viewable either individually or when combined with interactive widgets that control them. Being integrated with the HoloViz family, including GeoViews, Panel is especially good for handling geospatial data.

Panel objects are reactive, immediately updating to reflect changes to their state. This makes it easy to compose viewable objects and link them into simple one-off apps to do a specific exploratory task. You then can reuse the same objects in more complex combinations to build more ambitious apps. You can also share information between multiple pages so that you can build full-featured multipage apps. To see some example dashboards and how Panel works with multiple plotting libraries, visit *https://panel.holoviz.org/gallery/index.html*.

Choosing a Plotting Library

Even the simplest plotting libraries in Python require a bit of time and effort to learn, so you can't realistically learn them all. But with so many plotting choices available, how do you choose among them?

The throwaway answer is that it depends on what you're trying to do. But there's more to it than that. You need to look beyond your immediate needs. What will you be doing next year? What are your teammates and clients using? How do you position yourself for the long term, to reduce the number of libraries you need to learn?

The following sections are designed to help you choose the best library, or combination of libraries, for you. They include the libraries we've discussed so far and address the following criteria:

Size of dataset   The number of data points you need to plot

Types of plots   The types of plots you plan to make, from statistical charts to complex 3D visualizations

Format   The way you plan to present the data, such as static plots, Jupyter notebooks, interactive dashboards, and so on

Versatility   A library's range of capabilities, such as ease of use, the ability to make sophisticated plots, and dashboarding support

Maturity   The age of the library

For the first four criteria, we'll look at native, out-of-the-box functionality. Although it's always possible to extend the capabilities of a given library by using another library (for example, to enable interactivity), the assumption here is that the average user will want to avoid these types of complications.

And remember, we're only discussing a subset of the most popular plotting libraries. If you have highly specialized requirements, you'll need to perform an online search to find the most appropriate tool available.

Size of Dataset

The most important starting consideration for choosing a plotting library is the size of the datasets that you plan to use. In today's world of big data, you can't afford poor performance or memory issues during visualization. Although there are ways to decimate and otherwise manipulate large datasets so that they behave as smaller sets, you generally want to avoid this if possible.

Figure 16-21 presents a rough range of data sizes that you can practically plot with different libraries. These are more *relative* than *absolute*, as maximum limits can depend on the type of plot you're making, the hardware you're using, browser performance, whether you're working in a Jupyter notebook, and so on.

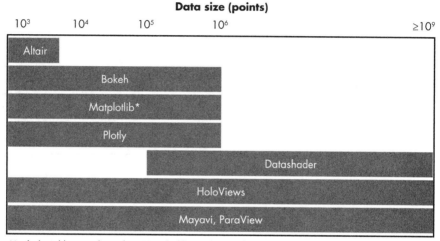

*Including libraries based on Matplotlib (seaborn, pandas, and so on)

Figure 16-21: InfoVis and SciVis libraries versus size of dataset (in number of samples)

Most of the InfoVis libraries we've discussed can plot somewhere between a hundred thousand and a million data points. Bokeh supports both Canvas- and WebGL-based plotting, and the default Canvas plotting limit may be in the hundreds of thousands. But if the WebGL JavaScript API (*https://get.webgl.org/*) is used for Bokeh, assuming it's supported for the particular type of plot involved, the limit should be similar to that for Matplotlib and Plotly.

Larger datasets require Datashader, which renders plots as images. The SciVis libraries Mayavi and ParaView can handle billions of samples using compiled data libraries and native GUI apps. Because HoloViews can use Matplotlib, Bokeh, or Plotly as its plotting backend, as well as use Datashader, it can theoretically cover the whole range shown in Figure 16-21.

Types of Plots

Knowing the types of plots that you plan to make, along with their degree of interactivity, will help you in selecting the most user-friendly tool for your needs. Figure 16-22 shows the capabilities of plotting libraries, with simple statistical plots on the left and complex 3D visualizations on the right.

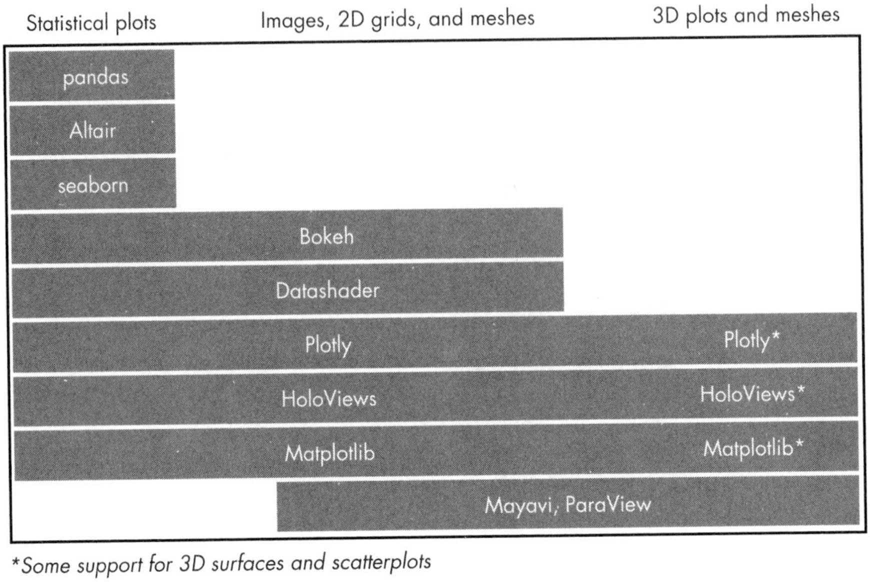

*Some support for 3D surfaces and scatterplots

Figure 16-22: InfoVis and SciVis libraries versus type of plot

All of the InfoVis libraries can handle statistical plotting. Even the SciVis tools Mayavi and ParaView have this capability to some extent, though they're hardly the best choice. Likewise, although several InfoVis libraries can generate 3D scatterplots (Figure 16-5) and meshes (Figures 16-2 and 16-6), you still need Mayavi or ParaView for high-performance visualization of large and complex 3D plots (such as Figures 16-17 and 16-18). Of the three major plotting libraries, only Bokeh has no built-in 3D capability, though it can be extended by installing other libraries.

Format

Knowing how you will present your visualizations will help you choose a library while keeping things as simple as possible. With the exception of the specialty products like Mayavi, ParaView, and the dashboarding tools, you can use most libraries to generate static plots and images to print or use in a report. You'll want to verify that you can output the smooth SVG format if you need it, though most support this option. Figure 16-23 shows more sophisticated options, ranging from Jupyter notebooks to highly interactive web applications viewed in a browser.

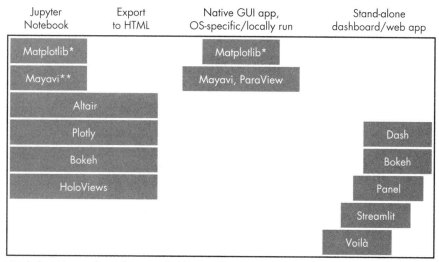

| | Jupyter Notebook | Export to HTML | Native GUI app, OS-specific/locally run | Stand-alone dashboard/web app |

*Including libraries based on Matplotlib (seaborn, pandas, and so on) **with limitations

Figure 16-23: The InfoVis and SciVis libraries versus publishing format

The dashboarding libraries are displayed so that the simplest, least flexible ones are shifted to the left and the more powerful and customizable are shifted to the right. Voilà, for example, works only with Jupyter Notebook, whereas Dash can produce enterprise-level visualizations. Bokeh operates over *WebSockets*, a library for maintaining a persistent connection between a client and server, allowing for constantly connected sessions that you can easily use for multiple back-and-forth interactions.

Versatility

Sometimes organically and sometimes by design, plotting libraries grow into "families" of a sort (Figure 16-24). The Plotly family, for example, has Plotly Express for quick and simple plotting, and Dash for dashboarding. In similar fashion, HoloViews has hvPlot and Panel, and pandas and seaborn make plotting with Matplotlib as easy as possible. With a truly versatile family, you can quickly produce plots using simple syntax, drop down into the core library to add sophisticated elements, and seamlessly share the result as a dashboard on the web.

Even though it's possible to mix and match these to a point, having to learn the syntax for multiple libraries is not very appealing. Both Plotly and HoloViews give you full built-in soup-to-nuts functionality, but that doesn't mean you're limited to just two options. The Matplotlib family can "adopt" a dashboarding library, such as Streamlit, Panel, or Voilà, whereas Chartify, Pandas-Bokeh, and hvPlot can serve as an "easy" option for Bokeh.

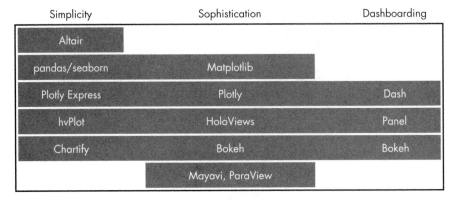

Simplicity	Sophistication	Dashboarding
Altair		
pandas/seaborn	Matplotlib	
Plotly Express	Plotly	Dash
hvPlot	HoloViews	Panel
Chartify	Bokeh	Bokeh
	Mayavi, ParaView	

Figure 16-24: Versatility of the InfoVis and SciVis libraries

Maturity

Figure 16-25 captures the relative age of the plotting libraries. The longer a library has been around, the more likely it is to be reliable, well documented, and have an established user base that produces helpful tutorials, example galleries, and extensions. Over time, users encounter bugs, learn usage patterns, and share their experiences. As a result, you'll be able to find answers to most questions at help sites like Stack Overflow (*https://stackoverflow.com/*).

Paraview, Matplotlib, and pandas have been around for a long time, whereas libraries like Voilà and Panel are more recent. Keep in mind that maturity is a somewhat scalable criterion. Wildly popular libraries will mature quickly. A good example of this is the newer dashboarding libraries Dash and Streamlit, with rapidly growing user bases constantly adding new features and supplementing the documentation.

Age ⟶

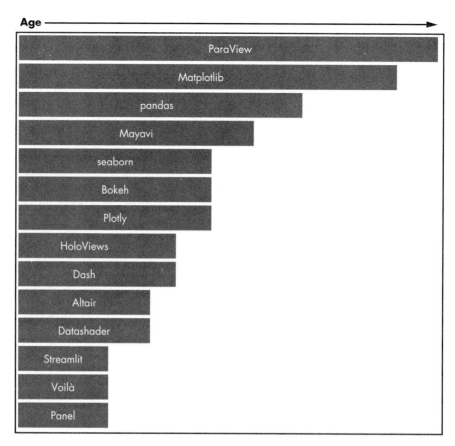

Figure 16-25: Relative age of the InfoVis and SciVis libraries

Making the Final Choice

Although it's true that the best plotting library might be dependent on your use case as well as your background and skill level, no one wants to jump from tool to tool with each new project. Still, there's a good chance you won't be able to get by with a single visualization library, especially if you need to do a range of things, including visualizing complicated 3D simulations.

If you expect to use Python *a lot*, you should look for a library, such as Matplotlib, Plotly, or the HoloViz family, that covers as much area as possible in Figures 16-21 through 16-25. These libraries may be more difficult to learn, but it will be worth it in the long run.

The case for learning Matplotlib is always strong due to its maturity, versatility, good integration with the ecosystem, and the fact that so many other libraries are built upon it. As a default plotting tool, it's a safe choice, but if you strongly favor a simpler library, all is not lost. As mentioned previously, Figures 16-21 through 16-24 assume that you're using the *native capability* of the posted libraries. They further assume that you want functionality, like zooming and panning, to work out of the box. But many other libraries exist that, with little effort, can *extend* their native capabilities. Earlier, you saw how, with one extra line of code, HoloViews could add interactivity to the static plots generated by the pandas plotting API.

With Anaconda, it's easy to install plotting libraries and play with them in Jupyter Notebook. You should take the time to experiment a little using online tutorials. If you find that you prefer a fairly simple library or one not discussed here, search for libraries that can add any missing capability. You may be able to cobble together a Frankenstein product that perfectly fits your needs.

As a final comment: the HoloViz concept is intriguing. Its goal is to provide a unified, consistent, and forward-looking plotting solution for Python. It's worth serious consideration, especially if you have a long career ahead of you.

> **NOTE** *After you choose a plotting library, you'll still need to pick a type of plot to use with your data. A great place to start is the From Data to Viz website at* https://www.data-to-viz.com/. *Here you'll find a decision tree that will help you determine the most appropriate chart based on the format of your dataset. You'll also find a Caveats page that will help you understand and avoid some of the most common data presentation mistakes.*

Summary

In this chapter, we reviewed the InfoVis libraries, used for 2D or simple 3D static or interactive representations of data, as well as the more sophisticated SciVis libraries, used for graphical representations of physically situated data. Because the InfoVis libraries address common displays such as bar charts and scattergrams, there are many libraries from which to choose.

The most popular InfoVis library is Matplotlib. Due to its maturity and flexibility, other plotting libraries, like seaborn, "wrap" Matplotlib to make it easier to use and to provide additional themes and styles. Newer plotting libraries such as Bokeh, Plotly, and Holoviews, provide much of the functionality of Matplotlib but also focus on web apps and the building of interactive dashboards. Other tools, like Datashader, address the need to efficiently plot large volumes of data.

The choice of a go-to plotting library is a personal one influenced by the tasks that you need to complete and the effort you're willing to apply. Because most users will want to focus on learning as few packages as possible, the best solution is to choose a plotting "family" that provides broad coverage of plot types, formats, dataset sizes, and so on. This will need to be weighed against the value of a mature (but possibly disjointed) solution that comes with lots of support versus newer, less well-documented libraries that try to provide a seamless, holistic approach that will stand the test of time.

17

THE GEOVIS LIBRARIES

Geospatial data is anything that includes a reference to geographical location, such as latitude and longitude, street address, and ZIP code. It's important to many fields of science, including geology, geography, meteorology, climatology, biology, archeology, anthropology, oceanography, economics, and sociology. As a result, there are lots of Python libraries dedicated to working with geospatial data.

Geospatial data comprises *vector* and *raster* data (Figure 17-1). With vector data, spatial elements (think polygons, lines, and points) are represented by x and y coordinates. Examples include road centerlines, country boundaries, and Starbucks locations. Raster data consists of a matrix of rows and columns with some information associated with each cell (think pixels). Examples include aerial photos and satellite images. These data types can be applied to maps as *layers*, letting you show just what you need for a given task, such as using only vector-based street maps for navigation. You can also use vector data to calculate distances and areas.

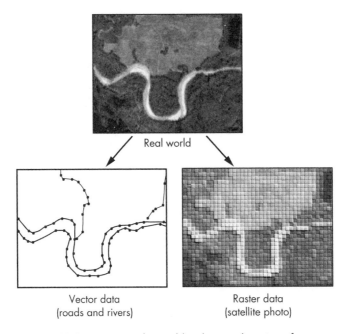

Real world

Vector data
(roads and rivers)

Raster data
(satellite photo)

Figure 17-1: Representing the world with a combination of vector and raster data

Geographic Information Systems (GIS), Global Positioning Systems (GPS), and remote sensing are examples of technology used to acquire, process, and store geospatial data. Python's flexibility makes it great for wrangling this data from a file or a database into something usable. Around 2008, major GIS platforms such as ArcGIS and QGIS adopted Python for scripting, tool-making, and analysis. As a result, Python is now the dominant computer language for performing geospatial analysis. And just as with statistical visualization, there are a daunting number of Python libraries designed to help you visualize geospatial data.

The Geospatial Libraries

The purpose of geospatial libraries is to keep track of and use spatial object types (like points and polygons), spatial reference systems (for projecting the Earth's curved surface onto a plane), geography and geometry formats (for measuring distances and areas accurately or quickly), common GIS data formats (for input/output), spatial indexing (to speed up processing), and map decorators (such as country borders and coastlines). Most will let you create animations, either by converting frames to MP4 or directly as live animation.

Table 17-1 lists some of the more important and popular geospatial libraries, along with a few specialty libraries. We'll take a high-level look at several of these in the sections that follow.

Table 17-1: Python's More Important Geospatial Libraries

Library	Description	Website
GeoPandas	GIS library meets "pandas with geometry"	https://geopandas.org/
Cartopy	Tools for projection-aware plots with Matplotlib	https://scitools.org.uk/cartopy/
geoplot	Cartopy extension ("seaborn for geospatial")	https://residentmario.github.io/geoplot/
Plotly	Easy interactive maps	https://plotly.com/python/maps/
folium	Easy interactive maps with low resource usage	https://python-visualization.github.io/folium/
ipyleaflet	Jupyter-LeafletJS bridge based on ipywidgets	https://github.com/jupyter-widgets/ipyleaflet/
GeoViews	Geographic plots with HoloViews and Cartopy.	http://geoviews.org/
KeplerGL	Tools to visualize large datasets in Jupyter	https://docs.kepler.gl/docs/keplergl-jupyter/
pydeck	Large-scale interactivity tools optimized for Jupyter	https://pydeck.gl/
PyGMT	Python wrapper for Generic Mapping tools	https://www.pygmt.org/
Bokeh	Reactive plots including on Google Maps	https://docs.bokeh.org/
EarthPy	Helper functions for working with spatial data	https://earthpy.readthedocs.io/
gmplot	Matplotlib-like interface to plot on Google Maps	https://github.com/gmplot/gmplot/
MovingPandas	Tools to track and analyze movement data	https://anitagraser.github.io/movingpandas/
cuSpatial	GPU acceleration tool for common spatial operations	https://github.com/rapidsai/cuspatial/

NOTE *The plotting examples in this chapter are intended to demonstrate the complexity of the code and the types of plots produced. You're not expected to run the code snippets, because most of the libraries discussed do not come preinstalled with Anaconda. If you do want to test them for yourself, you can find installation instructions in the product web page cited in each section. I recommend that you install them all in a dedicated conda environment (see Chapter 2) rather than dump them in your base environment.*

GeoPandas

GeoPandas is the most popular open source library for parsing geospatial data in Python. As you can guess from the name, it extends the data types used by pandas (see "pandas" on page 403) and makes working with geospatial vector data similar to working with tabular data. It also enables operations in Python that would otherwise require a dedicated spatial database such as PostGIS.

A *GeoDataFrame* in GeoPandas looks a lot like a tabular DataFrame in pandas but with a special "geometry" column for the location data (Figure 17-2).

	name	geometry
0	Vatican City	POINT (12.45339 41.90328)
1	San Marino	POINT (12.44177 43.93610)
2	Vaduz	POINT (9.51667 47.13372)
3	Luxembourg	POINT (6.13000 49.61166)
4	Palikir	POINT (158.14997 6.91664)

Figure 17-2: The geometry column (boxed) distinguishes a GeoDataFrame from a DataFrame.

This geometry column bundles together both the *type* of geometric object (Table 17-2) and the *coordinates* (as longitude and latitude) needed to draw it.

Table 17-2: Geometries Used in GeoPandas

Geometry type	Description
Point	A point
MultiPoint	A set of points
LineString	A line segment
MultiLineSting	A sequence of connected line segments
LinearRing	A closed collection of lines (zero-area polygon)
Polygon	A closed shape defined by a sequence of points
MultiPolygon	A collection of polygons

GeoPandas uses not only pandas but several other important open source libraries to produce a simple and convenient framework for handling geospatial data. It relies on the capabilities of *Shapely* (*https://pypi.org/project/Shapely/*) for working with planar geometric shapes (such as street centerlines or country boundary polygons), Fiona (*https://pypi.org/project/Fiona/*) for reading and writing geographic data file formats, pyproj (*https://pypi.org/project/pyproj/*) for handling projections, Matplotlib for plotting, and descartes (*https://pypi.org/project/descartes/*) for integrating Shapely geometry objects with Matplotlib.

As a result, you can plot a map from a GeoSeries or GeoDataFrame with only a couple of lines of code:

```
import geopandas as gpd

world = gpd.read_file(gpd.datasets.get_path('naturalearth_lowres'))
world.plot();
```

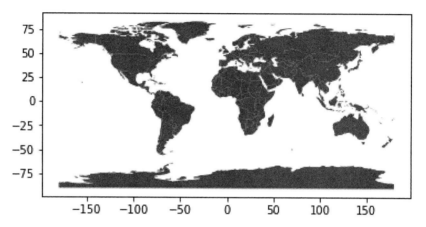

In this Jupyter Notebook example, the world variable represents a GeoDataFrame made from one of GeoPandas' internal global datasets. Of course, this simple plot can be customized further. Style options that you can pass to Matplotlib, especially those for lines, will work with the plot() method.

Here's an example of a choropleth map—where regions are shaded based on a data value—for population by country:

```
import geopandas as gpd

world = gpd.read_file(gpd.datasets.get_path('naturalearth_lowres'))
world = world[(world.name != 'Antarctica')]  # Omit Antarctica.
world.plot(column='pop_est',
           legend=True,
           legend_kwds={'label': "Population by Country in Billions",
                        'orientation': "horizontal"});
```

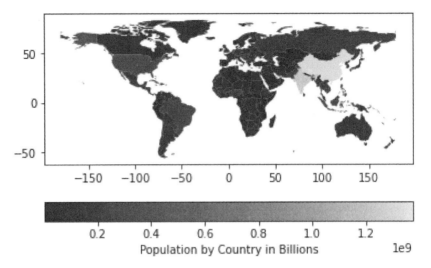

Population by Country in Billions   1e9

With the world data loaded as a GeoDataFrame, it's easy to filter the data and replot it. In the previous plot, we removed Antarctica, given that it has no permanent population. Now let's look at all the countries with a population greater than 300 million by changing one line of code:

```
world = world[(world.pop_est > 300000000) & (world.name != 'Antarctica')]
```

Rerunning the code block reveals only China, India, and the United States:

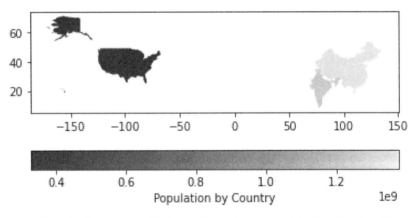

Population by Country   1e9

Historically, you could plot only static maps with GeoPandas. Now, thanks to Contextily (*https://github.com/geopandas/contextily*) for base maps, and IPYMPL (*https://github.com/matplotlib/ipympl*) for interactive Matplotlib plots in Jupyter, it's possible to make interactive maps with GeoPandas. Likewise, hvPlot, built on HoloViews (see Chapter 16), uses an interactive Bokeh-based plotting API to add zooming, panning, querying, sliders, and clickable legends to both pandas and GeoPandas output (Figure 17-3).

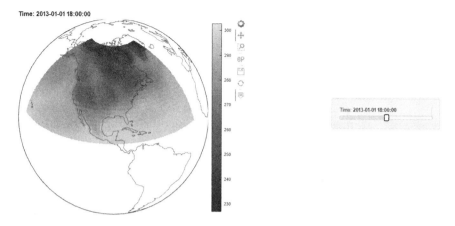

Time: 2013-01-01 18:00:00

Time: 2013-01-01 18:00:00

Figure 17-3: An interactive hvPlot with toolbar and slider widget (courtesy of holoviz.org*)*

With the Contextily library installed and imported, GeoPandas can support *tile-based* maps as well as the outline-based geo-maps shown previously. A tile map (or *web map tile*) is a map displayed in a browser by seamlessly joining dozens of individually requested image or vector data files over the Internet. The street and terrain layers in Google Maps are familiar examples of tile-based maps. Contextily provides easy access to popular tile sources like OpenStreetMap and Stamen, letting you add backgrounds similar to those in Google Maps (Figure 17-4).

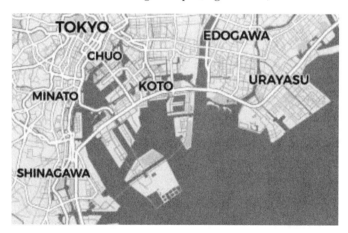

Figure 17-4: A tile map of part of Tokyo, Japan

Like pandas, GeoPandas operates on a single core, but it also supports *spatial indexing*, a technique that can significantly boost performance when querying large geospatial datasets. GeoPandas can generate spatial indexes, automatically in some cases and manually in others, by letting you call the sindex attribute on a GeoDataFrame. In addition, a new library, *geofeather* (*https://pypi.org/project/geofeather/*), can significantly speed-up reading and writing standard spatial file formats (such as *shapefile*).

GeoPandas is a good all-purpose tool if you're not planning to perform complex data transformations or work with millions of records. Plotting with this tool requires knowledge of the somewhat arcane Matplotlib syntax, and add-ons are needed to add interactivity. GeoPandas works best with vector data but you can also perform limited raster processing using *rasterio* (*https://rasterio.readthedocs.io/en/latest/*). Fortunately, many other geospatial libraries work well with GeoPandas, so you have the option of organizing your data in GeoPandas and plotting it with a different tool.

Cartopy

Cartopy is an open source library for producing maps and performing geospatial analyses. It's engineered for scientists and maintained by an active development community. Cartopy is an extension of Python's standard plotting library, Matplotlib, and makes use of other libraries, including NumPy, Shapely, and PROJ.4.

Cartopy prides itself on being very "projection aware." That is, it can handle a large number of projections (Figure 17-5) and transform points, lines, vectors, polygons, and images between these projections. It also pairs well with GeoPandas, allowing you to easily create cartographically accurate maps while using raster data more easily than in GeoPandas alone. If you use Matplotlib for basic plots, Cartopy lets you extend your skills into cartography with little extra effort.

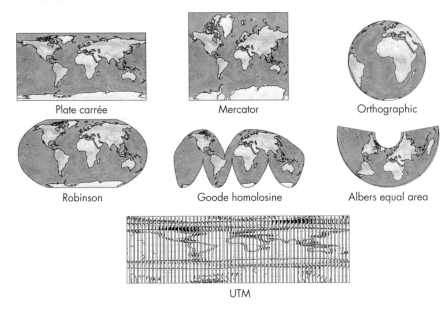

Figure 17-5: A few of the many map projections available in Cartopy

As with many other geospatial libraries, you can make a basic map using only a few lines of code:

```
import cartopy.crs as ccrs
import matplotlib.pyplot as plt

ax = plt.axes(projection=ccrs.Robinson())
ax.coastlines()
plt.show()
```

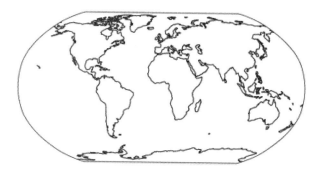

Depending on your setup, Cartopy can post up to a million points quickly but performance noticeably drags with larger datasets. You can see more Cartopy plot examples on its gallery page (*https://scitools.org.uk/cartopy/docs/latest/gallery/index.html*) and a list of supported map projections at *https://scitools.org.uk/cartopy/docs/latest/crs/projections.html*.

Geoplot

Geoplot is a fairly new, high-level, open source geospatial plotting library. As an extension to Cartopy and Matplotlib, it claims to be the "seaborn of geospatial," which means that it builds on the underlying libraries to make mapping easy.

Geoplot is designed to work well with GeoPandas input and provides a selection of easy-to-use geospatial visualizations (presumably, 90 percent of what you will ever need). And because geoplot is built on Cartopy, it can take advantage of Cartopy's extensive list of map projections.

A standout feature for geoplot is the *cartogram*, a thematic map of polygons, such as provinces or states, whose geographic size is warped to be proportional to a selected variable, like population, gross domestic product, or obesity level. In the following example from geoplot's plot reference page, you use geopandas and one of geoplot's native datasets, contiguous_usa, to easily generate a cartogram of the US population by state:

```
import geopandas as gpd
import geoplot as gplt
import geoplot.crs as gcrs

contiguous_usa = gpd.read_file(gplt.datasets.get_path('contiguous_usa'))
gplt.cartogram(contiguous_usa, scale='population',
               projection=gcrs.AlbersEqualArea(),
               color='black');
```

In this cartogram, California, the most populous state, is shown at its true size. The remaining states are reduced in size based on the relative size of their population.

Geoplot also lets you produce Sankey diagrams. These are a type of flow diagram in which the width of lines and arrows is proportional to the volume of movement being visualized, such as the flow of traffic over city streets (Figure 17-6). The most famous Sankey diagram depicts Napoleon's infamous Russian campaign and retreat from Moscow.

Figure 17-6: Sankey diagram of streets in Washington DC by average daily traffic (courtesy of geoplot)

Like GeoPandas, geoplot makes only static maps. With some extra work, however, such as writing your figure to HTML and using the mplleaflet library, you can enable interactivity like zooming and panning.

Geoplot lets you easily make maps if you're comfortable with giving up a lot of design control. To move beyond the basic functionality and produce highly customized maps, you'll need to be familiar with Matplotlib. And even though the core documentation is not bad, the immaturity of geoplot means you might have trouble finding tutorials or examples to match your specific use cases. Geoplot is also in "maintenance" state, with no new features planned.

Plotly

Plotly and Plotly Express, introduced in Chapter 16, have extensive geospatial data visualization capabilities. They offer many mapping options and the Plotly Express API is easy to use. You can make an animated choropleth map with a single line of code and deploy it to the web using Dash.

Plotly maps are useful for quickly exploring data, identifying outliers, and recognizing trends. You can use the convenience of GeoPandas, or if you have latitude and longitude in columns, plot straight from a pandas DataFrame. The following Jupyter Notebook example, using Plotly Express, turns a Plotly dataset on worldwide volcanoes into a highly interactive figure with only a few lines of code.

```
❶ import pandas as pd
  import plotly.express as px

  f = "https://raw.githubusercontent.com/plotly/datasets/master/volcano_db.csv"
  df = pd.read_csv(f, encoding="iso-8859-1")
❷ fig = px.scatter_geo(data_frame=df,
                       lat='Latitude',
                       lon='Longitude',
                       hover_name='Type',
                       hover_data={'Type':False,
                                   'Country':True,
                                   'Volcano Name':True},
                       symbol='Type',
                       color='Type',
                       projection='orthographic')
  fig.show()
```

Type
- ● Shield volcano
- ◆ Stratovolcano
- ▪ Maar
- ✕ Volcanic field
- + Submarine volcano
- ▪ Stratovolcanoes
- ◆ Caldera
- ▪ Complex volcano
- ✕ Fissure vent
- + Compound volcano
- ● Cinder cone
- ◆ Pyroclastic shield
- ▪ Unknown
- ✕ Pyroclastic cones
- + Scoria cone
- ● Scoria cones
- ◆ Cinder cones
- ▪ Lava dome
- ✕ Lava cone

Most of the code consists of importing libraries ❶ and loading the data before performing the actual plotting ❷. Within this plot, you can use the cursor to grab and rotate the map as if it were a real three-dimensional globe. You can hover the cursor over a volcano marker and get a pop-up window listing the type of volcano along with other information such as its location, country, and name. You also have the option to make markers clickable so that pop-up windows appear only when you deliberately use a mouse button.

If you look at the upper-right corner of this plot, you'll see a toolbar that lets you take screenshots, pan, zoom, and so on. These tools are hugely helpful, especially when you need to resolve closely packed data points, such as the numerous volcanoes in Iceland (Figure 17-7).

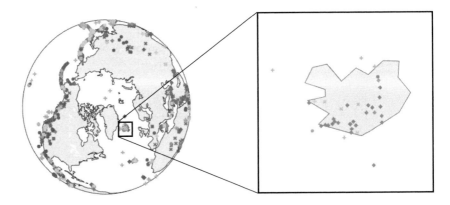

Figure 17-7: Plotly Express reposts data at the appropriate scale for the zoom level.

You can also make 3D surface plots with Plotly, whose automatic toolbars permit rotations about multiple axes. Here's an example for a single volcano:

```python
import pandas as pd
import plotly.graph_objects as go

df = pd.read_csv(
"https://raw.githubusercontent.com/plotly/datasets/master/volcano.csv")
fig = go.Figure(data=[go.Surface(z=df.values)])
fig.update_layout(title='Volcano',
                  autosize=False,
                  width=600, height=600,
                  margin=dict(l=65, r=50, b=65, t=90))
fig.show()
```

Volcano

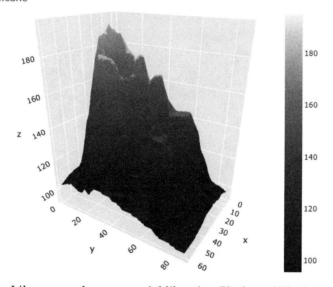

Like most other geospatial libraries, Plotly and Plotly Express support *tile-based* maps (see Figure 17-4) for adding streets, terrain, imagery, and so on. Unlike GeoPandas, you can access these directly, without the need for a separate library, like Contextily.

If you want to quickly build interactive plots in which you can query a map by hovering the cursor over a region, or position user-input widgets like sliders on the same screen as the map, Plotly and Plotly Express are sound choices. And with Plotly's Dash library (see "Dash" on page 446), you can seamlessly convert your work into a dashboard.

folium

The open source *folium* library lets you visualize maps using *Leaflet.JS*, a powerful JavaScript library for building interactive web-mapping applications on most mobile and desktop platforms. First released in 2013, folium is extremely popular, and as a result, you'll find a wealth of material on the internet to help you learn how to use it and customize it to your needs.

With folium, you can select from a number of *tilesets* from mapping services like OpenStreetMap, Mapbox, and Stamen. Tilesets are collections of raster or vector data broken up into a uniform grid of square tiles with up to 22 preset zoom levels. They let you produce beautiful leaflet maps with no effort at all:

```
import folium

map = folium.Map(location=[29.7, -95.2147])
map
```

This Jupyter Notebook example uses the OpenStreetMap tile by default. The location coordinates for the center of the map are in latitude and longitude (this can trip you up, as many libraries use the modern longitude-latitude order). You can look up these values for an address using tools like *LatLong.net* (*https://www.latlong.net/geo-tools*) or by simply doing an online search for a geographic feature's coordinates. It's also possible to query folium maps for this information using your cursor. This map is also scalable; when you zoom in, you get more and more detailed information until you exhaust the available tileset zoom levels.

Another strength of folium is its support of *markers*. You've probably seen these teardrop-shaped icons used to identify search locations on Google maps. Folium comes with several predefined markers and will also let you build a custom marker by using an image or by accessing free icon libraries. You can also include a pop-up window with content. Let's look at an example:

```
import folium

map = folium.Map(location=[37.15, -111.1], tiles='stamen terrain')
folium.Marker(location=[37.1, -111.17],
              popup="Water Sample #2",
              icon=folium.Icon(color="black")).add_to(map)
map
```

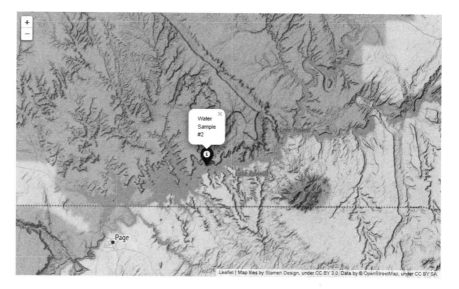

This code uses the "Stamen Terrain" tile depicting the area around Lake Powell in Utah. The marker represents the location of a water quality sample, and clicking it reveals the sample number.

Now let's revisit the volcanoes dataset used in "Plotly" on page 467. If you're running the code, you can download the volcano icon from sites such as Free onlinewebfonts.com (*https://onlinewebfonts.com/fonts*) or Iconfinder (*https://iconfinder.com/*).

```
import pandas as pd
import folium
from folium import plugins

f = "https://raw.githubusercontent.com/plotly/datasets/master/volcano_db.csv"
df = pd.read_csv(f, encoding="iso-8859-1")
map = folium.Map(tile='Stamen Terrain', control_scale=True)
for index, row in df.iterrows():
    volcano_icon = folium.features.CustomIcon('volcano_icon.png',
                                              icon_size=(25, 25))
    folium.Marker(location=(row['Latitude'], row['Longitude']),
                  popup=row['Type'],
                  icon=volcano_icon,
```

```
              tooltip=(row['Type'],
                       row['Country'],
                       row['Volcano Name'])
          ).add_to(map)
mini_map = folium.plugins.MiniMap(toggle_display=True)
map.add_child(mini_map)
map
```

This script produces another world map of volcano locations. Figure 17-8 is this map zoomed in to Iceland, similar to Figure 17-7. Note the custom volcano icons, terrain background, hover window, index map at lower right, and scale bar at lower left. All with just a few lines of code.

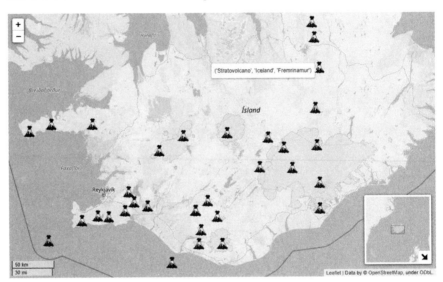

Figure 17-8: Icelandic volcanoes plotted with folium

Because folium embeds lots of information, file sizes can grow quite large. For example, the previous code produced a 138MB notebook file.

Combining folium with the popular GeoPandas library is a great way to visualize georeferenced data. Imagine that you're studying the urban "heat island" effect around Paris, France. You've recorded thousands of temperature measurements east of the city and you're using GeoPandas to manipulate this data. With a folium heatmap, the measurements will aggregate or separate depending on the map's zoom level (Figure 17-9). You can also add a time series, making it possible for you to see temperature variations throughout the day, month, year, and so on. And with folium's `MarkerCluster` plug-in, you can adapt this same technique to individual markers. Just don't try adding a legend; folium's support for this feature extends only to choropleth maps.

Figure 17-9: A heatmap of temperature data zoomed out (left) versus zoomed in (right)

The folium library is designed for simplicity, performance, and usability. By combining the data analysis capabilities of Python libraries like GeoPandas with the mapping strengths of LeafletJS, folium lets you generate maps with multiple layers of data representation. It's extremely easy to include useful backgrounds such as street maps and terrain maps, and there are lots of plug-ins available to extend folium's functionality (see *https://python-visualization.github.io/folium/plugins.html#folium-plugins/*).

ipyleaflet

The *ipyleaflet* open source interactive widgets library is based on ipywidgets (*https://github.com/jupyter-widgets/ipywidgets/*). Like folium, ipyleaflet wraps Leaflet.JS to bring mapping capabilities to both Jupyter Notebook and JupyterLab. Although folium is considered easier to use, ipyleaflet is considered more customizable and provides more avenues for interactivity.

Everything in ipyleaflet, such as tile maps and markers, is interactive, and you can dynamically update attributes from Python or the Notebook interface. And because ipyleaflet is built upon ipywidgets, you can write programs that use widgets to capture user input.

Suppose that you're compiling statistics on terrestrial impact craters. In this example, you use the measure control widget and your mouse to interactively find both the radius and area of Aorounga Crater in the Republic of Chad:

```
from ipyleaflet import Map, MeasureControl, basemaps

m = Map(basemap=basemaps.OpenTopoMap,center=(19.0933, 19.2431), zoom=11)
measure = MeasureControl(position='bottomleft',
                         active_color = 'black',
                         primary_length_unit = 'kilometers')
m.add_control(measure)
measure.completed_color = 'red'
m
```

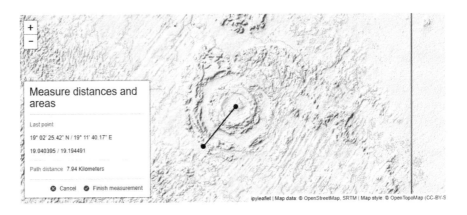

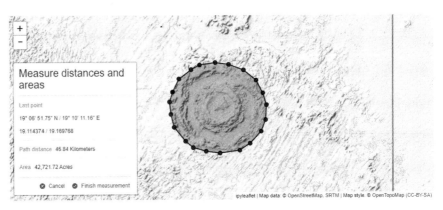

Clicking the square (■) icon on the map activates the Measure Distances and Areas tool. You can then click two locations to get the linear measurement between them or draw a polygon to get an area, as shown in the preceding example. You can even customize the units.

Another interesting control option is the *SplitMap*, which lets you compare a different set of layers at the same location. Imagine that you're studying a night view of Europe and you're curious about which city is causing a bright cluster of lights. With only a few lines of code, you can generate a dual-layer display to answer the question:

```
from ipyleaflet import Map, basemaps, basemap_to_tiles, SplitMapControl

m = Map(center=(42.6824, 365.581), zoom=5)
left_layer = basemap_to_tiles(basemaps.Esri.WorldStreetMap)
right_layer = basemap_to_tiles(basemaps.NASAGIBS.ViirsEarthAtNight2012)
control = SplitMapControl(left_layer=left_layer, right_layer=right_layer)
m.add_control(control)
m
```

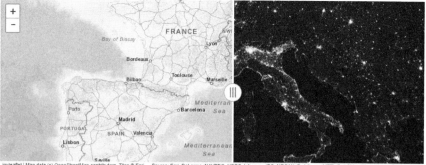

ipyleaflet | Map data (c) OpenStreetMap contributors, Tiles © Esri — Source: Esri, DeLorme, NAVTEQ, USGS, Intermap, iPC, NRCAN, Esri Japan, METI, Esri China (Hong Kong), Esri (Thailand), TomTom, 2012, Imagery provided by services from the Global Imagery Browse Services (GIBS), operated by the NASA/GSFC/Earth Science Data and Information System (ESDIS) with funding provided by NASA/HQ.

The previous code produces a "split" map with cities and streets on the left and the nighttime satellite view on the right. You can grab the round "|||" marker at the center of the screen and drag it to each side to extend one of the maps at the expense of the other (Figure 17-10). This lets you peek beneath the nighttime map to see the cities and roads, without the need to clutter one map by combining it with another or by adjusting the upper map's transparency. You can also zoom in to see smaller cities.

ipyleaflet | Map data (c) OpenStreetMap contributors, Tiles © Esri — Source: Esri, DeLorme, NAVTEQ, USGS, Intermap, iPC, NRCAN, Esri Japan, METI, Esri China (Hong Kong), Esri (Thailand), TomTom, 2012, Imagery provided by services from the Global Imagery Browse Services (GIBS), operated by the NASA/GSFC/Earth Science Data and Information System (ESDIS) with funding provided by NASA/HQ.

Figure 17-10: The SplitMap boundary dragged to the right

The *Magnifying Glass* is a particularly fun feature that lets you view details without changing the overall zoom level of a map. When it's active, you simply move a circle over a map with your cursor to get a zoomed-in view within the circle (Figure 17-11). It works at any zoom level and with all of the available base maps.

Figure 17-11: The Magnifying Glass option in ipyleaflet

Much of this functionality, along with things like marker clustering, is also available in folium, though you might need to use a plug-in (*https:// python-visualization.github.io/folium/plugins.html#folium-plugins/*) to replicate what you can do in ipyleaflet. However, this functionality overlap does not include ways to get user interactions such as selections back into Python for further processing, as folium provides only a one-way path from Python into a JavaScript map.

NOTE *Similar to ipyleaflet,* Jupyter-gmaps *(https://github.com/pbugnion/gmaps/) is also built upon the Jupyter interactive widgets framework but bridges between Jupyter and Google Maps rather than Leaflet.JS.*

GeoViews: The HoloViz Approach

The HoloViz-maintained libraries, discussed in Chapter 16 (see Figure 16-9), provide a unified solution for working with geospatial data. This includes dashboards and other types of interactive visualization. Within this collection of open source libraries, HoloViews provides a lot of support for geospatial data, including the ability to perform basic geoscience work.

For more advanced work, especially work involving map projections, HoloViz includes a dedicated geospatial library called *GeoViews*. Built on HoloViews, and with geographic plot types based on the Cartopy library, GeoViews can use either Matplotlib or Bokeh as a plotting backend.

GeoViews lets you work with large, multidimensional geographic datasets, quickly visualizing subsets or combinations with access to the underlying raw data. It's designed to work with Iris and xarray libraries and can accept multiple data formats including NumPy arrays, pandas DataFrames, and GeoPandas GeoDataFrames. In these cases, the data is wrapped in a HoloViews or GeoViews object that provides instant interactive visualizations (see "HoloViews" on page 436). Geographic projections use the extensive Cartopy coordinate reference system.

Like other geospatial libraries, GeoViews gives you access to all kinds of useful databases, polygon sets (such as for country boundaries), and tile maps of streets and terrain. Plots can be made from only a few lines of code, as in this Jupyter Notebook example from the official web page:

```
import geoviews as gv
import geoviews.feature as gf
from cartopy import crs

gv.extension('bokeh')
(gf.ocean + gf.land + gf.ocean * gf.land * gf.coastline * gf.borders).opts(
'Feature', projection=crs.Geostationary(), global_extent=True, height=325).
cols(3)
```

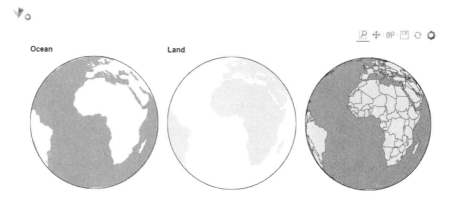

GeoViews' support for GeoPandas data structures allows for easy plotting of shapefiles and choropleths. Here's an example of plotting a human population choropleth map using a GeoPandas dataset:

```
import geopandas as gpd
import geoviews as gv
from cartopy import crs

gv.extension('bokeh')
gv.Polygons(gpd.read_file(gpd.datasets.get_path('naturalearth_lowres')),
            vdims=['pop_est', ('name', 'Country')]).opts(width=600,
            projection=crs.Robinson())
```

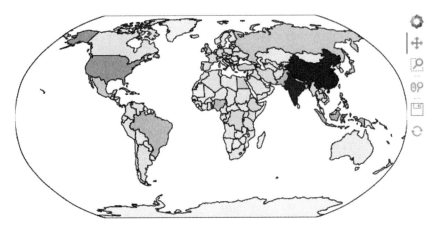

Finally, here's the volcanoes example with a twist. Because GeoViews is part of HoloViz, you have the option of plotting with hvPlot, which I personally find easier to use (much like Plotly Express versus Plotly):

```
import pandas as pd
import holoviews as hv
import hvplot.pandas

f = "https://raw.githubusercontent.com/plotly/datasets/master/volcano_db.csv"
df = pd.read_csv(f, encoding="iso-8859-1")

# Reassign the dataframe with only 3 volcano types:
❶ df = df[(df['Type'] == 'Cone') |
         (df['Type'] == 'Stratovolcano') |
         (df['Type'] == 'Shield volcano')]
❷ marker = hv.dim('Type').categorize({'Cone': 'triangle',
                                       'Shield volcano': 'circle',
                                       'Stratovolcano': 'square'})
size = hv.dim('Type').categorize({'Cone': 6,
                                   'Shield volcano': 5,
                                   'Stratovolcano': 4})
df.hvplot.points('Longitude', 'Latitude',
                 color='Type',
                 marker=marker,
                 size=size,
                 hover_cols=['Volcano Name'],
                 coastline=True)
```

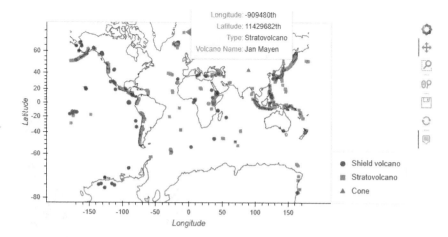

In this case, all the volcano types other than shield volcanoes, strato-volcanoes, and cones were dropped from the DataFrame ❶. The map was then customized to plot these volcano types with unique shapes ❷, sizes, and colors. Although not shown here, you also have the option of assigning a default shape and size.

Note the toolbar along the right side, with icons for panning, zooming, saving, and so on, and the customizable hover window. Unfortunately, there's no tool for rotating a globe in an orthographic projection as you can do with Plotly Express, as hvPlot uses only Bokeh rather than Plotly as a plotting backend.

A major selling point for GeoViews is that it's part of a holistic, forward-looking solution designed to satisfy all your plotting and mapping needs. On the downside, documentation is somewhat limited compared to other libraries.

KeplerGL

KeplerGL JupyterLab extension is an advanced open source geospatial library built on top of Mapbox GL (*https://www.mapbox.com/*) and deck.gl (*https://deck.gl/*). The latter is a WebGL (GPU)-powered framework for visually exploring large datasets using a layered approach. It has an extensive catalog of layer types for bitmaps, icons, point clouds, grids, contours, terrain, and more (see *https://deck.gl/docs/api-reference/layers/*).

Uber developed KeplerGL (*https://kepler.gl/*) as a web-based tool to make it easier for users with a variety of experience and skill levels to create meaningful data visualizations. It's designed for working with large geospatial datasets, especially those related to mobility. It includes impressive functionality including a GUI (Figure 17-12) that lets you drag and drop datasets, use built-in time–series animations, visualize in 3D, handle millions of data points, perform spatial aggregations on the fly, and customize maps by tweaking colors, changing sizes, filtering, and so on.

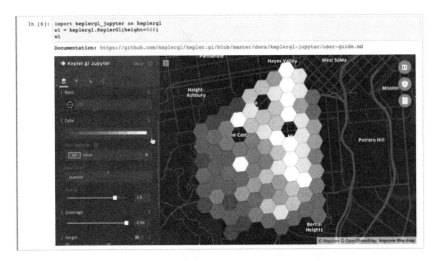

```
In [6]: import keplergl_jupyter as keplergl
        w1 = keplergl.KeplerGl(height=500)
        w1
```

Documentation: https://github.com/keplergl/kepler.gl/blob/master/docs/keplergl-jupyter/user-guide.md

Figure 17-12: The KeplerGL interface for customizing maps works in JupyterLab (courtesy of KeplerGL).

With the KeplerGL GUI running in Jupyter, you can eschew Python completely. You can drag and drop a data file into the browser, visualize it with different map layers, explore it by filtering and aggregating it, and eventually export the final visualization as a static map or an animated video. The website will walk you through the map-making workflow (*https:// docs.kepler.gl/docs/user-guides/b-kepler-gl-workflow/*) and show you how to use the GUI's friendly menus (Figure 17-13).

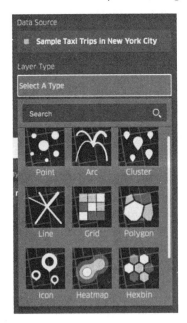

Figure 17-13: A KeplerGL interface menu for selecting a map layer type (courtesy of KeplerGL)

KeplerGL provides a set of Mapbox basemaps for backgrounds including ones for land, water, roads, building footprints, 3D buildings, and labels. You have to register with Mapbox, and the free plan comes with 50,000 map downloads per month, sufficient for most small applications. You're also restricted to using data in CSV, GeoJSON, pandas DataFrame, or GeoPandas GeoDataFrame formats, which rules out live streaming.

Setting up and using KeplerGL is a little more involved than other geospatial libraries. It works in JupyterLab and (currently) must be installed using Python's standard package manager (pip) rather than with conda or conda-forge.

pydeck

The *pydeck* graphics library is a set of Python bindings, optimized for a Jupyter Notebook environment, for making spatial visualizations using deck.gl. As mentioned in the previous section, deck.gl is a WebGL-powered framework for visually exploring large datasets using a layered approach.

The pydeck library grants you access to the full deck.gl layer catalog in Python. You can create beautiful deck.gl maps (Figure 17-14) without using a lot of JavaScript, and you can embed these maps in a Jupyter notebook or export them to a stand-alone HTML file. The library has been designed to work in tandem with popular JavaScript base map providers, especially Mapbox, but other map tile solutions, like OpenStreetMap, may come with different levels of compatibility.

Figure 17-14: Personal injury road accidents in Great Britain (https://pydeck.gl/gallery/ hexagon_layer.html)

Pydeck supports large-scale updates, such as color changes or data modification, to hundreds of thousands of visualized data points in 2D and 3D. And like ipyleaflet, there's support for two-way communication, by which data selected in a visualization can be passed back to the Jupyter Notebook kernel. For example, you can pass geometry data loaded into a map from a government source into a pandas DataFrame.

Let's visit our volcano database yet again. The following code snippet, entered in a Jupyter notebook, loads the data as a pandas DataFrame and then produces a global map zoomed-in on the Horn of Africa:

```
import pandas as pd
import pydeck as pdk

f = "https://raw.githubusercontent.com/plotly/datasets/master/volcano_db.csv"
df = pd.read_csv(f, encoding="iso-8859-1")
❶ layer = pdk.Layer('ScatterplotLayer',
                    df,
                    get_position=['Longitude', 'Latitude'],
                    auto_highlight=True,
                    get_radius=10_000,
                    radius_min_pixels=1,
                    radius_max_pixels=10_000,
                    get_fill_color='[255, 255, 255]',
                    pickable=True)
❷ view_state = pdk.ViewState(longitude=42.59, latitude=11.82,
                             zoom=5, min_zoom=1, max_zoom=8,
                             pitch=0, bearing=0)
r = pdk.Deck(layers=[layer], initial_view_state=view_state)
❸ r.to_html("scatterplot_layer.html")
```

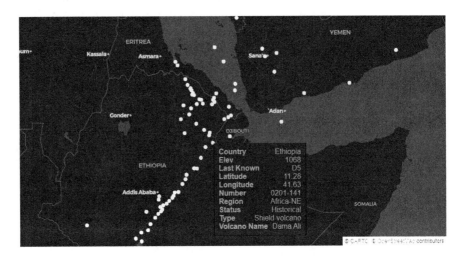

After importing the libraries and reading the CSV file in as a DataFrame, you call pydeck's Layer method and choose a ScatterplotLayer ❶. In the process, you also make the points 10 km in radius, color them white, and make them "pickable" so that you can hover the cursor over each point to see the associated data in the DataFrame (as shown on the map for the "Dama Ali" volcano).

Next, you need to set the view_state, which tells pydeck where to center the map, how far in to zoom, and the pitch and bearing ❷. These last two let you produce a tilted view, like the one in Figure 17-14. You end by telling pydeck how to render the map and save it as an HTML file ❸.

If you play with this example for a few minutes, some issues become apparent. To assign each type of volcano a unique color, you need to create a new column in the DataFrame using the following code:

```
color_lookup = pdk.data_utils.assign_random_colors(df['Type'])
df['color'] = df.apply(lambda row: color_lookup.get(row['Type']), axis=1)
```

Likewise, if you want a legend, you'll need to use an external library like Matplotlib to make one (search for `matplotlib.pyplot.colorbar`) and then render it beside your pydeck visualization. Compare this to the Plotly Express and hvPlot examples, in which both of these tasks were either extremely intuitive or completely automatic.

These issues are partly a function of pydeck's immaturity and might be addressed by the time you read this. However, the current takeaway is that pydeck is best reserved for data analytics use cases with large datasets—and that's where it excels.

With pydeck, you can use Python to access *Google Earth Engine* (*https:// earthengine.google.com/*), a cloud computing platform for processing satellite imagery and other Earth observation data. Earth Engine hosts a multi-petabyte catalog of geospatial datasets and satellite imagery that includes historical earth images going back more than 40 years. It ingests images on a daily basis, stores them in a public data archive, and then makes them freely available for global-scale data mining by academic, nonprofit, business, and government users.

In addition to allowing access to a large warehouse of geospatial data, Earth Engine provides the computational power, APIs, and other tools needed to analyze the large datasets. According to the website, these tools provide planetary-scale analysis capabilities that allow scientists, researchers, and developers to detect changes, map trends, and quantify differences on the Earth's surface.

The *pydeck-earthengine-layer* wrapper (*https://github.com/UnfoldedInc/ earthengine-layers/tree/master/py/*) connects pydeck to Google Earth Engine using deck.gl layers for the Earth Engine API (*https://earthengine-layers.com/*). This makes it possible to visualize enormous geospatial datasets with Python. The pydeck wrapper, released in 2020, can be easily installed through conda-forge. To use it, you'll need to authenticate with an Earth Engine–enabled Google Account (you can sign up at *https://earthengine .google.com/new_signup/*).

Whereas Earth Engine visualizations are typically raster based, pydeck gives you the ability to mix raster- and vector-based graphics to open up new visualization opportunities. You can add interactivity, such as hover-based tooltips, and you can interpret Earth Engine data as terrain elevations to display them in 3D. You can even upload and manipulate your own datasets using Earth Engine *platform* (*https://earthengine.google.com/platform/*).

To help you get started, Earth Engine comes with many prepackaged datasets (*https://developers.google.com/earth-engine/datasets/*) and example case studies (*https://earthengine.google.com/case_studies/*). With pydeck and Earth

Engine, you can monitor rainfall and floods, vegetation changes, forest fires and deforestation, urban sprawl, and more, without the need to download thousands of satellite images to your computer.

If you expect to work regularly with "planetary scale" datasets, pydeck is a great solution. It's also easier to install than KeplerGL because you can use conda-forge. Though it can't compete with Plotly Express or hvPlot for making quick and simple plots on smaller datasets, that gap should start to close as the product matures.

Bokeh

Bokeh, introduced in Chapter 16, is one of the major plotting libraries for Python. Like the Matplotlib and Plotly libraries, it comes with its own geospatial capabilities (*https://docs.bokeh.org/en/latest/docs/user_guide/geo.html*).

Bokeh can accept geospatial data from multiple sources, including GeoPandas and GeoJSON. It can also consume XYZ tile services which use the Web Mercator projection. With the gmap() method you can plot glyphs over a Google Map, though you must pass the method a Google API Key, and any use of Bokeh with Google Maps must be within Google's Terms of Service.

Although Bokeh lets you reproduce geospatial capabilities available in other libraries, such as choropleths, heatmaps, map tiles, and so on, you might find the process more difficult. A common user complaint is that the documentation and learning resources are limited. Beginners can also struggle with the "mid-level" API that is not exactly difficult, but it's not exactly easy, either. This can be mitigated somewhat by using a high-level API like hvPlot that uses Bokeh as its plotting backend.

Choosing a GeoVis Library

At this point, you've probably concluded that choosing any kind of visualization library in Python is like shopping for a new car. You'll never get all the features you want in one place, and for every really useful feature there's an offsetting limitation that forces you to compromise.

There is hope, however. Thanks to "bridging" libraries like Contextily, IPYMPL, hvPlot, and others, the lines between geospatial plotting libraries are becoming increasingly blurred. In addition, most libraries can work with GeoPandas, Python's workhorse for parsing geospatial data, and libraries like Datashader can help with plotting large datasets.

Nevertheless, there are still some important differences that can inform your decision of which library, or libraries, to use. As noted in the previous chapter, maturity can be a discriminating factor among plotting libraries. Figure 17-15 shows the age of the various GeoVis libraries as of the year 2022. This plot is at the same scale as Figure 16-26, and if you compare the two, you'll see that even the oldest GeoVis libraries are less than half the age of the oldest InfoVis and SciVis library.

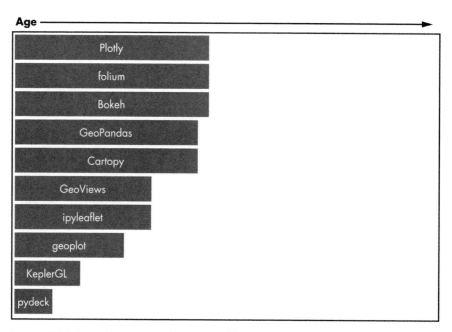

Age

Plotly

folium

Bokeh

GeoPandas

Cartopy

GeoViews

ipyleaflet

geoplot

KeplerGL

pydeck

Figure 17-15: The relative ages of the GeoVis libraries

Still, the volume of discussion around mature and widely used libraries such as GeoPandas, folium, and Plotly means you'll find abundant material on how to use them. They will be more battle-tested, and you're unlikely to be the first person to encounter a frustrating bug or show-stopping limitation. At the same time, some younger libraries have "old bones." For example, geoplot is built on Cartopy, and GeoViews is built on HoloViews and Cartopy, both of which have 10 times the number of users of GeoViews itself. Whether a library is mature and well used depends to some degree on the libraries on which it is built.

To further discriminate among libraries, let's focus on strongpoints. This book assumes that most scientists will want to abstract away as much programming as possible and learn only one API. To this end, the shaded cells in Figure 17-16 indicate an out-of-the-box distinguishing feature of a library, based on a combination of factors such as the developer's claims, online tutorials and reviews, and my own personal experience. The darker the shade the better, and qualifying factors are annotated. Lack of shading does not necessarily mean a feature is absent from a library but that 1) it's subordinate to what you can find in competing libraries, or 2) it requires the use of additional libraries for implementation.

Library	Age	Ease of use	Interactivity	MapTile	CRS accuracy	Large data
GeoPandas		Matplotlib users				
Cartopy		Matplotlib users				
geoplot		Matplotlib users				
Plotly		Plotly Express				
folium/ipyleaflet		folium				
GeoViews		hvPlot				
KeplerGL		Includes a GUI				
pydeck				Earth Engine		
Bokeh						

Figure 17-16: Strongpoints (shaded with qualifiers) of important Python geospatial libraries

As an example, one of Cartopy's main selling points is its powerful projection system, permitting highly accurate mapping together with the ability to perform complex data transformations between reference systems. This doesn't mean the other libraries will plot New York City in the middle of the Atlantic, it just means that they are subordinate to Cartopy, and libraries built on it, when it comes to handling projections. So, if this capability is very important to you, Cartopy, geoplot, and GeoViews should be on your radar.

The folium and ipyleaflet libraries come with a large selection of easily accessible map tiles. GeoPandas provides access to these through the Contextily library. Although this isn't a high hurdle to clear, it does break the premise of *science first, programming second.*

If you expect to do a lot of remote sensing work, the pydeck library comes with an easy connection to Google Earth Engine with its petabytes of satellite imagery.

When it comes to ease of use, Plotly Express and folium can't be beat. They represent plotting "sweet spots" that do many things well, as long as you're not using huge datasets. To appreciate this, try to reproduce the Plotly Express map in Figure 17-7 with other libraries and the same amount of code, as shown on page 467.

If you're already a seaborn and Matplotlib user, you should find GeoPandas, Cartopy, and geoplot somewhat intuitive. GeoViews suffers from limited documentation, but you can use hvPlot, also part of the HoloViz family, as an easy-to-use "Plotly Express-like" plotting option (see "GeoViews: The HoloViz Approach" on page 476).

GeoViews appears to check all the boxes to some degree. It's a single, do-it-all, cradle-to-grave library that, as part of the unified HoloViz family, may position you well for the future. First released in 2016, GeoViews has time to grow its popularity and, hopefully, its supporting documentation.

In terms of data size, most libraries have no trouble plotting hundreds of thousands of points, but many begin to choke on larger datasets. This can be mitigated somewhat with Datashader. Though not a geospatial library per se, it's a must-have library for scientists who deal with really large geospatial datasets. It breaks the visualization process into multiple steps and runs in parallel to quickly create displays for large datasets. Likewise, pydeck helps you manage the enormous datasets available through sites like Google Earth Engine.

Finally, just because GeoPandas doesn't tick many boxes doesn't mean you won't be using it. It's still the most popular way to wrangle geospatial data. There are just better ways to plot and explore the results.

Summary

Geospatial data comprises vector and/or raster data that includes a reference to geographical location. In this chapter, we reviewed the more important Python libraries for plotting this type of data.

The most popular open source Python library for parsing geospatial data is GeoPandas, which also comes with plotting capability built on top of Matplotlib. As many other packages work with GeoPandas, you might find yourself using this library for preparing data while plotting the results using a different tool.

As with the InfoVis libraries discussed in the Chapter 16, your personal choice for a geospatial plotting library will depend largely on what you need to plot—both now and in the future—and how much effort you want to expend. To help you choose, Figure 17-16 provided a summary of the out-of-the-box distinguishing features of the major geospatial libraries. Keep in mind, however, that it's always possible to cobble-together a custom suite of packages using "bridging" libraries that fill-in missing capabilities.

PART IV

THE ESSENTIAL LIBRARIES

Of all the Python libraries that are important to science, NumPy, Matplotlib, and pandas are arguably the most essential. These three libraries form a triptych, with each panel built on the one before. Together, they serve as an immense canvas for most scientific and data analysis work in Python.

What makes these libraries essential? For one thing, they're mature, reliable, and used across many disciplines. As established libraries, they come with enormous support networks consisting of online forums, tutorials, and example use cases as well as many excellent print and ebooks dedicated to each. Further, they form the basis for other important libraries. Some knowledge of NumPy, Matplotlib, and pandas is needed to proficiently use Python in most scientific and engineering endeavors.

In the following three chapters, we'll look at each of these packages in turn. As whole volumes can be dedicated to each library, we'll focus on the purpose of the libraries, the components that have proven difficult or frustrating for new users, and the basic functionality needed to begin applying the packages to your own projects.

18

NUMPY: NUMERICAL PYTHON

Short for *Numerical Python, NumPy* serves as Python's foundational library for numerical computing. It extends Python's mathematical capability and forms the basis of many scientific and mathematical packages. As a result, you'll need to understand NumPy in order to effectively use Python's scientific libraries such as Matplotlib (for plotting) and pandas (for data analysis).

NumPy is open source and comes preinstalled with Anaconda. It augments the built-in tools in the Python Standard Library, which can be too simple for many data analysis calculations. Using NumPy, you can perform fast operations, including mathematical, logical, shape manipulation, sorting, selecting, I/O, discrete Fourier transforms, basic linear algebra, basic statistical operations, random simulation, and more.

At the core of NumPy is the *array* data structure, which is basically a grid of values. By using precompiled C code, multidimensional arrays, and functions that operate on arrays, NumPy speeds up the running of slower algorithms and performs high-level mathematical calculations in a highly efficient manner. NumPy also makes it easier to work with large, uniform datasets with millions to billions of samples.

You can't understand NumPy if you don't understand arrays, so in this chapter, we'll focus on these features first and then look at some of the library's basic functionality. For further study, visit the official site (*https://numpy.org/*), which contains both "quickstart" and more detailed tutorials and guides.

Introducing the Array

In computer science, an array is a data structure that contains a group of elements (values or variables) of the same size and data type (referred to as *dytpes* in NumPy). An array can be indexed by a tuple of nonnegative integers, by Booleans, by another array, or by integers.

Here's an example of a two-dimensional array of integers, comprising a grid of two rows and three columns. Because arrays use square brackets, they look a lot like Python lists:

```
array([[0, 1, 2],
       [3, 4, 5]])
```

To select an element from this array, you can use standard indexing and slicing techniques. For example, to select the element 2, you would index first the row and then the column, using [0][2] (remember: Python starts counting at 0, not 1).

There are several reasons why you might want to work with arrays. Accessing individual elements by index is extremely efficient, making runtimes constant regardless of the array size. In fact, arrays let you perform complex computations on entire blocks of data without the need to loop through and access each element one at a time. As a result, NumPy-based algorithms run orders of magnitude faster than those in native Python.

In addition to being faster, arrays store data in contiguous memory blocks, resulting in a significantly smaller memory footprint than built-in Python sequences, like lists. A list, for example, is basically an array of pointers to (potentially) heterogeneous Python objects stored in non-contiguous blocks, making it much less compact than a NumPy array. Consequently, arrays are often the preferred data structure for storing data reliably and efficiently. The popular OpenCV computer vision library, for example, manipulates and stores digital images as NumPy arrays.

Describing Arrays Using Dimension and Shape

Understanding arrays requires knowledge of their layout. The number of *dimensions* in an array is the number of indexes needed to select an element from the array. You can think of a dimension as an array's *axis*.

The number of dimensions in an array, also called its *rank*, can be used to describe the array. Figure 18-1 is a graphical example of one-, two-, and three-dimensional arrays.

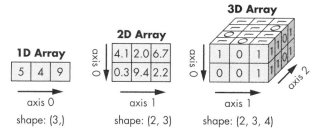

Figure 18-1: Graphical representations of arrays in one, two, and three dimensions

The *shape* of an array is a tuple of integers representing the size of the array along each dimension, starting with the first dimension (axis 0). Example shape tuples are shown below each array in Figure 18-1. The number of integers in these tuples equals the array's rank.

A one-dimensional array, also referred to as a *vector*, has a single axis. This is the simplest form of array and is the NumPy equivalent to Python's list data type. Here's an example of how the 1D array in Figure 18-1 looks in Python:

```
array([5, 4, 9])
```

Arrays with more than one dimension are basically arrays within arrays. An array with both rows and columns is called a 2D array. The 2D array in Figure 18-1 has a shape tuple of (2, 3) because the length of its first axis (0) is 2 and the length of its second axis (1) is 3.

A 2D array is used to represent a *matrix*. You might remember from math class that these are rectangular grids of elements such as numbers or algebraic expressions, arranged in rows and columns and enclosed by square brackets. Matrices store data in an elegant and compact manner, and despite containing many elements, each matrix is treated as one unit.

Here's an example of the 2D array in Figure 18-1 rendered in Python:

```
array([[4.1, 2.0, 6.7],
       [0.3, 9.4, 2.2]])
```

An array with three or more dimensions is called a *tensor*. As mentioned earlier, arrays can have any number of dimensions. Here's an example of the 3D array in Figure 18-1:

```
array([[[1, 0, 1, 1],
        [0, 1, 1, 1],
        [1, 1, 0, 1]],

       [[0, 0, 0, 0],
        [0, 0, 0, 0],
        [1, 1, 0, 1]]])
```

Tensors can be difficult to visualize in a two-dimensional display, but Python tries to help you out. Note how a blank line separates the two stacked matrices that comprise the 3D grid. You can also determine the rank of an array by counting the number of square brackets at the start of the output. Three square brackets in a row means that you're dealing with a 3D array.

Creating Arrays

NumPy handles arrays through its ndarray class, also known by the alias array. The ndarray name is short for *N-dimensional*, as this class can handle any number of dimensions. NumPy ndarrays have a fixed size at creation and can't grow like a Python list or tuple. Changing the size of an ndarray creates a new array and deletes the original.

NOTE *You should know numpy.array is not the same as array.array, found in the Python Standard Library. The latter is only a one-dimensional array and has limited functionality compared to NumPy arrays.*

NumPy comes with several built-in functions for creating ndarrays. These let you create arrays outright or convert existing sequence data types, like tuples and lists, to arrays. Table 18-1 lists some of the more common creation functions. We'll look at some of these in more detail in the sections that follow. You can find a complete listing of creation functions at *https:// numpy.org/doc/stable/reference/routines.array-creation.html*.

Table 18-1: Array Creation Functions

Function	Description
array	Convert (copy) input sequence to an ndarray by inferring or specifying a dtype
asarray	Like array but with fewer options and does not create a copy by default
arange	Like built-in range() function but returns an ndarray instead of a list
linspace	Return evenly spaced numbers over a specified interval
ones	Produce an ndarray of all 1s with a given shape and dtype
ones_like	Produce a ones ndarray of the same shape and dtype as an input array
zeros	Produce an ndarray of all 0s with a given shape and dtype
zeros_like	Produce a zeros ndarray of the same shape and dtype as an input array
empty	Allocate new memory for a new unpopulated ndarray of a given shape
empty_like	Allocate new memory for a new unpopulated ndarray based on an input array
full	Produce an ndarray of a given shape and dtype with all values set to a fill value
full_like	Take an input array and produce a filled array of the same shape and dtype
eye	Return a square 2D array with ones on the diagonal and zeros elsewhere
identity	Like eye but without the option to specify the index of the diagonal

Because arrays must contain data of the same type, the array needs to know the dtype that's being passed to it. You'll have the choice of letting the functions infer the most suitable dtype (although you'll want to check the result) or providing the dtype explicitly as an additional argument.

Some commonly used dtypes are listed in Table 18-2. The Code column lists the shorthand arguments you can pass to the functions in single quotes, such as dtype= 'i8 ', in place of dtype= 'int64 '. For a full list of supported data types, visit *https://numpy.org/doc/stable/user/basics.types.html*.

Table 18-2: Common NumPy Data Types

Type	Code	Description
bool	?	Boolean type (True and False).
object	0	Any Python object type.
string_	S*n*	Fixed-length ASCII string type with 1 byte per character. The *n* parameter represents the length of the longest string, such as 'S15'.
unicode_	U*n*	Fixed-length Unicode type with number of bytes platform specific. The *n* parameter represents the longest length, such as 'U12'.
int8, uint8	i1, u1	Signed and unsigned 8-bit (1 byte) integer types.
int16, uint16	i2, u2	Signed and unsigned 16-bit integer types.
int32, uint32	i4, u4	Signed and unsigned 32-bit integer types.
int64, uint64	i8, u8	Signed and unsigned 64-bit integer types.
float32	f4 or f	Single-precision floating-point type.
float64	f8 or d	Double-precision floating-point type compatible with Python float.
float128	f16 or g	Extended-precision floating-point type.
complex64	c8	Complex number represented by two 32-bit floats.
complex128	c16	Complex number represented by two 64-bit floats.
complex256	c32	Complex number represented by two 128-bit floats.

For string and Unicode dtypes, the length of the longest string or Unicode object must be included in the dtype argument. For example, if the longest string in a dataset has 12 characters, the assigned dtype should be 'S12'. This is necessary because all the ndarray elements should be of the same size. There's no way to create variable-length strings, so you must ensure that enough memory is allocated to hold every possible string in the dataset. When using *existing input*, such as when converting a list of strings to an array, NumPy can make this calculation for you.

Because the amount of memory used by the dtypes is automatically assigned (or can be input), NumPy knows how much memory to allocate when creating ndarrays. The choices in Table 18-2 give you plenty of control over how data is stored in memory, but don't let that intimidate you. Most of the time, all you'll need to know is the basic type of data you're using, such as a float or integer.

HOW NUMPY ALLOCATES MEMORY

The genius of NumPy is in how it allocates memory. The following figure shows information stored in a 3x4 2D array of numbers from 0 to 11, represented by the "Python View" diagram at the bottom of the figure. You're already familiar with parameters such as dtype, dimensions, and data, so let's focus on memory allocation and strides.

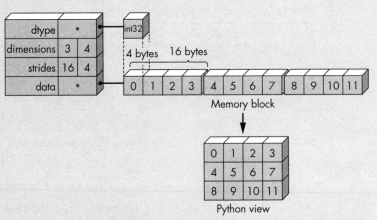

The values of an ndarray are stored as a contiguous block of memory in your computer's RAM, as shown by the Memory Block diagram in the figure. This is efficient, as processors prefer items in memory to be in chunks rather than randomly scattered about. The latter occurs when you store data in Python data types like lists, which keep track of *pointers* to objects in memory, creating "overhead" that slows down processing.

To help NumPy interpret the bytes in memory, the dtype object stores additional information about the *layout* of the array, such as the size of the data (in bytes) and the byte order of the data. Because we're using the int32 dtype in the example, each number occupies 4 bytes of memory (32 bits/8 bits per byte).

Ndarrays come with an attribute, strides, which is a tuple of the number of bytes to step in each dimension when traversing an array. This tuple informs NumPy on how to convert from the contiguous Memory Block to the Python View array shown in the figure.

In the figure, the memory block consists of 48 bytes (12 integers x 4 bytes each), stored one after the other. The array strides indicate how many bytes must be skipped in memory to move to the next position along a certain axis. For example, we must skip 4 bytes (1 integer) to reach the next column, but 16 bytes (4 integers) to move to the same position in the next row. Thus, the strides for the array are (16, 4).

Using the array() Function

The simplest way to create an array is to pass the NumPy array() function a sequence, such as a list, which is then converted into an ndarray. Let's do that now to create a 1D array. We'll begin by importing NumPy using the alias np (this is by convention and will reduce the amount of typing needed to call NumPy functions):

```
In [1]: import numpy as np

In [2]: arr1d = np.array([1, 2, 3, 4])

In [3]: type(arr1d)
Out[3]: numpy.ndarray

In [4]: print(arr1d)
[1 2 3 4]
```

You can also create an ndarray by passing the array() function a variable, like this:

```
In [5]: my_sequence = [1, 2, 3, 4]

In [6]: arr1d = np.array(my_sequence)

In [7]: arr1d
Out[7]: array([1, 2, 3, 4])
```

To create a multidimensional array, pass array() a nested sequence, where each nested sequence is the same length. Here's an example that uses a list containing three nested lists to create a 2D array:

```
In [8]: arr2d = np.array([[0, 1, 2], [3, 4, 5], [6, 7, 8]])

In [9]: print(arr2d)
[[0 1 2]
 [3 4 5]
 [6 7 8]]
```

Each nested list became a new row in the 2D array. To build the same array from tuples, you would replace all the square brackets [] in line In [8] with parentheses ().

NOTE *When you print an array, NumPy displays it with the following layout: the last axis is printed from left to right, the second-to-last is printed from top to bottom, the rest are also printed from top to bottom, with each slice separated from the next by an empty line. So, 1D arrays are printed as rows, 2D arrays as matrices, and 3D arrays as lists of matrices.*

Now, let's check some of the 2D array's attributes, such as its shape

```
In [10]: arr2d.shape
Out[10]: (3, 3)
```

its number of dimensions

```
In [11]: arr2d.ndim
Out[11]: 2
```

and its strides:

```
In [12]: arr2d.strides
Out[12]: (12, 4)
```

Although the items in an array must be the same data type, this doesn't mean that you can't pass these items to the array() function within a mixture of sequence types, such as tuples and lists:

```
In [13]: mixed_input = np.array([[0, 1, 2], (3, 4, 5), [6, 7, 8]])
```

```
In [14]: mixed_input
Out[14]:
array([[0, 1, 2],
       [3, 4, 5],
       [6, 7, 8]])
```

This worked because NumPy reads the data type of the elements in a sequence rather than the data type of the sequence itself.

You won't have the same luck, however, if you try to pass nested lists of different lengths:

```
In [15]: arr2d = np.array([[0, 1, 2], [3, 4, 5], [6, 7]])
```

```
C:\Users\hanna\AppData\Local\Temp/ipykernel_19556/570173853.py:1:
VisibleDeprecationWarning: Creating an ndarray from ragged nested sequences
(which is a list-or-tuple of lists-or-tuples-or ndarrays with different
lengths or shapes) is deprecated. If you meant to do this, you must specify
'dtype=object' when creating the ndarray.
  arr2d = np.array([[0, 1, 2], [3, 4, 5], [6, 7]])
```

You can avoid this warning by changing the dtype to object, as follows:

```
In [16]: arr2d = np.array([[0, 1, 2], [3, 4, 5], [6, 7]], dtype='object')
```

```
In [17]: print(arr2d)
[list([0, 1, 2]) list([3, 4, 5]) list([6, 7])]
```

Note that you now have a 1D array of list objects rather than the 2D array of integers you wanted. Just as with mathematical matrices, arrays need to have the *same number of rows and columns* if you plan to use them for mathematical calculations (there's some flexibility to this, but we'll save it for the section on "broadcasting").

Now, let's look at arrays with more than two dimensions. The array() function transforms sequences of sequences into two-dimensional arrays; sequences of sequences of sequences into three-dimensional arrays; and so on. So, to make a 3D array, you need to pass the function multiple nested sequences. Here's an example using nested lists:

```
In [18]: arr3d = np.array([[[0, 0, 0],
    ...:                     [1, 1, 1]],
    ...:                    [[2, 2, 2],
    ...:                     [3, 3, 3]]])

In [19]: arr3d
Out[19]:
array([[[0, 0, 0],
        [1, 1, 1]],

       [[2, 2, 2],
        [3, 3, 3]]])
```

In this example, we passed a list containing two nested lists that each contained two nested lists. Notice how the output array has a blank line in the middle. This visually separates the two stacked 2D arrays created by the function.

Keeping track of all those brackets when creating high-dimension arrays can be cumbersome and dangerous to your eyesight. Fortunately, NumPy provides additional methods for creating arrays that can be more convenient than the array() function. We'll look at some of these in the next sections.

Using the arange() Function

To create arrays that hold sequences of numbers, NumPy provides the arange() function, which works like Python's built-in range() function, only it returns an array rather than an immutable sequence of numbers.

The arange() function takes similar arguments to range(). Here, we make a 1D array of the integers from 0 to 9:

```
In [20]: arr1d = np.arange(10)

In [21]: arr1d
Out[21]: array([0, 1, 2, 3, 4, 5, 6, 7, 8, 9])
```

We can also add a start, stop, and step argument to create an array of the even numbers between 0 and 10:

```
In [22]: arr1d_step = np.arange(0, 10, 2)

In [23]: arr1d_step
Out[23]: array([0, 2, 4, 6, 8])
```

Next, we start the sequence at 5 and stop at 9:

```
In [24]: arr1d_start_5 = np.arange(5, 10)

In [25]: arr1d_start_5
Out[25]: array([5, 6, 7, 8, 9])
```

Whereas range() always produces a sequence of integers, arange() lets you specify the data type of the numbers in the array. Here, we use double-precision floating-point numbers:

```
In [26]: arr1d_float = np.arange(10, dtype='float64')

In [27]: arr1d_float.dtype
Out[27]: dtype('float64')
```

Interestingly, arange() accepts a float for the step parameter:

```
In [28]: arr1d_float_step = np.arange(0, 3, 0.3)

In [29]: arr1d_float_step
Out[29]: array([0. , 0.3, 0.6, 0.9, 1.2, 1.5, 1.8, 2.1, 2.4, 2.7])
```

NOTE: *When arange() is used with floating-point arguments, it's usually not possible to predict the number of elements obtained, due to the finite floating-point precision. For this reason, it's better to use the NumPy linspace() function, which receives as an argument the number of elements desired instead of the step argument. We'll look at linspace() shortly.*

With the arange() and reshape() functions, you can create a multidimensional array—and generate a lot of data—with a single line of code. The arange() function creates a 1D array, and reshape() divides this linear array into different parts as specified by a shape argument. Here's an example using the 3D shape tuple (2, 2, 4):

```
In [30]: arr3d = np.arange(16).reshape(2, 2, 4)
In [31]: print(arr3d)
[[[ 0  1  2  3]
  [ 4  5  6  7]]

 [[ 8  9 10 11]
  [12 13 14 15]]]
```

Because arrays need to be symmetrical, the product of the shape tuple must equal the size of the array. In this case, (8, 2, 1) and (4, 2, 2) will work, but (2, 3, 4) will raise an error because the resulting array has 24 elements, whereas you specified 16 (np.arange(16)):

```
In [32]: arr3d = np.arange(16).reshape(2, 3, 4)
Traceback (most recent call last):

File "C:\Users\hanna\AppData\Local\Temp/ipykernel_19556/3404575613.py", line 1, in <module>
arr3d = np.arange(16).reshape(2, 3, 4)

ValueError: cannot reshape array of size 16 into shape (2,3,4)
```

Using the linspace() Function

The NumPy linspace() function creates an ndarray of evenly spaced numbers within a defined interval. It's basically the arange() function with a num (number of samples) argument rather than a step argument. The num argument determines how many elements will be in the array, and the function calculates the intervening numbers so that the intervals between them are the same.

Suppose that you want an array of size 6 with values between 0 and 20. All you need to do is pass the function a start, stop, and num value, as follows, using keyword arguments for clarity:

```
In [33]: np.linspace(start=0, stop=20, num=6)
Out[33]: array([ 0., 4., 8., 12., 16., 20.])
```

This produced a 1D array of six floating-point values, with all the values evenly spaced. Note that the stop value (20) is included in the array.

You can force the function to not include the endpoint by setting the Boolean parameter endpoint to False:

```
In [34]: np.linspace(0, 20, 6, endpoint=False)
Out[34]:
array([ 0. , 3.33333333, 6.66666667, 10. , 13.33333333, 16.66666667])
```

If you want to retrieve the size of the intervals between values, set the Boolean parameter retstep to True. This returns the step value:

```
In [35]: arr1d, step = np.linspace(0, 20, 6, retstep=True)

In [36]: step
Out[36]: 4.0
```

By default, the linspace() function returns a dtype of float64. You can override this by passing it a dtype argument:

```
In [37]: np.linspace(0, 20, 6, dtype='int64')
Out[37]: array([ 0, 4, 8, 12, 16, 20], dtype=int64)
```

You'll need to be careful when changing the data type, however, as the result may no longer be a linear space due to rounding.

As with arange(), you can reshape the array on the fly. Here, we produce a 2D array with the same linspace() arguments:

```
In [38]: np.linspace(0, 20, 6).reshape(2, 3)
Out[38]:
array([[ 0.,  4.,  8.],
       [12., 16., 20.]])
```

NOTE *It's possible to create sequences with uneven spacing. The np.logspace() function, for example, creates a logarithmic space with numbers evenly spaced on a log scale.*

The linspace() function lets you control the number of elements in an array, something that can be challenging to do when using arange(). Arrays of evenly spaced numbers are useful when working with mathematical functions of continuous variables. Likewise, linear spaces come in handy when you need to evenly sample an object, such as a waveform. To see some useful examples of linspace() in action, visit *https://realpython.com/ np-linspace-numpy/*.

Along these lines, the meshgrid() function creates a rectangular grid out of two given 1D arrays. The resulting indexing matrix holds in each cell the x and y coordinates for each point in the 2D space. Whereas meshgrid() is useful when plotting and interpolating 2D arrays, the mgrid() function calls meshgrid() to produce a dense "meshgrid" with multiple dimensions.

Creating Prefilled Arrays

For convenience, NumPy lets you create ndarrays using prefilled ones, zeros, random values, or values of your own choosing. You can even create an empty array with no predefined values. These arrays are commonly used when you need a structure for holding computation results, for training machine learning applications, for creating image masks, for performing linear algebra, and so on.

To create a zero-filled array, simply pass the zero() function a shape tuple, as follows:

```
In [39]: np.zeros((3, 3))
Out[39]:
array([[0., 0., 0.],
       [0., 0., 0.],
       [0., 0., 0.]])
```

To create an array filled with ones, repeat the process with the ones() function:

```
In [40]: np.ones((3, 3))
Out[40]:
```

```
array([[1., 1., 1.],
       [1., 1., 1.],
       [1., 1., 1.]])
```

The np.eye() function creates an array where all items are equal to zero, except for the *k*th diagonal, whose values are equal to one:

```
In [41]: np.eye(N=3, M=3, k=0)
Out[41]:
array([[1., 0., 0.],
       [0., 1., 0.],
       [0., 0., 1.]])

In [42]: np.eye(N=3, M=3, k=1)
Out[42]:
array([[0., 1., 0.],
       [0., 0., 1.],
       [0., 0., 0.]])
```

By default, these functions return float64 values, but you can override this using a dtype argument, such as dtype=int.

To fill an array with a custom value and data type, use the full() function with following syntax:

```
In [43]: np.full((3, 3), fill_value=5, dtype='int64')
Out[43]:
array([[5, 5, 5],
       [5, 5, 5],
       [5, 5, 5]], dtype=int64)
```

The empty() function returns a new ndarray of a given shape and filled with uninitialized (arbitrary) data of the given data type:

```
In [44]: np.empty((2, 3, 2))
Out[44]:
array([[[2.20687562e-312, 2.05833592e-312],
        [5.73116149e-322, 0.00000000e+000],
        [2.35541533e-312, 2.07955588e-312]],

       [[2.05833592e-312, 2.44029516e-312],
        [2.35541533e-312, 2.33419537e-312],
        [0.00000000e+000, 0.00000000e+000]]])
```

According to the documentation, empty() does not set the array values to zero and may therefore be marginally faster than the zeros() function. On the other hand, it requires the user to manually set all the values in the array, thus you should use it with caution.

Finally, you can generate arrays of pseudo-random numbers using NumPy. For floating-point values between 0 and 1, just pass random() a shape tuple:

```
In [45]: np.random.random((3,3))
Out[45]:
array([[0.16666842, 0.54555604, 0.08931106],
       [0.14603673, 0.84008062, 0.67797898],
       [0.17353608, 0.34648653, 0.97878551]])
```

In addition, you can generate random integers, sample values from a "standard normal" distribution, shuffle an existing array's contents in place, and more. We'll look at some of these options later in the chapter, and you can find the official documentation at *https://numpy.org/doc/stable/reference/ random/generator.html*.

Accessing Array Attributes

As objects, ndarrays have attributes accessible through dot notation. We've looked at some of these already, and you can find more listed in Table 18-3.

Table 18-3: Important ndarray Attributes

Attribute	Description
ndim	The number of axes (dimensions) of the array
shape	A tuple of integers indicating the size of the array in each dimension
size	The total number of elements in the array
itemsize	The size in bytes of each element in the array
dtype	An object describing the data type of the elements in the array
strides	A tuple of bytes to step in each dimension when traversing an array

For example, to get the shape of the arr1d object, enter the following:

```
In [46]: arr1d = np.arange(0, 4)

In [47]: arr1d.shape
Out[47]: (4,)
```

As a 1D array, there's only one axis and thus only one index. Note the comma after the index, which tells Python that this is a tuple data type and not just an integer in parentheses.

The size of the array is the total number of elements it contains. This is the same as the product of the elements returned by shape. To get the array's size, enter the following:

```
In [48]: arr1d.size
Out[48]: 4
```

To get the array's dtype, enter:

```
In [49]: arr1d.dtype
Out[49]: dtype('int32')
```

Note that, even if you have a 64-bit machine, the default dtype for numbers *may be* 32-bit, such as int32 or float32. To ensure that you're using 64-bit numbers, you can specify the dtype when creating the array, as follows (for int64):

```
In [50]: test = np.array([5, 4, 9], dtype='int64')

In [51]: test.dtype
Out[51]: dtype('int64')
```

To get the array's strides, access the strides attribute with dot notation:

```
In [52]: arr1d.strides
Out[52]: (4,)
```

When using strings in arrays, the dtype needs to include the length of the longest string. NumPy can generally figure this out on its own, as follows:

```
In [53]: arr1d_str = np.array(['wheat', 'soybeans', 'corn'])

In [54]: arr1d_str.dtype
Out[54]: dtype('<U8')
```

Note how the unicode (U) dtype includes the number 8, which is the length of soybeans, the longest string item.

To see the data type and number of bits each item occupies, call the name attribute on dtype, as follows:

```
In [55]: arr1d_str.dtype.name
Out[55]: 'str256'
```

In this case, each item in the array is a string occupying 256 bits (8 characters x 32 bits). This is different from the itemsize attribute, which just displays the size of an individual character *in bytes*:

```
In [56]: arr1d_str.itemsize
Out[56]: 32
```

1. What is *not* a characteristic of an array?

   a. Enables fast computations with small memory footprint

   b. Composed entirely of elements of a single data type

   c. Can accommodate up to four dimensions

   d. Provides an efficient alternative to looping

2. A two-dimensional array is also known as a:

   a. Linear array

   b. Tensor

   c. Rank

   d. Matrix

3. A strides tuple tells NumPy:

   a. The number of different data types in the array

   b. The number of bytes to step in each dimension when traversing an array

   c. The step size when sampling an array

   d. The size of the array in bytes

4. You've been given a dataset of various-sized digital images and asked to take 100 evenly spaced samples of pixel intensity from each. Which NumPy function do you use to choose the sample locations?

   a. `arange()`

   b. `empty()`

   c. `empty_like()`

   d. `full()`

   e. `linspace()`

5. Write an expression to generate a square matrix of 100 zeros.

Indexing and Slicing Arrays

The elements within an ndarray can be accessed using indexes and slices. This lets you extract the value of elements as well as change the values using assignment statements. Array indexing uses square brackets [ ], just like Python lists.

Indexing and Slicing 1D Arrays

One-dimensional arrays are zero-indexed, so the first index is always 0. For indexing and slicing in reverse, the first value is -1. Figure 18-2 describes the indexes of five elements in an array.

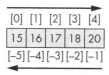

Figure 18-2: The indexes of a 1D ndarray

If you're familiar with list indexing, you won't have any problems indexing 1D arrays. Let's look at some examples in which we select elements using both positive and negative indexing:

```
In [57]: arr1d = np.array([15, 16, 17, 18, 19, 20])

In [58]: arr1d[0]
Out[58]: 15

In [59]: arr1d[-6]
Out[59]: 15

In [60]: arr1d[-1]
Out[60]: 20
```

To access every other element in the array, include a step value of 2:

```
In [61]: arr1d[::2]
Out[61]: array([15, 17, 19])
```

To access multiple elements at once, use an array of comma-separated indexes, as follows:

```
In [62]: arr1d[[0, 2, 4]]
Out[62]: array([15, 17, 19])
```

After you've selected these elements, you can assign them a new value and change the values in the underlying array, like this:

```
In [63]: arr1d[[0, 2, 4]] = 0

In [64]: arr1d
Out[64]: array([ 0, 16, 0, 18, 0, 20])
```

You can also assign new values to a group of array elements with array slices. In this next example, we use slicing to change the first three elements to a value of 100:

```
In [65]: arr1d[:3] = 100

In [66]: arr1d
Out[66]: array([100, 100, 100, 18, 0, 20])
```

In the previous example, the value of 100 was propagated across the entire slice. This process is known as *broadcasting*. Because array slices are views of the source array rather than copies, any changes to the view will modify the original array. This is advantageous when working with very large arrays, as it keeps NumPy from making memory-intensive copies on the fly.

Note that this assignment behavior persists even when array slices are assigned to a variable:

```
In [67]: arr1d = np.array([0, 1, 2, 3, 4])

In [68]: a_slice = arr1d[3:]

In [69]: a_slice
Out[69]: array([3, 4])

In [70]: a_slice[0] = 666

In [71]: arr1d
Out[71]: array([ 0, 1, 2, 666, 4])

In [72]: a_slice[:] = 42

In [73]: arr1d
Out[73]: array([ 0, 1, 2, 42, 42])
```

Because the slice itself is an array, it has its own set of indexes that are different from those of the source array. Thus, a_slice[:] corresponds to arr2d[3:].

To make an actual copy rather than a view, call the copy() method, as shown here:

```
In [74]: a_slice = arr1d[1:3].copy()

In [75]: a_slice[:] = 55

In [76]: a_slice
Out[76]: array([55, 55])

In [77]: arr1d
Out[77]: array([ 0, 1, 2, 42, 42])
```

Now, the a_slice array is separate from arr1d, and changing its elements does not affect the source array.

Alternatively, you can first call the array function on the slice and then mutate the result:

```
In [78]: a_slice = np.array(arr1d[:])

In [79]: a_slice[:] = 55

In [80]: arr1d
Out[80]: array([0, 1, 2, 42, 42])
```

Changing the a_slice array had no effect on arr1d, because the arrays represent separate objects.

Indexing and Slicing 2D Arrays

Two-dimensional arrays are indexed with a pair of values. These value pairs resemble Cartesian coordinates, except that the row index (the axis-0 value) comes before the column index (the axis-1 value), as shown in Figure 18-3. Square brackets are used again.

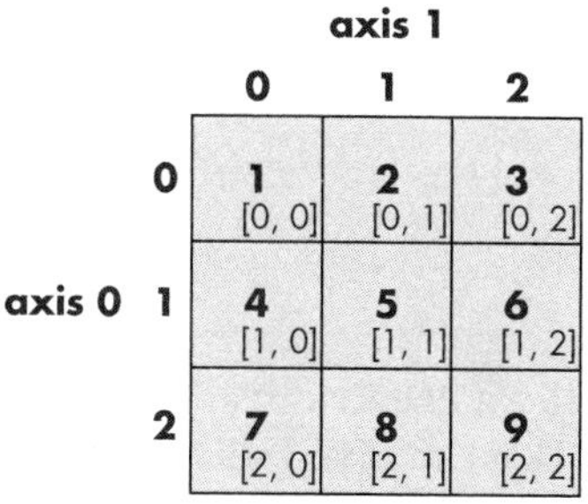

Figure 18-3: Indexes of a 2D ndarray

Let's create the 2D array in Figure 18-3 to study this further:

```
In [81]: arr2d = np.arange(1, 10).reshape(3, 3)

In [82]: arr2d
Out[82]:
array([[1, 2, 3],
       [4, 5, 6],
       [7, 8, 9]])
```

In a 2D array, each value in the index pair references a 1D array (a whole row or column) rather than a single element. For example, specifying an integer index of 1 outputs the 1D array that comprises the second row of the 2D array:

```
In [83]: arr2d[1]
Out[83]: array([4, 5, 6])
```

Slicing a 2D array also works along 1D arrays. Here we slice over rows, taking the last two:

```
In [84]: arr2d[1:3]
Out[84]:
array([[4, 5, 6],
       [7, 8, 9]])
```

This produced a 2D array of shape (2, 3).

To obtain a whole column in the 2D array, use the following syntax:

```
In [85]: arr2d[:, 1]
Out[85]: array([2, 5, 8])
```

The colon (:) tells NumPy to take all the rows; the 1 then selects only column 1, leaving you with only a 1D array from the center column of arr2d.

You can also extract a column with the following syntax, though in this case, rather than outputting a 1D array containing the column's values, you generate a 2D array of shape (3, 1):

```
In [86]: arr2d[:, 1:2]
Out[86]:
array([[2],
       [5],
       [8]])

In [87]: arr2d[:, 1:2].shape
Out[87]: (3, 1)
```

As a rule of thumb, if you slice a 2D array using a mixture of integer indexes and slices, you'll get a 1D array. If you slice along both axes, you'll get another 2D array. For a reference, see Figure 18-4, which shows the results of using various expressions to sample a 2D array.

As with 1D arrays, 2D slices are views of the array that you can use to modify the values in the source array. In this example, we select the middle column in the array in Figure 18-3 and change all of its elements to 42.

```
In [88]: a2_slice = arr2d[:, 1]

In [89]: a2_slice
Out[89]: array([2, 5, 8])

In [90]: a2_slice[:] = 42

In [91]: arr2d
Out[91]:
array([[ 1, 42,  3],
       [ 4, 42,  6],
       [ 7, 42,  9]])
```

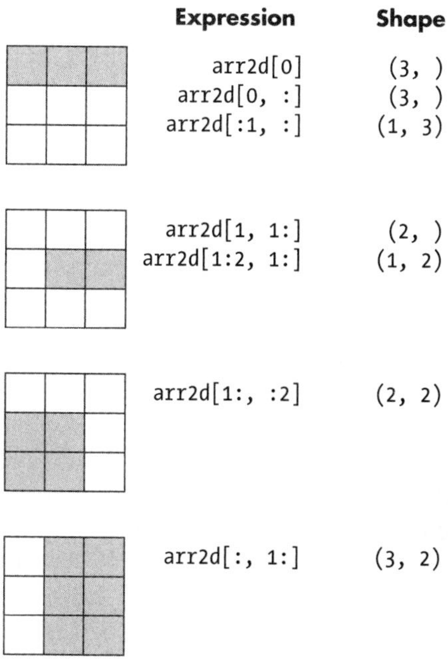

	Expression	Shape
	arr2d[0]	(3,)
	arr2d[0, :]	(3,)
	arr2d[:1, :]	(1, 3)
	arr2d[1, 1:]	(2,)
	arr2d[1:2, 1:]	(1, 2)
	arr2d[1:, :2]	(2, 2)
	arr2d[:, 1:]	(3, 2)

Figure 18-4: Example slices through a 2D ndarray

To select individual elements from 2D arrays, specify a pair of integers as the element's indexes. For example, to obtain the element from the intersection of the second row and second column, enter the following:

```
In [92]: arr2d[1, 1]
Out[92]: 42
```

Note that this syntax is a less cumbersome version of the more traditional nested list syntax in which each index is surrounded by brackets:

```
In [93]: arr2d[1][1]
Out[93]: 42
```

Indexing and Slicing Higher-Dimensional Arrays

The key to indexing and slicing arrays with more than two dimensions is to think of them as a *series of stacked arrays of a lower dimension*. We'll refer to these stacked arrays as *plans*. As with 2D arrays, the order in which you index 3D arrays is determined by their shape tuples.

Let's start by looking at a 3D array with a shape of (2, 3, 4). You can think of the first value in the shape tuple as the number of 2D arrays within that 3D array. The next two numbers are treated as the shape tuple for these 2D arrays, representing its rows and columns, respectively. Here's an example:

```
In [94]: arr3d = np.arange(24).reshape(2, 3, 4)

In [94]: arr3d
Out[94]:
array([[[ 0,  1,  2,  3],
        [ 4,  5,  6,  7],
        [ 8,  9, 10, 11]],

       [[12, 13, 14, 15],
        [16, 17, 18, 19],
        [20, 21, 22, 23]]])
```

When you look at the output, you should see two separate 2D arrays of shape (3, 4) stacked one atop the other. These are demarcated by a space in the output as well as by a new set of square brackets around the second 2D array.

Because the array contains two matrices, the 3D component to the shape tuple is 2. This number comes first, so you can think of the shape tuple as recording the number of plans, rows, and columns.

To see how this works, let's use indexes to retrieve the value 20 in the array. We can use the array's shape tuple (plans, rows, columns) to guide us:

```
In [95]: arr3d[1, 2, 0]
Out[95]: 20
```

First, we had to choose the second 2D array, which has an index of 1 because Python starts counting at 0. Next, we selected the third row using 2. Finally, we selected the first column using 0. The key is to work your way through the shape tuple in order. The dimension of the array will let you know how many indexes you'll need (three for a 3D array, four for a 4D array, and so on).

Slicing also follows the order of the shape tuple. For example, to get a view of the arr3d array's lower 2D array, you would enter 1 for the plan and then use the colon shorthand notation to select all of its rows and columns:

```
In [96]: arr3d[1, :, :]
Out[96]:
array([[12, 13, 14, 15],
       [16, 17, 18, 19],
       [20, 21, 22, 23]])
```

For reference, Figure 18-5 shows some example slices through a 3D array, along with the resulting shapes.

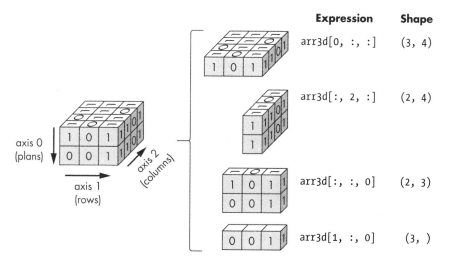

Expression	Shape
arr3d[0, :, :]	(3, 4)
arr3d[:, 2, :]	(2, 4)
arr3d[:, :, 0]	(2, 3)
arr3d[1, :, 0]	(3,)

Figure 18-5: Some example slices through a 3D ndarray

As always, changing the values of elements in a slice will change the source array, unless the slice is a copy:

```
In [97]: arr3d[0, :, :] = 0

In [98]: arr3d
Out[98]:
array([[[ 0,  0,  0,  0],
        [ 0,  0,  0,  0],
        [ 0,  0,  0,  0]],

       [[12, 13, 14, 15],
        [16, 17, 18, 19],
        [20, 21, 22, 23]]])
```

Before we move on, let's practice indexing and slicing an array with more than three dimensions. For example, look at the following 4D array:

```
In [99]: arr4d = np.arange(24).reshape(2, 2, 2, 3)

In [100]: arr4d
Out[100]:
array([[[[ 0,  1,  2],
         [ 3,  4,  5]],

        [[ 6,  7,  8],
         [ 9, 10, 11]]],


       [[[12, 13, 14],
         [15, 16, 17]],

        [[18, 19, 20],
         [21, 22, 23]]]])
```

Note how the array starts with four square brackets and uses two blank lines to separate the two stacked 3D arrays. Because we're dealing with a 4D array, to select the 20 element, you will need to enter four indexes:

```
In [101]: arr4d[1, 1, 0, 2]
Out[101]: 20
```

Here, from left to right, you indexed a 4D array to a 3D array; a 3D array to a 2D array; a 2D array to a 1D array; and a 1D array to a single element. This might be more obvious in Figure 18-6, which demonstrates stepping through these in order.

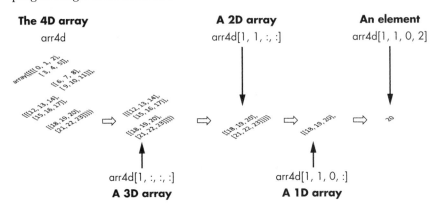

Figure 18-6: Indexing a 4D array down to a single element at [1, 1, 0, 2]

This style of ordering will hold true for any number of dimensions.

Finally, for 4D arrays, where the fourth dimension represents time, it can be useful to think of the arrays arranged horizontally, rather than vertically (Figure 18-7).

```
                    Time
─────────────────────────────────────────────►
        t₀                        t₁

[[[[ 0  1  2]              [[[12 13 14]
   [ 3  4  5]]                [15 16 17]]

  [[ 6  7  8]                [[18 19 20]
   [ 9 10 11]]]               [21 22 23]]]]]

└──────────┘               └──────────┘
   3D array at t₀              3D array at t₁
```

Figure 18-7: Each 4D slice can represent the same 3D array sampled at a different time.

In this case, each single 4D slice would represent the same dataset (the 3D array) measured at a different point in time. So, to look at the first set of measurements, you would enter arr4d[0, :, :], and for the last set of measurements, you could enter arr4d[-1, :, :].

Boolean Indexing

In addition to using numerical indexing and slicing, you can select elements in arrays using conditions and Boolean operators. This lets you extract elements without any prior knowledge of where they are in the array. For example, you might have hundreds of monitor wells around a landfill, and you want to find all the wells that detect the pollutant toluene above a certain threshold value. With Boolean indexing, not only can you identify these wells, but you can also create a new array based on the output.

To illustrate, the following condition searches an array for any elements that are integers greater than or equal to four:

```
In [102]: arr1d = np.array([1, 2, 3, 4, 5])

In [103]: print(arr1d >= 4)
[False False False True True]
```

As you can see, Python will return an array of Boolean values containing True values where the condition is satisfied. Note that this syntax works for ndarrays of any dimension.

NumPy can also use the Booleans behind the scenes, allowing you to slice an array based on a conditional:

```
In [104]: a_slice = arr1d[arr1d >= 4]

In [105]: a_slice
Out[105]: array([4, 5])
```

Comparing two arrays also produces a Boolean array. In this example we flag as True all the values in arr_2 that are greater than those in arr_1:

```
In [106]: arr_1 = np.random.randn(3, 4)

In [107]: arr_2 = np.random.randn(3, 4)

In [108]: arr_2 > arr_1
Out[108]:
array([[ True,  True, False,  True],
       [ True, False,  True, False],
       [False,  True,  True,  True]])
```

A common use of Boolean indexing is to partition a grayscale image into foreground and background segments, a process called *thresholding*. This produces a binary image based on a cutoff value. Here's an example in which we create a 2D image array and then threshold on values above 4:

```
In [109]: img = np.array([
          [12, 13, 14,  4, 16,  1, 11, 10,  9],
          [11, 14, 12,  3, 15,  1, 10, 12, 11],
          [10, 12, 12,  1, 14,  3, 10, 12, 12],
```

```
        [ 9, 11, 16,  0,  4,  2,  3, 12, 10],
        [12, 11, 16, 14, 10,  2, 16, 12, 13],
        [10, 15, 16, 14, 14,  4, 16, 15, 12],
        [13, 17, 14, 10, 14,  1, 14, 15, 10]])

In [110]: img_thresh = (img > 4).astype(int)
```

Remember that True evaluates to 1, and False evaluates to 0. This lets us convert a Boolean array to a numerical array by tacking on the astype() function and passing it the integer data type.

After thresholding, the 0 values in the new array should form the number 4:

```
In [111]: print(img_thresh)
[[1 1 1 0 1 0 1 1 1]
 [1 1 1 0 1 0 1 1 1]
 [1 1 1 0 1 0 1 1 1]
 [1 1 1 0 0 0 0 1 1]
 [1 1 1 1 1 0 1 1 1]
 [1 1 1 1 1 0 1 1 1]
 [1 1 1 1 1 0 1 1 1]]
```

To assign values based on a Boolean array, you index the source array based on a conditional and then assign a value. Here, we assign 0 to all the elements in the array with a value less than 5:

```
In [112]: img[img < 5] = 0

In [113]: img
Out[113]:
array([[12, 13, 14,  0, 16,  0, 11, 10,  9],
       [11, 14, 12,  0, 15,  0, 10, 12, 11],
       [10, 12, 12,  0, 14,  0, 10, 12, 12],
       [ 9, 11, 16,  0,  0,  0,  0, 12, 10],
       [12, 11, 16, 14, 10,  0, 16, 12, 13],
       [10, 15, 16, 14, 14,  0, 16, 15, 12],
       [13, 17, 14, 10, 14,  0, 14, 15, 10]])
```

Likewise, you can change entire rows, columns, and plans in a Boolean array using indexing. For example, img[0] = 0 changes all the elements in the first row of the img array to 0.

The use of Booleans in arrays involves a few quirks. Extracting elements from an array using Boolean indexing creates a copy of the data by default, meaning that there is no need to use the copy() function. Another idiosyncrasy of Boolean arrays is that you must replace the and and or keywords with the & and |characters, respectively, when writing comparison statements.

6. Create a 2D ndarray of size 30 and shape (5, 6). Then, slice the array to sample the values highlighted in gray:

0	1	2	3	4	5
6	7	8	9	10	11
12	13	14	15	16	17
18	19	20	21	22	23
24	25	26	27	28	29

7. Resample the array from Question 6 to retrieve the elements highlighted in gray:

0	1	2	3	4	5
6	7	8	9	10	11
12	13	14	15	16	17
18	19	20	21	22	23
24	25	26	27	28	29

8. Slicing an ndarray produces:

   a. A new array object

   b. A copy of the source array

   c. A view of the source array

   d. A Python list object

9. Slicing a 2D array with a combination of a scalar index and another slice produces:

   a. A 2D array

   b. A 1D array

   d. A single element (0D array)

   e. None of the above

10. What is the rank of this array?

```
array([[[[ 0,  1,  2,  3],
         [ 4,  5,  6,  7]],

        [[ 8,  9, 10, 11],
         [12, 13, 14, 15]]],


       [[[16, 17, 18, 19],
         [20, 21, 22, 23]],

        [[24, 25, 26, 27],
         [28, 29, 30, 31]]]])
```

Manipulating Arrays

NumPy provides tools for working with existing arrays. Common manipulations include reshaping arrays, swapping their axes, and merging and splitting arrays. These come in handy for jobs like rotating, enlarging, and translating images and for fitting machine learning models.

Shaping and Transposing

NumPy comes with functions to change the shape of arrays and to transpose arrays (invert columns with rows) and swap axes. You've already been working with one of these, the reshape() function.

One thing to be aware of with reshape() is that, like all NumPy assignments, it creates a *view* of an array rather than a *copy*. In the following example, reshaping the arr1d array produces only a temporary change to the array:

```
In [114]: arr1d = np.array([1, 2, 3, 4])

In [115]: arr1d.reshape(2, 2)
Out[115]:
array([[1, 2],
       [3, 4]])

In [116]: arr1d
Out[116]: array([1, 2, 3, 4])
```

This behavior is useful when you want to temporarily change the shape of the array for use in a computation, without copying any data.

Likewise, assigning an array to a new variable just creates another reference to the source array. In the following example, despite assigning the reshaped arr1d array to a new variable named arr2d, changing values in arr2d also changes the corresponding values in arr1d:

```
In [117]: arr2d = arr1d.reshape(2, 2)

In [118]: arr2d
Out[118]:
array([[1, 2],
       [3, 4]])

In [119]: arr2d[0] = 42

In [120]: arr2d
Out[120]:
array([[42, 42],
       [ 3,  4]])

In [121]: arr1d
Out[121]: array([42, 42, 3, 4])
```

Obviously, this type of behavior can really trip you up. As mentioned earlier, if you want to create a distinct ndarray object from an existing array, use the copy() function.

To modify an array in place rather than just create a view, use the shape() function and pass it a shape tuple:

```
In [122]: arr1d.shape = (2, 2)

In [123]: arr1d
Out[123]:
array([[42, 42],
       [3, 4]])
```

Compare this code to In [114] - Out [116]. Here, the source array is permanently changed.

Flattening an Array

There are times when you'll want to use 1D arrays as input to some process, even though your data is of a higher dimension. For example, standard plotting routines typically expect simple data structures, such as a list or single flat array. Likewise, image data is generally converted to 1D arrays before being fed to the input layer of a neural network.

Going from a higher dimension array to a 1D array is known as *flattening*. The ravel() function lets you do this while making a *view* of the array. Here's an example:

```
In [124]: arr2d = np.arange(8).reshape(2, 4)

In [125]: arr2d
Out[125]:
array([[0, 1, 2, 3],
       [4, 5, 6, 7]])

In [126]: arr1d = arr2d.ravel()

In [127]: arr1d
Out[127]: array([0, 1, 2, 3, 4, 5, 6, 7])
```

To create a copy of the array when flattening, you can use the flatten() method of the ndarray object. Because this produces a copy rather than a view, it's a bit slower than ravel(). Here's the syntax:

```
In [128]: arr2d.flatten()
Out[128]: array([0, 1, 2, 3, 4, 5, 6, 7])
```

You can also flatten the original array in place by using the shape()
function and passing it the number of elements in the array:

```
In [129]: arr2d.shape = (8)

In [130]: arr2d
Out[130]: array([0, 1, 2, 3, 4, 5, 6, 7])
```

Remember, you can get the size of an array by calling its size attribute
using dot notation.

Swapping an Array's Columns and Rows

When analyzing data, it's good to examine it in multiple ways. Figure 18-8
shows average temperature data by month for three Texas cities. How you
present the data, either *by month* or *by location,* can be beneficial depending
on the questions you're trying to answer as well as how much space you have
for printing the information in a report.

	Wichita Falls	Amarillo	Austin
Jan	42	37	51.1
Feb	45.9	40.3	55
Mar	54	47.8	61.7
Apr	62.6	56.3	69.3

	Jan	Feb	Mar	Apr
Wichita Falls	42	45.9	54	62.6
Amarillo	37	40.3	47.8	56.3
Austin	51.1	55	61.7	69.3

*Figure 18-8: The average monthly temperatures (°F) for three Texas cities displayed by
month and by city*

Just as Microsoft Excel lets you easily invert columns and rows, NumPy
provides the handy transpose() function for this operation:

```
In [131]: arr2d = np.arange(8).reshape(2, 4)

In [132]: arr2d
Out[132]:
array([[0, 1, 2, 3],
       [4, 5, 6, 7]])

In [133]: arr2d.transpose()
Out[133]:
array([[0, 4],
       [1, 5],
       [2, 6],
       [3, 7]])
```

This is still a view of the original array. To create a new array, you can
add the copy() function, like so:

```
In [134]: arr2d_transposed = arr2d.transpose().copy()
```

For higher-dimension arrays, you can pass transpose() a tuple of axis numbers in the order you desire. Let's transpose a 3D array so that the axes are reordered with the third axis first, the first axis second, and the second axis unchanged:

```
In [135]: arr3d = np.arange(12).reshape(2, 2, 3)

In [136]: arr3d
Out[136]:
array([[[ 0,  1,  2],
        [ 3,  4,  5]],

       [[ 6,  7,  8],
        [ 9, 10, 11]]])

In [137]: arr3d.transpose((2, 1, 0))
Out[137]:
array([[[ 0,  6],
        [ 3,  9]],

       [[ 1,  7],
        [ 4, 10]],

       [[ 2,  8],
        [ 5, 11]]])
```

Another method for swapping axes is swapaxes(). It takes a pair of axes and rearranges the array, returning a view of the array. Here's an example:

```
In [138]: arr3d
Out[138]:
array([[[ 0,  1,  2],
        [ 3,  4,  5]],

       [[ 6,  7,  8],
        [ 9, 10, 11]]])

In [139]: arr3d.swapaxes(0, 1)
Out[139]:
array([[[ 0,  1,  2],
        [ 6,  7,  8]],

       [[ 3,  4,  5],
        [ 9, 10, 11]]])
```

Joining Arrays

NumPy provides several functions that let you merge, or *stack*, multiple existing arrays into a new array. Let's begin by making two 2D arrays, the first composed of zeros, and the second composed of ones:

```
In [140]: zeros = np.zeros((3, 3))

In [141]: ones = np.ones((3, 3))
```

Now let's vertically stack the two arrays using the vstack() function. This will add the second array to the first as new rows along axis 0:

```
In [142]: np.vstack((zeros, ones))
Out[142]:
array([[0., 0., 0.],
       [0., 0., 0.],
       [0., 0., 0.],
       [1., 1., 1.],
       [1., 1., 1.],
       [1., 1., 1.]])
```

The hstack() function adds the second array as new columns on the first:

```
In [143]: np.hstack((zeros, ones))
Out[143]:
array([[0., 0., 0., 1., 1., 1.],
       [0., 0., 0., 1., 1., 1.],
       [0., 0., 0., 1., 1., 1.]])
```

The row_stack() and column_stack() functions stack 1D arrays to form new 2D arrays. For example:

```
In [144]: x = np.array([1, 2, 3])

In [145]: y = np.array([4, 5, 6])

In [146]: z = np.array([7, 8, 9])

In [147]: np.row_stack((x, y, z))
Out[147]:
array([[1, 2, 3],
       [4, 5, 6],
       [7, 8, 9]])

In [148]: np.column_stack((x, y, z))
Out[148]:
array([[1, 4, 7],
       [2, 5, 8],
       [3, 6, 9]])
```

You also can accomplish column stacking along axis 2 using the depth stacking function (dstack((x, y, z))). This function is like hstack(), except that it first converts 1D arrays to 2D column vectors.

Splitting Arrays

NumPy also lets you divide, or *split*, arrays. As with joining, you can perform splitting both vertically and horizontally.

Here's an example using the vsplit() function. First, let's create an array:

```
In [149]: source = np.arange(24).reshape((4, 6))

In [150]: source
Out[150]:
array([[ 0,  1,  2,  3,  4,  5],
       [ 6,  7,  8,  9, 10, 11],
       [12, 13, 14, 15, 16, 17],
       [18, 19, 20, 21, 22, 23]])
```

To split the source array in half vertically (axis=0), pass the vsplit() function the array and 2 as arguments:

```
In [151]: split1, split2 = np.vsplit(source, 2)

In [152]: split1
Out[152]:
array([[ 0, 1, 2, 3,  4,  5],
       [ 6, 7, 8, 9, 10, 11]])

In [153]: split2
Out[153]:
array([[12, 13, 14, 15, 16, 17],
       [18, 19, 20, 21, 22, 23]])
```

To split the source array in half horizontally (axis=1), pass hsplit() the array and 2 as arguments:

```
In [154]: split1, split2 = np.hsplit(source, 2)

In [155]: split1
Out[155]:
array([[ 0,  1,  2],
       [ 6,  7,  8],
       [12, 13, 14],
       [18, 19, 20]])

In [156]: split2
Out[156]:
array([[ 3,  4,  5],
       [ 9, 10, 11],
       [15, 16, 17],
       [21, 22, 23]])
```

In the previous examples, the array split must result in an *equal division*. With the split() function, you can split an array into multiple arrays along an axis. You pass the function the original array and the indexes for the parts to be split, along with an optional axis number (the default is axis 0). For example, to divide the source array into three arrays of two, three, and one columns, you would enter the following:

```
In [157]: a, b, c = np.split(source, [2, 5], axis=1)

In [158]: a
Out[158]:
array([[ 0,  1],
       [ 6,  7],
       [12, 13],
       [18, 19]])

In [159]: b
Out[159]:
array([[ 2,  3,  4],
       [ 8,  9, 10],
       [14, 15, 16],
       [20, 21, 22]])

In [160]: c
Out[160]:
array([[ 5],
       [11],
       [17],
       [23]])
```

The indexes[2, 5] told NumPy where along axis 1 to split the array. To repeat this over the rows, just change the axis argument to 0.

Doing Math Using Arrays

Now that you know how to create and manipulate arrays, it's time to apply them to their main purpose: mathematical operations. NumPy uses two internal implementations to efficiently perform math on arrays: *vectorization* and *broadcasting*. Vectorization supports operations between equal-sized arrays, and broadcasting extends this behavior to arrays with different shapes.

Vectorization

One of the most powerful features of ndarrays, *vectorization* lets you perform batch operations on data without the need for explicit for loops. This means that you can apply an operation on a entire array at once without selecting each individual element from it.

For equal-sized arrays, arithmetic operations are applied elementwise, as shown in Figure 18-9.

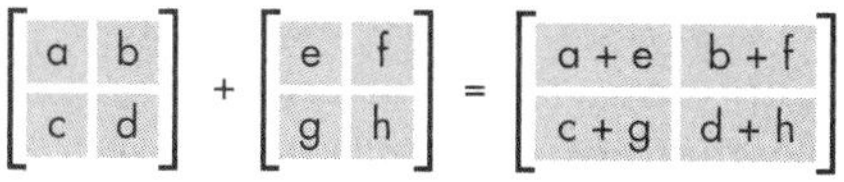

$$\begin{bmatrix} a & b \\ c & d \end{bmatrix} + \begin{bmatrix} e & f \\ g & h \end{bmatrix} = \begin{bmatrix} a+e & b+f \\ c+g & d+h \end{bmatrix}$$

Figure 18-9: Mathematical operations involving equal-sized arrays are performed on corresponding elements.

Because looping takes place behind the scenes with code implemented in C, vectorization leads to faster processing. Let's look at an example in which we compare looping in Python to vectorization in NumPy.

Start by creating two datasets of 100,000 randomly-selected integers between 0 and 500:

```
In [161]: data_a = np.random.randint(500, size=100_000)

In [162]: data_b = np.random.randint(500, size=100_000)
```

Now, make an empty list and then loop through the two datasets, appending each item in data_a to the list if it also occurs in data_b:

```
In [163]: shared_list = []

In [164]: for item in data_a:
     ...:     if item in data_b:
     ...:         shared_list.append(item)
```

Note that this can also be written as shared_list = [item for item in data_a if item in data_b] using list comprehension.

Depending on your hardware, you'll need to wait around five seconds or more for this loop to complete.

Here's the first three values in the list (yours may differ, as these were randomly generated):

```
In [165]: shared_list[:3]
Out[165]: [326, 159, 155]
```

Let's repeat this exercise using the NumPy isin() function. This optimized function compares each element in a target array to another array and returns a Boolean. We can combine this with indexing to return the elements with values of True:

```
In [166]: data_a[np.isin(data_a, data_b)]
Out[166]: array([326, 159, 155, ..., 136, 416, 307])
```

This computation ran almost instantly compared to the previous standard Python loop.

Vectorization also permits more concise and readable code that can resemble mathematical expressions. For example, to multiply two arrays together, you can forgo writing nested loops and just state arr1 * arr2, as follows:

```
In [167]: arr1 = np.array([[1, 1, 1], [2, 2, 2]])

In [168]: arr1
Out[168]:
array([[1, 1, 1],
       [2, 2, 2]])

In [169]: arr2 = np.array([[3, 3, 3], [4, 4, 4]])

In [170]: arr2
Out[170]:
array([[3, 3, 3],
       [4, 4, 4]])

In [171]: arr1 * arr2
Out[171]:
array([[3, 3, 3],
       [8, 8, 8]])
```

This behavior applies to all basic arithmetic operations, such as adding, subtracting, multiplying, and dividing.

Broadcasting

The technique of *broadcasting* allows operations on arrays of different shapes. To understand how it works, consider Figure 18-10, in which a 1D array of four elements is multiplied by a 1D array of a single element.

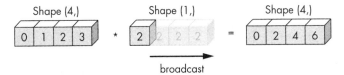

Figure 18-10: An example of broadcasting when multiplying a 1D ndarray by a scalar

As you can see, the smaller array is stretched across the larger array until they have compatible shapes. The array of shape (1,) becomes an array of shape (4,) with its single value repeated so that element-by-element multiplication can occur. This same behavior applies to operations between scalars and arrays.

For broadcasting to work, the dimensions of the two arrays must be compatible. Two dimensions are compatible when they are equal or one of them is 1. NumPy determines this compatibility by comparing the array shape tuples, starting with the trailing (rightmost) dimension and moving left. For example, to check whether different 24-element 3D arrays are broadcastable, NumPy would compare their shape tuples, as shown in Figure 18-11.

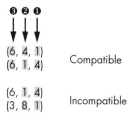

(6, 4, 1)
(6, 1, 4)   Compatible

(6, 1, 4)
(3, 8, 1)   Incompatible

Figure 18-11: Checking 3D array dimensions for compatibility (gray-shaded values)

Starting with the trailing dimension ❶, NumPy determines that both pairs of arrays are compatible, as at least one is equal to 1. This holds true for the next comparison ❷, but the bottom pair fails in the last comparison ❸, because 6 and 3 are not equal. Consequently, we can't perform any mathematical operations between these two arrays.

By contrast, in Figure 18-12, a 2D and 1D array are compatible, so the 1D array can broadcast down to fill in the missing rows.

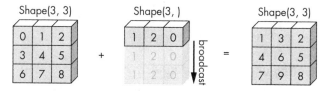

Figure 18-12: An example of broadcasting when adding a 2D array to a 1D array

This allows for element-by-element addition. Broadcasting can occur along rows, columns, or plans, as needed. For more on broadcasting, including a practical example, visit *https://numpy.org/doc/stable/user/basics .broadcasting.html*.

The Matrix Dot Product

In NumPy, basic multiplication between arrays is executed element for element. In other words, each element in one array is multiplied by the corresponding element in a second array. This includes the multiplication of 2D arrays, also known as matrices.

You might remember from math class, however, that proper matrix multiplication involves performing operations on rows and columns, not elements. This is the *matrix dot product*, in which the horizontals in the first matrix are multiplied by the verticals in the second matrix. The results are then added, as shown by the gray-shaded values in Figure 18-13. Not only is this process not *by element*, but it's also noncommutative, as `arr1 * arr2` is not equal to `arr2 * arr1`.

$$\underset{\text{arr1}}{\begin{bmatrix} 0 & 1 \\ 2 & 3 \end{bmatrix}} \cdot \underset{\text{arr2}}{\begin{bmatrix} 4 & 5 \\ 6 & 7 \end{bmatrix}} = \begin{bmatrix} 0 \times 4 + 1 \times 6 & 0 \times 5 + 1 \times 7 \\ 2 \times 4 + 3 \times 6 & 2 \times 5 + 3 \times 7 \end{bmatrix} = \begin{bmatrix} 6 & 7 \\ 26 & 31 \end{bmatrix}$$

Figure 18-13: The matrix dot product

For multiplying two matrices in this way, NumPy provides the `dot()` function. Here's an example using the matrices in Figure 18-13:

```
In [172]: arr1 = np.array([[0, 1], [2, 3]])

In [173]: arr2 = np.array([[4, 5], [6, 7]])

In [174]: np.dot(arr1, arr2)
Out[174]:
array([[ 6,  7],
       [26, 31]])
```

You can also use the alternate syntax `arr1.dot(arr2)` to compute the dot product.

In addition to the dot product, NumPy comes with other functions for performing linear algebra. To see the full list, visit *https://numpy.org/doc/stable/reference/routines.linalg.html.*

Incrementing and Decrementing Arrays

You can use augmented operators such as `+=` to change the values in an array without creating a new array. Here are some examples using a 1D array:

```
In [175]: arr1d = np.array([0, 1, 2, 3])

In [176]: arr1d += 10

In [177]: arr1d
Out[177]: array([10, 11, 12, 13])

In [178]: arr1d -= 10

In [179]: arr1d
Out[179]: array([0, 1, 2, 3])

In [180]: arr1d *= 2

In [181]: arr1d
Out[181]: array([0, 2, 4, 6])
```

In these cases, the scalar value is applied to every element in the array.

Using NumPy Functions

Like Python's standard math module, NumPy comes with its own set of mathematical functions. These include universal functions and aggregate functions. A *universal* function, also known as a *ufunc*, acts in an element-by-element fashion and generates a new array with the same size as the input. *Aggregate* functions act on a whole array and produce a single value, such as the sum of the elements in the array.

Universal Functions

Universal functions that perform simple element-by-element transformations, such as taking the log or squaring an element, are referred to as *unary* ufuncs. To use them, call the function and pass it an ndarray, as follows:

```
In [182]: arr1d = np.array([10, 20, 30, 40])

In [183]: np.log10(arr1d)
Out[183]: array([1. , 1.30103 , 1.47712125, 1.60205999])

In [184]: np.square(arr1d)
Out[184]: array([ 100, 400, 900, 1600], dtype=int32)
```

Some of the more useful unary ufuncs are listed in Table 18-4. You can find a complete list at *https://numpy.org/doc/stable/reference/ufuncs.html#ufuncs/*.

Table 18-4: Useful NumPy Unary Universal Functions

Function	Description
abs	Compute absolute value of each element
fabs	Compute absolute value of each element and return float
all	Test whether all array elements along an axis evaluate to True
any	Test whether any array element along an axis evaluates to True
ceil	Compute smallest integer greater than or equal to each element
floor	Compute largest integer less than or equal to each element
clip	Limit values in array to a specified min, max range
round	Round values in array to a specified number of decimals
exp	Compute the exponent (ex) of each element
log, log10, log2	Compute the natural, base 10, or base 2 log per element
rint	Round elements to the nearest integer preserving the dtype
sign	Compute sign of each element (positive=1, zero=0, negative=-1)
sqrt	Compute square root of each element

(continued)

Table 18-4: Useful NumPy Unary Universal Functions *(continued)*

Function	Description
square	Compute the square of each element
modf	Return the fractional and integral parts of array as a new array
isnan	Return Boolean array indicating NaN (Not a Number) values
degrees	Convert elements representing radians to degrees
radians	Convert elements representing degrees to radians
cos, sin, tan	Compute cosine, sine, or tangent for each element
cosh, sinh, tanh	Compute hyperbolic cosine, sine, or tangent for each element
arccos, arcsin, arctan	Compute inverse trigonometric functions per element
arccosh, arcsinh, arctanh	Compute inverse hyperbolic trigonometric functions per element
sort	arr.sort() sorts in-place; np.sort() returns a sorted copy

Universal functions that accept two arrays as input and return a single array are called *binary* ufuncs. The following binary functions find the maximum and minimum values in two arrays and return them in new arrays:

```
In [185]: a = np.array([1, 2, 500])

In [186]: b = np.array([0, 2, -1])

In [187]: np.maximum(a, b)
Out[187]: array([ 1, 2, 500])

In [188]: np.minimum(a, b)
Out[188]: array([ 0, 2, -1])
```

Some other binary functions are listed in Table 18-5.

Table 18-5: Useful NumPy Binary Universal Functions

Function	Description
add	Add arrays element by element
subtract	Subtract second argument array from first argument array by element
multiply	Multiply arrays element by element
divide	Divide arrays element by element
floor_divide	Divide arrays and truncate the remainder
power	Raise elements in first array to powers in second array
maximum, fmax	Return the maximum value by element, ignoring NaN values for fmax
minimum, fmin	Return the minimum value by element, ignoring NaN values for fmax

Function	Description
mod	Return the modulus by element
copysign	Copy sign of values in second array to values in first array
greater	Return Boolean array for elementwise *greater than* comparison
greater_equal	Return Boolean array for elementwise *greater than or equal to* comparison
less	Return Boolean array for elementwise *less than* comparison
less_equal	Return Boolean array for elementwise *less than or equal to* comparison
equal	Return Boolean array for elementwise *equality* comparison
not_equal	Return Boolean array for elementwise negative equality comparison

For more on universal functions visit: *https://numpy.org/doc/stable/user/basics.ufuncs.html.*

Statistical Methods

NumPy also comes with methods that compute statistics for an entire array or for data along an axis. Reducing the elements in an array to a single value can be referred to as *aggregation* or *reduction.*

Let's try out some of these using a 2D array of randomly generated integers:

```
In [189]: arr = np.random.randint(100, size=(3, 5))
```

```
In [190]: arr
Out[190]:
array([[85, 77,  0, 10, 24],
       [16, 39, 94, 11, 21],
       [71, 54,  8, 73, 98]])
```

To calculate the mean value for all the elements in this array, call mean() on the array using dot notation:

```
In [191]: arr.mean()
Out[191]: 45.4
```

You can also pass the array to the mean() function, like so:

```
In [192]: np.mean(arr)
Out[192]: 45.4
```

The optional axis argument lets you specify the axis over which to compute the statistics. For example, specifying axis 1 means that the calculation is performed *across the columns*, producing a 1D array with the same number of elements as rows in the array:

```
In [193]: arr.mean(axis=1)
Out[193]: array([39.2, 36.2, 60.8])
```

Specifying axis 0 tells the function to compute the *down the rows*. In the following example, this yields a 1D array of five elements, equal to the number of columns:

```
In [194]: arr.sum(axis=0)
Out[194]: array([172, 170, 102, 94, 143])
```

These functions can also be called without the axis keyword:

```
In [195]: arr.mean(1)
Out[195]: array([39.2, 36.2, 60.8])
```

Table 18-6 lists some useful statistical methods for arrays. You can use the whole array or specify an axis.

Table 18-6: Useful NumPy Statistical Methods

Function	Description
argmin	Index of the element with the minimum value
argmax	Index of the element with the maximum value
count_nonzero	Counts the number of non-zero values in an array
cumprod	Cumulative product of elements starting with index 1
cumsum	Cumulative sum of elements starting with index 0
mean	Arithmetic mean of elements
min	Minimum value of elements
max	Maximum value of elements
std	Standard deviation of elements
sum	Sum of the elements
var	Variance of the elements

Note that NumPy also comes with the apply_along_axis() aggregate function that lets you supply the statistical function, axis, and array as arguments. Here's an example using the previous array:

```
In [196]: np.apply_along_axis(np.mean, axis=1, arr=arr)
Out[196]: array([37.4, 31. , 74.4])
```

You can also define your own functions and pass them to apply_along _axis():

```
In [197]: def cube(x):
     ...:     return x**3

In [198]: np.apply_along_axis(cube, axis=1, arr=arr)
Out[198]:
array([[614125, 456533,      0,   1000,  13824],
       [  4096,  59319, 830584,   1331,   9261],
       [357911, 157464,    512, 389017, 941192]], dtype=int32)
```

Notice how, in these examples, you were able to work with the array without explicitly iterating over every element. Again, this is one of the great strengths of NumPy.

Generating Pseudorandom Numbers

NumPy comes with functions for creating arrays from different types of probability distributions. These are useful for tasks such as generating randomized data to test machine learning models, creating data distributions with a known shape or distribution, randomly drawing data for a Monte Carlo simulation, and so on. They're also at least an order of magnitude faster than similar functions in Python's built-in `random` module.

Table18-7 lists some of the functions you can find in `np.random`. For the full list, visit *https://numpy.org/doc/stable/reference/random/index.html.*

Table 18-7: Useful NumPy Pseudorandom Functions

Function	Description
beta	Draw samples from a Beta distribution
binomial	Draw samples from a binomial distribution
chisquare	Draw samples from a chi-square distribution
gamma	Draw samples from a Gamma distribution
normal	Draw random samples from a normal (Gaussian) distribution
permutation	Return a permuted range or random permutation of a sequence
power	Draw from a power function distribution
rand	Create an array of a given shape populated with random samples from a uniform distribution over (0, 1)
randint	Return random integers from low (inclusive) to high (exclusive)
randn	Return a sample (or samples) from the "standard normal" distribution
random	Return random floats in the half-open interval (0.0, 1.0)
seed	Change the seed for the random number generator
shuffle	Randomly permute a sequence in-place
uniform	Draw samples from uniform distribution over half-open interval (low, high)

Reading and Writing Array Data

NumPy can load and save data from and to disk in both binary and text format. Supported text formats are *.txt* and *.csv*. Generally, you will want to use the pandas library, built on NumPy, to work with text or tabular data. We look at pandas in Chapter 20.

For storing and retrieving data in binary format, NumPy provides the save() and load() functions. To save an array to disk, just pass a filename and the array as arguments, as shown here:

```
In [199]: arr = np.arange(8).reshape(2, 4)

In [200]: arr
Out[200]:
array([[0, 1, 2, 3],
       [4, 5, 6, 7]])

In [201]: np.save('my_array', arr)
```

This will produce the binary file *my_array.npy* (the *.npy* extension is added automatically).

To reload this file, enter the following:

```
In [202]: np.load('my_array.npy')
Out[202]:
array([[0, 1, 2, 3],
       [4, 5, 6, 7]])
```

The np.savez() function lets you save several arrays into a single file in uncompressed *.npz* format. Providing keyword arguments lets you store them under the corresponding name in the output file:

```
In [203]: arr1 = np.arange(5)

In [204]: arr2 = np.arange(4)

In [205]: np.savez('arr_arch.npz', a=arr1, b=arr2)

In [206]: archive = np.load('arr_arch.npz')

In [207]: archive['a']
Out[207]: array([0, 1, 2, 3, 4])
```

If arrays are specified as *positional* arguments (no keywords), their names will be *arr_0*, *arr_1*, and so on, by default.

To compress data when archiving, use the savez_compressed() function:

```
In [208]: np.savez_compressed('arr_arch_compressed.npz', a=arr1, b=arr2)
```

In the event you do want to read-in a text file, NumPy provides the genfromtxt() (generate from text) function. To load a *.csv* file, for example, you would pass the function the file path, the character (comma) that separates the values, and whether the data columns have headers, as follows:

```
In [209]: arr = np.genfromtxt('my_data.csv', delimiter=',', names=True)
```

This will produce a *structured* array that contains records rather than individual items. We haven't discussed structured arrays, because they are a low-level tool and we'll be using pandas for operations such as loading *.csv* files. However, you can read more about structured arrays at *https://numpy .org/doc/stable/user/basics.rec.html*.

TEST YOUR KNOWLEDGE

11. Why is there so much whitespace in the first two elements in this output array: ([0,   2, -10000])?

12. Which function would you use to flatten a higher-dimension array to a 1D array?

    a. meshgrid()

    b. vsplit()

    c. ravel()

    d. thresh()

13. For the array [[0, 1, 2], [3, 4, 5], [6, 7, 8]], what does the slice arr2d[:2, 2] produce?

    a. array([1])

    b. array([2, 5])

    c. array([6, 7])

    d. array([3, 4, 5])

14. In NumPy, array multiplication is done:

    a. Row by column

    b. Column by row

    c. Element by element

    d. Row by row then column by column

15. Which array is broadcastable with an array of shape (4, 3, 6, 1)?

    a. (4, 6, 6, 1)

    b. (1, 6, 3, 1)

    c. (4, 1, 6, 6)

    d. (6, 3, 1, 6)

Summary

When working with uniform datasets, NumPy's ndarrays represent a faster, more efficient alternative to competing data structures such as Python lists. Complex computations can be performed without the use of for loops, and ndarrays require significantly less memory than other Python data types.

This chapter touched on a lot of NumPy basics, but there's still more to learn. To expand your knowledge of NumPy, I recommend NumPy's "Beyond the Basics" page (*https://numpy.org/doc/stable/user/c-info.beyond-basics .html*) and Wes McKinney's *Python for Data Analysis: Data Wrangling with Pandas, NumPy, and IPython*, 2nd edition (O'Reilly, 2018).

Before you run off and start applying NumPy, you'll want to check out the next two chapters on Matplotlib and pandas. These libraries are built on top of NumPy and provide higher-level wrappers for performing data analysis and plotting.

19

DEMYSTIFYING MATPLOTLIB

Even among the large number of plotting packages available in Python, Matplotlib stands out. Launched in 2003 to provide a MATLAB-like graphing interface for science and engineering, it now dominates plotting in Python. It has spawned numerous visualization add-ons, like seaborn, and provides the underlying plotting functionality for popular analytical tools like pandas. With knowledge of Matplotlib, you can generate quick and simple plots as well as elaborate, complex charts while controlling every aspect of the display.

The Matplotlib library comes preinstalled with Anaconda. Thanks to its maturity, popularity, and open source status, it has a large supporting community ready to offer you advice and code samples. The best resource is the famous Matplotlib gallery (*https://matplotlib.org/stable/gallery/index.html*), which contains code "recipes" for making just about any kind of plot you can imagine.

Like any powerful piece of software, Matplotlib can be, as one author put it, "syntactically tedious." The simplest plots are easy, but difficulty ramps up quickly. And even though resources like the Matplotlib gallery provide helpful code examples, if you want something slightly different than what's provided, you might find yourself scratching your head. In fact, many people use Matplotlib by copying and pasting other people's code and then hacking at the edges until they get something they like. As a user once told me, "No matter how many times I use Matplotlib, it always feels like the first time!"

Fortunately, you can greatly alleviate this pain by taking the time to learn some key aspects of the package. So, in this chapter, we'll study the fundamentals of Matplotlib plots, including its two plotting interfaces and methods for making multipanel, animated, and customized plots. Armed with this knowledge, you may find Matplotlib a tool to embrace instead of to avoid or use reluctantly.

However, if you don't aspire to be a plotting warrior, take a look at the easier seaborn wrapper in the next chapter. And if seaborn is more than you need, there's also the easier—though less flexible—pandas plotting option.

Anatomy of a Plot

The first step in understanding Matplotlib is mastering the sometimes-awkward nomenclature used for its plots. To that end, let's dissect a plot and its components.

Plots in Matplotlib are held within a `Figure` object (on left in Figure 19-1). This is a blank canvas that represents the top-level container for all plot elements. Besides providing the canvas on which the plot is drawn, the `Figure` object also controls things like the size of the plot, its aspect ratio, the spacing between multiple plots drawn on the same canvas, and the ability to output the plot as an image.

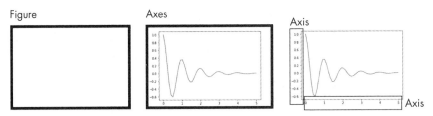

Figure 19-1: The Figure*,* Axes*, and* Axis *components of a Matplotlib plot*

The plots themselves—that is, the things that you and I think of as *figures*—are represented by the Axes class (Figure 19-1, center). This class includes most of the figure *elements*, such as lines, polygons, markers (points), text, titles, and so on, as well as the methods that act *on* them. It also sets the coordinate system. A Figure can contain multiple Axes objects, but each Axes object can belong to only one Figure.

The Axes object should not be confused with the Axis element that represents the numerical values on, say, the x- or y-axis of a chart (Figure 19-1, right). This includes the tick marks, labels, and limits. All these elements are contained within the Axes class.

Each of the components in Figure 19-1 exists within a hierarchical structure (Figure 19-2). The lowest layer includes elements in Figure 19-1 such as each axis, the axis tick marks, and labels, and the curve (Line2D). The highest level is the Figure object, which serves as a container for everything below it.

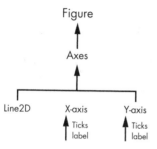

Figure 19-2: The hierarchy of the plot components in Figure 19-1

Because a Figure object can hold multiple Axes objects, you could have more than one Axes object point to the Figure in Figure 19-2. The common example is subplots, in which one Figure canvas holds two or more different plots side by side.

The pyplot and Object-Oriented Approaches

There are two primary interfaces for plotting with Matplotlib. Using the first, referred to as the *pyplot approach*, you rely on Matplotlib's internal pyplot module to *automatically* create and manage Figure and Axes objects, which you then manipulate with pyplot methods for plotting. Designed mainly for dealing with single plots, the pyplot approach reduces the amount of code that you need to know and write. It's a MATLAB-like API that can be very convenient for quick, interactive work.

Using the second approach, called the *object-oriented style*, you *explicitly* create Figure and Axes objects and then call methods on the resulting objects. This gives you the most control over customizing your plots and keeping track of multiple plots in a large program. It's also easier to understand interactions with other libraries if you first create an Axes object.

In the sections that follow, we'll look at both approaches. However, according to the Matplotlib documentation, to maintain consistency you should *choose one approach and stick to it*. They suggest using the object-oriented style, particularly for complicated plots as well as for methods and scripts that are intended to be reused as part of a larger project.

It can certainly be argued that one of the reasons beginners find Matplotlib intimidating is that they see a mixture of these approaches in existing code, such as on question-and-answer sites like Stack Overflow. Because this is unavoidable, I suggest that you read over the descriptions for both approaches so that you can make an informed decision on which one to choose for yourself and you'll have an awareness of the alternate approach when you encounter it in legacy code or in tutorials.

Using the pyplot Approach

To generate a simple plot using the `pyplot` approach, let's use the Jupyter Qt console. To launch a console from your base environment, open an Anaconda prompt (in Windows) or a terminal (in macOS or Linux).

First, run the following (you can ignore this command if your prompt includes "base" in the name):

```
conda activate base
```

Next, enter the following:

```
jupyter qtconsole
```

Now, import Matplotlib's `pyplot` module into the console. For convenience and by convention, you should use the alias `plt`:

```
In [1]: import matplotlib.pyplot as plt
```

By default, plots in the console will display *inline* (within the console). To enable plot interactivity, such as zooming and panning, you can use the magic command `%matplotlib qt`. Subsequent plots will render in an external Qt window, which comes with a toolbar. To restore inline plotting, use the `%matplotlib inline` magic command.

NOTE *In Jupyter Notebook, you can also use `%matplotlib notebook` to enable in-cell interactivity. This can cause some latency in drawing plots, however, as rendering is done on the server side.*

Now, import NumPy and use it to generate a simple 1D array for plotting:

```
In [2]: import numpy as np

In [3]: data = np.arange(5, 10)
```

```
In [4]: data
Out[4]: array([5, 6, 7, 8, 9])
```

To plot the data, pass it to the aptly named plot() method:

```
In [5]: plt.plot(data);
```

The semicolon at the end of the line suppresses display of the Figure object's name, which you don't need. You should now see Figure 19-3 in the console.

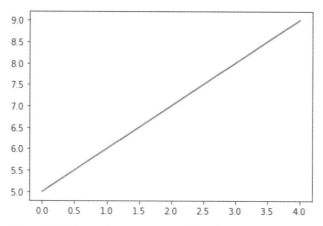

Figure 19-3: A simple autogenerated line plot

Two things are worth noting here: we didn't explicitly refer to Figure or Axes objects in the code, as pyplot took care of these behind the scenes. Nor did we specify what elements to show in the plot, including the ticks and values displayed along the x- and y-axes. Instead, Matplotlib looked at your data and made intelligent choices about the type of plot you wanted and how to annotate it.

Along these lines, the plot() method makes line charts, scatter() makes scatterplots, bar() makes bar charts, hist() makes histograms, pie() makes pie charts, and so on. We'll look at many of these in the sections to come, and you can also visit *https://matplotlib.org/stable/plot_types/index*.

The automatic nature of these methods is useful when you want to quickly explore a dataset, but the resulting plots are generally too plain for presentations or reports. One issue is that the default configuration of methods like plt.plot() assumes that you want the size of each axis to match the range of the input data (such as x from 5 to 8, rather than 0 to 10, if the data is limited to values between 5 and 8). It also assumes that you don't want a legend, title, or axis labels, and that you want lines and markers drawn in blue. This isn't always the case, so pyplot provides many methods to embellish charts with titles, axis labels, grids, and so on. We'll look at these next.

Creating and Manipulating Plots with pyplot Methods

Despite being considered a simpler approach than the object-oriented style, pyplot can still produce some very elaborate plots. To demonstrate, let's use some pyplot methods to create a more sophisticated plot than the one shown in Figure 19-3.

A *catenary* is the shape that a chain assumes when it's hung from both of its ends. It's a common shape in nature and architecture, examples being a square sail under wind pressure and the famous Gateway Arch in St. Louis, Missouri. You can generate a catenary by entering the following code in the console window, where cosh(x) represents the hyperbolic cosine of the x values:

```
In [6]: import numpy as np

In [7]: x = np.arange(-5, 5, 0.1)

In [8]: y = np.cosh(x)
```

Now, let's plot the catenary using a black line with a width of 3 and add a title, axis labels, limits to the axis values, and a background grid. Be sure to use CTRL-ENTER after the first six lines to prevent premature generation of the plot. After the last line, you can press ENTER.

```
In [9]: plt.title('A Catenary')
   ...: plt.xlabel('Horizontal Distance')
   ...: plt.ylabel('Height')
   ...: plt.xlim(-8, 8)
   ...: plt.ylim(0, 60)
   ...: plt.grid()
   ...: plt.plot(x, y, lw=3, color='k');
```

For the line color, 'k' represents the single character shorthand notation for "black." You can see more color choices at *https://matplotlib.org/stable/tutorials/colors/colors.html* and more on plot() parameters at *https://matplotlib.org/stable/api/_as_gen/matplotlib.pyplot.plot.html*. Your output should look like Figure 19-4.

In Matplotlib, the elements rendered on a figure canvas, such as a title, legend, or line, are called Artist objects. Standard graphical objects, like rectangles circles, and text, are referred to as *primitive* Artists. The objects that hold the primitives, like the Figure, Axes, and Axis objects, are called *container* Artists.

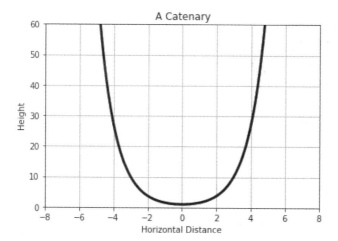

Figure 19-4: A line plot of a catenary

Some of the more common pyplot methods for making plots and working with Artists are listed in Tables 19-1 and 19-2, respectively. To see the full list, visit *https://matplotlib.org/stable/api/pyplot_summary.html*. Clicking the method names in this online list will take you to detailed information on the method parameters, along with example applications. To read more about Artists in general, visit *https://matplotlib.org/stable/tutorials/intermediate/artists.html*.

Table 19-1: Useful pyplot Methods for Creating Plots

Method	Description	Example
bar	Make a bar chart	plt.bar(x, height, width=0.8)
barh	Make a horizontal bar chart	plt.barh(x, height)
contour	Draw a contour map	plt.contour(X, Y, Z)
contourf	Draw a filled contour map	plt.contourf(X, Y, Z, cmap='Greys')
hist	Make a 2D histogram	plt.hist(x, bins)
pie	Display a pie chart	plt.pie(x=[8, 80, 9], labels=['A', 'B', 'C'])
plot	Plot data as lines/markers	plt.plot(x, y, 'r+') # Red crosses
Polar	Make a polar plot	plt.polar(theta, r, 'bo') # Blue dots
Scatter	Make a scatterplot	plt.scatter(x, y, marker='o')
stem	Plot vertical lines to y coordinate	plt.stem(x, y)

Table 19-2: Useful pyplot Methods for Manipulating Plots

Method	Description	Example
annotate	Add text, arrows to Axes	`plt.annotate('text', (x, y))`
axis	Set axis properties (min, max)	`plt.axis([xmin, xmax, ymin, ymax])`
axhline	Add a horizontal line	`plt.axhline(y_loc, lw=5)`
axvline	Add a vertical line	`plt.axvline(x_loc, lw=3, c='red')`
close	Close a plot	`plt.close()`
draw	Update if interactive mode off	`plt.draw()`
figure	Create or activate a figure	`plt.figure(figsize=(4.0, 6.0))`
grid	Add grid lines	`plt.grid()`
imshow	Display data as an image	`pic = plt.imread('img.png')` `plt.imshow(pic, cmap='gray'))`
legend	Place a legend on the Axes	`plt.plot(data, label='Data')` `plt.legend()`
loglog	Use log scaling on each axis	`plt.loglog()`
minorticks_off	Remove minor ticks from axis	`plt.minorticks_off()`
minorticks_on	Display minor ticks on axis	`plt.minorticks_on()`
savefig	Save as *.jpg*, *.png*, *.pdf*, and so on	`plt.savefig('filename.jpg')`
semilogx	Use log scaling on x-axis	`plt.semilogx()`
semiology	Use log scaling on y-axis	`plt.semilogy()`
set_cmap	Set colormap	`plt.set_cmap('Greens')`
show	Show plot run from terminal or when interactive mode is off	`plt.show()`
subplot	Create subplots on a figure	`plt.subplot(nrows, ncols, index)`
text	Add text to the Axes	`plt.text(x, y, 'text')`
tight_layout	Adjust padding in subplots	`plt.tight_layout(pad=3)`
title	Add a title to the Axes	`plt.title('text')`
xkcd	Turn on xkcd sketch-style*	`plt.xkcd()`
xlabel	Set the x-axis label	`plt.xlabel('text')`
xlim	Set x-axis limits	`plt.xlim(xmin, xmax)`
xticks	Set tick information	`plt.xticks([0, 2], rotation=30)`
ylabel	Set the y-axis label	`plt.ylabel('text')`
ylim	Set y-axis limits	`plt.ylim(ymin, ymax)`
yticks	Set tick information	`plt.yticks([0, 2], rotation=30)`

*For best results, the Humor Sans font should be installed.

Note that the code examples in the tables represent simple cases. Most methods take many arguments, letting you fine tune your plots with respect to properties like font style and size, line widths and colors, rotation angles, exploded views, and much more (see *https://matplotlib.org/stable/api/pyplot_summary.html*).

Working with Subplots

So far, we've been working with single figures, but there'll be times when you'll want to compare two plots side by side, or bundle several charts into a summary display. For these occasions, Matplotlib provides the `subplot()` method.

To see how this works, let's begin by generating data for two different sine waves:

```
In [10]: time = np.arange(-12.5, 12.5, 0.1)

In [11]: amplitude = np.sin(time)

In [12]: amplitude_halved = np.sin(time) / 2
```

One way to compare these waveforms is to plot them in the same Axes, like so:

```
In [13]: plt.plot(time, amplitude, label='sine1')
    ...: plt.plot(time, amplitude_halved, lw=3, ls='--', label='sine2')
    ...: plt.legend();
```

This produces the output in Figure 19-5. By default, the curves plot with different colors in the Qt console, but because this is a black-and-white book, we used a different line width (`lw`) and line style (`ls`) for the amplitude_halved data to distinguish it from the amplitude data. The `label` parameter in `plt.plot()` also permits the use of a legend. For the list of characters available for marker and line styles, visit *https://matplotlib.org/stable/api/_as_gen/matplotlib.pyplot.plot.html*.

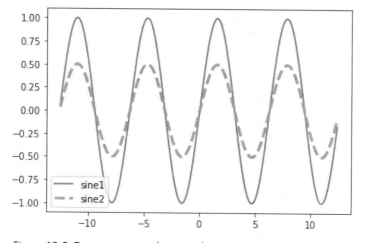

Figure 19-5: Two sine waves drawn in the same Axes object

If you're comparing more than a few curves, a single plot can become cluttered and difficult to read. In those cases, you'll want to use separate stacked plots created by the subplot() method. Figure 19-6 describes the syntax for this method, in which four subplots (Axes) are placed in a single Figure container.

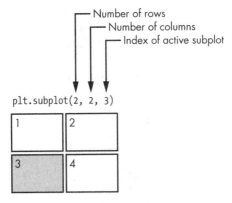

Figure 19-6: Understanding the subplot() method

The subplots will be arranged in a grid, and the first two arguments passed to the subplot() method specify the dimensions of this grid. The first argument represents the number of rows in the grid, the second, the number of columns, and the third argument is the index of the *active* subplot (highlighted in gray in the figure).

The active subplot is the one you are currently plotting in when you call a method like plot() or scatter(). Unlike most things in Python, the first index is 1, not 0.

Matplotlib uses a concept called the *current figure* to keep track of which Axes is currently being worked. For example, when you call plt.plot(), pyplot creates a new "current figure" Axes to plot on. That's why you must press CTRL-ENTER in the console when working on a plot. As soon as you press ENTER, the plot is complete, and a new "current figure" is queued up. When you're working with multiple subplots, the index argument tells pyplot which subplot represents the "current figure."

NOTE *For convenience, you don't need to use commas with the subplot() arguments. For example, plt.subplot(223) works the same as plt.subplot(2, 2, 3), although it's arguably less readable.*

Now, let's plot our sine waves as two separate stacked plots. The process will be to call the subplot() method and alter its active subplot argument to change the current subplot. For each current subplot, the plot() method will post the data specific to that subplot, as follows:

```
In [14]: plt.subplot(2, 1, 1)
    ...: plt.plot(time, amplitude, label='sine1')
    ...: plt.legend(loc='upper right')
    ...:
```

```
...: plt.subplot(2, 1, 2)
...: plt.ylim(-1, 1)
...: plt.plot(time, amplitude_halved, label='sine2')
...: plt.legend(loc='best');
```

Note that if you don't set the y limits on the second plot, pyplot will automatically scale the graph so that the two subplots look identical. Because we manually set the scale using the ylim() method, it's clear that the second sine wave has half the amplitude of the first (Figure 19-7).

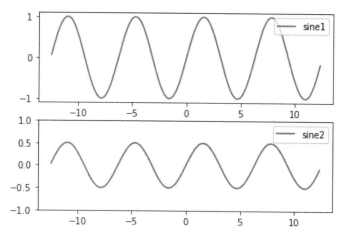

Figure 19-7: Sine waves displayed in two horizontal subplots

These plots appear a bit cramped. Let's give them some breathing room by calling the tight_layout() method and passing it a pad value. The larger the pad value, the larger the space between plots, though there is a limit to how much space can be accommodated. Additional arguments are available that let you fine tune the display; for example, by padding the height and width between edges of adjacent subplots using h_pad and w_pad.

Use the arrow keys to bring up the previous code and add the tight _layout() method, as follows:

```
In [15]: plt.subplot(2, 1, 1)
    ...: plt.plot(time, amplitude, label='sine1')
    ...: plt.legend(loc='upper right')
    ...:
    ...: plt.subplot(2, 1, 2)
    ...: plt.ylim(-1, 1)
    ...: plt.plot(time, amplitude_halved, label='sine2')
    ...: plt.legend(loc='best')
    ...:
    ...: plt.tight_layout(pad=2)
```

This produces the plot in Figure 19-8. Now it's clear which x-axis goes with which subplot.

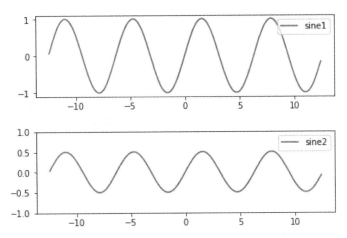

Figure 19-8: The result of calling `tight_layout()` on the figure

You've just seen how the `subplot()` method lets you subdivide a figure into different drawing areas and then focus the plotting commands on a single subplot. To help you manage even more sophisticated plots, Matplotlib provides the `GridSpec` class, which we'll look at next.

TEST YOUR KNOWLEDGE

1. An Axes object represents:

    a. The x-, y-, and z-axes of a plot

    b. Individual elements of a plot, such as titles and legends

    c. The container for individual figure elements

    d. A blank canvas

2. True or False: For complicated plots, and for methods and scripts that are intended to be reused as part of a larger project, the Matplotlib documentation recommends that you use the object-oriented style.

3. Which code produces a grid of subplots four columns wide, three rows high, and with the second subplot active?

    a. `plt.subplot(3, 4, 1)`

    b. `plt.subplots(3, 4, 2)`

    c. `plt.subplot(4, 3, 2)`

    d. `plt.subplot(342)`

4. The %matplotlib qt magic command is used to:

    a. Enable graphics within the console

    b. Allow interactive graphics within Jupyter Notebook

    c. Open an external window with interactive controls

    d. Restore inline graphics after using an external window

5. Make a Python dictionary of rocket heights using this data: Atlas: 57, Falcon9: 70, Saturn V: 111, Starship: 120. Plot a bar chart of the data, label the y-axis to indicate height in meters, and set the bar width to 0.3.

Building Multipanel Displays Using GridSpec

The matplotlib.gridspec module includes a GridSpec class that lets you split a Figure into a grid of subareas. This, in turn, helps you to create subplots that have different widths and heights. The resulting *multipanel* displays are useful for summarizing information in presentations and reports.

Constructing a Martian Multipanel Display

Let's work through an example: imagine that you're studying an ancient lakebed on Mars. You want to summarize some of your findings about hematite, goethite, and jarosite, three iron-bearing minerals associated with aqueous environments. You've sketched a layout for a compilation figure (Figure 19-9), and now you want to create it using Matplotlib.

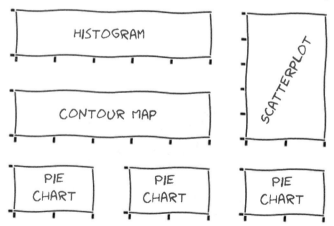

Figure 19-9: A sketch of the summary figure for a Mars study (generated with the xkdc() method from Table 19-2)

NOTE *If you want to save the code, you can create this project in the Spyder text editor or Jupyter Notebook rather than in the console. You've already played with Matplotlib in these applications in Chapters 4, 5, and 6. If you're working in Jupyter Notebook, all of the code that defines a plot should be contained in the same cell.*

To start, if you haven't done so already, import NumPy and Matplotlib:

```
In [16]: import numpy as np

In [17]: import matplotlib.pyplot as plt
```

Now, call `GridSpec` to create a 3×3 grid and assign the resulting object to a variable named gs, for *grid spec*. In the console, use CTRL-ENTER after this statement, as we'll now start defining the subplots:

```
In [18]: gs = plt.GridSpec(3, 3)
```

The previous code creates a grid with three rows and three columns. To place a subplot within this grid, you index the gs object. Unlike with the `subplot()` method, indexes start with 0, rather than 1.

Figure 19-10 shows the subplot locations along with their grid indexes. To place the upper-left subplot (to hold the histogram, as illustrated in Figure 19-9), use gs[0, :2]. This references the first row [0] and columns one and two [:2]. Thus, the subplot spans the first two columns of row one. Likewise, gs[:2, 2] spans the first two rows of column three, and gs[2, 1] places that subplot in the center of row three.

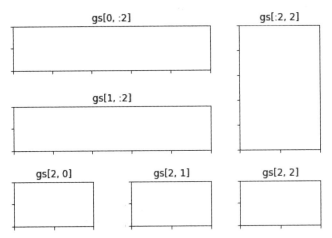

Figure 19-10: Subplots for the sketch in Figure 19-9 with their GridSpec indexes annotated

Before building a subplot in the summary figure, you'll need to specify its grid location using the indexes in Figure 19-10. Let's do this now for the histogram. Because we don't have any real Mars data, we'll use a dummy dataset drawn from a normal distribution (using the NumPy `random.normal()` method):

```
...:
...: plt.subplot(gs[0,:2])
...: plt.title('Goethite Distribution Location 1')
...: plt.hist(np.random.normal(0.22, 0.02, size=500), bins=5)
...:
```

The arguments for the `np.random.normal()` method are the mean, standard deviation, and number of draws from the normal distribution. The `plt.hist()` method takes this output, along with the number of bins for the histogram.

This will produce the chart in Figure 19-11, though you won't see this until the entire plot is finished. Your view might look slightly different given that the histogram data is generated randomly.

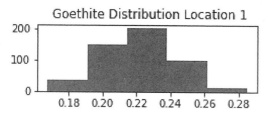

Figure 19-11: The histogram subplot

Next, we'll build the contour map below the histogram. Note that we can build the subplots in whatever order we want, but following a logical order makes it easier to go back and edit the code later. As always, start by locating the subplot on the grid using `plt.subplot()`:

```
...: plt.subplot(gs[1, :2])
...: plt.title('Goethite Concentration Location 1')
...: plt.text(1.3, 1.6, ❶ 'o--Sample A')
...: x, y = np.arange(0, 3, 0.1), np.arange(0, 3, 0.1)
...: X, Y = ❷ np.meshgrid(x, y)
...: Z = np.absolute(np.cos(X * 2 + Y) * 2 + np.sin(Y + 3))
...: plt.contourf(X, Y, Z, cmap='Greys')
...: plt.colorbar()
...:
```

To demonstrate placing text on plots, add an annotation identifying the location of Sample A ❶. The `text()` method used to do this needs at least an x, y location and a text string. The circle and line part of the string (o--) represents a pointer to the sample location. Many other arguments are available for the `text()` method, including ones for `fontsize`, `color`, and `rotation`.

Next, generate some dummy coordinates and a *mesh grid* ❷ using NumPy. The meshgrid() method creates a rectangular grid out of two given one-dimensional arrays representing Cartesian or matrix indexing. From this grid we can use an equation to generate corresponding Z values. Calling the pyplot contourf() method and passing it the coordinates and a gray color-map generates filled contours. Finish by posting the colorbar.

This code will produce a map like the one presented in Figure 19-12. If you want to get fancy, you can use an arrow artist to point to the sample location (see *https://matplotlib.org/stable/tutorials/text/annotations .html#annotating-with-arrow/*).

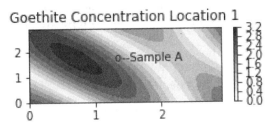

Figure 19-12: The contour map subplot

Next, we'll generate the scatterplot in the upper-right corner of Figure 19-9. This will plot the concentration of hematite versus goethite at Location 1. Start by assigning the grid location, and then add a title and labels for the x-axis and y-axis.

```
...: plt.subplot(gs[:2, 2])
...: plt.title('Loc1 Goe-Hem Ratio')
...: plt.xlabel('Hematite mg')
...: plt.ylabel('Goethite mg')
...: plt.scatter(np.random.normal(3, 1, 30), np.random.uniform(1, 30, 30))
...:
```

To generate scatterplots, you pass the plt.scatter() method a sequence of x, y values. In this case, we'll randomly generate these on the fly using NumPy's normal and uniform distribution methods, respectively. For the normal method, the arguments are the mean, standard deviation, and number of draws. For the uniform method, they represent the low and high values and number of draws.

This will ultimately produce the subplot in Figure 19-13. Again, because the data is randomly generated, every scatterplot will look different.

NOTE *For datasets with more than a few thousand points, passing a marker type to plt .plot() can be much more efficient than using plt.scatter(). The reason for this is because plt.plot() renders points as clones, whereas plt.scatter() renders each point individually, to permit altering marker sizes to reflect data values or to differentiate datasets.*

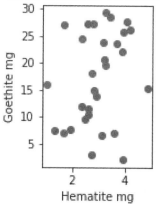

Figure 19-13: The scatterplot subplot

Now, let's build three pie charts that record the percentage of hematite, goethite, and jarosite in samples A, B, and C. We'll string these along the bottom of the summary figure.

Each pie chart will use the same wedge labels (representing the categories in the chart), so these should be assigned at the outset to avoid repeating code. In addition, we'll use the plt.pie() method's explode parameter to separate the pie wedges. To specify the size of the gap between the wedges, we'll use a list named explode, which emphasizes the jarosite wedge by pulling it slightly out of the rest of the pie:

```
...: labels = 'Goethite', 'Hematite', 'Jarosite'
...: explode = [0.1, 0.1, 0.2]
...:
```

To make a pie chart, pass the plt.pie() method the labels (representing the categories in the chart), the sizes (representing the percentages of each category), and the explode list:

```
...: plt.subplot(gs[2, 0])
...: plt.title('Sample A')
...: sizes = [35, 55, 10]
...: plt.pie(sizes, labels=labels, explode=explode)
...:
...: plt.subplot(gs[2, 1])
...: plt.title('Sample B')
...: sizes = [35, 45, 20]
...: plt.pie(sizes, labels=labels, explode=explode)
...:
...: plt.subplot(gs[2, 2])
...: plt.title('Sample C')
...: sizes = [35, 35, 30]
...: plt.pie(sizes, labels=labels, explode=explode)
...:
```

Complete the figure by calling the tight_layout() method to add some space between the subplots. After this last line, if you're in the console, press ENTER or SHIFT-ENTER to generate the final multipanel plot, which you can see in Figure 19-14.

```
...: plt.tight_layout();
```

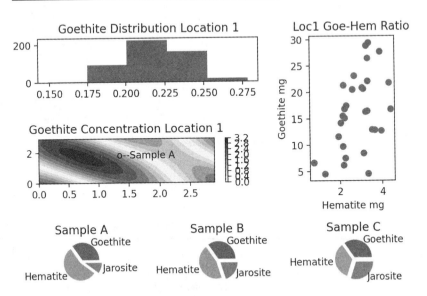

Figure 19-14: The final multipanel summary plot

Thanks to GridSpec, the summary display contains subplots that span multiple rows and columns.

Changing the Width and Height of the Subplots

Within certain bounds, you can set the width and height of the rows and columns produced by GridSpec. You can do this through the width_ratios and height_ratios parameters, which each accept a list of numbers. Only the *ratios* between these numbers matter. For example, to set the width ratios for each column in our 3×3 grid, [1, 2, 4] is the same as [2, 4, 8].

To demonstrate, enter the following code to alter our Martian multipanel display:

```
In [19]: widths = [2, 3, 2]

In [20]: heights = [2, 10, 3]
```

The widths list addresses column widths, starting at index 0. The heights list repeats this for row heights.

Now, bring up the code from the previous section (using the arrow key if in the console) and edit the call to plt.GridSpec, as follows:

```
In [21]: gs = plt.GridSpec(3, 3, width_ratios=widths, height_ratios=heights)
```

Rerun the code and you should see the plot in Figure 19-15. Note the changes such as the shorter histogram and the taller contour map.

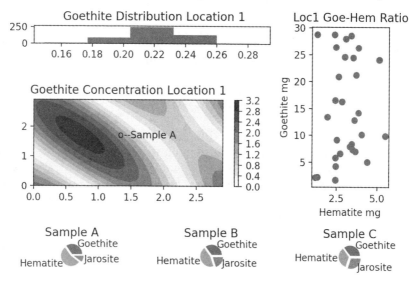

Figure 19-15: The multipanel display with new row and column widths and heights

To read more about GridSpec and see some example use cases, visit *https://matplotlib.org/stable/api/_as_gen/matplotlib.gridspec.GridSpec.html.* For a tutorial on the pyplot approach, see *https://matplotlib.org/stable/tutorials/introductory/pyplot.html.*

Using the Object-Oriented Style

The object-oriented plotting style generally requires a bit more code than the previously described pyplot approach, but it lets you get the absolute most out of Matplotlib. By explicitly creating Figure and Axes objects, you'll be able to more easily control your plots, better understand interactions with other libraries, create plots with multiple x- and y-axes, and more.

NOTE   *You'll appreciate the object-oriented style more if you're familiar with object-oriented programming. This programming paradigm is covered in Chapter 13.*

To familiarize ourselves with the object-oriented style, let's re-create the simple plot from Figure 19-3. If you're using the Qt console, restart it now.

The Matplotlib `import` statement stays the same regardless of the plotting approach:

```
In [22]: import matplotlib.pyplot as plt
```

```
In [23]: import numpy as np
```

Now, regenerate the dataset with NumPy:

```
In [24]: data = np.arange(5, 10)
```

To start using the object-oriented style, enter the following and press CTRL-ENTER in the console:

```
In [25]: fig, ax = plt.subplots()
```

As soon as you see this line of code in a program, you know you're dealing with the object-oriented style. The `plt.subplots()` method creates a `Figure` instance and a set of subplots (a NumPy array of `Axes` objects). If a number of subplots is not specified, a single subplot is returned by default. Because two objects are returned, you need to unpack the results to two variables, called `fig` and `ax` by convention. Remember that, with the `pyplot` approach, these two entities are created behind the scenes.

To display the plot, add the following line and then press ENTER:

```
    ...: ax.plot(data);
```

This produces the plot in Figure 19-16, which is identical to the one in Figure 19-3.

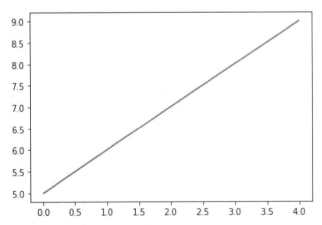

Figure 19-16: A simple line plot generated using the object-oriented style

Because you assigned the plot to a fig variable, you can regenerate it by simply entering fig in the console:

```
In [26]: fig
```

The object-oriented plotting style really isn't mysterious. The key is to assign the Figure and Axes objects created by pyplot to variables. You'll no longer get the benefit of the automated features of pyplot, but in return you open the door to a host of object attributes and methods for customizing plots.

Creating and Manipulating Plots with the Object-Oriented Style

To better understand the object-oriented style, let's use it to re-create the catenary example from "Creating and Manipulating Plots with pyplot Methods" on page 542. To demonstrate some of the style's enhanced functionality, we'll force the y-axis to pass through the center of the plot.

If you're using the Qt console, restart the kernel now. Then, import NumPy and Matplotlib and regenerate the catenary data:

```
In [27]: import numpy as np

In [28]: import matplotlib.pyplot as plt

In [29]: x = np.arange(-5, 5, 0.1)

In [30]: y = np.cosh(x)
```

To create a single plot, enter the following and then press CTRL-ENTER (in the console):

```
In [31]: fig, ax = plt.subplots()
```

Next, call the AXES object's set() method and pass it keyword arguments for a title, axis labels, and axis limits. This is a convenience method that lets you set multiple properties at once rather than calling specific methods for each. You can use either a single line that wraps, or press ENTER after each comma to produce a more readable vertical stack, as follows:

```
   ...: ax.set(title='A Catenary',
   ...:        xlabel='Horizontal Distance',
   ...:        ylabel='Height',
   ...:        xlim=(-8, 8.1),
   ...:        ylim=(0, 60))
```

Now let's move the y-axis to the center of the chart instead of along the side. In Matplotlib, *spines* are the lines connecting the axis tick marks and noting the boundaries of the area containing the plotted data. The default position for these is around a plot with the ticks and labels along the left and bottom margins (see Figure 19-16). But spines can also be placed at arbitrary positions. With the object-oriented style, we can accomplish this using the set_position() method of the Spine subclass.

The following code first moves the left (y) axis to the 0 value on the x-axis. Then, the line width is set to 2 so that the axis stands out a bit from the background grid that we're going to use later:

```
...: ax.spines.left.set_position('zero')
...: ax.spines.left.set_linewidth(2)
```

The following line turns off the right boundary of the plot by setting its color to none:

```
...: ax.spines.right.set_color('none')
```

The next three lines repeat this overall process for the bottom axis and top axis, respectively:

```
...: ax.spines.bottom.set_position('zero')
...: ax.spines.bottom.set_linewidth(2)
...: ax.spines.top.set_color('none')
```

To finish the plot, add the background grid and call the plot method, passing it the x and y data and setting the line width to 3 and the color to black ('k'):

```
...: ax.grid()
...: ax.plot(x, y, lw=3, color='k');
```

This produces the plot in Figure 19-17.

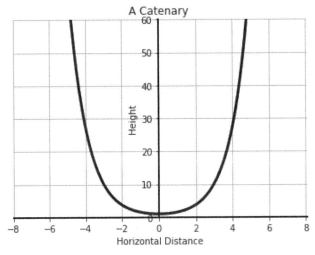

Figure 19-17: The line plot of a catenary built using the object-oriented style

If you omit the code related to the spines, you can reproduce the plot in Figure 19-4 with essentially the same amount of code as used by the pyplot approach. Thus, the verbosity of the object-oriented style has much to do with the fact that you can do more with it, and people generally take advantage of this.

Methods available in the `pyplot` approach have an equivalent in the object-oriented style. Unfortunately, the method names are often different. For example, `title()` in `pyplot` becomes `set_title()`, and `xticks()` becomes `set_xticks()`. This is one reason why it's good to pick one approach and stick with it.

Some of the more common methods for making object-oriented plots are listed in Table 19-3. You can find additional methods, such as for making box plots, violin plots, and more, at *https://matplotlib.org/stable/plot_types/index.html* and in the Matplotlib gallery, referenced previously in this chapter on page 538.

Table 19-3: Useful Object-Oriented Methods for Creating Plots

Method	Description	Example
bar	Make a bar chart	`ax.bar(x, height)`
barh	Make a horizontal bar chart	`ax.barh(x, height)`
contour	Draw a contour map	`ax.contour(X, Y, Z)`
contourf	Draw a filled contour map	`ax.contourf(X, Y, Z, cmap='Greys')`
hist	Make a 2D histogram	`ax.hist(x, bins)`
pie	Display a pie chart	`ax.pie(x=[8, 80, 9], labels=['A', 'B', 'C'])`
plot	Plot data as lines/markers	`ax.plot(x, y, 'r+') # Red crosses`
polar	Make a polar plot	`fig, ax = plt.subplots(subplot_kw={'projection': 'polar'})` `ax.plot(theta, r, 'bo') # Blue dots`
scatter	Make a scatterplot	`ax.scatter(x, y, marker='o')`
stem	Plot vertical lines to y coordinate	`ax.stem(x, y)`

Common methods for working with `Figure` and `Axes` objects are listed in Tables 19-4 and 19-5, respectively. In many cases, these work like the `pyplot` methods in Table 19-2, though the method names might be different.

Table 19-4: Useful Object-Oriented Methods for Working with `Figure` Objects

Method	Description	Example
add_subplot	Add or retrieve an Axes	`ax = fig.add_subplot(2, 2, 1)`
close()	Close a figure	`plt.close(fig2)`
colorbar	Add a colorbar to an Axes	`fig.colorbar(image, ax=ax)`
constrained_layout	Auto-adjust fit of subplots	`fig, ax = plt.` `subplots(constrained_layout=True)`

(continued)

Table 19-4: Useful Object-Oriented Methods for Working with Figure Objects *(continued)*

Method	Description	Example
gca	Get the current Axes instance on the current figure	`fig.gca()`
savefig	Save as *.jpg, .png, .pdf,* and so on	`fig.savefig('filename.jpg')`
set_size_inches	Set Figure size in inches	`fig.set_size_inches(6, 4)`
set_dpi	Set Figure dots per inch	`fig.set_dpi(200) # Default is 100.`
show	Show plot run from terminal or when interactive mode is off	`plt.show()`
subplots	Create Figure with Axes	`fig, ax = plt.subplots(2, 2)`
suptitle	Add a super title to a Figure	`fig.suptitle('text')`
tight_layout	Auto-adjust subplots fit	`fig.tight_layout()`

Table 19-5: Useful Object-Oriented Methods for Working with Axes Objects

Method	Description	Example
annotate	Add text and arrows to Axes	`ax.annotate('text', xy=(5, 2))`
axis	Get or set axis properties	`ax.axis([xmin, xmax, ymin, ymax])`
axhline	Add a horizontal line	`ax.axhline(y_loc, lw=5)`
axvline	Add a vertical line	`ax.axvline(x_loc, lw=3, c='red')`
grid	Add grid lines	`ax.grid()`
imshow	Display data as an image	`pic = plt.imread('img.png')` `ax.imshow(pic, cmap='gray'))`
legend	Place a legend on the Axes	`ax.plot(data, label='Data')` `ax.legend()`
loglog	Use log scaling on each axis	`ax.loglog()`
minorticks_on	Display minor ticks on axis	`ax.yaxis.get_ticklocs(minor=True)` `ax.minorticks_on()`
minorticks_off	Remove minor ticks from axis	`plt.minorticks_off()`
semilogx	Use log scaling on x-axis	`ax.semilogx()`
semiology	Use log scaling on y-axis	`ax.semilogy()`
set	Set multiple properties at once	`ax.set(title, ylabel, xlim, alpha)`
set_title()	Set the Axes title	`ax.set_title('text', loc='center')`
set_xticks()	Set x-axis tick marks	`xticks = np.arange(0, 100, 10)` `ax.set_xticks(xticks)`
set_yticks()	Set y-axis tick marks	`yticks = np.arange(0, 100, 10)` `ax.set_yticks(yticks)`

Method	Description	Example
set_xticklabels	Set x-axis labels after calling set_xticks()	labels = ['a', 'b', 'c', 'd'] ax.set_xticklabels(labels)
set_yticklabels	Set y-axis labels after calling set_yticks()	ax.set_yticklabels([1, 2, 3, 4])
tick_params	Change ticks, labels, and grid	ax.tick_params(labelcolor= 'red')
twinx	New y-axis with shared x-axis	ax.twinx()
twiny	New x-axis with shared y-axis	ax.twiny()
set_xlabel()	Set label for x-axis	ax.set_xlabel('text', loc='left')
set_ylabel()	Set label for y-axis	ax.set_ylabel('text', loc='top')
set_xlim()	Set limits of x-axis	ax.set_xlim(-5, 5)
set_ylim()	Set limits of y-axis	ax.set_ylim(0, 10)
set_xscale()	Set the x-axis scale	ax.set_xscale('log')
set_yscale()	Set the y-axis scale	ax.set_yscale('linear')
text	Add text to the Axes	ax.text(x, y, 'text')
xaxis.grid()	Add x-axis grid lines	ax.xaxis.grid(True, which='major')
yaxis.grid()	Add y-axis grid lines	ax.yaxis.grid(True, which='minor')

As mentioned in the pyplot section, the code examples in all these tables represent simple cases. Most methods take many arguments, letting you fine tune your plots with respect to properties like font style and size, line widths and colors, rotation angles, exploded views, and much more. To learn more, visit the Matplotlib documentation at *https://matplotlib.org/*.

Working with Subplots

Like the pyplot approach, the object-oriented style supports the use of subplots (see "Working with Subplots" on page 545). Although there are multiple ways to assign subplots to Figure and Axes objects, the plt.subplots() method is convenient and returns a NumPy array that lets you select subplots using standard indexing or with unique names such as axs[0, 0] or ax1. Another benefit is that you can preview the subplots' geometry prior to plotting any data.

NOTE *The object-oriented method for creating subplots is spelled subplots, whereas the pyplot approach uses subplot.*

Calling plt.subplots() with no arguments generates a single empty plot (Figure 19-18). Technically, this produced a 1×1 AxesSubplot object.

```
In [32]: fig, ax = plt.subplots()
```

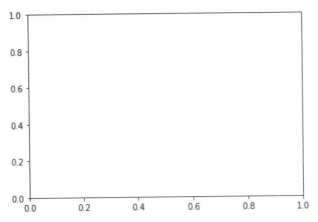

Figure 19-18: Empty plot produced using the subplots()
method of the object-oriented style

Producing multiple subplots is like the plt.subplot() method, only without an index argument for the active subplot. The first argument indicates the number of rows; the second specifies the number of columns. By convention, multiple Axes are given the plural name, axs, rather than axes so as to avoid confusion with a single instance of Axes.

Passing the plt.subplots() method two arguments lets you control the number of subplots and their geometry. The following code generates the 2×2 grid of subplots shown in Figure 19-19 and stores a list of two AxesSubplot objects in the axs variable:

```
In [33]: fig, axs = plt.subplots(2, 2)
    ...: axs
Out[33]:
array([[<AxesSubplot:>, <AxesSubplot:>],
       [<AxesSubplot:>, <AxesSubplot:>]], dtype=object)
```

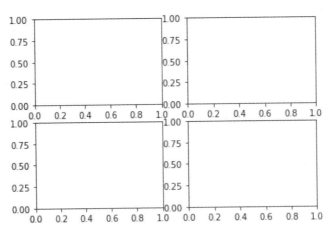

Figure 19-19: Four subplots in a 2×2 arrangement

To activate a subplot, you can use its index. In this example, we plot on the second subplot in the first row, producing Figure 19-20:

```
In [34]: fig, axs = plt.subplots(2, 2)
    ...: axs[0, 1].plot([1, 2, 3]);
```

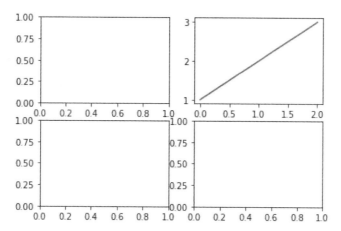

Figure 19-20: Plotting using subplot index [0, 1]

Alternatively, you can name and store the subplots individually by using tuple unpacking for multiple Axes. Each row of subplots will need to be in its own tuple. You can then select a subplot using a name, versus a less-readable index. The following code reproduces Figure 19-20:

```
In [35]: fig, ((ax1, ax2), (ax3, ax4)) = plt.subplots(2, 2)
    ...: ax2.plot([1, 2, 3]);
```

Finally, the subplots() method takes additional keywords, including figure keywords, that let you do things like share an axis among plots, adjust the figure size and layout, and so on (Figure 19-21):

```
In [36]: fig, axs = plt.subplots(ncols=2,
    ...:                         nrows=2,
    ...:                         sharex=True,
    ...:                         sharey=True,
    ...:                         figsize=(6, 4),
    ...:                         tight_layout=True)
```

For more on these keywords, see the method's documentation at *https://matplotlib.org/stable/api/_as_gen/matplotlib.pyplot.subplots.html*.

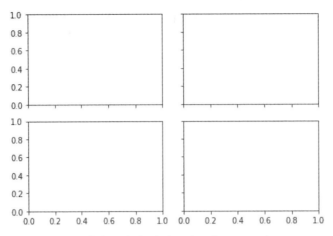

Figure 19-21: A 2×2 grid of subplots that share x- and y-axes

Building Multipanel Displays Using GridSpec

The `matplotlib.gridspec` module (described in "Building Multipanel Displays Using GridSpec" on page 549) also works with the object-oriented style. Let's use it now to reproduce the Martian multipanel display in Figure 19-14. This will let you directly compare the `pyplot` and object-oriented approaches.

Reconstructing the Martian Multipanel Display

To start fresh, restart the kernel in an open console (**Kernel ▸ Restart Current Kernel**) or exit and reopen the console. If you're restarting, use CTRL-L to clear the window. To restart in Jupyter Notebook, use **Kernel ▸ Restart & Clear Output**.

Now, import NumPy and Matplotlib and set up a 3×3 grid using GridSpec. In the console, press CTRL-ENTER after line In [39] to prevent generation of the plot (in Notebook, use ENTER):

```
In [37]: import numpy as np

In [38]: import matplotlib.pyplot as plt

In [39]: fig = plt.figure()
    ...: gs = fig.add_gridspec(3, 3)
    ...:
```

Next, build the histogram subplot. Name it `ax1` and position it using the grid indexes in Figure 19-10:

```
    ...: ax1 = fig.add_subplot(gs[0, :2])
    ...: ax1.set_title(' Goethite Distribution Location 1')
    ...: ax1.hist(np.random.normal(0.22, 0.02, size=500), bins=5)
```

Continuing to use CTRL-ENTER in the console, build the contour map, as follows:

```
...: ax2 = fig.add_subplot(gs[1, :2])
...: ax2.set_title('Goethite Concentration Location 1')
...: ax2.annotate('o--Sample A', xy=(1.3, 1.6))
...: x, y = np.arange(0, 3, 0.1), np.arange(0, 3, 0.1)
...: X, Y = np.meshgrid(x, y)
...: Z = np.absolute(np.cos(X * 2 + Y) * 2 + np.sin(Y + 3))
...: contour_map = ax2.contourf(X, Y, Z, cmap='Greys')
...: fig.colorbar(contour_map)
...:
```

Next, we'll build the scatterplot located in the upper-right corner of the display:

```
...: ax3 = fig.add_subplot(gs[:2, 2])
...: ax3.set_title('Loc1 Goe-Hem Ratio')
...: ax3.set_xlabel('Hematite mg')
...: ax3.set_ylabel('Goethite mg')
...: ax3.scatter(np.random.normal(3, 1, 30),
...:             np.random.uniform(1, 30, 30))
...:
```

Add the code for the pie wedge labels and the gaps between pie wedges. This reduces code duplication because these variables are the same for all the charts:

```
...: labels = 'Goethite', 'Hematite', 'Jarosite'
...: explode = [0.1, 0.1, 0.2]
...:
```

Finish the charts and then call the Figure object's tight_layout() method to prevent the plots from crowding one another. Generate the display by pressing ENTER or SHIFT-ENTER in the console, and CTRL-ENTER in Jupyter Notebook:

```
...: ax4 = fig.add_subplot(gs[2, 0])
...: ax4.set_title('Sample A')
...: sizes = [35, 55, 10]
...: ax4.pie(sizes, labels=labels, explode=explode)
...:
...: ax5 = fig.add_subplot(gs[2, 1])
...: ax5.set_title('Sample B')
...: sizes = [35, 45, 20]
...: ax5.pie(sizes, labels=labels, explode=explode)
...:
...: ax6 = fig.add_subplot(gs[2, 2])
...: ax6.set_title('Sample C')
...: sizes = [35, 35, 30]
...: ax6.pie(sizes, labels=labels, explode=explode)
...:
...: fig.tight_layout();
```

Note that the main changes from the pyplot approach are some method names, such as set_title() for title(), and the use of subplot names. The resulting display should be identical to Figure 19-14, except for some variations in the randomly generated data.

To change the width and height of the subplots, refer to the section "Changing the Width and Height of the Subplots" on page 554. This task works the same for both plotting approaches.

The matplotlib.gridspec module gives you a lot of control over the placement of subplots in multipanel displays. As always with Python, however, there are multiple ways to do the same thing, and we'll look at one of these alternatives next.

Higher-Level Alternatives to GridSpec

The Matplotlib library includes some higher-level alternatives to using GridSpec. With the subplot_mosaic() method, for example, you can lay out your grid using logical names like upper_left and right. You then can use these to index the axs object, as follows:

```
In [40]: fig, axs = plt.subplot_mosaic([['left', 'upper right'],
    ...:                                ['left', 'lower right']],
    ...:                                figsize=(4.5, 3.5),
    ...:                                tight_layout=True)
    ...: axs['upper right'].set_title('upper right');
```

This produces the display in Figure 19-22. The subplots are laid out in the order in which they are assigned in line In [40].

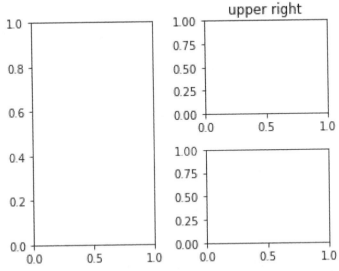

Figure 19-22: Subplots created using the plt.subplot_mosaic() method

To read more about `subplot_mosaic()` and other multipaneling options, see the sections "Working with Multiple Figures and Axes" at *https://matplotlib.org/stable/tutorials/introductory/usage.html#sphx-glr-tutorials -introductory-usage-py/* and "Arranging Multiple Axes in a Figure" at *https:// matplotlib.org/stable/tutorials/intermediate/arranging_axes.html.*

Insetting Plots

An *inset* plot—that is, a plot within a plot—is useful for showing greater detail in part of an enclosing plot, a different treatment of the same data, the geographical location of the data, and so on. An inset plot is like a subplot, but it's built using a different technique.

To make an inset, you first make a `Figure` object and then add `Axes` to it using the `add_axes()` method. Enter the following code in the console or in Notebook; ignore the imports if you've already executed them in your current session:

```
In [41]: import numpy as np

In [42]: import matplotlib.pyplot as plt

In [43]: %matplotlib inline

In [44]: x = np.arange(0, 25)

In [45]: y = x**3
```

Now, set up the `Figure` and `Axes` objects. In this case, `ax2` represents the inset plot:

```
In [46]: fig = plt.figure()
    ...: ax1 = fig.add_axes([0, 0, 1.0, 1.0])
    ...: ax2 = fig.add_axes([0.1, 0.6, 0.4, 0.3])
    ...:
```

The arcane-looking list passed to the `add_axes()` method represents the `Axes` rect parameter. This defines the dimensions of the rectangular `Axes` object. The values range from 0 to 1 and represent, respectively, the left, bottom, width, and height of the rectangle.

Now make the main plot and the inset:

```
    ...: # Main plot
    ...: ax1.plot(x, y, 'k*-')
    ...: ax1.set_xlabel('x')
    ...: ax1.set_ylabel('y')
    ...:
    ...: # Inset plot
    ...: ax2.plot(x, np.sin(y), 'r*-')
    ...: ax2.set_xlabel('x')
    ...: ax2.set_ylabel('y')
    ...: ax2.set_title('Sine of Y')
```

You should get a plot like the one depicted in Figure 19-23.

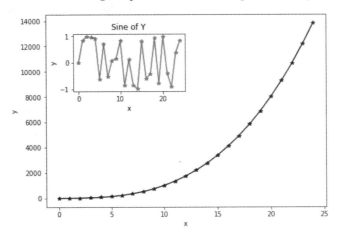

Figure 19-23: A plot with an inset plot

Plotting in 3D

Although designed primarily for 2D plotting, Matplotlib includes an Axes3D class that supports 3D scatterplots, histograms, surfaces, contour maps, and more. Here's an example:

```
In [47]: import numpy as np
    ...: import matplotlib.pyplot as plt
    ...:
    ...: z = np.arange(0, 200, 1)
    ...: x = z * np.cos(25 * z)
    ...: y = z * np.sin(25 * z)
    ...:
❶   ...: ax = plt.figure().add_subplot(projection='3d')
    ...:
    ...: ax.plot(x, y, z, 'black');
```

The key is to pass the projection='3d' keyword when creating an Axes object ❶. You can also use this alternate two-line syntax if you find it easier to read:

```
    ...: fig = plt.figure()
    ...: ax = plt.axes(projection ='3d')
```

Both will yield the plot shown in Figure 19-24.

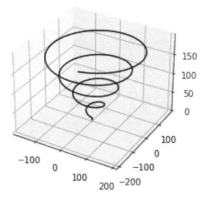

Figure 19-24: A 3D line plot

To read more about 3D plotting, visit *https://matplotlib.org/stable/tutorials/ toolkits/mplot3d.html*.

Animating Plots

Scientists commonly study dynamic phenomena such as ocean currents and caribou migrations. Whether based on actual observations or simulated behavior, the ability to visualize movement in plots, a process called *animation*, can lead to insights and better understanding of the phenomena. Animations also enhance presentations, helping your audience to better understand the points you're trying to convey.

As you might expect, Matplotlib provides numerous ways for animating plots. For simple animations, you can manually update and plot variables by iterating in a loop. For convenience and for working with more complicated animations, you can use the `matplotlib.animation` module (*https://matplotlib.org/ stable/api/animation_api.html*).

The `animation` module contains the `FuncAnimation` class, which animates a visualization by repeatedly calling a function. The `ArtistAnimation` class makes an animation using a fixed set of `Artist` objects such as a precomputed list of images. In general, `FuncAnimation` is simpler to use and more efficient. We won't cover `ArtistAnimation` here.

Animating Plots Using a for Loop

Perhaps the simplest way to animate a plot is to use a `for` loop. Let's give this a try using the sine wave example from "Working with Subplots" on page 545. In the console, enter the following code, using CTRL-ENTER after the first line and ENTER after the last line:

```
In [48]: import numpy as np
    ...: import matplotlib.pyplot as plt
❶   ...: import time
    ...: %matplotlib qt
    ...:
```

```
    ...: t = np.arange(-12.5, 12.5, 0.1)
    ...: amplitude = np.sin(t)
    ...:
    ...: fig, ax = plt.subplots()
❷  ...: line, = ax.plot(t, amplitude)
    ...: for i in range(30):
    ...:     updated_amp = np.sin(t + i)
❸  ...:     line.set_ydata(updated_amp)
    ...:     fig.canvas.draw()
    ...:     fig.canvas.flush_events()
    ...:     time.sleep(0.1)
```

Start by importing NumPy and Matplotlib, as usual, but this time, add the Standard Library's time module ❶. The time.sleep() method will let us control the speed of the animation later.

We'll show the animation in the external Qt window, so call the %matplotlib qt magic command. If you're working in Jupyter Notebook, you can use the %matplotlib notebook command to show the animation within the notebook.

Next, reproduce the time (t) and amplitude data from before and then assign the fig and ax variables. To animate plots using a for loop, you need to update the displayed data before each iteration of the loop. Because we're plotting a line, assign a line variable to the plot ❷. Note the comma after line, which indicates that this is a tuple unpacking process.

Start a for loop that runs 30 times. With each loop, shift the time series one second by adding the loop number (i) to the y data using the equation np.sin(t + i). Assign the result to the updated_amp variable. To update the line object prior to plotting, call its set_ydata()method and pass it the updated_amp variable ❸.

To update a Figure object that has been altered but not automatically redrawn, call canvas.draw(). Follow this with the canvas_flush_events() method, which clears the plot so that the next iteration can start with a blank screen.

Finish by calling the time.sleep() method and passing it 0.1. This is the number of seconds to suspend program execution. Feel free to play around with this number to see the effect on the animation; the larger the number, the slower it will proceed.

To run the animation, in the console, press ENTER; in Jupyter Notebook, press CTRL-ENTER. To return to inline plotting, remember to use the %matplotlib inline magic command.

You also can accomplish this type of real-time animation by using the pyplot approach. Here's an example in which we continuously update a scatterplot by adding a new point calculated with the quadratic equation. Run it in the Qt console or the Spyder text editor:

```
In [49]: import numpy as np
    ...: import matplotlib.pyplot as plt
    ...: %matplotlib qt
    ...:
    ...: x = 0
```

```
...: for i in range(30):
...:     x = x + 1
...:     y = x**2
...:     plt.scatter(x, y)
...:     plt.title("Quadratic Function")
...:     plt.xlabel("x")
...:     plt.ylabel("x-squared")
...:     plt.pause(0.1)
```

Note the use of `plt.pause()` in place of `time.sleep()`. The `plt.pause()` method takes seconds as an argument and runs the GUI event loop for this time interval. An active figure will be updated and displayed before the pause, and the GUI event loop (if any) will run during the pause.

As the animation runs, the x- and y-axes automatically adjust to accommodate the expanding plot limits. When the animation finishes, you should see a plot like the one shown in Figure 19-25.

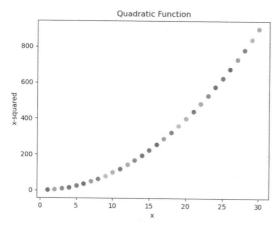

Figure 19-25: The finished pyplot animation

For complex animations, the Matplotlib documentation recommends using the `matplotlib.animation` module rather than a for loop. We'll cover this technique next.

Animating Plots Using the FuncAnimation Class

The `FuncAnimation` class makes an animation by repeatedly calling a function. It provides a more formal and flexible approach than the for loop process used in the previous section.

Let's animate two lines in the same plot using the object-oriented style. Enter the following in the console or Jupyter Notebook (if you're using Notebook, replace the `%matplotlib qt` magic with `%matplotlib notebook`):

```
In [50]: import numpy as np
    ...: import matplotlib.pyplot as plt
    ...: from matplotlib.animation import ❶ FuncAnimation
    ...: %matplotlib qt
    ...:
```

```
        ...: x = np.arange(-6, 6, 0.02)
        ...: y = np.sin(2 * x) / x
❷   ...: scaler = np.arange(1, 10, 0.1)
        ...:
        ...: fig, ax = plt.subplots()
        ...: line1 = ax.plot(x, y, color='k', lw=2) ❸[0]
        ...: line2 = ax.plot(x, y, color='r', ls='--')[0]
        ...:
        ...: def animate(frame):
        ...:     line1.set_ydata(y / frame)
        ...:     line2.set_ydata(y / frame * -0.2)
        ...:
        ...: animated = FuncAnimation(fig, animate, frames=scaler, interval=20)
```

Add FuncAnimation from the matplotlib.animation module to the imports ❶. Next, use NumPy to generate some data for plotting. The scaler array will let you alter the x and y data, giving you something new to print as the animation runs ❷.

Set up the fig and ax objects and then make a plot for each line, setting the color for the first to black and the second to red. Also set the line width of the first line to 2, and the line style for the second to dashed.

For both lines, add a zero index [0] to the end of the plotting code ❸. The plot command returns a sequence of line objects, and we want only the first item in the sequence. This represents an alternative to the tuple unpacking approach (line, = ax.plot(t, amplitude)) used to animate the sine wave in the previous section.

Now it's time to define a function that will update the data to create each frame of the animation. We'll call this function animate, with a parameter named frame. The argument for this parameter will be the scalar array, which will be passed by the frames parameter in the FuncAnimation() class.

Use the set_ydata() method on each line and pass it the y data divided by the scaler array. For the second line, multiply scaler by a negative scalar so that line2 will look different than line1.

To complete the code, call FuncAnimation() and pass it the Figure object on which it will draw (fig), the user-defined function (animate), and a frames and interval argument. The frames argument represents the source of the data passed to the user-defined function for each frame of the animation. It can be either an iterable, an integer, a generator function, or None. The interval argument sets the delay time between frames in milliseconds. Increasing this number will slow down the animation.

NOTE *You can assign the scalar NumPy array directly to the frames parameter, like so: FuncAnimation(fig, animate, frames=np.arange(1, 10, 0.1), interval=20). Although this removes the need for the scalar variable, the code is arguably less readable.*

Run the code by pressing ENTER in the console, or CTRL-ENTER in Jupyter Notebook. You should see two animated line plots, as shown in

Figure 19-26. To stop the animation, click the **Close** button at the upper right of the plot window. Otherwise, when calling `FuncAnimation()`, set `repeat=False` to stop the animation after a single run-through.

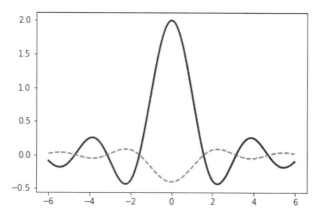

Figure 19-26: A screen capture from the functional animation

An optional parameter in `FuncAnimation()` worth mentioning is `fargs`. Short for *functional arguments*, you use this when your user-defined function takes multiple arguments. The first parameter is always reserved for the `frames` parameter in `FuncAnimation()`, but you can pass subsequent parameters (those that follow `frames`) as an ordered tuple of arguments, such as the following:

```
...: ani = FuncAnimation(fig, func, frames=param1,fargs=(param2, param3))
```

Finally, to save the animation as a *.gif*, use the save() method with optional frames per second (`fps`) and dots-per-inch (`dpi`) arguments, as follows:

```
...: animated.save('animation.gif', fps=20, dpi=150)
```

Other supported file formats include *.avi*, *.mp4*, *.mov*, and other save options include the methods to_html5_video() and to_jshtml(). To learn more about the methods and parameters of `FuncAnimation`, visit *https://matplotlib.org/stable/api/_as_gen/matplotlib.animation.FuncAnimation.html*.

Styling Plots

Up until now, you've changed the default settings for a plot, such as the line width or marker color, by passing new values as you made the plot. But what if you want to set these values for multiple plots *at the same time* so that all your lines are colored black? Or what if you'd like to *cycle through* a defined order of colors?

Well, one way to do this is to set the parameters at runtime, using an instance of the RcParams class. The name of this class stands for *runtime configuration parameters*, and you run it from a notebook, script, or console using either the pyplot approach or the object-oriented style. It stores settings in the matplotlib.rcParams variable, which is a dictionary-like object.

There's a very long list of configurable parameters, which you can view in multiple ways. To see a list of valid parameters, visit *https://matplotlib.org/ stable/api/matplotlib_configuration_api.html?highlight=rcparams/*. To see more details about the parameters, run import matplotlib as mpl followed by print(mpl.matplotlib_fname()). This will reveal the path to the *matplotlibrc* file on your computer, which you then can open and view.

Changing Runtime Configuration Parameters

Let's look at a pyplot example in which we standardize the size of figures, use black for all plotted lines, and cycle through two different line styles. This means that the first line plotted will always have a certain consistent style, and that the second plotted will have another consistent style. In the console, enter the following:

```
In [51]: import numpy as np
    ...: import matplotlib.pyplot as plt
    ...: import matplotlib as mpl
    ...: from cycler import cycler
    ...: %matplotlib inline
    ...:
```

Notice here that we import Matplotlib as mpl. Importing Matplotlib in this manner gives us access to more features than in the pyplot module alone. We also import cycler. The Cycler class will let us specify which colors and other style properties we want to cycle through when making multidata plots. You can read about it at *https://matplotlib.org/stable/tutorials/intermediate/ color_cycle.html*.

To access a property in rcParams, treat it like a dictionary key. You can find the valid parameter names by entering mpl.rcParams.keys() or by visiting the sources listed in the previous section. In the next three lines, we set the figure size, line color, and line styles:

```
    ...: mpl.rcParams['figure.figsize'] = (5, 4)
    ...: mpl.rcParams['lines.color'] = 'black'
    ...: mpl.rcParams['axes.prop_cycle'] = cycler('linestyle', ['-', ':'])
    ...:
```

NOTE   *You can also set parameters through pyplot, using syntax like plt.rcParams['lines .color'] = 'black'.*

To cycle through the line styles, use the `axes.prop_cycle` key and then pass the cycler factory function the parameter (`'linestyle'`) and a list of the styles (solid and dotted). These defaults have now been reset for all plots that you will make in the current session.

Finish by generating some data and plotting it:

```
...: x = np.arange(0, 15, 0.1)
...: y = np.sin(x)
...:
...: plt.plot(x, y)
...: plt.plot(x + 1, y - 2);
```

Normally, this code would produce a plot with two solid lines, one blue and one orange. Now, however, you get two black lines distinguished by different line styles (Figure 19-27).

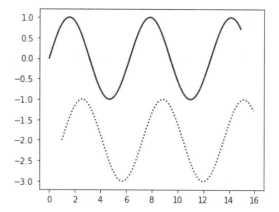

Figure 19-27: A plot built with global figure size, line color, and line style parameters

Note that if you were to plot *three* lines in the previous plot, the third line would cycle back to using the solid line style, and you'd have one dotted and two solid lines. If you want three different styles, you'll need to add the extra style to the cycler.

For convenience, Matplotlib comes with functions for simultaneously modifying multiple settings in a single group using keyword arguments. Here's an example, using the previous plotting data, in which we start by resetting the Matplotlib "factory defaults":

```
In [52]: mpl.rcParams.update(mpl.rcParamsDefault)
    ...:
```

Now, let's use the `rc()` convenience function to change the default line width to 5 and the line style to dash-dot:

```
In [53]: mpl.rc('lines', lw=5, ls='-.')
    ...: plt.plot(x, y);
```

This produces the plot in Figure 19-28.

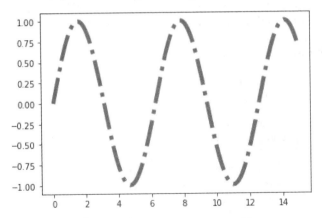

Figure 19-28: The new plotting parameters set with a convenience function

If you want to use a style for only a specific block of code, the style package provides a context manager for limiting your changes to a specific scope. For more on this, see "Temporary Styling" at *https://matplotlib.org/ stable/tutorials/introductory/customizing.html.*

Creating and Using a Style File

You can save changes to the Matplotlib default style in a file. This lets you standardize plots for a report or presentation and share the customization within a project team. It also reduces code redundancy and complexity by letting you preset certain plot parameters and encapsulate them in an external file.

Let's create a simple style file that sets some standards for plots, such as the figure size and resolution, use of a background grid, and the typeface and size to use for titles, axes labels, and tick labels. In the Spyder text editor, or any text editor, enter the following:

```
# scientific_style.mplstyle

figure.figsize:    4, 3  # width & height in inches
figure.dpi:        200   # dots per inch
axes.grid:         True
font.family:       Times New Roman
axes.titlesize:    24
axes.labelsize:    20
xtick.labelsize:   16
ytick.labelsize:   16
```

 For guidance on creating style files, use the matplotlibrc *file on your computer, mentioned previously. You can also find a copy at* https://matplotlib.org/stable/ tutorials/introductory/customizing.html.

For Matplotlib to easily find this file, you need to save it in a specific location. First, find the location of the *matplotlibrc* file by entering the following in the console:

```
In [54]: import matplotlib as mpl

In [55]: mpl.matplotlib_fname()
Out[55]: 'C:\\Users\\hanna\\anaconda3\\lib\\site-packages\\matplotlib\\mpl-data\\matplotlibrc'
```

This shows you the path to the *mpl-data* folder, which contains the *matplotlibrc* file and a folder named *stylelib*, among others. Save your style file into the *stylelib* folder as *scientific_style.mplstyle* (replacing the *.txt* extension).

NOTE   *If Matplotlib has trouble finding this file later, you might need to restart the kernel. In the console, click **Kernel ▶ Restart Current Kernel**. In Jupyter Notebook, click **Kernel ▶ Restart**.*

Now, let's use this file to create a standardized plot. After importing pyplot, use its style.use() method to load the style file *without* its file extension:

```
In [56]: import matplotlib.pyplot as plt

In [57]: plt.style.use('scientific_style')
```

Next, generate an empty figure using the object-oriented style. You should see a plot like Figure 19-29.

```
In [58]: fig, ax = plt.subplots()
    ...: ax.set_title('Standardized Title')
    ...: ax.set_xlabel('Standardized X-labels')
    ...: ax.set_ylabel('Standardized Y-labels');
```

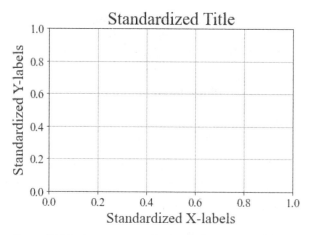

Figure 19-29: An empty standardized plot generated by the style file

When you saved your style file, you might have noticed that the *stylelib* folder was full of preexisting *mplstyle* files. These files create many different plot formats, and you can look through them for clues on how to write your own style files. In the next section, we'll use one of these files to override some of Matplotlib's default values.

Applying Style Sheets

Besides letting you customize your own plots, Matplotlib provides predefined *style sheets* that you can import by using `style.use()`. Style sheets look the same as the *matplotlibrc* file, but within one, you can set only `rcParams` that are related to the actual style of the plot. This makes style sheets portable between different machines because there's no need to worry about uninstalled dependencies. Only a few `rcParams` can't be reset, and you can view a list of these at *https://matplotlib.org/stable/api/style_api .html#matplotlib.style.use/*.

You can see examples of the available style sheets at *https://matplotlib .org/stable/gallery/style_sheets/style_sheets_reference.html*. These take the form of a strip of thumbnails, as shown in Figure 19-30. Some of the style sheets emulate popular plotting libraries like seaborn and ggplot.

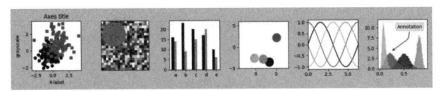

Figure 19-30: Example of the grayscale style sheet

NOTE *An important style sheet to be aware of is the* seaborn-colorblind *sheet. This style sheet uses "colorblind-safe" colors designed for the 5 to 10 percent of the population that suffers from color blindness.*

Let's try out a scatterplot using the grayscale style sheet that ships with Matplotlib. First, in either the console or Jupyter Notebook, import NumPy and Matplotlib and then call the grayscale file:

```
In [59]: import numpy as np
    ...: import matplotlib.pyplot as plt
    ...:
    ...: plt.style.use('grayscale')
    ...:
```

Now, generate some dummy data for making two different point clouds:

```
    ...: x = np.arange(0, 20, 0.1)
    ...: noise = np.random.uniform(0, 10, len(x))
    ...: y = x + (noise * x**2)
    ...: y2 = x + (noise * x**3)
    ...:
```

Finish by setting up and executing the plot using the `pyplot` approach. Use log scales for both axes.

```
...: plt.title('Grayscale Style Scatterplot')
...: plt.xlabel('Log X')
...: plt.ylabel('Log Y')
...: plt.loglog()
❶ ...: plt.scatter(x, y2, alpha=0.4, label='X Cubed')
...: plt.scatter(x, y, marker='+', label='X Squared')
...: plt.legend(loc=(1.01, 0.7));
```

You should see a plot similar to the one in Figure 19-31. The point locations might differ due to the use of randomly generated data.

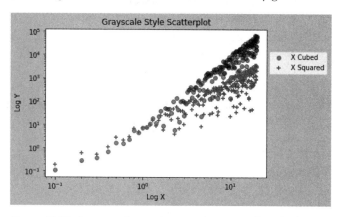

Figure 19-31: A scatterplot made using the grayscale style sheet

Note the use of the `alpha` keyword when calling `plt.scatter()` ❶. The `alpha` attribute controls opacity, letting you regulate the transparency of a line or marker. A value of `1` is completely opaque.

Making one dataset slightly transparent helps to resolve *over-posting*, wherein markers from one dataset plot on top of markers from other datasets, obscuring the over-posted markers. Semi-transparent markers also become darker as they stack on top of one another, letting you visualize data density (such as the blacker circles in Figure 19-31).

NOTE *To control the plot order of markers, use the `zorder` parameter (such as `zorder=2`) when calling `plt.scatter()`. Artists with higher `zorder` values will post over those with lower values.*

Back to our style sheet: if you open the *grayscale.mplstyle*, you'll see that it looks a lot like the *scientific_style.mplstyle* file that we made in "Creating and Using a Style File" on page 576. So, if an existing style sheet is not quite right for your purposes, you can always copy the file, edit it, and save it as a new style sheet!

6. True or False: The ability to manipulate spines is an advantage of the pyplot approach.

7. Add the summary title "Martian Goethite, Hematite, and Jarosite Distributions" to the display in Figure 19-14. Use whichever plotting approach you prefer.

8. Use the following code to produce three datasets for plotting: np.random .normal(0, 1, 50).cumsum(). Generate three subplots in a row and use a for loop to populate each with a different dataset. Give each subplot a unique title and plot the data using black crosses.

9. Generate a 2D NumPy array of randomized data using np.random.rand (4, 4). Then, plot a heatmap using heat = ax.imshow(data). Animate the heatmap using a for loop and a range of 30.

10. Use the equation velocity = 9.81 * time to calculate the speed of a falling object. Let the object fall for 15 seconds, and for every second, post its position and velocity in a single plot, using a different y-axis for each.

Summary

The goal of this chapter was to introduce the powerful Matplotlib plotting library and (hopefully) address some of its more frustrating aspects. A major source of confusion is that there are two main interfaces for making plots; for consistency, you should choose one and stick with it.

The pyplot approach works with implicit, "currently active," Figure and Axes objects, in which the Figure is a blank canvas, and the Axes holds plot elements like lines, legends, titles, and so on. To simplify plotting, pyplot creates these objects behind the scenes.

The pyplot approach works well when using Matplotlib interactively and in small scripts, but when building larger applications, the object-oriented style is preferred. Explicitly assigning Figure and Axes objects to variables will help you keep track of multiple plots and ensure that the code producing them is as clear as possible. You'll also have more control over certain plot elements.

For simpler, more automated plotting than the pyplot approach, you can use the seaborn package, which is a wrapper around Matplotlib. Additionally, the pandas data analysis package wraps Matplotlib for even easier, though less sophisticated, plotting. Chapter 16 included overviews of seaborn and pandas plotting, and we'll look at them again in the next chapter.

For further study and to learn advanced Matplotlib features, check out the tutorials and user guide at the official website (*https://matplotlib.org/*) and at Real Python (*https://realpython.com/python-matplotlib-guide/*). You can find useful cheat sheets at *https://matplotlib.org/cheatsheets/*.

20

PANDAS, SEABORN, AND SCIKIT-LEARN

A common scientific practice is evaluating data and using it to generate predictive models. In this chapter, we'll use three of Python's most popular open source libraries to solve a zoological classification problem. Using this hands-on, project-based approach will showcase the functionality and synergy among the libraries and demonstrate what's involved in doing science with Python.

For data loading, analysis, and manipulation, we'll use the pandas package (*https://pandas.pydata.org/*). Built on NumPy and Matplotlib, pandas uses array-based computing under the hood but has simpler syntax, making coding and plotting faster, easier, and more error free. Unlike native Python, pandas can intelligently read tabular text-file data, recognizing columns, rows, headers, and so on. And unlike NumPy, on which it's built, pandas can handle heterogeneous data types such as mixtures of text and numbers.

We'll also use the seaborn library (*https://seaborn.pydata.org/*), which wraps Matplotlib to produce more attractive and easier visualizations. It represents a nice plotting compromise between the highly customizable but verbose syntax of Matplotlib and the bare-bones simplicity of pandas. Even better, seaborn is tightly integrated with pandas for seamless plotting and effective data exploration.

Lastly, the scikit-learn library (*https://scikit-learn.org/*) is Python's primary general-purpose machine learning toolkit. It provides algorithms for classification, regression, clustering, dimensionality reduction, preprocessing, and model selection.

In the sections that follow, you'll apply these libraries to a real-world problem and observe how they work together. But, due to their enormous size and scope, we won't be able to study them in-depth. Whole books have been dedicated to each, and the admittedly non-exhaustive pandas overview in Wes McKinney's *Python for Data Analysis*, 2nd edition requires no less than 270 pages!

If you'd like a complete picture, I list some additional resources in the "Summary" section at the end of this chapter. You can also find useful tutorials and examples in the official websites, cited previously.

Introducing the pandas Series and DataFrame

The pandas library contains data structures designed for working with common data sources such as Excel spreadsheets and SQL relational databases. Its two primary data structures are series and DataFrames. Other libraries, like seaborn, are designed to integrate well with these data structures and supply additional functionality, making pandas a great foundation to any data science project.

The Series Data Structure

A *series* is a one-dimensional labeled array that can hold any type of data such as integers, floats, strings, and so on. Because pandas is based on NumPy, a series object is basically two associated arrays. One array contains the data point values, which can have any NumPy data type. The other array contains labels for each data point, called *indexes* (Table 20-1).

Table 20-1: A Series Object

Index	Value
0	42
1	549
2	' Steve '
3	−66.6

Unlike the indexes of Python list items, the indexes in a series don't need to be an integer. In Table 20-2, the indexes are the names of elements, and the values are their atomic numbers.

Table 20-2: A Series Object with Meaningful Indexes

Index	Value
Silicon	14
Sodium	11
Argon	18
Cobalt	27

A series acts much like a Python dictionary in so much as indexes represent keys. It can thus serve as a replacement for a dictionary in many contexts. Another useful feature is that different series will align by index label when doing arithmetic operations between them even if the labels don't occur in the same order.

As with a list or NumPy array, you can slice a series or select individual elements by specifying an index. You can manipulate the series many ways, such as filtering it, performing mathematical operations on it, and merging it with other series. To see the many attributes and methods available for a series object, visit *https://pandas.pydata.org/pandas-docs/stable/reference/api/pandas.Series.html.*

The DataFrame Data Structure

A *DataFrame* is a more complex structure made up of two dimensions. It's a collection of objects organized using a tabular structure, like a spreadsheet, with columns, rows, and data (Table 20-3). You can think of it as an ordered collection of columns with two indexing arrays. Each column represents a pandas series.

Table 20-3: A DataFrame Object

	Columns			
Index	Country	State	County	Population
0	USA	Alabama	Autauga	54,571
1	USA	Alabama	Baldwin	182,265
2	USA	Alabama	Barbour	27,457
3	USA	Alabama	Bibb	22,915

The first index, for the rows, works much like the index array in a series. The second keeps track of the series of labels, with each label representing a column header. DataFrames also resemble dictionaries; the column names form the keys, and the series of data in each column forms the values. Like series, DataFrames come with many attributes and methods. For more information on these, see *https://pandas.pydata.org/pandas-docs/stable/reference/api/pandas.DataFrame.html.*

By integrating index objects and labels into their structure, you can easily manipulate DataFrames. We'll look at some of this functionality as we work through the classification problem. You can also get up to speed on the basics by visiting the "10 Minutes to pandas" tutorial at *https://pandas.pydata.org/docs/user_guide/10min.html.*

The Palmer Penguins Project

The *Palmer Penguins dataset* consists of 342 observations of Antarctic penguins from three islands in the Palmer Archipelago (Figure 20-1). It was made available through the *Palmer Station Antarctica LTER (https://pallter.marine.rutgers.edu/).*

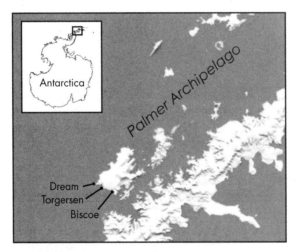

Figure 20-1: The location of Dream, Torgersen, and Biscoe Islands, Palmer Archipelago, Antarctica

Three different penguin species were sampled in the study. In order of decreasing body size, these are the Gentoo, Chinstrap, and Adélie (Figure 20-2).

Figure 20-2: The three penguin species in the Palmer Penguins dataset, as drawn by Charles Joseph Hullmandel (courtesy of Wikimedia Commons)

The goal of this project will be to generate a model to predict the species of penguin from a combination of morphological features such as flipper length and body mass. In machine learning, this is considered a *classification* problem. We'll use pandas to load, explore, validate, and clean the data, seaborn to plot the data, and scikit-learn to produce the predictive model.

The Project Outline

Data science projects like this one follow a series of logical steps, as listed here:

1. Frame the problem (the single most important step).
2. Collect raw data and set up the project.
3. Process the data (through cleaning, merging, infilling, reducing, and so on).
4. Explore the data.
5. Perform in-depth analysis and develop models and algorithms.
6. Apply the models and present the project results.

Jupyter Notebook is ideal for this process, as it can handle all the steps in order and is basically self-documenting. It can also be turned into a slideshow for presentations (as discussed in Chapter 5).

Setting Up the Project

For this project, we'll use Jupyter Notebook in a dedicated project folder. We'll install Notebook and the scientific and plotting libraries using the naive approach, in other words, directly in a conda environment within the project folder (see Chapter 5). In general, you use the naive approach when you want to work with *specific* and *persistent* versions of a library or application. We're using it here for practice, as we've previously been focusing on the modular approach, in which Notebook is installed in the *base* environment.

Start by making a folder named *penguins* under your user directory. Although you can do this through Anaconda Navigator, the command line is more succinct, so we'll use that going forward.

To make the directories for the project, open Anaconda Prompt (in Windows) or a terminal (in macOS or Linux) and enter the following (using your own directory path):

```
mkdir C:\Users\hanna\penguins
mkdir C:\Users\hanna\penguins\notebooks
```

This makes a *penguins* directory with a *notebooks* subdirectory. Next, create a conda environment named *penguins_env* under the project directory, activate it, and install the libraries we'll use (substituting your own path where needed):

```
conda create --prefix C:\Users\hanna\penguins\penguins_env
conda activate C:\Users\hanna\penguins\penguins_env
conda install python notebook pandas seaborn
conda install -c conda-forge scikit-learn
```

You now have a conda environment for the project that contains the notebook, pandas, python, scikit-learn, and seaborn packages. Remember from Chapter 2 that this environment is isolated and can't "see" other packages on your system, such as those in the *base* environment.

At this point, your *penguins_env* should be active, and your project directory structure should look like Figure 20-3. We'll be loading the dataset straight from seaborn, so there's no need for a *data* folder.

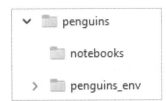

Figure 20-3: Directory
structure for the penguins
project

To create a notebook for the project, first navigate to the *notebooks* folder using Anaconda Prompt or the terminal:

```
cd C:\Users\hanna\penguins\notebooks
```

To launch Notebook, enter the following:

```
jupyter notebook
```

You should now see the Jupyter dashboard in your browser. Click the **New** button and choose **Python[conda env:penguins_env]** to create a new notebook. A new notebook should appear in your browser. Click **Untitled**, name it *penguins_project*, and then click the **Save** button. You're ready to go!

NOTE *If you want to open the notebook in the future using Anaconda Navigator, launch Navigator, use the Environments tab to activate* penguins_env, *and then click the Launch button on the Jupyter Notebook tile. This will open the dashboard, where you can navigate to the* notebook *folder and launch* penguins_project.ipynb. *If you want to use Notebook in JupyterLab, see the instructions in Chapter 6 for installing JupyterLab and launching Notebook.*

Importing Packages and Setting Up the Display

In the first notebook cell, import Matplotlib, seaborn, and pandas. Enter the following code and execute it using SHIFT-ENTER, which automatically moves you to a new blank cell:

```
import matplotlib.pyplot as plt
import seaborn as sns
import pandas as pd

# Enable multiple outputs per cell:
%config InteractiveShell.ast_node_interactivity = 'all'

# Set plotting styles
sns.set_style('whitegrid')
sns.set_palette(['black', 'red', 'grey'])
```

NOTE   *Normally, we would perform all the imports here, but for the sake of the narrative, we'll import the scikit-learn components later, so we can discuss them just before applying them.*

By default, Notebook displays only one output per cell. The %config magic command overrides this, allowing us to see multiple outputs, such as a data table plus a bar chart, in a single output cell.

The default seaborn color palette is undeniably beautiful (see *http:// seaborn.pydata.org/tutorial/function_overview.html*), but it loses its charm somewhat in a black-and-white book. As a compromise, we'll use the whitegrid stylesheet and reset the palette to black, red, and gray, one for each of the three penguin species in the dataset.

Loading the Dataset

Seaborn comes with a few practice datasets that are automatically downloaded during installation. These are all comma-separated values (*.csv*) files stored in a repository at *https://github.com/mwaskom/seaborn-data/*. If you ever need to get the dataset names, you can retrieve them by running sns.get _dataset_names() in a notebook or console (after importing seaborn, of course).

As a data analysis tool, pandas can read and write data stored in many types of media, such as files and databases (Table 20-4). Example syntax is df = pd.read_excel('*filename.xlsx*') and df.to_excel('*filename.xlsx*'), where df stands for *DataFrame*. For more options, visit *https://pandas.pydata.org/docs/ reference/io.html*.

In addition to the methods in Table 20-4, the read_table() method reads tabular data, such as text (*.txt*) files, in which the values are separated by spaces or tabs. Python can generally detect the separator in use, but you can also pass it as an argument, for example, sep='\t' for a tab.

Table 20-4: Useful pandas I/O Methods

Input (Reader)	Output (Writer)
read_csv()	to_csv()
read_excel()	to_excel()
read_hdf()	to_hdf()
read_sql()	to_sql()
read_json()	to_json()
read_html()	to_html()
read_stata()	to_stata()
read_clipboard()	to_clipboard()
read_pickle()	to_pickle()

Besides loading external sources, you can create a DataFrame from many different types of input. These include 2D ndarrays, lists of lists or tuples, list of dictionaries or series, an existing DataFrame, and more.

In spite of all these choices, we'll use seaborn's load_dataset() method to load the penguins dataset. This specialized method reads a CSV-format dataset from the seaborn repository and returns a pandas DataFrame object. Enter the following in the new cell and press SHIFT-ENTER:

```
# Load penguins dataset:
df = sns.load_dataset('penguins')
```

NOTE   *In this notebook, I'm using simple comments, such as # Load penguins dataset, as cell headers. To make proper headers, you can add a Markdown cell before each code cell, as described in Chapter 5.*

In the previous code, we assigned the DataFrame to a variable named df. This is handy, as the name reflects the datatype. There's no reason why you couldn't use another name, however, such as penguins_df.

Displaying the DataFrame and Renaming Columns

The first thing you'll want to do after loading the data is look at it. For larger datasets such as penguins, pandas will show you part of the top and part of the bottom of a DataFrame by default. To see an example, enter the following, and then press CTRL-ENTER to execute the cell without leaving it:

```
# View dataframe head and tail:
df
```

To see the entire DataFrame in a scrollable output cell, place this command at the top of the cell and rerun it: pd.set_option('display.max _rows', None).

Calling the DataFrame displays all the columns along with the first five rows and last five rows (Figure 20-4). When possible, column names should be descriptive and short. However, this isn't always an option, so let's practice changing a column header.

	species	island	bill_length_mm	bill_depth_mm	flipper_length_mm	body_mass_g	sex
0	Adelie	Torgersen	39.1	18.7	181.0	3750.0	Male
1	Adelie	Torgersen	39.5	17.4	186.0	3800.0	Female
2	Adelie	Torgersen	40.3	18.0	195.0	3250.0	Female
3	Adelie	Torgersen	NaN	NaN	NaN	NaN	NaN
4	Adelie	Torgersen	36.7	19.3	193.0	3450.0	Female
...	...	...	...	...	...	...	...
339	Gentoo	Biscoe	NaN	NaN	NaN	NaN	NaN
340	Gentoo	Biscoe	46.8	14.3	215.0	4850.0	Female
341	Gentoo	Biscoe	50.4	15.7	222.0	5750.0	Male
342	Gentoo	Biscoe	45.2	14.8	212.0	5200.0	Female
343	Gentoo	Biscoe	49.9	16.1	213.0	5400.0	Male

344 rows × 7 columns

Figure 20-4: DataFrame head and tail display

In the same cell, add the following code to rename the sex header to gender. The inplace argument tells pandas to alter the current DataFrame rather than return a copy. Press CTRL-SHIFT to execute the code and move to a new cell.

```
# Rename sex column to 'gender' and verify change:
df.rename(columns={'sex': 'gender'}, inplace=True)
df.head()
```

The head() method displays the first five rows in a DataFrame, as shown in Figure 20-5. To see more, just pass it the number of rows you want to see as an argument.

	species	island	bill_length_mm	bill_depth_mm	flipper_length_mm	body_mass_g	gender
0	Adelie	Torgersen	39.1	18.7	181.0	3750.0	Male
1	Adelie	Torgersen	39.5	17.4	186.0	3800.0	Female
2	Adelie	Torgersen	40.3	18.0	195.0	3250.0	Female
3	Adelie	Torgersen	NaN	NaN	NaN	NaN	NaN
4	Adelie	Torgersen	36.7	19.3	193.0	3450.0	Female

Figure 20-5: The head of the DataFrame after changing the sex column header to gender

In Figure 20-4 the number of rows and columns is included at the bottom of the output. You might immediately notice an issue: there are 344 rows, but earlier, I stated that the dataset has 342 observations. The discrepancy could be due to one of two common dataset problems: duplicate or missing values.

Checking for Duplicates

It's not uncommon for data rows to become duplicated by accident. This can happen during the initial creation of a dataset, in later edits, or during data transfers and transformations. You should remove this redundant data before you begin an analysis because it takes up memory, slows processing speeds, and distorts statistics due to the overweighting of the duplicate values.

Fortunately, pandas comes with the `duplicated()` method for finding duplicate rows. In the new cell, enter the following and then press CTRL-ENTER:

```
# Check for duplicate rows:
duplicate_rows = df[df.duplicated(keep=False)]
print(f'Number of duplicate rows = {len(duplicate_rows)}')
```

You should get the following output, as there are no duplicate rows:

```
Number of duplicate rows = 0
```

Had there been any duplicates in the dataset, we could have removed them using the `drop_duplicates()` method, like so:

```
df.drop_duplicates(inplace=True)
```

You can also look across specific columns for duplicate values. At the bottom of the current cell, enter the following and execute it by pressing SHIFT-ENTER:

```
# Check for duplicates across specified columns:
df[df.duplicated(['bill_length_mm', 'bill_depth_mm', 'flipper_length_mm', 'body_mass_g'])]
```

Note that the inner square brackets define a Python list with column names, whereas the outer brackets represent "selection brackets" used to select data from a pandas DataFrame. We specified four of the seven columns, producing the output in Figure 20-6.

	species	island	bill_length_mm	bill_depth_mm	flipper_length_mm	body_mass_g	gender
339	Gentoo	Biscoe	NaN	NaN	NaN	NaN	NaN

Figure 20-6: Row with duplicate values across the four columns with the float data type

As you'll see in a moment, row 339 is a duplicate of row 3 (for the four columns specified). But even though there are duplicate values here, they're not the kind that we need to treat as duplicates. Instead, they represent *missing values*, which we'll cover in the next section.

Handling Missing Values

The duplicate values in Figure 20-6 are represented by the *Not a Number* (NaN) value. This is a special floating-point value recognized by all systems that use the standard IEEE floating-point representation. For computational speed and convenience, it serves as the default missing value marker for both NumPy and pandas. NaN and Python's built-in None value are essentially interchangeable. By default, these null values are not included in computations.

Finding Missing Values

Missing data values reduce statistical power and can cause bias when estimating parameters and making predictions. To find the missing values in the penguins DataFrame, enter the following in a new cell and then press SHIFT-RETURN:

```
# Find null values
df.isnull().sum()
df[df.isnull().any(axis=1)]
```

The first method sums the missing values and displays the results as a table (Figure 20-7). The penguins dataset is missing 11 gender calls and a total of 8 morphological measurements.

```
species              0
island               0
bill_length_mm       2
bill_depth_mm        2
flipper_length_mm    2
body_mass_g          2
gender              11
dtype: int64
```

Figure 20-7: The output of
df.isnull().sum()

The second call indexes the DataFrame where a value in *any* column is missing (as opposed to *all*). Remember, pandas is built on NumPy, so axis 1 refers to columns and axis 0 refers to rows. You should get the result depicted in Figure 20-8.

	species	island	bill_length_mm	bill_depth_mm	flipper_length_mm	body_mass_g	gender
3	Adelie	Torgersen	NaN	NaN	NaN	NaN	NaN
8	Adelie	Torgersen	34.1	18.1	193.0	3475.0	NaN
9	Adelie	Torgersen	42.0	20.2	190.0	4250.0	NaN
10	Adelie	Torgersen	37.8	17.1	186.0	3300.0	NaN
11	Adelie	Torgersen	37.8	17.3	180.0	3700.0	NaN
47	Adelie	Dream	37.5	18.9	179.0	2975.0	NaN
246	Gentoo	Biscoe	44.5	14.3	216.0	4100.0	NaN
286	Gentoo	Biscoe	46.2	14.4	214.0	4650.0	NaN
324	Gentoo	Biscoe	47.3	13.8	216.0	4725.0	NaN
336	Gentoo	Biscoe	44.5	15.7	217.0	4875.0	NaN
339	Gentoo	Biscoe	NaN	NaN	NaN	NaN	NaN

Figure 20-8: All the DataFrame rows containing missing data

Filling and Removing Missing Values

Missing values must be addressed before you try to conduct analyses or build models from a dataset. Although ignoring the issue is a possibility, it's much better to either fill in the missing values or remove (drop) them completely. Methods for doing this are listed in Table 20-5.

Table 20-5: Useful Methods for Handling Missing Data

Method	Description
dropna	Depending on arguments, remove row or column that contains missing data based on whether any or all values are null.
fillna	Fill in missing value with a constant or an interpolation method. Arguments include the ffill and bfill methods.
ffill	"Forward fill" by propagating the last valid observation forward.
bfill	"Back fill" by replacing missing values with values from the next row or column, as specified.
isnull	Return Boolean indicating missing/NA values.
notnull	Negate isnull.

Options for filling in the missing values with fillna() include replacing them with the mean, median, or most frequent values in the dataset so that the overall statistics aren't skewed. For example, to use the mean value of a column, you would use this syntax (don't add this to your project code):

```
df['col1'] = df['col1'].fillna(df['col1'].mean())
```

NOTE *The pandas library tries to mimic the R programming language, and the* na *in the* fillna() *method stands for the* NA *(not* available*) marker used for missing data in R.*

Filling in the missing data is important when a dataset is small and you need to take into account every observation. And if only a single column among many is missing a value, you might not want to "throw away" all the other useful data in the row.

Because we have a robust dataset and can't easily impute and replace missing gender data, we'll *drop* the rows with missing data. In the new cell, enter the following and then press SHIFT-ENTER:

```
# Drop Null Values
df = df.dropna(how='any')
df.isnull().sum()
```

Using an assignment statement when calling dropna() causes the current DataFrame (df) to be overwritten. This allows the DataFrame to evolve over time, but be aware that to erase changes and restore the DataFrame to its previous state, you'll need to run all the cells above the current cell. Passing a how argument of any to dropna() means that any row with at least one missing value will be deleted.

To check the results, rerun the `isnull()` method. You should get the output in Figure 20-9.

```
species              0
island               0
bill_length_mm       0
bill_depth_mm        0
flipper_length_mm    0
body_mass_g          0
gender               0
dtype: int64
```

Figure 20-9: A summary of null values after dropping nulls

The DataFrame no longer includes missing values.

Reindexing

Reindexing refers to the process of making data conform to a given set of labels along a particular axis. Missing value markers will be automatically inserted in label locations where no data for the label exists.

When we dropped the rows with null values in the previous section, we also deleted their corresponding indexes. To see the result, run the following code in the new cell and press SHIFT-ENTER:

```
# Check index values after dropping rows.
df.head()
```

As you can see in Figure 20-10, there is a gap in the DataFrame index (leftmost column) where row 3, which contained a null value, was dropped.

	species	island	bill_length_mm	bill_depth_mm	flipper_length_mm	body_mass_g	gender
0	Adelie	Torgersen	39.1	18.7	181.0	3750.0	Male
1	Adelie	Torgersen	39.5	17.4	186.0	3800.0	Female
2	Adelie	Torgersen	40.3	18.0	195.0	3250.0	Female
4	Adelie	Torgersen	36.7	19.3	193.0	3450.0	Female
5	Adelie	Torgersen	39.3	20.6	190.0	3650.0	Male

Figure 20-10: Dropping rows results in missing DataFrame indexes.

To restore the indexes, run the following and then execute the cell using SHIFT-ENTER.

```
# After dropping nulls, reindex:
df.reset_index(drop=True, inplace=True)
df.head()
```

In the `reset_index()` method, `drop=True` signals that the old index isn't preserved as a new column in the DataFrame, because there's no need to keep that information. The inplace=True argument means that the method adjusts the current DataFrame rather than returning a copy. As an alternative, you could simply reassign the DataFrame, like so:

```
df = df.reset_index(drop=True).
```

Calling the head() method shows that the indexes are now consecutively ordered (Figure 20-11).

	species	island	bill_length_mm	bill_depth_mm	flipper_length_mm	body_mass_g	gender
0	Adelie	Torgersen	39.1	18.7	181.0	3750.0	Male
1	Adelie	Torgersen	39.5	17.4	186.0	3800.0	Female
2	Adelie	Torgersen	40.3	18.0	195.0	3250.0	Female
3	Adelie	Torgersen	36.7	19.3	193.0	3450.0	Female
4	Adelie	Torgersen	39.3	20.6	190.0	3650.0	Male

Figure 20-11: The DataFrame head after reindexing

Pandas includes several other reindexing functions, such as reindex() and reindex_like(). You can find these, and other DataFrame functions, at *https://pandas.pydata.org/pandas-docs/stable/reference/frame.html*. For more on missing values, see *https://pandas.pydata.org/docs/user_guide/missing_data.html*.

Exploring the Dataset

At this point, you've cleaned the data by checking for duplicates, removing missing values, and reindexing the DataFrame. Of course, there might still be problems, such as incorrect values (a penguin body mass of one million grams, for example). Catching and correcting these requires an exploration of the dataset, and pandas and seaborn provide several methods to aid you in this process. These same methods will help you to understand the dataset so that you can formulate a plan for addressing the project goal.

Describing the DataFrame

Let's explore the DataFrame using a combination of tables and graphs. To begin, we'll look at the data types in play and overall statistics. In a new cell, enter the following and then press SHIFT-ENTER:

```
# Display datatypes and data statistics:
df.dtypes
df.describe(include='all')
```

This produces the output shown in Figure 20-12.

```
species                 object
island                  object
bill_length_mm          float64
bill_depth_mm           float64
flipper_length_mm       float64
body_mass_g             float64
gender                  object
dtype: object
```

	species	island	bill_length_mm	bill_depth_mm	flipper_length_mm	body_mass_g	gender
count	333	333	333.000000	333.000000	333.000000	333.000000	333
unique	3	3	NaN	NaN	NaN	NaN	2
top	Adelie	Biscoe	NaN	NaN	NaN	NaN	Male
freq	146	163	NaN	NaN	NaN	NaN	168
mean	NaN	NaN	43.992793	17.164865	200.966967	4207.057057	NaN
std	NaN	NaN	5.468668	1.969235	14.015765	805.215802	NaN
min	NaN	NaN	32.100000	13.100000	172.000000	2700.000000	NaN
25%	NaN	NaN	39.500000	15.600000	190.000000	3550.000000	NaN
50%	NaN	NaN	44.500000	17.300000	197.000000	4050.000000	NaN
75%	NaN	NaN	48.600000	18.700000	213.000000	4775.000000	NaN
max	NaN	NaN	59.600000	21.500000	231.000000	6300.000000	NaN

Figure 20-12: Output of the dtypes() and describe() methods

The describe() method returns a quick-look statistical overview of the DataFrame. Passing it all produces a statistical summary of *all* the columns. If you omit the include argument, you'll see only a summary of the *numeric* columns.

The NaN values present in the table represent *not applicable* values rather than missing values. For example, you can't take the mean of a categorical feature like species, so the result is presented as NaN.

The stats table doesn't tell you if every value in the dataset is valid, but it does help bracket how good or bad things can be. If the minimum, maximum, and mean values appear to be reasonable, the dataset is probably reliable.

Counting Observations Using countplot

Tables of data, although useful, can be dense and difficult to interpret. For example, is the data skewed toward male or female penguins? The information is there, but you must work to back it out.

In these cases, it's beneficial to create a visualization of the data. The seaborn library provides many statistical plot types for data exploration (Table 20-6). You can see examples of these in the seaborn gallery (*https://seaborn.pydata.org/examples/index.html*), and you can find a plotting tutorial at *https://seaborn.pydata.org/tutorial.html*.

Table 20-6: Useful seaborn Plotting Methods

Method	Description
barplot()	Categorical data presented with bars whose heights or lengths are proportional to the values that they represent.
boxplot()	Graphical representation of the locality, spread, and skewness groups of numerical data through their quartiles.
countplot()	A visualization of the counts of observations in each categorical bin using bars.
histplot()	Series of bars used to bin and display continuous data in a categorical form.
jointgrid()	Grid for drawing a bivariate plot with marginal univariate plots.
jointplot()	jointgrid() wrapper for drawing jointgrid() of two variables using canned bivariate and univariate graphs.
lineplot()	Graphical display of data along a number line where markers recorded above the responses indicate the number of occurrences.
pairgrid()	Subplot grid for plotting pairwise relationships in a dataset.
pairplot()	Easier-to-use wrapper for pairgrid().
relplot()	Function for visualizing statistical relationships using scatter plots and line plots
scatterplot()	Graph that uses Cartesian coordinates to display values for two variables. Additional variables can be incorporated through marker coding (color/size/shape).
stripplot()	A scatterplot for which one variable is categorical.
swarmplot()	A stripplot with non-overlapping points.
violinplot()	A combination of boxplot and kernel density estimate showing the distribution of quantitative data across several levels of one (or more) categorical variables.

Let's look at an example in which we plot the number of penguins and their gender. In a new cell, enter the following and then press SHIFT-ENTER:

```
# Plot species and gender counts:
sns.countplot(data=df, x='species', hue='gender')
plt.xticks(rotation=45)
plt.legend(loc='best');
```

This produces the output in Figure 20-13.

By visualizing the data, we can instantly see that the Chinstrap species is a bit underrepresented, and the division between genders is close to equal.

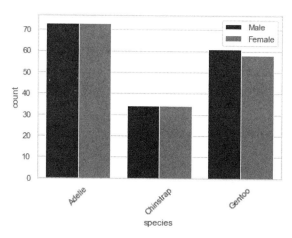

Figure 20-13: Bar chart of penguin species and gender counts

What about the distribution of penguins *per island*? Are they one big happy family, or do some prefer one island over another? To check, in a new cell, enter the following code and then press SHIFT-ENTER:

```
# Count and plot penguin species per island:
sns.countplot(data=df, x='island', hue='species')
plt.legend(loc='best');
```

This code counts the penguins per island, presents the results in a bar chart, and colors the bars based on species (Figure 20-14).

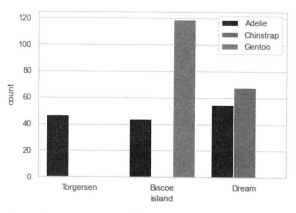

Figure 20-14: Bar chart of the number of penguins sampled per island, colored by species

Based on Figure 20-14, we can see that the Adélie penguins live on all the islands, but the Chinstraps are found only on Dream Island, and the

Gentoos only on Biscoe Island (see Figure 20-1 for island locations). So, if you have measurements from Torgersen Island, you know you're dealing with an Adélie. And the dimensional space is reduced for the other two islands, as you need to choose between only two species on both islands.

An assumption here is that each island was thoroughly sampled. We're saying that if a penguin species is not present in the dataset for a particular island, that species doesn't live on that island. You'd want to verify this assumption in a real study, as absence of evidence is not evidence of absence.

Another way to count each species per island is to use the pandas get_dummies() method in combination with the groupby() method. The first method converts categorical variables to *dummy variables*, which are numeric variables used to represent categorical data. The second method is used to group large amounts of data and compute operations on these groups.

In this case, we want to *sum* the penguin species per island, so we chain the methods and pass them the species column grouped by the island column, followed by the sum() method. In a new cell, enter the following code and then press SHIFT-ENTER:

```
# Count penguins per species per island
count_df = (pd.get_dummies(data=df, columns=['species']).groupby(
    'island', as_index=False).sum())
print(count_df.columns)
count_df[['island', 'species_Adelie', 'species_Gentoo', 'species_Chinstrap']]
```

The call to print() lets you see the names of the new "dummy" columns (highlighted in bold):

```
Index(['island', 'bill_length_mm', 'bill_depth_mm', 'flipper_length_mm',
       'body_mass_g', 'species_Adelie', 'species_Chinstrap', 'species_
       Gentoo'],
      dtype='object')
```

The final line of code displays the new columns in the count_df DataFrame (Figure 20-15).

	island	species_Adelie	species_Gentoo	species_Chinstrap
0	Biscoe	44	119	0
1	Dream	55	0	68
2	Torgersen	47	0	0

Figure 20-15: The count_df DataFrame that sums columns per island per penguin species

An advantage of checking tabular data is that low values are just as apparent as high values. With a bar chart, very low values may be mistaken for 0, due to the shortness of the bars.

You can read more about the get_dummies() and groupby() methods at *https://pandas.pydata.org/pandas-docs/stable/reference/api/pandas.get_dummies .html* and *https://pandas.pydata.org/pandas-docs/stable/reference/api/pandas .DataFrame.groupby.html*, respectively.

Getting the Big Picture with pairplot

Because visualizations are so effective for understanding data, seaborn provides the pairplot() method for plotting pairwise relationships in a dataset. This method creates a grid of axes where each variable shares the y-axis across a single row and the x-axis across a single column. This lets you quickly spot patterns in the data.

To make a pairplot, in a new cell, enter the following code and then press SHIFT-ENTER:

```
sns.pairplot(df, hue='species', markers=['o', '*', 'd']);
```

The arguments here are the name of the DataFrame, the column used to color the plots, and the marker types. You can find a list of marker types at *https://matplotlib.org/stable/api/markers_api.html*.

Because the dataset contains only four numeric columns, the pairplot (Figure 20-16) is very accessible and easy to consume.

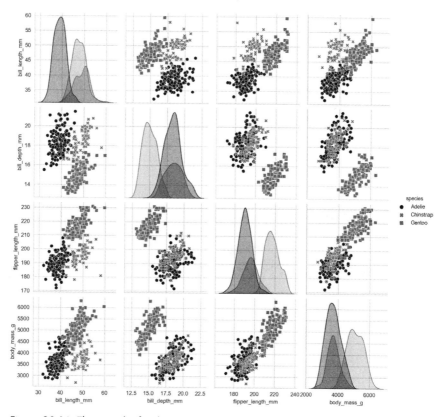

Figure 20-16: The pairplot for the penguins dataset

The pairplot makes it easy to see data distributions and relationships. For example, the scatterplots where the points cluster into separate groups are important because they indicate that classification strategies such as principal component analysis (PCA) and *k*-nearest neighbors should be able to distinguish one species from another. Scatterplots with linear relationships, like flipper length versus body mass, suggest that regression techniques could predict one of these features when the other is known.

Digging into Details with scatterplot

Despite being packed with information, even a pairplot can't tell the whole story. For example, what role does gender play in determining the body mass and bill length of each species? To explore this, you'll need more detailed plots. In a new cell, enter the following and press SHIFT-ENTER:

```
# Investigate bill length vs. body mass by species by gender:
sns.scatterplot(data=df,
                x='body_mass_g',
                y='bill_length_mm',
                hue='species',
                style='species',
                size='gender')

plt.legend(bbox_to_anchor=(1.3, 1.0));
```

In the call to scatterplot(), the hue, style, and size arguments control marker color, shape, and size, respectively. The first two are based on species, and the latter on gender; thus, data points representing female penguins are sized differently than males of the same species. Calling legend() with the bbox_to_anchor argument prevents the legend from posting over the plot and obscuring some of the data. You should get the results in Figure 20-17.

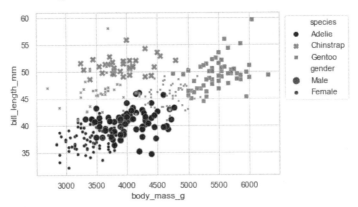

Figure 20-17: A scatterplot of bill length versus body mass, colored by species and sized by gender

This plot shows that the female of each species tends to be smaller, with shorter bills and lower body mass than the males. Bill length also appears to be more strongly correlated with body mass for the Adélie and Gentoo species, regardless of gender.

You can learn more about the scatterplot at *https://seaborn.pydata.org/generated/seaborn.scatterplot.html.*

Investigating Categorical Scatter Using boxplot and stripplot

We can explore the gender relationships further using different types of plots such as *box plots* and *strip plots.* To make a box plot, in a new cell, enter the following code and then press SHIFT-ENTER:

```
# Plot body mass by species by gender:
box = sns.boxplot(x="body_mass_g",
                  y="gender",
                  orient='h',
                  hue='species',
                  data=df)
```

This produces the plot in Figure 20-18.

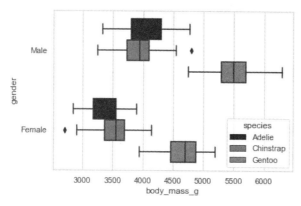

Figure 20-18: A box-and-whisker plot of penguin body mass by species by gender

Box plots provide insight on the symmetry, grouping, and skewness of data. Each box encompasses the first through third quartiles of the data distribution, with the vertical line within the box marking the median value. The "whiskers" extend to show the rest of the distribution, except for points that are considered "outliers," which are represented by diamonds.

Based on the box plot in Figure 20-18, Adélie and Chinstrap penguins are similar in size and smaller than Gentoo penguins, and the females tend to be smaller for all species. There is overlap between the genders, however, meaning body mass alone cannot positively differentiate males from females.

The seaborn strip plot posts the actual data points rather than summarizing them, as in the box plot. Let's examine bill length measurements in both species and genders. In a new cell, enter the following code and then press SHIFT-ENTER:

```
# Plot bill length by species by gender:
strip = sns.stripplot(data=df,
                      x='bill_length_mm',
                      y='gender',
                      hue='species',
                      dodge=True)
```

The dodge argument shifts points for each species to reduce overlap, making the plot easier to read (Figure 20-19).

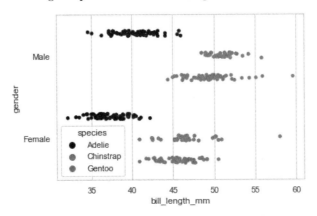

Figure 20-19: A strip plot of penguin bill length by species by gender

Based on the plot, we can see that the Adélie penguins have markedly shorter bills than the other two species. Gender differences are less distinct, though female penguins of all species tend to have shorter bills, on average.

Combining Views Using jointplot

Another potentially characteristic feature of the penguins is the vertical thickness of the bill, referred to as its *depth*. You can see in Figure 20-2 that Gentoos have narrow, pointed bills, whereas the other two species have more bulbous bills. Although there are numerous ways to compare these graphically, let's try out a joint plot using a *kernel density estimation (KDE)*.

A KDE plot is a method for visualizing the distribution of observations in a dataset, much like a histogram. But whereas a histogram approximates the underlying probability density of the data by counting observations in discrete bins, a KDE plot smooths the observations using a Gaussian kernel, producing a continuous density estimate. This results in a less cluttered and more interpretable plot when drawing multiple distributions. The joinplot() method lets you plot two variables using bivariate and univariate KDE graphs.

In a new cell, enter the following and then press SHIFT-ENTER:

```
# Plot bill depth vs. bill length by species:
sns.jointplot(data=df,
              x="bill_length_mm",
              y="bill_depth_mm",
              kind="kde",
              hue="species",
              alpha=0.75);
```

This produces the chart in Figure 20-20.

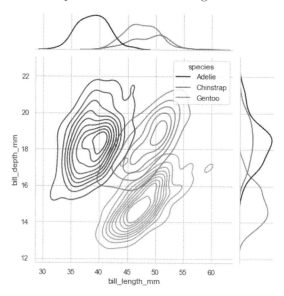

Figure 20-20: A joint plot of bill depth versus bill
length by species

From the Gaussian curves along the edges of the joint plot, it's clear that the Adélie is distinguished by its shorter bill length, and the Gentoo by its shallower bill depth.

You can customize joint plots in many ways. To see some examples, check out the documentation at *http://seaborn.pydata.org/generated/seaborn .jointplot.html*.

Visualizing Multiple Dimensions Using radviz

The pandas library comes with its own plotting capability built on Matplotlib. This includes the radviz() (radial visualization) method for plotting multidimensional datasets in a 2D format (Figure 20-21).

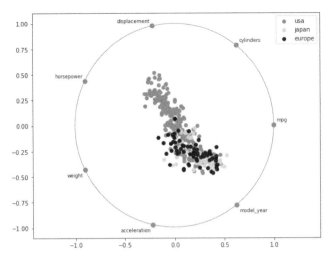

Figure 20-21: An example radviz plot for an automotive dataset

In a radial visualization, the dimensions in a DataFrame, such as a penguin's body mass or bill length, are evenly spaced around the circumference of a circle. Data in these numerical columns are normalized to values between 0 and 1 so that all dimensions have equal weights. These are then projected into the circular 2D space as if imaginary springs anchor them to the column labels along the circumference. A point is plotted where the sum of the "spring" forces acting on it equals zero.

Radial visualizations are intuitive by nature. Points with similar dimension values will plot near the center, as will points with similar values whose dimensions are opposite each other on the circle. Points with dimension values greater than others will be "pulled" toward that side of the circle. The penguins dataset has only four dimensions, but the radviz() method can handle many more.

To make a radial visualization, in a new cell, enter the following and then press SHIFT-ENTER:

```
# Make radial visualization:
❶ sns.set_theme(context='talk')
  plt.figure(figsize=(7, 7))
❷ pd.plotting.radviz(df.drop(['island', 'gender'], axis=1),
                  class_column='species',
                  color=['black', 'red', 'grey'],
                  marker='+',
                  alpha=0.7)
  plt.legend(loc=(1.01, 0.7));
```

For a better-looking radviz plot, reset the default seaborn plotting parameters using the set_theme() method and set the context to 'talk' ❶. The context parameter controls the scaling of plot elements like label size and line thickness. The base context is notebook, and the other contexts are paper, `talk`, and `poster`, which are just versions of the `notebook` parameter scaled by different values. Using the `talk` argument ensures that the plot labels are easy to read. To better increase readability, manually set the figure size to 7" × 7".

Next, call pandas' `plotting.radviz()` method ❷. This method accepts only one categorical column, called the `class_column`, which in this case will be `species`. The rest of the DataFrame columns are assumed to be numerical, so we must remove the `island` and `gender` columns, which don't contain numerical data. You could do this by creating a copy of the DataFrame, but because we need only this revised DataFrame for plotting, we'll temporarily delete the columns, using the `drop()` method, while passing the DataFrame to the `radviz()` method.

The `drop()` method takes two arguments: the column names as a list, and the axis number, where 0 = row and 1 = column. Unless you pass it an `inplace=True` argument, the DataFrame will be changed for only the current operation.

Because we're not plotting with seaborn, we need to remind pandas of the color scheme we're using and then change the marker style and transparency to make over-posted points easier to see. Moving the legend to the side also helps. Notice how we're able to mix in seaborn (`sns`) and Matplotlib's pyplot (`plt`) with pandas `plotting`. This is because both seaborn and pandas are built on top of Matplotlib.

You should see the plot depicted in Figure 20-22.

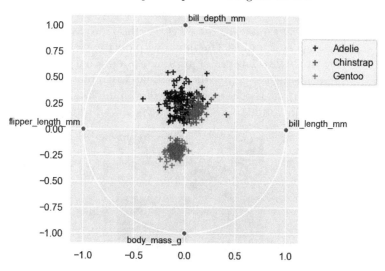

Figure 20-22: Radviz plot for the penguins dataset

In the plot, the Gentoo data points form a distinct cluster skewed toward body mass and flipper length. The similarly sized Chinstrap and Adélie penguins are distinguished mainly by bill length, which "pulls" the Chinstrap points to the right of center.

The radviz plot is another way of exploring the data, and it becomes more useful with more dimensions. To read more about the pandas implementation, visit *https://pandas.pydata.org/docs/reference/api/pandas.plotting .radviz.html*.

It's worth noting here that we changed the plotting style. I'll be continuing with this new look, but if you want to return to the previous 'whitegrid' style, you'll need to enter the following code in a new cell before making more plots:

```
# Restore theme and palette:
sns.set_theme(context='notebook')
sns.set_style("whitegrid")
sns.set_palette(['black', 'red', 'grey'])
```

Quantifying Correlations Using corr()

The pandas DataFrame class comes with a corr() method that quantifies data correlations by computing a pairwise correlation of columns, excluding NA/null values. This is useful when you plan to use regression techniques to make predictions.

In a new cell, enter the following and then press SHIFT-ENTER:

```
correlations = df.corr()
sns.heatmap(correlations, center=1, annot=True);
```

The first line calls the corr() method and assigns the results to the correlations variable. The next line plots the results as a seaborn heatmap (Figure 20-23). The center argument is optional and tells the method the value at which to center the colormap when plotting divergent data. With a value of 1, the best correlations will plot in black. The annot argument turns on the plot annotations within each colored square.

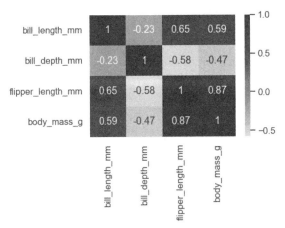

Figure 20-23: The correlation heatmap

The heatmap confirms and quantifies correlations that we noted in the pairplot (Figure 20-16). Flipper length and body mass are the most closely correlated, followed by flipper length and bill length.

For more on the corr() method and seaborn heatmap, visit *https://pandas.pydata.org/pandas-docs/stable/reference/api/pandas.DataFrame.corr.html* and *https://seaborn.pydata.org/generated/seaborn.heatmap.html*.

TEST YOUR KNOWLEDGE

1. Which of the following are advantages of pandas series or DataFrames over NumPy arrays?

   a. Ability to use heterogeneous data types

   b. Ability to use either numbers *or* labels as indexes

   c. Ability to load Python dictionaries

   d. Ease of use with tabular data

2. True or False: Reindexing is required after renaming columns in a DataFrame.

3. Convert the following dictionary into a DataFrame and rename the last column to "whales":

```
animals = {'canines': ['husky', 'poodle', 'bulldog'],
           'felines': ['Siamese', 'Persian', 'Maine Coon'],
           'cetaceans': ['humpback', 'sperm', 'right']}
```

4. Display the first row of the animals DataFrame from the previous question.

5. Flip the rows and columns in the animals DataFrame (hint: look up the pandas transpose() method).

Predicting Penguin Species Using k-Nearest Neighbors

The goal of this project is to develop a model that classifies penguins based on the Palmer Archipelago dataset. Our data exploration has revealed that four morphological features (bill length and depth, flipper length, and body mass) form separate but overlapping clusters in numerous plots. This implies that a machine learning classification algorithm should be able to handle the problem.

It's always best to start simple, and if you do a little research, you'll find that *k-Nearest Neighbors* (*k*-NN) is one of the most basic, beginner-friendly, yet important classification algorithms in machine learning. It uses distance measures to intuitively find the *k* nearest neighbors to a new, unknown data point and then uses those neighbors to make a prediction.

In Figure 20-24, numerical data points for two categorical classes (A and B) are plotted in a scattergram. A new and unlabeled data point (★) falls between the two clusters. To classify this new point, the algorithm's *k* parameter has been set to 7. As most of the closest points belong to Class B, the new data point will be assigned to B. Because this is a "voting" algorithm, *k* should always be set to an odd number to avoid a tie.

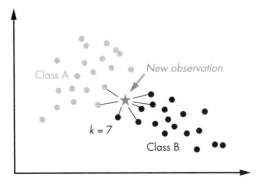

Figure 20-24: Example of the k-NN algorithm choosing the seven nearest neighbors to a new data point

Besides being intuitive and easy to explain, *k*-NN runs quickly, works well with small datasets, is robust to noise, and can be tuned for greater accuracy. It's also versatile, given that it can be applied to both classification and regression problems.

The algorithm needs a dense dataset, however, so that points aren't too far apart. The more data *dimensions* you have, like flipper length and body mass, the more data you need for *k*-NN to work properly.

In addition, *k*-NN, like other machine learning algorithms, requires its own data preparation routines. Because the algorithm works only with numerical data, you'll commonly need to convert categorical values to integers and normalize numerical values between 0 and 1. Normalization is needed so that dimensions with larger values don't skew the distance calculations.

Converting Categorical Data to Numerical Data

As stated previously, the *k*-NN algorithm uses *numerical* data. To take advantage of important non-numerical data, such as the island of origin and gender, you need to convert these values into numbers.

Let's do this first for the island column, using the pandas get_dummies() method that we used previously when counting penguins per island. Next, we'll repeat the exercise manually for gender so that you can practice DataFrame *indexing*. In a new cell, enter the following and then press SHIFT-ENTER:

```
# Prepare for k-NN.
# Add numerical columns for island and gender labels:
knn_df = pd.get_dummies(data=df, columns=['island'])
```

❶
```
knn_df['male'] = 0
knn_df.loc[knn_df['gender'] == 'Male', 'male'] = 1

knn_df['female'] = 0
knn_df.loc[knn_df['gender'] == 'Female', 'female'] = 1
```

❷
```
knn_df.iloc[:300:30]
```

To use get_dummies(), pass it the DataFrame and the column label that you want to convert. Assign the result to a new DataFrame named knn_df. The method will create three new columns—one for each island—with values of either 0 or 1, depending on the value in the island column (either Biscoe, Dream, or Torgersen).

Next, for demonstration purposes, we'll use a different approach to convert the gender column to new male and female columns. We'll create a new column for each class and fill it with zeros. Then, using conditional statements, we'll find the rows where the targeted class exists and change the column values to ones. For example, the column for male penguins will contain 1 for rows in the gender column containing a male designation; for all other rows, the column will contain 0.

Start by creating a new column named male. Assign to that column a value of 0 ❶. Next, use the pandas loc indexing operator to select a *subset* of the knn_df DataFrame. Because pandas can use both label-based and integer-based indexing for rows and columns, it comes with two indexing operators. The loc operator is for strictly label-based indexing, and the iloc operator handles cases in which the labels are strictly integers. In our case, the columns use labels (such as "species") and the rows use integers.

The current operation will convert Male values in the gender column to 1 values in the male column. So, select the gender column (knn_df['gender']) and then use a conditional to overwrite the 0 values that we set in the previous line. What you're saying here is, "get the gender column and, if its value is Male, put a 1 in the male column."

Repeat this code for the female column and then check the results by using the iloc operator to sample rows throughout the DataFrame ❷. This works like indexing a list, for which you start at the beginning, go up to index 300, and use a step of 30 to select every 30th row.

You should get the output in Figure 20-24. Note the five new columns on the right side of the DataFrame. The categorical island and gender columns can now be used by the *k*-NN algorithm, so you can make use of all the data at your disposal.

	species	bill_length_mm	bill_depth_mm	flipper_length_mm	body_mass_g	gender	island_Biscoe	island_Dream	island_Torgersen	male	female
0	Adelie	39.1	18.7	181.0	3750.0	Male	0	0	1	1	0
30	Adelie	39.2	21.1	196.0	4150.0	Male	0	1	0	1	0
60	Adelie	35.5	16.2	195.0	3350.0	Female	1	0	0	0	1
90	Adelie	38.1	18.6	190.0	3700.0	Female	0	1	0	0	1
120	Adelie	38.8	17.6	191.0	3275.0	Female	0	0	1	0	1
150	Chinstrap	52.7	19.8	197.0	3725.0	Male	0	1	0	1	0
180	Chinstrap	49.7	18.6	195.0	3600.0	Male	0	1	0	1	0
210	Chinstrap	43.5	18.1	202.0	3400.0	Female	0	1	0	0	1
240	Gentoo	47.8	15.0	215.0	5650.0	Male	1	0	0	1	0
270	Gentoo	45.5	15.0	220.0	5000.0	Male	1	0	0	1	0

Figure 20-25: A sample of the new knn_df DataFrame with new numerical columns for islands and gender

By converting the gender and island data to numbers, we've supplemented our morphological data with two more dimensions.

If your goal was to predict the species of penguins sampled at sea, you'd want to drop the island-related columns because you couldn't be sure of a penguin's point of origin.

Setting Up the Training and Testing Data

The *k*-NN classifier is a *supervised* learning algorithm. This means that you show it what the answer *should* look like by providing a set of examples known as the "training" dataset. You can't use all the available data for the training set, however, as you'll have no objective way to test the results. Consequently, you need to randomly split out a smaller subset of the data that you can use to test the model's accuracy.

As a lazy learning algorithm, *k*-NN doesn't have an actual training phase in which it "learns" a discriminative function to apply to new data. Instead, it loads, or memorizes, the data and performs calculations with it during the prediction phase.

For convenience, scikit-learn provides the train_test_split() method as part of the sklearn.model_selection module. In a new cell, enter the following and then press SHIFT-ENTER:

```
# Break out numerical and target data and split off train and test sets:
from sklearn.model_selection import train_test_split

❶ X = knn_df.select_dtypes(include='number')  # Use numerical columns.
y = knn_df['species']  # The prediction target.

# Split out training and testing datasets:
❷ X_train, X_test, y_train, y_test = train_test_split(X, y,
                                          test_size=0.25,
                                          random_state=300)
```

After importing the module, call the pandas select_dtypes() method on the new knn_df DataFrame ❶. This method returns a subset of a DataFrame including or excluding columns based on their data type. We want the numerical columns for use with the *k*-NN algorithm, so set include equal to 'number' and assign the result to a variable named X.

Next, assign the `species` column to a variable named y. This represents the categorical class you're trying to predict. Note that the uppercase "X," lowercase "y" format follows the convention in the scikit-learn documentation.

Split out the training and testing data using the `train_test_split()` method ❷. You'll need to unpack four variables for both X and y training and testing. Because we passed the method DataFrames, it will return DataFrames.

A key argument here is `test_size`, expressed as a proportion of the complete dataset. By default, this is 0.25, or 25 percent. So, for our penguins dataset, this represents 83 samples (332 × 0.25).

To avoid biasing, the `train_test_split()` method randomly shuffles the data before splitting it. For reproducible output across multiple function calls, you can pass an integer to the `random_state` argument. As written, this code lets you produce one set of repeatable training and testing data. To generate a new random set, you'll need to either change the `random_state` value or not use it at all.

Although we don't need it here to get a good result, the `train_test_split()` method comes with a `stratify` parameter that ensures the split preserves the proportions of samples of each target class as observed in the original dataset. So, if the original dataset sampled 25 percent of Class A and 75 percent of Class B, the training and testing sets would reflect this proportion. This helps you avoid sampling bias, wherein a sample is not representative of the true population.

To read more about the `train_test_split()` method, visit *https://scikit -learn.org/stable/modules/generated/sklearn.model_selection.train_test_split.html*.

Normalizing the Data

Each numerical data column in the training and testing sets should be normalized to values between 0 and 1. This prevents columns with large numerical values from biasing the *k*-NN distance measurement.

Because column transformations, such as normalization, are a common operation in machine learning, scikit-learn comes with two modules, `compose` and `preprocessing`, to simplify the task. In a new cell, enter the following and then press SHIFT-ENTER:

```
# Normalize numerical columns to 0-1:
from sklearn.compose import make_column_transformer
from sklearn.preprocessing import MinMaxScaler

❶ column_transformer = make_column_transformer((MinMaxScaler(),
                                                ['bill_depth_mm',
                                                 'bill_length_mm',
                                                 'flipper_length_mm',
                                                 'body_mass_g']),
                                               remainder='passthrough')
X_train = column_transformer.fit_transform(X_train)
```

```
❷ X_train = pd.DataFrame(data=X_train,
                        columns=column_transformer.get_feature_names_out())
  X_train.head()

  X_test = column_transformer.fit_transform(X_test)
  X_test = pd.DataFrame(data=X_test,
                        columns=column_transformer.get_feature_names_out())
  X_test.head()
```

Start by importing the `make_column_transformer()` method and the `MinMaxScaler()` method. The first method lets us transform columnar data; the second method specifies how to do it.

To normalize the data, scale it so that the minimum and maximum values fall between 0 and 1. Pass the `make_column_transformer()` method the `MinMaxScaler()` method ❶. Next, pass it the columns that you want to transform; in this case, the numerical columns that aren't already scaled to 0 and 1. By default, the transformer *drops* columns that you didn't specify in the previous argument. To prevent this, set the `remainder` argument to `passthrough`.

Now you need to apply the transformer by calling its `fit_transform()` method and passing it the `X_train` variable you made in the previous section. This method transforms the data and concatenates the results, returning an array. To convert this array back into a DataFrame, call pandas' `DataFrame` class and pass it the `X_train` array ❷. The column transformer renames the columns as well as transforming them, so for the `columns` argument, call the `get_feature_names_out()` method on the `column_transformer` object.

Call `X_train.head()` to see the results, and then repeat this code for the testing set. The two DataFrame heads are shown in Figure 20-26. In both, the columns should have new names, all the columns should contain numerical data, and values should fall between 0.0 and 1.0.

	minmaxscaler__bill_depth_mm	minmaxscaler__bill_length_mm	minmaxscaler__flipper_length_mm	minmaxscaler__body_mass_g	remainder__island_Biscoe
0	0.321429	0.766038	0.796610	0.777778	1.0
1	0.250000	0.607547	0.830508	1.000000	1.0
2	0.345238	0.173585	0.186441	0.104167	1.0
3	0.607143	0.607547	0.389831	0.472222	0.0
4	0.226190	0.664151	0.864407	0.791667	1.0

	minmaxscaler__bill_depth_mm	minmaxscaler__bill_length_mm	minmaxscaler__flipper_length_mm	minmaxscaler__body_mass_g	remainder__island_Biscoe
0	0.5875	0.756303	0.230769	0.172414	0.0
1	0.9375	0.836134	0.615385	0.655172	0.0
2	0.0625	0.634454	0.692308	0.698276	1.0
3	0.5750	0.285714	0.057692	0.250000	0.0
4	0.5125	0.432773	0.057692	0.241379	0.0

Figure 20-26: The head of the normalized X_train and X_test DataFrames (shown horizontally truncated)

To read more about the scikit-learn column transformer, visit *https://scikit-learn.org/stable/modules/generated/sklearn.compose.make_column_transformer.html.*

You now have numbers-only DataFrames that can be used for training and testing. The x_test and y_test DataFrames let you relate these numerical DataFrames back to a species call. With only seven dimensions, this is a *low-dimensional* dataset. A *high-dimensional* dataset, common in machine learning, could have 100,000-plus features!

Running k-NN and Checking the Accuracy of the Prediction

At this point, the data is ready to be used with the *k*-NN classifier. It's time to name that penguin!

In a new cell, enter the following code and then press SHIFT-ENTER:

```
# Run k-NN and check accuracy of prediction:
from sklearn.neighbors import KNeighborsClassifier
from sklearn.metrics import accuracy_score

knn = KNeighborsClassifier(n_neighbors=5, p=2)
knn.fit(X_train, y_train)
predictions = knn.predict(X_test)

accuracy = accuracy_score(y_test, predictions)
print(f"Model accuracy = {accuracy}")
```

To run *k*-NN, you need to import the KNeighborsClassifier class from the scikit-learn neighbors module (*https://scikit-learn.org/stable/modules/generated/sklearn.neighbors.KNeighborsClassifier.html*). To check the results, import the accuracy_score() method from the metrics module.

Call the KNeighborsClassifier and pass it a k value of 5, and a p value of 2. The p value tells the classifier to use Euclidian, or straight-line, distance measurements. This is usually appropriate for low-dimensional datasets with few outliers.

Next, call the classifier's fit() method to train it and then run the predict() method on the X_test dataset. This will take the measurement data you *withheld* from training and predict the species.

Finish by calling the accuracy_score() method and passing it the y_test and predictions variables. This method will compare the two datasets and store the accuracy measure in the accuracy variable, which you then print (Figure 20-27).

```
KNeighborsClassifier()

Model accuracy = 0.9880952380952381
```

Figure 20-27: The accuracy of the k-NN model

Right out of the gate, the model correctly matched about 99 percent of the samples in the X_test dataset. But don't get too excited. It's only matched samples in a *randomly chosen* subset of the penguins dataset. If you generate a new test set by changing the train_test_split() method's random _state argument from 300 to 500 and rerun the cells, you'll get an accuracy of 0.9642857142857143. Although this is about as good as it gets for a real-world dataset, let's use this discrepancy to see how you might handle a larger mismatch in a different project.

Optimizing the k Value Using Cross-Validation

The goal of supervised machine learning is to generalize *beyond* what we see in training samples so that we can reliably classify new data. The *k*-NN classifier uses numerous *hyperparameters* such as k, p, and weights to control the learning process. In addition, the test_size and other parameters in the train_test_split() method can have a big impact on model results. You can think of these parameters as knobs that you can turn to "tune" or "dial in" the model fit.

You must be careful however, not to tune things *too* finely. *Overfitting* is a common problem that can lurk behind an apparently accurate model (Figure 20-28).

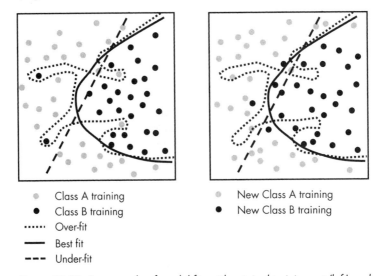

○ Class A training	○ New Class A training
● Class B training	● New Class B training
⋯⋯ Over-fit	
── Best fit	
─ ─ Under-fit	

Figure 20-28: An example of model fits with original training set (left) and superimposed with new, unmatched training set (right)

In Figure 20-28, the chart on the left shows three fits (over, under, and best) to a randomized training dataset. Any points that fall to the right of these lines will be classified as belonging to Class B.

If the left-hand chart represents all the data we would ever have, the overfitted model would be the most accurate. But look what happens when we choose and post a new training set (right side of Figure 20-28). Some points stay the same and others change, and the overfitted model no longer matches the data as well. On the other hand, the underfitted model can neither model the training data nor generalize sufficiently to match new data.

Typically, the smaller the k value, the "tighter" the fit of the model to the data and the more likely that it is overfit; the larger the k value, the more likely that the model is underfit.

In Figure 20-28, the more generalized "Best fit" model does a good job of matching the two datasets. To achieve this generalized model, we'll need to find the best values for the important hyperparameters. But this isn't something that can be intuited when working in multidimensional space. It requires iterative investigation, where parameters are changed multiple times and the results are tabulated and scored.

In the code that follows, we'll investigate a range of k values using *cross-validation* (*cv* for short). This is a model validation technique for assessing how the results of a statistical analysis will generalize to a new, unknown dataset. It also helps flag issues like overfitting.

Cross-validation resamples different parts of the training set to create testing sets. To ensure that the entire training set is evaluated, it repeats this sampling multiple times. As it iterates, it changes the value of a hyperparameter, like k, and scores the result based on model accuracy. When the optimal parameters are identified, you input them in the k-NN classifier and perform a final evaluation against the test dataset, which has been kept separate from the cv process (Figure 20-29).

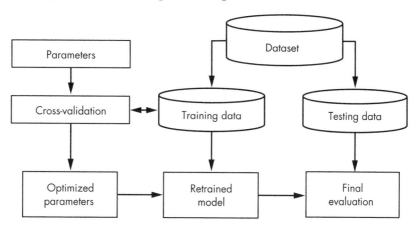

Figure 20-29: Building a predictive model using cross-validation (modified from scikit-learn.org)

To use cross-validation on our penguins dataset model, in a new cell, enter the following and then press SHIFT-ENTER:

```
# Run cross-validation on k:
import numpy as np
from sklearn.model_selection import cross_validate

cv_metrics = {'train_score_ave': [],
              'cv_score_ave': []}

num_neighbors = {'k': np.arange(1, 25)}
```

```
for k in num_neighbors['k']:
❶  knn = KNeighborsClassifier(n_neighbors=k, p=2)
    scores = cross_validate(knn, X_train, y_train, return_train_score=True)
    cv_metrics['cv_score_ave'].append(np.mean(scores['test_score']))
    cv_metrics['train_score_ave'].append(np.mean(scores['train_score']))

❷ cv_metrics_df = pd.DataFrame(cv_metrics)
  cv_metrics_df.insert(loc=0, column='k', value=num_neighbors['k'])
  cv_metrics_df.head(10)

  best_k = cv_metrics_df.loc[cv_metrics_df['cv_score_ave'].idxmax()]
  print(f"Best k for current training and testing set: {int(best_k['k'])}")
```

We'll use NumPy to make a range of *k* values to evaluate, and average, the results of each cv iteration. The cross_validate() method is found in the scikit-learn model_selection module.

Python dictionaries are great for storing data like test results, and they can easily be turned into pandas DataFrames. After the imports, create a dictionary named cv_metrics with keys for the average training set and cross-validation score per iteration. The initial values for these dictionary keys are empty lists.

Next, make a num_neighbors dictionary with one key-value pair: *k* and a 1D ndarray from 1 to 25. These represent the range of *k* values you'll be testing.

Loop through the num_neighbors dictionary and pass the current *k* value to the KNeighborsClassifier ❶. Then, call the cross_validate() method, pass it the knn model and the training data, and set the return_train_score argument to True. Finish each loop by appending the score results to the appropriate key in the cv_metrics dictionary. Use the NumPy mean() method to average the scores for each data point during the process.

Outside the loop, turn the cv_metrics dictionary into a DataFrame ❷ and then add the num_neighbors dictionary as a new column on the leftmost side of the DataFrame. Do this by calling the insert() method on the DataFrame and passing it the first column position (loc=0), the column name, and the value, obtained by passing the num_neighbors dictionary the k key. Finish by calling head(10) to display the first 10 rows.

Rather than scroll through the DataFrame looking for the *k* value with the best score, let pandas find it using the idxmax() method, which returns the index of the first occurrence of a maximum value over a requested axis. This is axis 0 (row) by default. When you print the result, you should see the output in Figure 20-30.

Comparing cross-validation and training scores can be insightful with respect to model overfitting and underfitting. The process is computationally expensive for high-dimensionality datasets, however, and the training scores are not required to select the best parameters.

	k	train_score_ave	cv_score_ave
0	1	1.000000	0.984
1	2	0.996990	0.988
2	3	0.998995	0.988
3	4	1.000000	0.996
4	5	1.000000	0.996
5	6	0.996990	0.996
6	7	0.997995	0.996
7	8	0.996990	0.996
8	9	0.997995	0.996
9	10	0.997995	0.996

```
Best k for current training and testing set:  4
```

Figure 20-30: The first 10 rows in the cv_metrics DataFrame and the cv choice for best k value

To plot the cv results, in a new cell, enter the following and then press SHIFT-ENTER:

```
# Plot cross-validation results:
sns.set_palette(['black', 'red', 'grey'])
df_melt = cv_metrics_df.melt('k', var_name='cols',  value_name='vals')
sns.lineplot(x='k', y="vals", hue='cols', data=df_melt);
```

The first line resets the seaborn color palette that we've been using for a consistent look. The next line prepares the DataFrame for plotting. In seaborn, plotting multiple columns against the same y-axis requires a call to the pandas melt() method. This method returns a new DataFrame reshaped into a long table with one row for each column. To learn about wide-form and long-form data, see *https://seaborn.pydata.org/tutorial/data_structure.html.*

With the new df_melt DataFrame, you can call seaborn's lineplot() method to get the plot in Figure 20-31. The top curve represents the average training score.

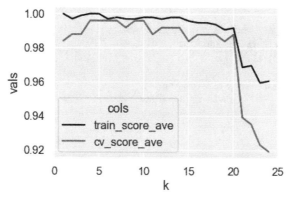

Figure 20-31: A comparison of the average training scores and cross-validation scores with the k value

If you expect to work extensively with pandas, consider learning its plotting syntax. For example, you can re-create Figure 20-31 with a single line: `cv_metrics_df.plot.line(x='k')`. The plots are less customizable than with seaborn or Matplotlib, but they're more than suitable for data exploration. To learn more, visit *https://pandas.pydata.org/docs/getting_started/intro_tutorials/04_plotting.html*.

A model that is underfit will have both low training and low testing accuracy, whereas an overfit model will have a high training accuracy but a low testing accuracy. On the left side of Figure 20-31, where $k = 1$, the training set is perfectly accurate because the training data point is compared only to itself. The cv results, however, are less accurate. This indicates slight overfitting, which we would expect with a low value of k.

On the right side of the figure, where k is greater than 20, accuracy falls off for both curves. The model is trying to accommodate too many data points and is becoming underfit.

When $k = 4$, the cv score reaches its highest average accuracy, and the two curves begin to meet and run parallel to each other. In this case, values of k between 5 and 10 will only add computational burden without increasing the model accuracy.

If you change the `random_state` or the `test_size` parameters in the `train_test_split()`method and rerun the code, you'll see variations in the choice of best k value. This is because the model starts out with such high accuracy that subtle stochastic effects can have a large relative impact with essentially no absolute impact.

Optimizing Multiple Hyperparameters Using GridSearchCV

Running cross-validations can take time, so scikit-learn comes with a convenience class called `GridSearchCV`. It takes a dictionary of parameter names and values, cross-validates them, and reports the outcome. You can find the documentation at *https://scikit-learn.org/stable/modules/generated/sklearn.model_selection.GridSearchCV.html*.

In a new cell, enter the following and then press SHIFT-ENTER:

```
from sklearn.model_selection import GridSearchCV

params = {'n_neighbors': np.arange(1, 20),
          'weights': ['uniform', 'distance'],
          'p': [1, 2]}

grid = GridSearchCV(estimator=knn,
                    param_grid=params,
                    scoring='accuracy',
                    verbose=1)
grid_results = grid.fit(X_train, y_train)
print(f"Best parameter values: {grid_results.best_params_}")
```

A dictionary named `params` holds the hyperparameter ranges. In this example, the k (`n_neighbors`) range is a NumPy array, and the `weights` and `p` parameters use lists.

The GridSearchCV class needs to know the DataFrame you're using (knn), the name of the parameters dictionary (params), what to score on (accuracy), and how much detail you want it to report. By passing it verbose=1, we suppressed most of the extraneous output.

After fitting the model, you can print the best_params_ attribute to see the results (Figure 20-32).

```
Fitting 5 folds for each of 76 candidates, totalling 380 fits
Best parameter values: {'n_neighbors': 4, 'p': 2, 'weights': 'uniform'}
```

Figure 20-32: Results of running GridSearchCV

Next, in a new cell, pass the best parameters identified by the grid search to the KNeighborsClassifier, fit the model, predict against the testing dataset, and evaluate the accuracy. This corresponds to the "retrained model" and "final evaluation" steps shown in Figure 20-29.

```
knn = KNeighborsClassifier(n_neighbors=4, p=2, weights='uniform')
knn.fit(X_train, y_train)
predictions=knn.predict(X_test)

accuracy = accuracy_score(y_test, predictions)
print(f"Model accuracy = {accuracy}")
```

You should get the output shown in Figure 20-33.

```
KNeighborsClassifier(n_neighbors=4)

Model accuracy = 0.9761904761904762
```

Figure 20-33: Model accuracy after applying the optimized hyperparameters

You might notice that this score is worse than the one we got initially using k=5 (Figure 20-27). That's because this initial run on a *single* train-test dataset was equivalent to a lucky roll of dice. The current model built with k=4 has been tested over *multiple* datasets and should, when used repeatedly, yield the same average accuracy as k=5 (see Figure 20-31) but with better computational efficiency.

Along these lines, we've used only 75 percent of the penguins dataset to train the model. How do we know that 80 percent wouldn't produce better results? To find out, you could use a loop to run multiple combinations of the test_size and random_state parameters of the train_test_split() method and model each.

The need to test many parameter and dataset combinations, along with heavy memory use, renders the *k*-NN algorithm inappropriate for very large datasets. Otherwise, it has numerous benefits, including being simple to use, easy to understand, fast to train, versatile, and agnostic to assumptions about the underlying data distributions.

TEST YOUR KNOWLEDGE

6. For a big picture overview of a dataset, call the:

    a. seaborn `relplot()` method

    b. pandas `radviz()` method

    c. seaborn `pairplot()` method

    d. seaborn `jointplot()` method

7. The *k*-NN algorithm is appropriate for:

    a. Classification in high-dimensional datasets

    b. Projects for which computer memory is at a premium

    c. Classification in a noisy, low-dimensional dataset

    d. All of the above

8. Using a very low *k* value with the *k*-NN algorithm can result in:

    a. Excessive run times

    b. An underfit model

    c. An overfit model

    d. A generalized model

9. In machine learning, a hyperparameter is:

    a. A parameter chosen automatically by the algorithm

    b. A parameter set at the top level of an algorithm

    c. An adjustable parameter used to control the learning process

    d. An overly excitable parameter

10. Cross-validation is used to:

    a. Check the accuracy of a model against an independent test set

    b. Find the best hyperparameters from an input range of hyperparameters

    c. Check a dataset for duplicate samples

    d. Gain insight on model underfitting and overfitting

Summary

The Palmer penguins project provided a good overview of how pandas, seaborn, and scikit-learn work, how they work together, and what you can accomplish using them. At this point, though, you've barely glimpsed the enormity of these packages. To expand your knowledge, I recommend the official library documentation cited in the introduction to this chapter as well as the following books:

Python for Data Analysis: Data Wrangling with Pandas, NumPy, and IPython, 2nd edition, by Wes McKinney (O'Reilly Media, 2018), is an indispensable guide by the creator of the pandas library.

Python Data Science Handbook: Essential Tools for Working with Data, by Jake VanderPlas (O'Reilly Media, 2016), is a thorough reference for important Python data science tools, including pandas.

Hands-on Machine Learning with Scikit-Learn, Keras, & TensorFlow: Concepts, Tools, and Techniques to Build Intelligent Systems, 2nd edition, by Aurélien Géron (O'Reilly Media, 2019), provides practical instruction for machine learning novices.

Although the penguins project covered a lot of ground, it didn't address one of the most important forms of structured data used by scientists: time series data. In the next and final chapter, we'll look at methods for incorporating dates and times in your programs and plots.

21

MANAGING DATES AND TIMES
WITH PYTHON AND PANDAS

In mathematics, a *time series* is a series of data points indexed in chronological order. They are common components in scientific datasets where observations are made over periods of time.

Although you and I recognize "11/11/1918" as a date, a computer sees this value as a string. To intelligently work with calendar dates as well as hours, minutes, seconds, and so on, Python and pandas treat them as special objects. These objects are "aware" of the mechanics of the Gregorian calendar, the sexagesimal (base 60) time system, time zones, daylight-saving time, leap years, and more.

Native Python supports times series through its datetime module, and pandas is oriented toward using arrays of dates, such as for an index or column in a DataFrame. In addition to its built-in tools and algorithms for working with both fixed-frequency and irregular time series, pandas also uses the datetime module. Observations in *fixed frequency* time series are those recorded at regular intervals such as once a day. Otherwise, the time series is said to be *irregular* in nature.

We'll look at both the Python and pandas approaches here, with the goal of introducing you to the basics of working with time series and making you conversant in the subject. For more detail, you can visit *https://docs.python.org/3/library/datetime.html* for Python's datetime module and *https://pandas.pydata.org/pandas-docs/stable/user_guide/timeseries.html* for the pandas tools.

Python datetime Module

Python's built-in datetime module includes the date, time, and combined datetime types. By treating time information as specific data types, Python knows how to manipulate it properly and efficiently. This includes working with time zones, daylight saving time (DST), leap years, and different international formatting methods.

In this brief introduction, we'll look at marking time series data with *timestamps*, for specific time instants; time *intervals*, delineated by a starting and ending timestamp; and fixed *periods*, such as a year. You can keep track of *elapsed* time, too, such as the time relative to the start of an experiment. We'll also look at converting datetime objects to strings and back again.

Getting the Current Date and Time

The datetime.now() method returns the current date and time based on your computer's clock. In an environment where you have NumPy, pandas, and Matplotlib installed, start the Jupyter Qt console and enter the following (you will see a different date, for obvious reasons):

```
In [1]: from datetime import date, time, datetime

In [2]: now=datetime.now()

In [3]: now
Out[3]: datetime.datetime(2022, 10, 27, 17, 51, 26, 382489)

In [4]: type(now)
Out[4]: datetime.datetime
```

The now() method returns dates in ISO 8601 format (year-month-day). ISO 8601 is the global standard format for numeric dates.

The now variable represents the datetime data type. Other types for storing date and time information are shown in Table 20-1.

Table 21-1: Data Types in the Python datetime Module

Datatype	Description
date	Gregorian calendar date in year, month, day format
datetime	Combined date and time types
time	Twenty-four-hour (military) time in hours, minutes, seconds, and microseconds
timedelta	Difference between two datetime objects in days, seconds, and microseconds
tzinfo	Time zone information

To access date and time data in the now object, or any other timestamp, use its datetime attributes, called with dot notation:

```
In [5]: now.day
Out[5]: 27

In [6]: now.hour
Out[6]: 17

In [7]: now.minute
Out[7]: 51

In [8]: now.microsecond
Out[8]: 382489
```

To extract the date and time *objects*, call datetime methods with the same name:

```
In [9]: now.date()
Out[9]: datetime.date(2022, 10, 27)

In [10]: now.time()
Out[10]: datetime.time(17, 51, 26, 382489)
```

Assigning Timestamps and Calculating Time Delta

To assign a timestamp to a variable, pass datetime() the date and time in the year-month-day-hour-minute-second-microsecond format:

```
In [11]: ts = datetime(1976, 7, 4, 0, 0, 1, 1)
```

To view it as a string, pass the variable to Python's built-in str() function:

```
In [12]: str(ts)
Out[12]: '1976-07-04 00:00:01.000001'
```

If you're not interested in time data, just pass datetime() the date:

```
In [13]: ts = datetime(1976, 7, 4)

In [14]: str(ts)
Out[14]: '1976-07-04 00:00:00'
```

A timedelta object represents a *duration*, or the difference between two dates or times. Subtracting two datetime objects yields the elapsed time. To demonstrate, let's calculate Python creator Guido van Rossum's age on October 28, 2022:

```
In [15]: delta = datetime(2022, 10, 28) - datetime(1956, 1, 31)

In [16]: delta
Out[16]: datetime.timedelta(days=24377)

In [17]: age = delta.days / 365.2425

In [18]: int(age)
Out[18]: 66
```

If you include both date and time information, the timedelta object will present days, seconds, and microseconds, which are the only units stored internally:

```
In [19]: dt1 = datetime(2022, 10, 28, 10, 36, 59, 3)

In [20]: dt2 = datetime(1956, 1, 31, 0, 0, 0, 0)

In [21]: delta = dt1 - dt2

In [22]: delta
Out[22]: datetime.timedelta(days=24377, seconds=38219, microseconds=3)
```

The timedelta object supports arithmetic operations like addition, subtraction, multiplication, division, modulus, and more. To see the complete list of supported operations, visit *https://pandas.pydata.org/pandas-docs/stable/reference/api/pandas.Timedelta.html.*

Formatting Dates and Times

As you've seen, the datetime output isn't very human friendly. Converting it to a string using the str() function helps, but you can accomplish even more by using the datetime strftime() method.

The strftime() method uses C programming language (ISO C89)–compatible specifications, or *directives*, preceded by the % sign. Some of the most useful directives are listed in Table 21-2.

Table 21-2: Selected Datetime Format Specifications

Directive	Description	Examples
%a	Weekday as abbreviated name	Sun, So, Mon, Mo, Sat, Sa
%A	Weekday as full name	Sunday, Sonntag
%d	Two-digit weekday	01, 02, . . ., 05
%b	Month as abbreviated name	Jan, Feb, Dec, Dez
%B	Month as full name	February, Februar
%m	Two-digit month	01, 02, . . ., 31
%y	Two-digit year	00, 01, . . ., 99
%Y	Year with century as a decimal number	0001, . . . 2022, . . . 9999
%H	Twenty-four-hour clock hour	01, 02, . . ., 23
%I	Twelve-hour clock hour	01, 02, . . ., 12
%p	AM or PM	AM, am, PM, pm
%M	Two-digit minute	01, 02, . . ., 59
%S	Two-digit second (60, 61 account for leap seconds)	01, 02, . . ., 59
%f	Microsecond as decimal number (zero-padded six digits)	000000, 000001, . . ., 999999
%w	Integer weekday with 0 for Sunday	0, 1, . . ., 6
%W	Week number of year (Monday = 1st day of week; days before 1st Monday are week 0)	00, 01, . . . 53
%U	Week number of year (Sunday = 1st day of week; days before 1st Sunday are week 0)	00, 01, . . . 53
%Z	Time zone name (empty is naive object)	(empty), UTC, GMT
%c	Appropriate date and time representation for locale	Wed Mar 30 09:14:12 2022
%x	Appropriate date representation for locale	07/31/1984, 31.07.1984
%X	Appropriate time representation for locale	13:30:15
%F	Shortcut for %Y-%m-%d format	2022-03-30
%D	Shortcut for %m/%d/%y format	03/30/22

Bold = locale-specific date formatting

In the console, enter the following to see some example formats:

```
In [23]: now = datetime.now()

In [24]: now.strftime('%m/%d/%y %H:%M')
Out[24]: '10/30/22 09:14'
```

```
In [25]: now.strftime('%x')
Out[25]: '10/30/22'

In [26]: now.strftime('%A, %B %d, %Y')
Out[26]: 'Sunday, October 30, 2022'

In [27]: now.strftime('%c')
Out[27]: 'Sun Oct 30 09:14:00 2022'
```

You can find more formatting directives at *https://pandas.pydata.org/
pandas-docs/stable/reference/api/pandas.Period.strftime.html*.

Converting Strings to Dates and Times

Sometimes, you might need to import date and time information from a
file rather than creating it yourself. If the input data is in string format,
you'll need to convert the strings into dates. This is basically the opposite
of the operation we performed in the previous section, and you can do this
using either the dateutil.parser.parse() method or the datetime.strptime()
method. The third-party dateutil date utility package extends the datetime
module and is automatically installed with pandas.

For common datetime formats, use the parse() method for convenience.
In the console, enter the following:

```
In [28]: from dateutil.parser import parse
```

The parse() method can handle most date representations. If you enter
the month before the day, as in the US, it will honor this convention in the
datetime object:

```
In [29]: parse('Oct 31, 2022, 11:59 PM')
Out[29]: datetime.datetime(2022, 10, 31, 23, 59)
```

For locales where the day comes before the month, set the dayfirst
argument to True:

```
In [30]: parse('2/10/2022', dayfirst=True)
Out[30]: datetime.datetime(2022, 10, 2, 0, 0)
```

Let's look at a real-world example. Suppose that you've recorded, by
date, the type of animal captured in a trail camera. You've loaded the data
as a list and want to replace the dates in string format with datetime objects:

```
In [31]: data = ['2022/10/31', 'bobcat',
                 '2022/11/1', 'fox',
                 '2022/11/2', ['bobcat', 'opposum']]

In [32]: data_dt = data.copy()
```

```
In [33]: for i, date in enumerate(data):
   ...:     if i % 2 == 0:
   ...:         data_dt[i] = parse(date)

In [34]: data_dt
Out[34]:
[datetime.datetime(2022, 10, 31, 0, 0),
 'bobcat',
 datetime.datetime(2022, 11, 1, 0, 0),
 'fox',
 datetime.datetime(2022, 11, 2, 0, 0),
 ['bobcat', 'opposum']]
```

In this example, we made a copy (date_dt) of the initial data list and then looped through the data list using the built-in enumerate() function to get both the item in the list and its index. If the index is an even number, which corresponds to the locations of the dates, we parsed the date at that location in the data_dt list. Now we have the original data plus a version for which the dates are datetime objects.

Although parse() is useful for common, known date formats, it can't handle every situation. For edge cases, you'll need to use the datetime module's strptime() method and pass it the proper format specification. For example, Suppose that your lab assistant input a bunch of dates using an underscore to separate the date components:

```
In [35]: date = '2022_10_31'
```

The parse() method is unable to recognize this format and will raise an error (ParserError: Unknown string format: 2022_10_31). To handle this non-standard format, use strptime() with directives from Table 21-2. Be sure to place the underscores in the same relative position:

```
In [36]: datetime.strptime(date, '%Y_%m_%d')
Out[36]: datetime.datetime(2022, 10, 31, 0, 0)
```

As with parse(), you can convert a sequence of dates using strptime(). Here's an example using list comprehension, in place of a for loop:

```
In [37]: dates = ('8/11/84', '9/11/84', '10/11/84')

In [38]: dates_dt = [datetime.strptime(date, '%m/%d/%y') for date in dates]

In [39]: dates_dt
Out[39]:
[datetime.datetime(1984, 8, 11, 0, 0),
 datetime.datetime(1984, 9, 11, 0, 0),
 datetime.datetime(1984, 10, 11, 0, 0)]
```

For more on the dateutil package, visit *https://pypi.org/project/python -dateutil*.

Plotting with datetime Objects

Plotting dates can be messy due to the tendency of long date labels to overlap. In addition, standard plotting defaults don't consider proper time intervals when displaying ticks. To see an example using Matplotlib, in the console, enter the following:

```
In [40]: import datetime as dt
    ...: import numpy as np
    ...: import matplotlib.pyplot as plt

In [41]: dates = [dt.date(2022, 1, 31),
    ...:          dt.date(2022, 2, 28),
    ...:          dt.date(2022, 3, 31),
    ...:          dt.date(2022, 4, 30)]

In [42]: obs = [5, 12, 25, 42]

In [43]: plt.plot(dates, obs);
```

This produces the unreadable results in Figure 21-1.

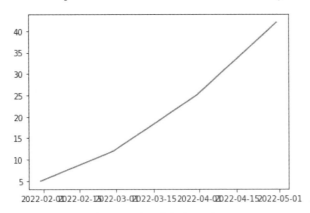

Figure 21-1: Overlapping date labels on the x-axis

To handle dates when plotting, you must inform Matplotlib that it's dealing with datetime objects by importing the matplotlib.dates module. This specialized module is built on the datetime and third-party dateutil modules. Among its sophisticated plotting capabilities, it helps you define time scales using *locator* methods, which find and comprehend the types of dates you're using, such as months and years.

Let's rebuild the previous plot using matplotlib.dates. Remember to use CTRL-ENTER to prevent early execution when entering the code in line In [48]:

```
In [44]: import matplotlib.dates as mdates

In [45]: months = mdates.MonthLocator()
```

```
In [46]: days = mdates.DayLocator()

In [47]: date_fmt = mdates.DateFormatter('%Y-%m')

In [48]: fig, ax = plt.subplots()
    ...: plt.plot(dates, obs)
    ...: ax.xaxis.set_major_locator(months)
    ...: ax.xaxis.set_major_formatter(date_fmt)
    ...: ax.xaxis.set_minor_locator(days)
```

This produces the results in Figure 21-2.

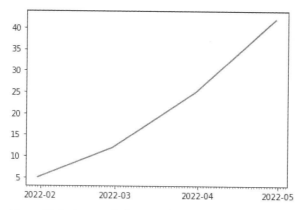

Figure 21-2: A properly formatted plot of dates

The date labels are now readable and, if your eyesight is good enough, you'll be able to count the proper number of tick marks for each day of the month. Locator functions are also available for other units such as hours, minutes, seconds, and weekdays. To learn more, visit the module documentation at *https://matplotlib.org/stable/api/dates_api.html*.

Creating Naive vs. Aware Objects

Python datetime objects may be categorized as either *naive* or *aware* depending on whether they include time zone information. A naive object does not contain time zone information and can't locate itself relative to other datetime objects. With knowledge of metadata such as time zone and DST information, an aware object represents a specific and unambiguous moment in time that can be located with respect to other aware objects.

To generate an aware object, datetime and time objects have an optional time zone attribute, tzinfo, that is used to capture information about the offset from the *coordinated universal time (UTC)*, the time zone name, and whether DST is in effect.

UTC is the successor to Greenwich Mean Time (GMT) and represents the primary time standard by which the world regulates clocks and time. Precision is usually in milliseconds, but submicrosecond precision is possible when using satellite signals. UTC does not change with the seasons, nor is it affected by DST. By working in UTC, you can confidently share your work and remove the need for fiddly time zone and similar corrections.

Although it's possible for the `tzinfo` attribute to hold detailed, country-specific time zone information, the `datetime` module's `timezone` class can represent only simple time zones with fixed offsets from UTC, such as UTC itself or North American EST and EDT time zones.

To access more detailed time zone information, you can use the third-party pytz library (*https://pypi.org/project/pytz/*), which is wrapped by pandas. To make an aware timestamp, import pytz and pass the datatime method the pytz library's name for a time zone. You can find these names using the `common_timezones` attribute, shown here sliced to the last 10 items in the list:

```
In [49]: import pytz

In [50]: pytz.common_timezones[-10:]
Out[50]:
['Pacific/Wake',
 'Pacific/Wallis',
 'US/Alaska',
 'US/Arizona',
 'US/Central',
 'US/Eastern',
 'US/Hawaii',
 'US/Mountain',
 'US/Pacific',
 'UTC']
```

First, let's make an *aware* timestamp in UTC:

```
In [51]: aware = datetime(2022, 11, 2, 21, 15, 19, 426910, pytz.UTC)

In [52]: aware
Out[52]: datetime.datetime(2022, 11, 2, 21, 15, 19, 426910, tzinfo=<UTC>)
```

Note that the aware timestamp has time zone metadata (`tzinfo=<UTC>`).

To convert an existing unaware timestamp to an aware timestamp, call the `localize()` method on the pytz time zone and pass the method the datetime object, like so:

```
In [53]: unaware = datetime(2022, 11, 3, 0, 0, 0)

In [54]: aware = pytz.timezone('Europe/London').localize(unaware)

In [55]: aware
Out[55]: datetime.datetime(2022, 11, 3, 0, 0, tzinfo=<DstTzInfo 'Europe/
London' GMT0:00:00 STD>)
```

To convert from one time zone to another, you can use the `astimezone()` method:

```
In [56]: here = datetime(2022, 11, 3, 14, 51, 3,
                         tzinfo=pytz.timezone('US/Central'))

In [57]: there = here.astimezone(pytz.timezone('Europe/London'))
```

```
In [58]: there
Out[58]: datetime.datetime(2022, 11, 3, 20, 42, 3, tzinfo=<DstTzInfo 'Europe/
London' GMT0:00:00 STD>)
```

The pytz library will consider the idiosyncrasies of your local area, such as DST, when making the conversion.

NOTE *Because naive* datetime *objects are treated by many* datetime *methods as local times, it's preferable to use* aware *datetimes to represent times in UTC. As such, the recommended way to create an object representing the current time in UTC is by calling* datetime.now(timezone.utc).

TEST YOUR KNOWLEDGE

1. Which date is written in the global standard numeric date format?

   a. 23-2-2021

   b. 2021-2-23

   c. 2-23-2021

   d. 23/2/21

2. Which methods convert a string representation of a date into a datetime object?

   a. strftime()

   b. str()

   c. strptime()

   d. parse()

3. Which directive produces the format '03/30/2022 21:09'?

   a. '%m/%d/%y %H:%M'

   b. '%M/%D/%Y %H:%m'

   c. '%m/%d/%Y %H:%M'

   d. '%m/%d/%y %H:%M'

4. What is the global time standard?

   a. pytz

   b. US/Eastern

   c. UTC

   d. GMT

(continued)

5. Which method lets you convert naive datetime objects to a new time zone?

   a. `mdates()`

   b. `parse()`

   c. `timedelta()`

   d. `localize()`

Time Series and Date Functionality with pandas

As you might expect, pandas has extensive capabilities for working with time series. This functionality is based on the NumPy `datetime64` and `timedelta64` data types with nanosecond resolution. In addition, features have been consolidated from many other Python libraries, and new functionality has been developed. With pandas, you can load time series; convert data to the proper datetime format; generate ranges of datetimes; index, merge, and resample both fixed- and irregular-frequency data; and more.

The pandas library uses four general time-related concepts. These are date times, time deltas, time spans, and date offsets (Table 21-3). Except for date offsets, each time concept has a *scalar* class, for single observations, along with an associated *array* class, which serves as an index structure.

Table 21-3: Time-Related Concepts in pandas

Concept	Scalar class	Array class	Data type	Creation method
Date times	Timestamp	DatetimeIndex	datetime64[ns] datetime64[ns, tz]	to_datetime or date_range
Time deltas	Timedelta	TimedeltaIndex	timedelta64[ns]	to_timedelta or timedelta_range
Time spans	Period	PeriodIndex	period[freq]	Period or period_range
Date offsets	DateOffset	None	None	Dateoffset

A *date time* represents a specific date and time with time zone support. It's similar to `datetime.datetime` from the Python standard library. A *time delta* is an absolute time duration, similar to `datetime.timedelta` from the standard library. *Time spans* are a period defined by a point in time and its associated frequency (daily, monthly, and so on). A *date offset* represents a relative time duration that respects calendar arithmetic.

In the sections that follow, we'll look at these various concepts and the methods used to create them. For more detail, you can visit the official documentation at *https://pandas.pydata.org/pandas-docs/stable/user_guide/timeseries.html.*

Parsing Time Series Information

To create a timestamp representing the time for a particular event, use the Timestamp class:

```
In [59]: import pandas as pd
In [60]: ts = pd.Timestamp('2021, 2, 23 00:00:00')

In [61]: ts
Out[61]: Timestamp('2021-02-23 00:00:00')
```

Likewise, to create a DatetimeIndex object, use the DatetimeIndex class:

```
In [62]: dti = pd.DatetimeIndex(['2022-03-31 14:39:00',
                                 '2022-04-01 00:00:00'])

In [63]: dti
Out[63]: DatetimeIndex(['2022-03-31 14:39:00', '2022-04-01 00:00:00'],
dtype='datetime64[ns]', freq=None)
```

For existing data, the pandas to_datetime() method converts scalar, array-like, dictionary-like, and pandas series or DataFrame objects to pandas datetime64[ns] objects. This lets you easily parse time series information from various sources and formats.

To see what I'm talking about, in the console, enter the following:

```
In [64]: import numpy as np
    ...: from datetime import datetime
    ...: import pandas as pd

In [65]: dti = pd.to_datetime(["2/23/2021",
    ...:                       np.datetime64("2021-02-23"),
    ...:                       datetime(2022, 2, 23)])

In [66]: dti
Out[66]: DatetimeIndex(['2021-02-23', '2021-02-23', '2022-02-23'],
dtype='datetime64[ns]', freq=None)
```

In this example, we passed a list of dates in three different formats to the to_datetime() method. These included a string, a NumPy datetime64 object, and a Python datetime object. The method returned a pandas DatetimeIndex object that consistently stores the dates as datetime64[ns] objects in ISO 8601 format (year-month-day).

The method can accommodate times as well as dates:

```
In [67]: dates = ['2022-3-31 14:39:00',
   ...:           '2022-4-1 00:00:00',
   ...:           '2022-4-2 00:00:20',
   ...:           '']

In [68]: dti = pd.to_datetime(dates)

In [69]: dti
Out[69]:
DatetimeIndex(['2022-03-31 14:39:00', '2022-04-01 00:00:00',
'2022-04-02 00:00:20', 'NaT'],
dtype='datetime64[ns]', freq=None)
```

In this example, we passed a list of both dates and times, which were all correctly converted. Note that we included an empty item ('') at the end of the list. The to_datetime() method converted this entry into a NaT (Not a Time) value, which is the timestamp equivalent of the pandas NaN (Not a Number) value that you learned about in the previous chapter.

The to_datetime() method also works with pandas DataFrames. Let's look at an example in which you have recorded (in Microsoft Excel) the date and time a trail camera captured an image of an animal. You've exported the spreadsheet as a *.csv* file that you now want to load and parse using pandas.

To create the *.csv* file, in a text editor such as Notepad or TextEdit, enter the following and then save it as *camera_1.csv*:

```
Date,Obs
3/30/22 11:43 PM,deer
3/31/22 1:05 AM,fox
4/1/22 2:54 AM,cougar
```

Back in the console, enter the following to read the file in as a Data-Frame (substitute your path to the *.csv* file):

```
In [70]: csv_df = pd.read_csv('C:/Users/hanna/camera_1.csv')

In [71]: csv_df
Out[71]:
            Date     Obs
0 3/30/22 11:43 PM   deer
1  3/31/22 1:05 AM    fox
2   4/1/22 2:54 AM cougar
```

To convert the Date column to ISO 8601 format, enter the following:

```
In [72]: csv_df['Date'] = pd.to_datetime(csv_df['Date'])

In [73]: csv_df
Out[73]:
```

```
          Date      Obs
0 2022-03-30 23:43:00    deer
1 2022-03-31 01:05:00     fox
2 2022-04-01 02:54:00  cougar
```

These datetimes were recorded in the Eastern US time zone, but that information is not encoded. To make the datetimes aware, first make the following imports:

```
In [74]: import pytz
```

Next, assign a variable to a pytz tzfile object and then pass the variable to the localize() method:

```
In [75]: my_tz = pytz.timezone('US/Eastern')
In [76]: csv_df['Date'] = csv_df['Date'].dt.tz_localize(my_tz)
```

You can do all this in one line, but using a my_tz variable makes the code more readable and less likely to wrap. To check the results, print the Date column:

```
In [77]: print(csv_df['Date'])
0 2022-03-30 23:43:00-04:00
1 2022-03-31 01:05:00-04:00
2 2022-04-01 02:54:00-04:00
Name: Date, dtype: datetime64[ns, US/Eastern]
```

Even though it's a good idea to work in UTC, it's also important to have *meaningful* time data. For example, you'll probably want to study when these animals are on the prowl in *local* time, so you'll want to preserve the times recorded in the Eastern US. In this case, you'll want to make a new "UTC-aware" column based on the Date column so that you can have the best of both worlds. Because the Date column is now aware of its time zone, you must use tz_convert() instead of tz_localize():

```
In [78]: csv_df['Date_UTC'] = csv_df['Date'].dt.tz_convert(pytz.utc)
```

Print the columns to verify the conversion:

```
In [79]: print(csv_df[['Date', 'Date_UTC']])
                      Date                 Date_UTC
0 2022-03-30 23:43:00-04:00 2022-03-31 03:43:00+00:00
1 2022-03-31 01:05:00-04:00 2022-03-31 05:05:00+00:00
2 2022-04-01 02:54:00-04:00 2022-04-01 06:54:00+00:00
```

NOTE *To remove time zone information from a datetime so that it becomes naive, pass the* tz_convert() *method* None, *like so:* csv_df['Date'] = csv_df['Date'].dt.tz_convert(None).

Finally, if you look at the previous printout of the `csv_df` DataFrame, you'll see that the index values range from 0 to 2. This is by default, but there's no reason why you can't use `datetime` values as the index instead. In fact, `datetime` indexes can be helpful when doing things like plotting. So, let's make the `Date_UTC` column the index for the DataFrame. In the console, enter the following:

```
In [80]: csv_df = csv_df.set_index('Date_UTC')
Out[80]:
                                            Date Obs
Date_UTC
2022-03-31 03:43:00+00:00  2022-03-31 03:43:00 deer
2022-03-31 05:05:00+00:00  2022-03-31 05:05:00 fox
2022-04-01 06:54:00+00:00  2022-04-01 06:54:00 cougar
```

To read more about the `to_datetime()` method, visit *https://pandas.pydata .org/pandas-docs/stable/reference/api/pandas.to_datetime.html*. You can find the documentation for `dt.tz_localize()` and `dt.tz_convert()` at *https://pandas .pydata.org/pandas-docs/stable/reference/api/pandas.Series.dt.tz_localize.html* and *https://pandas.pydata.org/pandas-docs/stable/reference/api/pandas.Series.dt.tz _convert.html*, respectively.

Creating Date Ranges

Time series with a *fixed* frequency occur often in science for jobs as diverse as sampling waveforms in signal processing, observing target behaviors in psychology, recording stock market movements in economics, and logging traffic flow in transportation engineering. Not surprisingly, pandas ships with many standardized frequencies and tools that generate them, resample them, and infer them.

The pandas `date_range()` method returns a `DatetimeIndex` object with a fixed frequency. To generate an index composed of days, pass it a start and end date, as follows:

```
In [81]: day_index = pd.date_range(start='2/23/21', end='3/1/21')

In [82]: day_index
Out[82]:
DatetimeIndex(['2021-02-23', '2021-02-24', '2021-02-25', '2021-02-26',
'2021-02-27', '2021-02-28', '2021-03-01'],
dtype='datetime64[ns]', freq='D')
```

You can also pass it either a start date or an end date, along with the number of periods to generate (such as a number of days). In the following example, we start with a timestamp for a certain observation and ask for six periods:

```
In [83]: day_index = pd.date_range(start='2/23/21 12:59:59', periods=6)

In [84]: day_index
Out[84]:
```

```
DatetimeIndex(['2021-02-23 12:59:59', '2021-02-24 12:59:59',
'2021-02-25 12:59:59', '2021-02-26 12:59:59',
'2021-02-27 12:59:59', '2021-02-28 12:59:59'],
dtype='datetime64[ns]', freq='D')
```

Note that the six datetimes represent days starting at 12:59:59. Normally, you want the days to start at midnight, so, pandas provides a handy `normalize` parameter to make this adjustment:

```
In [85]: day_index_normal = pd.date_range(start='2/23/21 12:59:59',
                                           periods=6,
                                           normalize=True)

In [86]: day_index_normal
Out[86]:
DatetimeIndex(['2021-02-23', '2021-02-24', '2021-02-25', '2021-02-26',
'2021-02-27', '2021-02-28'],
dtype='datetime64[ns]', freq='D')
```

After they're normalized to days, the output `datetime64` objects no longer include a time component.

By default, the date_range() method assumes that you want a *daily* frequency. Other frequencies are available, however, with many designed for business applications (such as the end of a business month, end of a business year, and so on).

Table 21-4 lists some of the time series frequencies more relevant for science. For the complete list, including financial frequencies, see "DateOffset objects" at *https://pandas.pydata.org/pandas-docs/stable/user_guide/timeseries.html*.

Table 21-4: Useful Time Series Frequencies

Freq string	Offset type	Description
N	Nano	By nanosecond
U	Micro	By microsecond
L or ms	Milli	By millisecond
S	Second	By second
T or min	Minute	By minute
H	Hour	By hour
D	Day	By calendar day
W-MON, W-TUE, . . .	Week	Weekly, optionally anchored on a day of the week
MS	MonthBegin	By first calendar day of month
M	MonthEnd	By last calendar day of month
Q	Quarter	By quarter year
AS-JAN, AS-FEB, . . .	YearBegin	Yearly, anchored on first calendar day of given month
A-JAN, A-FEB, . . .	YearEnd	Yearly, anchored on last calendar day of given month

To specify an offset type, pass a frequency string alias from Table 21-4 as the freq argument. You can also specify a time zone using the tz argument. Here's how to make an hourly frequency referenced to UTC:

```
In [87]: hour_index = pd.date_range(start='2/23/21',
                                     periods=6,
                                     freq='H',
                                     tz='UTC')

In [88]: hour_index
Out[88]:
DatetimeIndex(['2021-02-23 00:00:00+00:00', '2021-02-23 01:00:00+00:00',
'2021-02-23 02:00:00+00:00', '2021-02-23 03:00:00+00:00',
'2021-02-23 04:00:00+00:00', '2021-02-23 05:00:00+00:00'],
dtype='datetime64[ns, UTC]', freq='H')
```

For an existing time series, you can retrieve its frequency by using the freq attribute, as shown here:

```
In [89]: hour_index.freq
Out[89]: <Hour>
```

The frequencies shown in Table 21-4 represent *base* frequencies. Think of these as building blocks for alternative frequencies, such as bi-hourly. To make this new frequency, just place the integer 2 before the H in the freq argument, as follows:

```
In [90]: bi_hour_index = pd.date_range(start='2/23/21', periods=6, freq='2H')

In [91]: bi_hour_index
Out[91]:
DatetimeIndex(['2021-02-23 00:00:00', '2021-02-23 02:00:00',
'2021-02-23 04:00:00', '2021-02-23 06:00:00',
'2021-02-23 08:00:00', '2021-02-23 10:00:00'],
dtype='datetime64[ns]', freq='2H')
```

You can also combine offsets by passing frequency strings like '2H30min', like this:

```
In [92]: pd.date_range(start='2/23/21', periods=6, freq='2H30min')
Out[92]:
DatetimeIndex(['2021-02-23 00:00:00', '2021-02-23 02:30:00',
'2021-02-23 05:00:00', '2021-02-23 07:30:00',
'2021-02-23 10:00:00', '2021-02-23 12:30:00'],
dtype='datetime64[ns]', freq='150T')
```

Creating Periods

Timestamps associate data with points in time. Sometimes, however, data remains constant through a certain *time span*, such as a month, and you want to associate the data with that interval.

In pandas, regular intervals of time such as a day, month, year, and so on are represented by `Period` objects. With the `period_range()` method, `Period` objects can be collected into a sequence to form a `PeriodIndex`. You can specify a period's time span using the `freq` keyword with frequency aliases from Table 21-4.

Suppose that you want to keep track of a daily observation for the month of September 2022. First, use the `period_range()` method to create a time span with a frequency of days:

```
In [93]: p_index = pd.period_range(start='2022-9-1',
                                    end='2022-9-30',
                                    freq='D')
```

Next, create a pandas series and use the NumPy `random.randn()` method to generate some fake data on the fly. Note that the number of data points must equal the number of days in the index:

```
In [94]: ts = pd.Series(np.random.randn(30), index=p_index)

In [95]: ts
Out[95]:
2022-09-01  0.412853
2022-09-02  0.350678
2022-09-03  0.086216
--snip--
2022-09-28  1.944123
2022-09-29  0.311337
2022-09-30  0.906780
Freq: D, dtype: float64
```

You now have a time series, organized by day, for the month of September.

To shift a period by its own frequency, just add or subtract an integer. Here's an example using a yearly time span:

```
In [96]: year_index = pd.period_range(2001, 2006, freq='A-DEC')

In [97]: year_index
Out[97]: PeriodIndex(['2001', '2002', '2003', '2004', '2005', '2006'],
dtype='period[A-DEC]')

In [98]: year_index + 10
Out[98]: PeriodIndex(['2011', '2012', '2013', '2014', '2015', '2016'],
dtype='period[A-DEC]')
```

Using a frequency of `'A-DEC'` means that each year represents January 1 through December 31. Adding 10 shifted the periods up by 10 years. You can only perform arithmetic in this manner between `Period` objects with the *same* frequency.

Here's an example of making monthly periods:

```
In [99]: month_index = pd.period_range('2022-01-01', '2022-12-31', freq='M')

In [100]: month_index
Out[100]:
PeriodIndex(['2022-01', '2022-02', '2022-03', '2022-04', '2022-05', '2022-06',
'2022-07', '2022-08', '2022-09', '2022-10', '2022-11', '2022-12'],
dtype='period[M]')
```

With the asfreq() method, you can convert an existing period to another frequency. Here's an example in which we convert the month_index variable's period to hours, anchored on the first hour of each month:

```
In [101]: hour_index = month_index.asfreq('H', how='start')

In [102]: hour_index
Out[102]:
PeriodIndex(['2022-01-01 00:00', '2022-02-01 00:00', '2022-03-01 00:00',
'2022-04-01 00:00', '2022-05-01 00:00', '2022-06-01 00:00',
'2022-07-01 00:00', '2022-08-01 00:00', '2022-09-01 00:00',
'2022-10-01 00:00', '2022-11-01 00:00', '2022-12-01 00:00'],
dtype='period[H]')
```

To read more about the pandas Period class and the asfreq() method, visit *https://pandas.pydata.org/docs/reference/api/pandas.Period.html* and *https://pandas.pydata.org/docs/reference/api/pandas.Period.asfreq.html*, respectively.

Creating Time Deltas

The timedelta_range() method creates TimedeltaIndex objects. It behaves similarly to date_range() and period_range():

```
In [103]: pd.timedelta_range(start='1 day', periods = 5)
Out[103]: TimedeltaIndex(['1 days', '2 days', '3 days', '4 days', '5 days'],
dtype='timedelta64[ns]', freq='D')
```

In the television drama *Lost*, a character had to enter a code and push a button every 108 minutes to avert some unknown catastrophe. With the timedelta_range() method and a frequency argument, he could schedule his day around this requirement. Assuming he last pushed the button at midnight, he won't be getting much uninterrupted sleep:

```
In [104]: pd.timedelta_range(start="1 day", end="2 day", freq="108min")
Out[104]:
TimedeltaIndex(['1 days 00:00:00', '1 days 01:48:00', '1 days 03:36:00',
'1 days 05:24:00', '1 days 07:12:00', '1 days 09:00:00',
'1 days 10:48:00', '1 days 12:36:00', '1 days 14:24:00',
'1 days 16:12:00', '1 days 18:00:00', '1 days 19:48:00',
'1 days 21:36:00', '1 days 23:24:00'],
dtype='timedelta64[ns]', freq='108T')
```

Shifting Dates with Offsets

In addition to working with frequencies, you can import offsets and use them to shift `Timestamp` and `DatetimeIndex` objects. Here's an example in which we import the `Day` class and use it to shift a famous date:

```
In [105]: from pandas.tseries.offsets import Day

In [106]: apollo_11_moon_landing = pd.to_datetime('1969, 7, 20')

In [107]: apollo_11_splashdown = apollo_11_moon_landing + 4 * Day()
In [108]: print(f"{apollo_11_splashdown.month}/{apollo_11_splashdown.day}")
7/24
```

You can also import `DateOffset` class and then pass it the time span as an argument:

```
In [109]: from pandas.tseries.offsets import DateOffset

In [110]: ts = pd.Timestamp('2021-02-23 09:10:11')

In [111]: ts + DateOffset(months=4)
Out[111]: Timestamp('2021-06-23 09:10:11')
```

A nice thing about `DateOffset` objects is that they honor DST transitions. You just need to import the appropriate class from `pandas.tseries.offsets`. Here's an example of shifting one hour across the vernal DST transition in the US Central time zone:

```
In [112]: from pandas.tseries.offsets import Hour

In [113]: pre_dst_date = pd.Timestamp('2022-03-13 1:00:00', tz='US/Central')

In [114]: pre_dst_date
Out[114]: Timestamp('2022-03-13 01:00:00-0600', tz='US/Central')

In [115]: post_dst_date = pre_dst_date + Hour()

In [116]: post_dst_date
Out[116]: Timestamp('2022-03-13 03:00:00-0500', tz='US/Central')
```

Note that the final datetime (`03:00:00`) is *two hours* later than the starting datetime (`01:00:00`), even though you shifted it *one hour*. This is due to crossing the DST transition.

Along these lines, you can combine two time series even if they are in different time zones. The result will be in UTC, as pandas automatically keeps track of the equivalent UTC timestamps for each time series.

To see the long list of available offsets, visit *https://pandas.pydata.org/ pandas-docs/stable/reference/api/pandas.tseries.offsets.DateOffset.html.*

Indexing and Slicing Time Series

When you're working with time series data, it's conventional to use the time component as the index of a series or DataFrame so that you can perform manipulations with respect to the time element. Here, we make a series whose index represents a time series and whose data is the integers 0 through 9:

```
In [117]: ts = pd.Series(range(10), index=pd.date_range('2022',
                                                         freq='D',
                                                         periods=10))

In [118]: ts
Out[118]:
2022-01-01 0
2022-01-02 1
2022-01-03 2
2022-01-04 3
2022-01-05 4
2022-01-06 5
2022-01-07 6
2022-01-08 7
2022-01-09 8
2022-01-10 9
Freq: D, dtype: int64
```

Even though the indexes are now dates, you can slice and dice the series, just as with integer indexes. For example, to select every other row, enter the following:

```
In [119]: ts[::2]
Out[119]:
2022-01-01 0
2022-01-03 2
2022-01-05 4
2022-01-07 6
2022-01-09 8
Freq: 2D, dtype: int64
```

To select the data associated with the January 5, index the series using that date:

```
In [120]: ts['2022-01-05']
Out[120]: 4
```

Conveniently, you don't need to enter the date in the same format that it was input. Any string interpretable as a date will do:

```
In [121]: ts['1/5/2022']
Out[121]: 4

In [122]: ts['January 5, 2022']
Out[122]: 4
```

Duplicate dates will produce a *slice* of the series showing all values associated with that date. Likewise, you will see all the rows in a DataFrame indexed by the same date using the syntax: `dataframe.loc['datetime_index']`.

Additionally, if you have a time series with multiple years, you can index based on the year and retrieve all the indexes and data that include that year. This also works for other timespans, such as months.

Slicing works the same way. You can use timestamps not explicitly included in the time series, such as 2021-12-31:

```
In [123]: ts['2021-12-31':'2022-1-2']
Out[123]:
2022-01-01 0
2022-01-02 1
Freq: D, dtype: int64
```

In this case, we started indexing with December 31, 2021, which precedes the dates in the time series.

NOTE *Remember that pandas is based on NumPy, so slicing creates* views *rather than copies. Any operation you perform on a view will change the* source *series or DataFrame.*

If you want the datetime component to be the data instead of the index, leave off the index argument when creating the series:

```
In [124]: pd.Series(pd.date_range('2022', freq='D', periods=3))
Out[124]:
0 2022-01-01
1 2022-01-02
2 2022-01-03
dtype: datetime64[ns]
```

The result is a pandas series with an integer index and the dates treated as data.

Resampling Time Series

The process of converting the frequency of a time series to a different frequency is called *resampling*. This can involve *downsampling*, by which you aggregate data to a lower frequency, perhaps to reduce memory requirements or see trends in the data; *upsampling*, wherein you move to a higher frequency, perhaps to permit mathematical operations between two datasets with different resolutions; or simple resampling, for which you keep the same frequency but change the anchor point from, say, the start of the year (AS-JAN) to the year end (A-JAN).

In pandas, resampling is accomplished by calling the `resample()` method on a pandas object using dot notation. Some of its commonly used parameters are listed in Table 21-5. To see the complete list, visit *https://pandas .pydata.org/pandas-docs/stable/reference/api/pandas.DataFrame.resample.html*. Both series and `dataframe` objects use the same parameters.

Table 21-5: Useful Parameters of the pandas `resample()` Method

Parameter	Description
freq	DateOffset or Timedelta object, or string, indicating resampling frequency (such as 'D', 'Q', '10min').
axis	Axis on which to resample (0 or 'index', 1 or 'columns'). Defaults to 0.
closed	When downsampling, indicates which interval end is inclusive, either 'right' or 'left'. The default value changes depending on freq type.
label	When downsampling, which bin edge to use to label the result, either 'right' or 'left'. The default value changes depending on freq type.
convention	For PeriodIndex only, controls whether to use the start or end of freq when converting frequencies from low to high. Defaults to 'start'.
kind	Pass 'timestamp' to convert the resulting index to a DateTimeIndex or 'period' to convert it to a PeriodIndex. By default, the input representation is retained.
on	For a DataFrame, specifies the column to use instead of index for resampling. The column must be datetime-like.

Upsampling

Upsampling refers to resampling to a *shorter* time span, such as from daily to hourly. This creates bins with NaN values that must be filled; for example, as with the forward-fill and backfill methods `ffill()` and `bfill()`. This two-step process can be accomplished by chaining together the calls to the resample and fill methods.

To illustrate, let's make a toy dataset with yearly values and expand it to quarterly values. This might be necessary when, say, production targets go up every year, but progress must be tracked against quarterly production. In the console, enter the following:

```
In [125]: import pandas as pd

In [126]: dti = pd.period_range('2021-02-23', freq='Y', periods=3)

In [127]: df = pd.DataFrame({'value': [10, 20, 30]}, index=dti)

In [128]: df.resample('Q').ffill()
```

After importing pandas, establish an annual PeriodIndex named dti. Next, create the DataFrame and pass it a dictionary with the values in list form. Then, set the index argument to the dti object. Call the resample() method and pass it Q, for quarterly, and then call the ffill() method, chained to the end.

The results of this code are broken down in Figure 21-3, which, from left to right, shows the original DataFrame, the resampling results, and the fill results. The original annual values are shown in bold.

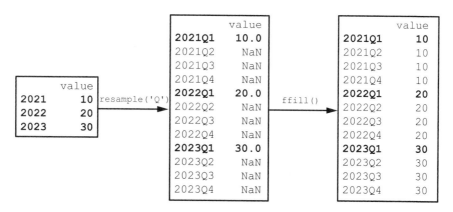

Figure 21-3: Resampling a DataFrame with a yearly range to a quarterly range using resample() followed by ffill()

The resample() method builds the new quarterly index and fills the new rows with NaN values. Calling ffill() fills the empty rows going "forward." What you're saying here is, "The value for the first quarter of each year (Q1) is the value to use for all quarters within that year."

Backfilling does the opposite and assumes that the value at the start of each new year (Q1) should apply to the quarters in the previous year *excluding* the previous first quarter:

```
In [129]: df.resample('Q').bfill()
```

The execution of this code is described by Figure 21-4. Again, original annual values are shown in bold.

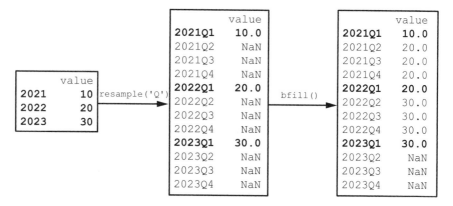

Figure 21-4: Resampling a DataFrame with a yearly range to a quarterly range using resample() followed by bfill()

In this case, the values associated with the first quarter are "back filled" to the previous three quarters. You must be careful, however, as "leftover" NaNs can occur. You can see these at the end of the value column in the right-hand DataFrame in Figure 21-4. The last three values are unchanged because no 2024Q1 data was available to set the value.

To fill the missing data, let's assume that the values keep increasing by 10 each quarter and rerun the code using the `fillna()` method chained to the end. Pass it 40 to fill the remaining holes:

```
In [130]: df.resample('Q').bfill().fillna(40)
Out[130]:
        value
2021Q1  10.0
2021Q2  20.0
2021Q3  20.0
2021Q4  20.0
2022Q1  20.0
2022Q2  30.0
2022Q3  30.0
2022Q4  30.0
2023Q1  30.0
2023Q2  40.0
2023Q3  40.0
2023Q4  40.0
```

Both `bfill()` and `ffill()` are synonyms for the `fillna()` method. You can read more about it at *https://pandas.pydata.org/pandas-docs/stable/reference/api/pandas.DataFrame.fillna.html*.

Downsampling

Downsampling refers to resampling from a higher frequency to a lower frequency, such as from minutes to hours. Because multiple samples must be combined into one, the `resample()` method is usually chained to a method for *aggregating* the data (see Table 21-6).

Table 21-6: Useful Aggregation Methods in pandas

Method	Description
count()	Returns the number of non-null values
max()	Returns the maximum value
mean()	Returns the arithmetic mean of the values
median()	Returns the median of the values
min()	Returns the minimum value
std()	Returns the standard deviation of the values
sum()	Returns the sum of the values
var()	Returns the variance of the values

To practice downsampling, let's use a real-world dataset from "The COVID Tracking Project" at *The Atlantic*. This dataset includes COVID-19 statistics from March 3, 2020, to March 7, 2021.

To reduce the size of the dataset, we'll download the data for just the state of Texas. Navigate to *https://covidtracking.com/data/download/*, scroll down, and then click the link for **Texas**. For convenience, I recommend moving this file to the same folder from which you launched Jupyter Qt console; this prevents the need for a file path when loading the data.

To begin, load the data as a pandas DataFrame. The input file has many columns of data that we don't need, so we'll select only the date and deathIncrease columns. The latter column is the number of COVID-related deaths for the day.

```
In [131]: df = pd.read_csv("texas-history.csv",
     ...:                   usecols=['date','deathIncrease'])

In [132]: df.head()
Out[132]:
        date deathIncrease
0 2021-03-07            84
1 2021-03-06           233
2 2021-03-05           256
3 2021-03-04           315
4 2021-03-03           297
```

It's good to keep an eye on what's happening to the data by calling the head() method on the DataFrame, which returns the first five rows by default. Here, we see that dates are organized in *descending* order, but we generally use and plot datetime data in *ascending* order. So, call the pandas sort_values() method, pass it the column name, and set the ascending argument to True:

```
In [133]: df = df.sort_values('date', ascending=True)

In [134]: df.head()
Out[134]:
          date deathIncrease
369 2020-03-03             0
368 2020-03-04             0
367 2020-03-05             0
366 2020-03-06             0
365 2020-03-07             0
```

Next, the dates look like dates, but are they? Check the DataFrame's dtypes attribute to confirm one way or the other:

```
In [135]: df.dtypes
Out[135]:
date object
deathIncrease int64
dtype: object
```

They're not. This is important because the resample() method works only with objects that have a datetime-like index, such as DatetimeIndex, PeriodIndex, or TimedeltaIndex. We'll need to change their type and set them as the DataFrame's index, replacing the current integer values. We'll also drop the date column because we no longer need it:

```
In [136]: df = df.set_index(pd.DatetimeIndex(df['date'])).drop('date',
    ...:                                                        axis=1)
```

```
In [137]: df.head()
Out[137]:
            deathIncrease
date
2020-03-03              0
2020-03-04              0
2020-03-05              0
2020-03-06              0
2020-03-07              0
```

At this point, we've wrangled the data so that our DataFrame uses a DatetimeIndex with dates in ascending order. Let's see how it looks by making a quick plot using pandas plotting, which is quick and easy for data exploration:

```
In [138]: df.plot();
```

This returns the plot depicted in Figure 21-5.

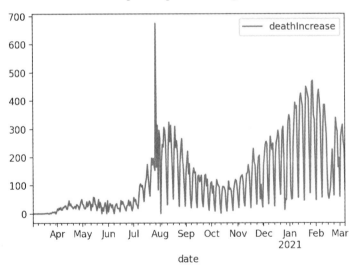

Figure 21-5: Texas COVID-19-related daily deaths for the period March 3, 2020, through March 7, 2021

One aspect of Figure 21-5 that really stands out is the spike in values near the start of August 2020. Because this is clearly a *maximum* value, you can easily retrieve the value and its date index by using the max() and idxmax() methods, respectively:

```
In [139]: print(df.max(), df.idxmax())
deathIncrease 675
dtype: int64 deathIncrease 2020-07-27
dtype: datetime64[ns]
```

This is most likely an anomalous value, especially given that the CDC records only 239 deaths on this date, which is more consistent with the adjacent data (see *https://covid.cdc.gov/covid-data-tracker/#trends_dailydeaths/*). Let's use the CDC value going forward. To change the DataFrame, apply the .loc indexer, passing it the date (index) and column name, as follows:

```
In [140]: df.loc['2020-7-27', 'deathIncrease'] = 239

In [141]: df.plot();
```

The spike is gone now, and the plot looks more reasonable, as demonstrated in Figure 21-6.

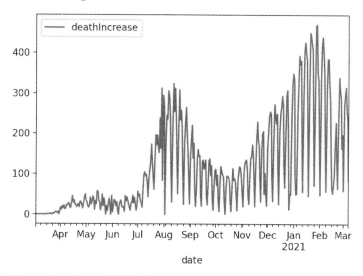

Figure 21-6: Texas COVID-19-related daily deaths with the anomalous spike removed

Another thing that's noticeable is the "sawtooth" nature of the curve caused by periodic oscillations in the number of deaths. These oscillations have a high frequency, and it's doubtful that the disease progressed in this manner.

To investigate this anomaly, make a new DataFrame that includes a column for weekdays:

```
In [142]: df_weekdays = df.copy()
```

```
In [143]: df_weekdays['weekdays'] = df.index.day_name()
```

Now, print out multiple weeks' worth of data using pandas' iloc[] indexing:

```
In [144]: print(df_weekdays.iloc[90:115])
          deathIncrease    weekdays
date
2020-06-01             6       Monday
2020-06-02            20      Tuesday
2020-06-03            36    Wednesday
2020-06-04            33     Thursday
2020-06-05            21       Friday
2020-06-06            31     Saturday
2020-06-07            11       Sunday
2020-06-08             0       Monday
2020-06-09            23      Tuesday
2020-06-10            32    Wednesday
2020-06-11            35     Thursday
2020-06-12            19       Friday
2020-06-13            18     Saturday
2020-06-14            19       Sunday
2020-06-15             7       Monday
2020-06-16            46      Tuesday
2020-06-17            33    Wednesday
2020-06-18            43     Thursday
2020-06-19            35       Friday
2020-06-20            25     Saturday
2020-06-21            17       Sunday
2020-06-22            10       Monday
2020-06-23            28      Tuesday
2020-06-24            29    Wednesday
2020-06-25            47     Thursday
```

As I've highlighted in gray, the lowest reported number of deaths consistently occurs on a Monday, and the Sunday results also appear suppressed. This suggests a reporting issue over the weekend, with a one-day time lag. You can read more about this reporting phenomenon at *https://www.ncbi.nlm.nih.gov/pmc/articles/PMC7363007/*.

NOTE *If you were on social media during the pandemic, you might have noticed people questioning the veracity of COVID data based on plots like Figure 21-5. This is a good example of how, with a simple application of data science, you can easily solve mysteries and quickly quell conspiracy theories.*

Because the oscillations occur *weekly*, downsampling from daily to weekly should merge the low and high reports and smooth the curve. Enter the following to test the hypothesis:

```
In [145]: df.resample('W').sum().plot();
```

This produces the plot in Figure 21-7. The high-frequency oscillations are gone.

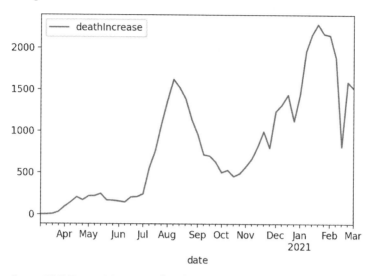

Figure 21-7: Texas COVID-19-related weekly deaths for the period March 3, 2020, through March 7, 2021

Now, let's downsample to a monthly period:

```
In [146]: df.resample('m').sum().plot();
```

This produces the even smoother plot in Figure 21-8.

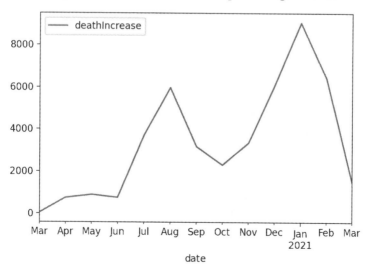

Figure 21-8: Texas COVID-19-related monthly deaths for the period March 3, 2020, through March 7, 2021

Note that you can also downsample to custom periods, such as `'4D'`, for every four days.

Changing the Start Date When Resampling

So far, we've been taking the *default* origin (start date) when aggregating intervals, but this can lead to unwanted results. Here's an example:

```
In [147]: raw_dict = {'2022-02-23 09:00:00': 100,
     ...:             '2022-02-23 10:00:00': 200,
     ...:             '2022-02-23 11:00:00': 100,
     ...:             '2022-02-23 12:00:00': 300}

In [148]: ts = pd.Series(raw_dict)

In [149]: ts.index = pd.to_datetime(ts.index)

In [150]: ts.resample('2H').sum()
Out[150]:
2022-02-23 08:00:00 100
2022-02-23 10:00:00 300
2022-02-23 12:00:00 300
Freq: 2H, dtype: int64
```

Despite the first data point being recorded *at* 9 AM, the resampled sums *start* at 8 AM. This is because the default for aggregated intervals is 0, causing the two-hour (`'2H'`) frequency timestamps to be 00:00:00, . . . 08:00:00, 10:00:00, and so on, skipping 09:00:00.

To force the output range to start at 09:00:00, pass the method's origin argument `'start'`. Now it should use the actual start of the time series:

```
In [151]: ts.resample('2H', origin='start').sum()
Out[151]:
2022-02-23 09:00:00 300
2022-02-23 11:00:00 400
Freq: 2H, dtype: int64
```

The aggregation starts at 9 AM, as desired.

Resampling Irregular Time Series Using Interpolation

Scientific observations are often irregular in nature. After all, wildebeests don't show up at waterholes on a fixed schedule. Fortunately, resampling works the same whether a time series has an irregular or fixed frequency.

As with upsampling, *regularizing* a time series will generate new timestamps with empty values. Previously, we filled these blank values using backfilling and front filling. In the next example, we'll use the pandas interpolate() method.

Let's begin by generating a list of irregularly spaced datetimes with a resolution measured in seconds:

```
In [152]: raw = ['2021-02-23 09:46:48',
     ...:        '2021-02-23 09:46:51',
```

```
...:        '2021-02-23 09:46:53',
...:        '2021-02-23 09:46:55',
...:        '2021-02-23 09:47:00']
```

Next, in a single line, create a pandas series object where the index is the datetime string converted to a DatetimeIndex:

```
In [153]: ts = pd.Series(np.arange(5), index=pd.to_datetime(raw))

In [154]: ts
Out[154]:
2021-02-23 09:46:48 0
2021-02-23 09:46:51 1
2021-02-23 09:46:53 2
2021-02-23 09:46:55 3
2021-02-23 09:47:00 4
dtype: int32
```

Now, resample this time series at the same resolution ('s') and call interpolate() using 'linear' for the method argument:

```
In [155]: ts_regular = ts.resample('s').interpolate(method='linear')

In [156]: ts_regular
Out[156]:
2021-02-23 09:46:48 0.000000
2021-02-23 09:46:49 0.333333
2021-02-23 09:46:50 0.666667
2021-02-23 09:46:51 1.000000
2021-02-23 09:46:52 1.500000
2021-02-23 09:46:53 2.000000
2021-02-23 09:46:54 2.500000
2021-02-23 09:46:55 3.000000
2021-02-23 09:46:56 3.200000
2021-02-23 09:46:57 3.400000
2021-02-23 09:46:58 3.600000
2021-02-23 09:46:59 3.800000
2021-02-23 09:47:00 4.000000
Freq: S, dtype: float64
```

You now have timestamps for every second, and new values have been interpolated between the original data points. The method argument comes with other options, including nearest, pad, zero, spline, and more. You can read about them at *https://pandas.pydata.org/pandas-docs/stable/reference/api/pandas.Series.interpolate.html*.

Resampling and Analyzing Irregular Time Series: A Binary Example

Let's look at a realistic example of working with irregular time series. Imagine that you've attached a sensor to the compressor of a refrigeration unit to see how often it's on (1) and off (0) during a day.

To build the toy dataset, enter the following in the console:

```
In [157]: import pandas as pd

In [158]: raw_dict = {'2021-2-23, 06:00:00': 0,
     ...:             '2021-2-23, 08:05:09': 1,
     ...:             '2021-2-23, 08:49:13': 0,
     ...:             '2021-2-23, 11:23:21': 1,
     ...:             '2021-2-23, 11:28:14': 0}

In [159]: ts = pd.Series(raw_dict)

In [160]: ts.index = pd.to_datetime(ts.index)

In [161]: ts.plot();
```

This produces the plot in Figure 21-9. Note that it doesn't reflect the binary (0 or 1) nature of the data.

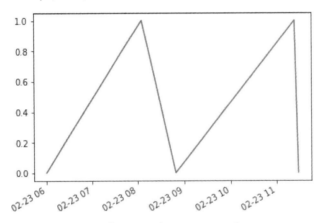

Figure 21-9: A plot of an irregular time series of compressor on-off data

In its raw, irregular form, the data is difficult to visualize and work with. For example, if you try to check the state of the compressor at 11 AM, you'll get an error:

```
In [162]: ts['2021-02-23 11:00:00']
KeyError --snip--
```

The problem is that series indexing doesn't interpolate on the fly. We need to first resample the data to a "working resolution," in this case, seconds:

```
In [163]: ts_secs = ts.resample('S').ffill()

In [164]: ts_secs
```

```
Out[164]:
2021-02-23 06:00:00    0
2021-02-23 06:00:01    0
2021-02-23 06:00:02    0
2021-02-23 06:00:03    0
2021-02-23 06:00:04    0
                      ..
2021-02-23 11:28:10    1
2021-02-23 11:28:11    1
2021-02-23 11:28:12    1
2021-02-23 11:28:13    1
2021-02-23 11:28:14    0
Freq: S, Length: 19695, dtype: int64
```

In [165]: **ts_secs.plot();**

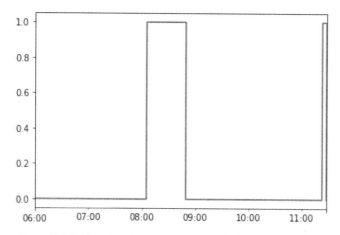

Figure 21-10: The plot of time series resampled to one-second frequency

Now the plot reflects the binary "on-off" nature of the data, and you can extract the state at 11 AM:

In [166]: **ts_secs['2021-2-23 11:00:00']**
Out[166]: 0

To determine how many seconds the compressor was off and on during the time period, call the value_counts() method on the series:

In [167]: **ts_secs.value_counts()**
Out[167]:
0   16758
1   2937
dtype: int64

To determine the fraction of the day that the compressor was on, just divide the value_counts() output at index 1 by the number of seconds in a day:

```
In [168]: num_secs_per_day = 60 * 60 * 24

In [169]: print(f"On = {ts_secs.value_counts()[1] / num_secs_per_day}")
On = 0.033993055555555554
```

The compressor ran for only three percent of the day. That's some good insulation!

Sliding Window Functions

The pandas library comes with functions for transforming time series using a *sliding window* or with exponentially decaying weights. These functions smooth raw data points so that long-term trends are more apparent.

A *moving average* is a commonly used time series technique for smoothing noise and gaps and revealing underlying data trends. Well-known examples are the 50- and 200-day moving averages used to analyze stock market data.

To make a moving average, a "window" of a specified length is used to average rows in a DataFrame column. The window starts at the earliest date and slides down the column one time unit at a time, and then it repeats this process. Here's an example for a three-day moving average, with the averaged values in bold:

```
             value
date
2020-06-01     6   |
2020-06-02    20   | |
2020-06-03    36   | | |--> (6 + 20 + 36)/3 = 20.67
2020-06-04    33   | |--> (20 + 36 + 33)/3 = 29.67
2020-06-05    21   |--> (36 + 33 + 21)/3 = 30.0
```

To make a "monthly" 30-day moving average of our COVID data from the section "Downsampling" on page 650, let's first reimport it as a new DataFrame named df_roll and replace the anomalous value. (If you still have the data in memory, you can use df_roll = df.copy() in place of the next five lines):

```
In [170]: df_roll = pd.read_csv("texas-history.csv",
     ...:                        usecols = ['date','deathIncrease'])

In [171]: df_roll = df_roll.sort_values('date', ascending=True)

In [172]: df_roll = df_roll.set_index(pd.DatetimeIndex(df_roll['date']))
```

```
In [173]: df_roll = df_roll.drop('date', axis=1)
```

```
In [174]: df_roll.loc['2020-7-27', 'deathIncrease'] = 239
```

Next, make a 30_day_ma column for this DataFrame and calculate the values by calling the rolling() method on the deathIncrease column, passing it 30 and then tacking on the mean() method. Finish by calling plot():

```
In [175]: df_roll['30_day_ma'] = df_roll.deathIncrease.rolling(30).mean()
```

```
In [176]: df_roll.plot();
```

As you can see in Figure 21-11, the moving average curve is smoother than the curve produced by monthly resampling (Figure 21-8) but retains some of the periodic oscillations.

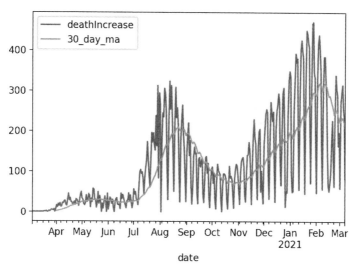

Figure 21-11: Texas COVID-related deaths with 30-day moving average curve

By default, the averaged values are posted at the *end* of the window, which makes the average curve look offset relative to the daily data. To post at the *center* of the window, pass True to the rolling() method's center argument:

```
In [177]: df_roll['30_day_ma'] = df_roll.deathIncrease.rolling(30,
                                             center=True).mean()
```

```
In [178]: df_roll.plot();
```

Now, the peaks and valleys in the averaged curve and the raw data are better aligned, as illustrated in Figure 21-12.

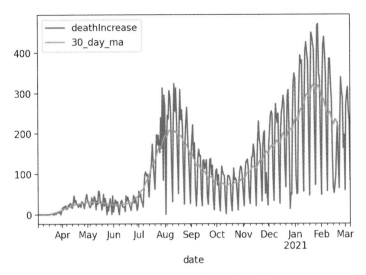

Figure 21-12: Texas COVID-related deaths with 30-day moving average curve posted at the center of the window interval

You can call other aggregation methods in Table 21-6 with `rolling()`. Here, we call the standard deviation method on the same 30-day sliding window and display the new column with the others (Figure 21-13):

```
In [179]: df_roll['30_std'] = df_roll.deathIncrease.rolling(30,
                                                center=True).std()

In [180]: df_roll.plot();
```

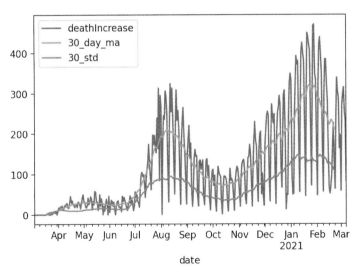

Figure 21-13: A 30-day sliding window standard deviation and moving average for COVID-related daily deaths in Texas

In addition to rolling averages with a fixed-sized window, pandas has methods for using expanding windows (`expanding()`), binary moving windows (`corr()`), exponentially weighted functions (`ewm()`), and user-defined moving window functions (`apply()`). You can read about these at *https://pandas.pydata.org/pandas-docs/stable/reference/frame.html* for DataFrames, and *https://pandas.pydata.org/pandas-docs/stable/reference/series.html* for series.

TEST YOUR KNOWLEDGE

6. What is the index structure associated with the pandas `Timestamp` class?

   a. `datetime64`

   b. `datetime64[ns]`

   c. `TimedeltaIndex`

   d. `DatetimeIndex`

7. Convert `'2021-2-23 00:00:00'` to a pandas `Timestamp`.

8. Localize the previous timestamp to the Europe/Warsaw time zone.

9. Create a PeriodIndex for every hour in May 1, 2021.

10. Which of the following are examples of downsampling?

   a. Minutes to seconds

   b. Minutes to hours

   c. Years to weeks

   d. Days to months

Summary

Time series represent data indexed to a time reference. Both native Python and pandas provide special "time-aware" data types and tools. These let you easily handle issues like sexagesimal arithmetic, time zone transitions, daylight saving time, leap years, datetime plotting, and more. With the ability to manipulate time series, you can gain insights into your data and solve otherwise imponderable problems.

Well, that does it for *Python Tools for Scientists*. This book had a simple goal: to get you up and running as a scientist using Python.

If you've finished the book, you've learned how to set your computer up for doing science with the Anaconda distribution, organize your projects using conda environments and dedicated project folders, and become familiar with coding tools like the Jupyter Qt console, Spyder, Jupyter Notebook, and JupyterLab. If you were new to Python, you've now learned the basics of the language. You're aware of many of the important scientific and visualization packages, and you should have some ideas about how to choose among them. Finally, you've gotten some hands-on experience with key libraries like NumPy, Matplotlib, pandas, seaborn, and scikit-learn.

Moving forward, the absolute best way to progress your programming knowledge and skill is to *do projects*, either for pay or for play. Projects let you reduce the enormous Python universe into manageable chunks, force you to focus your full attention on a select group of tasks, and grow your confidence. They'll lead to questions that you never knew you had, and seeking the answers will help you further your own education. Onward!

APPENDIX

ANSWERS TO THE "TEST YOUR KNOWLEDGE" CHALLENGES

Chapter 7

1. False

2. c

3. c

4.
```
In [22]: (42**0.5)**4
Out[22]: 1764.0000000000002
```

5.
```
In [1]: 42 **= 2
  File "C:\Users\hanna\Local\Temp/ipykernel_24188/3589881457.py", line 1
    42 **= 2
       ^
SyntaxError: cannot assign to literal
```

6. a

7. ```
In [1]: import math

In [2]: round(math.pi, 5)
Out[2]: 3.14159
```

8. ```
In [1]: type((1, 2, 3))
Out[1]: tuple
```

9. False

10. String

11. a

12. ```
print (
    """

      ^-------^
    | >   < |
    | V |
     \   /

    Hooty Hoot!

    """)
```

13. ```
In [35]: secs = 1824
In [36]: minutes, seconds = ((int(secs / 60)), (int(secs % 60)))
In [37]: print(f'{secs} seconds = {minutes} minutes, {seconds} seconds')
1824 seconds = 30 minutes, 24 seconds
```

14. b (supposedly the longest continuous string of consonants in English)

15. d

16. c

17. ```
In [1]: from string import punctuation
In [2]: punc = punctuation.replace('-','')
In [3]: caesar_said = 'Tee-hee, Brutus.'
In [4]: hyph_only = caesar_said.translate(str.maketrans('', '',punc))
In [5]: hyph_only
Out[5]: 'Tee-hee Brutus'
```

18. ```
In [1]: 'impractical python projects'.title()
Out[1]: 'Impractical Python Projects'
```

Chapter 8

1. a, b, d

2. b

3. c

4. Change its name

5.
```
In [1]: x = 42
In [2]: del x
```

---

6. False

7.
```
In [1]: print(input("Enter your first name: ")[::-1])

Enter your first name: Lee
eeL
```

8.
```
In [1]: john == 'Harry'
Traceback (most recent call last):

File "<ipython-input-112-97bd679db026>", line 1, in <module>
john == 'Harry'

NameError: name 'john' is not defined
```

9.
```
In [1]: a = 'Alice'

In [2]: b = 42

In [3]: c = a / b
Traceback (most recent call last):

File "C:\Users\hanna\AppData\Local\Temp/ipykernel_24188/2258966649.py",
line 1, in <module>
c = a / b

TypeError: unsupported operand type(s) for /: 'int' and 'str'
```

---

10. a

Chapter 9

1. IndexError: tuple index out of range
2. '!'

3.
```
In [1]: tup = ('Rust', 'R', 'Go', 'Julia'), ('Python')
In [2]: tup[1][1]
Out[2]: 'y'
```

---

4. a, b, c

5.
```
In [1]: field_trip = 'pith helmet',
   ...:              'rock hammer',
   ...:              'hand lens',
   ...:              'hiking boots',
   ...:              'sunglasses'
```

```
In [2]: field_trip
Out[2]: ('pith helmet', 'rock hammer', 'hand lens', 'hiking boots',
'sunglasses')
In [3]: field_trip = field_trip[1:]
In [4]: field_trip
Out[4]: ('rock hammer', 'hand lens', 'hiking boots', 'sunglasses')
```

6.
```
In [1]: patroni = []
In [2]: patroni.extend(['tiger', 'shark', 'weasel'])
In [3]: patroni
Out[3]: ['tiger', 'shark', 'weasel']
```

7.
```
In [1]: patroni.clear()
In [2]: patroni
Out[2]: []
```

8. c

9. c

10. c

11. Sets remove duplicate values in a dataset; each unique value will occur only once.

12. c

13. True

14. c

15. a

16. b

17.
```
In [1]: jokes = {"Did you hear about the kidnapping?":
   ...:         "He slept for 4 hours!",
   ...:         "You shot your dog? Was he mad?":
   ...:         "He wasn't too happy about it!"}

In [2]: jokes["Did you hear about the kidnapping?"]
Out[2]: 'He slept for 4 hours!'
```

18. c

Chapter 10

1. Four spaces

2. False

3.
```
In [1]: while True:
   ...:     print('Heeellllllpppppppp!!!!')
```

4. a, b

5.  ```python
    print('English to Pig Latin Translator')

    VOWELS = 'aeiouy'

    while True:
        word = input("Enter a word else enter '0' to stop: ")
        if word == '0':
            break
        if word[0] in VOWELS:
            pig_latin = word + 'way'
        else:
            pig_latin = word[1:] + word[0] + 'ay'
        print(f'\n{pig_latin}')
    ```

6.  ```python
    In [1]: while True:
       ...:        name = input('Enter your username: ')
       ...:        if name != 'Alice':
       ...:            continue
       ...:        while True:
       ...:            pwd = input('Enter your password: ')
       ...:            if pwd == 'Star Lord':
       ...:                break
       ...:            else:
       ...:                print('That password is incorrect')
       ...:        break
    ```

7.  ```python
    In [1]: count = 0

    In [2]: while count < 5:
       ...:        print('Python')
       ...:        count += 1
    Python
    Python
    Python
    Python
    Python
    ```

8.  ```python
    In [1]: print([i for i in range(1, 10) if i % 2 == 0])
    [2, 4, 6, 8]
    ```

9.  ```python
    In [1]: for i in range(10, -1, -1):
       ...:        print(i)
       ...:
    10
    9
    8
    7
    6
    5
    4
    3
    2
    1
    0
    ```
    ```

10.
```
In [1]: words = ['age', 'moody', 'knock', 'adder', 'project', 'stoop',
'blubber']

In [2]: for word in words:
   ...:         middle = int(len(word) / 2)
   ...:         print(word[middle])

g
o
o
d
j
o
b
```

11.
```
import random

answer = random.randint(1, 100)
guess = int(input('Guess a number between 1 and 100: '))
attempts = 1

while guess != answer:
    if guess > answer:
        print('You guessed too high.')
    else:
        print('You guessed too low.')
    guess = int(input('Guess again: '))
    attempts += 1

print('\nYou got it!')
print(f'It only took you {attempts} tries.')
```

12.
```
import random

fortunes = ['Dogogone it, people LIKE you!',
            'You will learn a new coding skill today.',
            'You are a quick learner!',
            'Your wisdom makes you superior to others.']

misfortunes = ['Your eyes are like pools. Cesspools.',
               'Your ears are like flowers. Cauliflowers.',
               'Your breath would kill a thousand camels.',
               'Run up an alley and holler fish!']
print("""
    0 - Quit
    1 - A fortune cookie
    2 - A misfortune cookie
    """)

while True:
    choice = input('Choose a number from the menu: ')
    if choice.isdigit():
        choice = int(choice)
```

```
            if choice == 0:
                print('Thanks for playing!')
                break
            if choice == 1:
                print(random.choice(fortunes))
            elif choice == 2:
                print(random.choice(misfortunes))
        else:
            print('Choose from the menu options.')
```

Chapter 11

1. c

2. c

3. b

4.
```
In [1]: def vowel_voider():
   ...:         name = input("Enter your last name: ")
   ...:         new_name = ''
   ...:         vowels = 'aeiouy'
   ...:         for char in name:
   ...:             if char not in vowels:
   ...:                 new_name += char
   ...:             else:
   ...:                 continue
   ...:         return new_name
```

5.
```
In [1]: def calc_momentum(*, mass, velocity):
   ...:         return mass * velocity

In [2]: calc_momentum(mass=10, velocity=50)
Out[2]: 500
```

6. c

7.
```
In [1]: from random import uniform

In [2]: samples = [round(uniform(0, 50), 1) for x in range(10)]
In [3]: samples
Out[3]: [42.7, 37.8, 30.2, 35.0, 0.4, 35.1, 22.4, 9.8, 23.4, 30.0]
```

8.
```
In [1]: nums = [3, 10, 16, 25, 88, 75]
In [2]: filtered = filter(lambda x: x % 5 == 0, nums)
In [3]: print(list(filtered))
[10, 25, 75]
```

9. False. Calling main() at the end grants it access.

10. c

11.
```
In [1]: G = 0.0000000000667
In [2]: def calc_force_gravity(mass1, mass2, radius):
   ...:     global G
   ...:     f = (G * mass1 * mass2) / radius**2
   ...:     return f
```

NOTE *You can also use G = 6.67e-11.*

12.
```
In [1]: import math

In [2]: dir(math)
```

13. b

14. d

15.
```
In [1]: x = 25

In [2]: def use_x(x):
   ...:     print(x**2)

In [3]: use_x(x)
625

In [4]: def use_x():
   ...:     global x
   ...:     print(x**2)

In [5]: use_x()
625
```

Chapter 12

1. c

2. d

3. False

4. c

5. Objects

6. e

7.
```
In [1]: from pathlib import Path

In [2]: p = Path('.\\test1\\another_haiku.txt')
In [3]: p.rename('.\\test1\\haiku_2.txt')
Out[3]: WindowsPath('test1/haiku_2.txt')
```

8. Remember, Python starts counting at 0:

```
In [93]: with open('haiku.txt') as f:
    ...:         f.seek(14)
    ...:         print(f.read())

Contemplating cherry trees
Strangers are like friends
                  --Issa
```

9. c

10. True

11. True

12. c

13.
```
In [1]: import json

In [2]: crew = dict(Mercury=1, Gemini=2, Apollo=3)
In [3]: capsules_data = json.dumps(crew)

In [4]: with open('capsules_data.json', 'w') as f:
    ...:         f.write(capsules_data)

In [5]: with open('capsules_data.json', 'r') as f:
    ...:         crew = json.load(f)

In [6]: for key in crew:
    ...:         if crew[key] == 1:
    ...:             seat = 'seat'
    ...:         else:
    ...:             seat = 'seats'
    ...:         print(f"The {key} capsule had {crew[key]} {seat}.")
The Mercury capsule had 1 seat.
The Gemini capsule had 2 seats.
The Apollo capsule had 3 seats.
```

14.
```
In [1]: test = ["don't", "do"]
In [2]: test_json = json.dumps(test)
In [3]: test_json
Out[3]: '["don\'t", "do"]'

In [4]: test = ['don\'t', 'do']
In [5]: test
Out[5]: ["don't", 'do']
In [6]: test_json = json.dumps(test)
In [7]: test_json
Out[7]: '["don\'t", "do"]'
```

15. d

16. Assumes that cwd is `file_play`:

```
In [1]: import shutil

In [2]: shutil.move('lines.txt', 'test1')
Out[2]: 'test1\\lines.txt'
In [3]: shutil.make_archive('.\\test1\\lines.txt', 'zip')
Out[3]: '.\\test1\\lines.txt.zip'
```

Chapter 13

1. b

2. c

3. True

4. c

5.
```
class Parrot():
    def __init__(self, name, color, age):
        self.name = name
        self.color = color
        self.age = age

    def squawk(self):
        print("\nSQUAWK!\n")

    def parroting(self):
        phrase = input("Enter something for parrot to repeat: ")
        print(f"\nSquawk! {phrase} Squawk!")

polly = Parrot('Polly', 'green', 80)
polly.squawk()
polly.parroting()
```

Output:

```
SQUAWK!

Enter something for parrot to repeat: Polly wants a cracker!

Squawk! Polly wants a cracker! Squawk!
```

6. c

7. e

8. b

9. True

10. b

11. d

12. New code in the *ship_display.py* program is highlighted in gray:

```python
from math import dist
from dataclasses import dataclass
import matplotlib.pyplot as plt


@dataclass
class Ship:
    '''Object for tracking a ship on a grid.'''
    name: str
    classification: str
    registry: str
    location: tuple
    obj_type = 'ship'
    obj_color = 'black'

    def distance_to(self, other):
        distance = round(dist(self.location, other.location), 2)
        return str(distance) + ' ' + 'km'


garcia = Ship('Garcia', 'frigate', 'USA', (20, 15))
ticonderoga = Ship('Ticonderoga', 'destroyer', 'USA', (5, 10))
kobayashi = Ship('Kobayashi', 'maru', 'Federation', (10, 22))

VISIBLE_SHIPS = [garcia, ticonderoga, kobayashi]

def plot_ship_dist(ship1, ship2):
    sep = ship1.distance_to(ship2)
    for ship in VISIBLE_SHIPS:
        plt.scatter(ship.location[0], ship.location[1],
                    marker='d',
                    color=ship.obj_color)
        plt.text(ship.location[0], ship.location[1], ship.name)
        plt.plot([ship1.location[0], ship2.location[0]],
                 [ship1.location[1], ship2.location[1]],
                 color='gray',
                 linestyle="--")
    plt.text((ship2.location[0]), (ship2.location[1] - 2), sep, c='gray')
    plt.xlim(0, 30)
    plt.ylim([0, 30])
    plt.show()

for i in range(30):
    garcia.location = (20, i)
    plot_ship_dist(kobayashi, garcia)i)
```

Before running the script in Spyder, go to the Plots pane and select **Mute inline plotting** (Figure A-1). This will force plots to appear in the Plots pane rather than inline within the console.

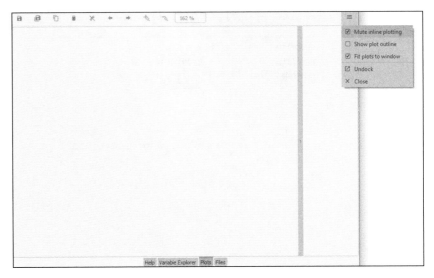

Figure A-1: Selecting Mute inline plotting from the Plots pane in Spyder

To close all the plots, click the large **X** icon on the Plots pane toolbar. As a challenge, see if you can make the Garcia move diagonally across the screen.

Chapter 14

1. b
2. True
3. a, c
4. c
5. d

6.
```
In [1]: import itertools
In [2]: help(itertools.product)
```

7. b, d
8. a, c
9. d

10.
```
class Frigate():
    """A frigate class warship for use in a war game simulation.

    Attributes
        name (str): Name of the ship without a designation/registry.
        crew (int): Number of crew members.
        length_ft(int): Length of the ship in feet.
        tonnage (int): Weight of the ship in short tons (US).
        fuel_gals(int): Fuel tank capacity in US gallons.
        guns (int): Number of big guns.
        ammo (int): Number of rounds of ammo available.
```

```
        heading (int): The compass direction in which the bow is pointed.
        max_speed (float): Maximum speed of ship in knots.
        speed (float): Current speed of ship in knots.

    Methods defined here:
        __init__(self, name)
            Constructs all the necessary attributes for the ship object.

            Parameters
                name (str):
                    Name of the ship without a designation/registry.

        helm(self, heading, speed)
            Sets and displays ship's current heading and speed.

            Parameters
                heading (int):
                    The compass direction the bow is pointed.
                speed (float):
                    The current speed of the ship in knots.

            Returns
                None

        fire_guns(self)
            Prints "BOOM!" and decrements and prints remaining ammo.

            Parameters
                None

            Returns
                None
    """
```

Chapter 18

1. c (arrays can hold any number of dimensions)

2. d

3. b

4. e

5. In [1]: **import numpy as np**

   In [2]: **np.zeros((10, 10))**

6. In [1]: **import numpy as np**

   In [2]: **arr2d = np.arange(30).reshape(5, 6)**

   In [3]: **arr2d[::2]**

7.
```
In [4]: arr2d[1::2, 1::2]
In [5]: # also:
In [6]: arr2d[1:5:2, 1:6:2]
```

8. c

9. b

10. 4

11. Because the byte size for each element is set by the largest element (–10000)

12. c

13. b

14. c

15. c

Chapter 19

1. c

2. True

3. d

4. c

5. Note: This solution uses the pyplot approach.

```
In [1]: rockets = {'Atlas': 57, 'Falcon9': 70, 'SaturnV': 111, 'Starship': 120}

In [2]: plt.ylabel('Height (m)')
   ...: plt.bar(rockets.keys(), rockets.values(), width=0.3);
```

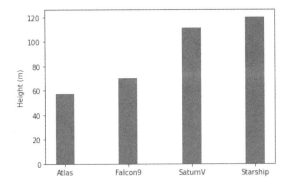

6. False

7. Use the suptitle() method, like so, for the pyplot approach:

```
plt.suptitle('Martian Goethite, Hematite, and Jarosite Distributions')
```

and like this for the object-oriented style:

```
fig.suptitle('Martian Goethite, Hematite, and Jarosite Distributions')
```

8. This solution uses the object-oriented style:

```
In [64]: # Create dummy datasets:
    ...: x = np.random.normal(0, 1, 50).cumsum()
    ...: y = np.random.normal(0, 1, 50).cumsum()
    ...: z = np.random.normal(0, 1, 50).cumsum()
    ...:
    ...: # Make list of datasets and titles:
    ...: data = [x, y, z]
    ...: titles = ['Data X', 'Data Y', 'Data Z']
    ...:
    ...: # Create subplots:
    ...: fig, axs = plt.subplots(1, 3)
    ...: fig.tight_layout()
    ...:
    ...: # Loop through subplots and plot data using black crosses:
    ...: for i, ax in enumerate(axs):
    ...:     ax.set_title(titles[i])
    ...:     ax.plot(data[i], 'k+')
```

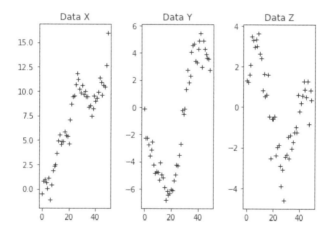

NOTE *Your plot will look different because the data is randomly generated.*

9. This solution uses the object-oriented style:

```
import numpy as np
import matplotlib.pyplot as plt
%matplotlib notebook

fig, ax = plt.subplots()

for _ in range(30):
```

```
data = np.random.rand(4, 4)
heat = ax.imshow(data)
fig.canvas.draw()
fig.canvas.flush_events()
```

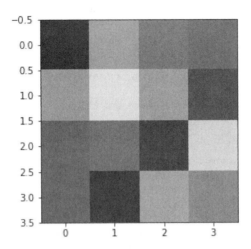

NOTE *Your plot will look different because the data is randomly generated.*

10. This solution uses the object-oriented style:

```
import matplotlib.pyplot as plt
%matplotlib inline

def calc_data(t, pos, vel, dt):
    """Return time, position, and velocity of object falling in a
vacuum."""
    time = []   # seconds
    position = []  # meters
    velocity = []  # meters per second
    for _ in range(15):  # Duration of fall in seconds.
        pos = pos + vel * dt
        vel = vel + -9.81 * dt  # 9.81 m/s**2 for Earth gravity.
        t += dt
        position.append(pos)
        velocity.append(abs(vel))  # Convert to absolute value.
        time.append(t)
    return time, position, velocity

time, position, velocity = calc_data(t=0, pos=0, vel=0, dt=1)

# Set up plot:
fig, ax1 = plt.subplots()
ax2 = ax1.twinx()  # Share the x-axis with ax.
ax1.set_xlabel('Time (sec)')
ax1.set_ylabel('Distance (m)', color='green')
ax2.set_ylabel('Velocity (m/s)', color='red')
```

```
ax2.invert_yaxis()  # So larger numbers plot toward bottom.

# Plot data:
ax1.plot(time, position, 'go', label='Position')
ax1.legend()
ax2.plot(time, velocity, 'red', label='Velocity')
ax2.legend(loc='lower left');
```

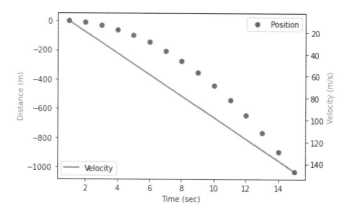

Chapter 20

1. a, b, d

2. False

3.
```
import pandas as pd

animals = {'canines': ['husky', 'poodle', 'bulldog'],
           'felines': ['Siamese', 'Persian', 'Maine Coon'],
           'cetaceans': ['humpback', 'sperm', 'right']}

df = pd.DataFrame(animals)
df.rename(columns={'cetaceans': 'whales'}, inplace=True)
df.head()
```

	canines	felines	whales
0	husky	Siamese	humpback
1	poodle	Persian	sperm
2	bulldog	Maine Coon	right

4. `df.head(1)`

	canines	felines	whales
0	husky	Siamese	humpback

5.
```
df_t = df.T
df_t
```

	0	1	2
canines	husky	poodle	bulldog
felines	Siamese	Persian	Maine Coon
whales	humpback	sperm	right

6. b and c

7. c

8. c

9. c

10. b and d

Chapter 21

1. b

2. c and d

3. c

4. c

5. d

6. d

7.
```
In [1]: import pandas as pd

In [2]: date = '2021-2-23 00:00:00'

In [3]: dt = pd.to_datetime(date)

In [4]: dt
Out[4]: Timestamp('2021-02-23 00:00:00')
```

8.
```
In [1]: import pandas as pd
In [2]: dt_warsaw = dt.tz_localize('Europe/Warsaw')
In [3]: dt_warsaw
Out[3]: Timestamp('2021-02-23 00:00:00+0100', tz='Europe/Warsaw')
```

9.
```
In [1]: import pandas as pd

In [2]: hours = pd.period_range(start='2021-5-1',
   ...:                         periods=24,
   ...:                         freq='H')

In [3]: hours
Out[3]:
PeriodIndex(['2021-05-01 00:00', '2021-05-01 01:00', '2021-05-01 02:00',
```

```
                          '2021-05-01 03:00', '2021-05-01 04:00', '2021-05-01 05:00',
                          '2021-05-01 06:00', '2021-05-01 07:00', '2021-05-01 08:00',
                          '2021-05-01 09:00', '2021-05-01 10:00', '2021-05-01 11:00',
                          '2021-05-01 12:00', '2021-05-01 13:00', '2021-05-01 14:00',
                          '2021-05-01 15:00', '2021-05-01 16:00', '2021-05-01 17:00',
                          '2021-05-01 18:00', '2021-05-01 19:00', '2021-05-01 20:00',
                          '2021-05-01 21:00', '2021-05-01 22:00', '2021-05-01 23:00'],
              dtype='period[H]')
```

10. b and d

INDEX

Symbols

= (assignment) operator, 177

\ (backslash escape character), 190–191

() (call operator), 284

** (combining dictionaries) operator, 252

/ (division) operator, 176

// (floor division) operator, 176

(hash) symbol, 378

\ (line continuation character), 189

% (modulo) operator, 176

+ (plus sign)

    addition operator, 176

    string concatenation operator, 192

** (power) operator, 177

* (splat) operator, 226, 228

* (string replication) operator, 192

- (subtraction) operator, 176

@ (syntactic sugar) symbol, 363–364

A

absolute paths, 319–320

abstraction, 283

accessing

    array attributes, 504–506

    data types, 185

access modes, 325–326

    for binary files, 335

    for shelves, 337

accuracy of prediction in *k*-NN, 615–616

accuracy_score() method, 615

add_axes() method, 567

adding

    consoles in JupyterLab, 152–153

    images in notebooks, 107–109

    key-value pairs, 251

    list items, 231–232

    text in notebooks, 102–104

addition operator (+), 176

aggregate functions in NumPy, 529, 531–533

aggregation methods in pandas, 650

allocating memory in NumPy, 496

Altair, 429–456

Anaconda

    components of, 7–9

    ecosystem, 397–398

    installing

        on Linux, 12–13

        on macOS, 11–12

        space requirements, 9

        on Windows, 9–11

Anaconda, Inc., 9

Anaconda Navigator

    Community tab, 17–18

    conda environments, 24–34

        backing up, 33

        creating, 25–26

        duplicating, 33

        package management in, 27–33

        removing, 34

    Environments tab, 15–17

    File menu, 18–19

    Home tab, 13–14

    JupyterLab, installing and launching, 140–142

    Jupyter Notebook, installing and launching, 95

    Jupyter Qt console, installing and launching, 51–52

    launching, 13, 25

    Learning tab, 17

    purpose of, 13

    Spyder, installing and launching, 62–63

Anaconda.org, 9

Anaconda Prompt. *See* CLI
animating plots, 569–573
ANNs (artificial neural networks), 406
anonymous functions. *See* lambda
    functions
Apache Spark, 418
append() method, 231–232
apply_along_axis() function, 532
applying style sheets, 578–580
arange() function, 499–501
arguments
    in exceptions, 276
    of functions, 285–286
    positional and keyword, 286–287
Armstrong, Joe, 348
array() function, 497
arrays, 491–492
    accessing attributes, 504–506
    broadcasting, 526–527
    creating, 494–504
        arange() function, 499–501
        array() function, 497
        functions for, 494
        linspace() function, 501–502
        prefilled, 502–504
    describing with dimension and
        shape, 492–494
    flattening, 519–520
    incrementing and decrementing,
        528
    indexing and slicing
        Boolean indexing, 515–517
        multidimensional arrays,
            511–514
        1D arrays, 507–509
        2D arrays, 509–511
    joining, 521–522
    matrix dot product, 527–528
    printing, 497
    purpose of, 492
    reading and writing data, 533–535
    shaping, 518–519
    splitting, 522–524
    transposing, 520–521
    vectorization, 524–526
artificial intelligence, branches of, 405
artificial neural networks, 406
Artist objects (Matplotlib), 542

asfreq() method, 644
assigning variables, 177, 203–206
assignment operator (=), 177
astimezone() method, 634
attributes, 347
    of arrays, accessing, 504–506
    defining classes, 350–352
augmented assignment operators, 178
autocompleting text in Editor pane
    (Spyder), 84
aware objects in datetime module,
    633–636
Axes object (Matplotlib), 538–539,
    560–561

B

backing up conda environments in
    Anaconda Navigator, 33
backslash escape character (\), 190–191
base classes, 355
base condition, 297
base environment, 23
Beautiful Soup, 414
Bednar, James, 93, 422
best practices, naming variables,
    209–211
bfill() method, 649
binary files, access modes, 335
binary universal functions in NumPy,
    530–531
Binder, 129–130, 448
binding. *See* assigning variables
block comments, 380
blocks of code, 259–260
Bokeh, 430–456
    dashboards, 446
    geospatial data, 484
Boolean data type, 215
Boolean indexing, 515–517
Boolean operators, 263–264
Bowtie, 446
boxplot() method, 603–604
branching, 258
break statement, 270
broadcasting, 508, 526–527
built-in data types, 184–185
built-in functions, 290–292
built-in modules, 311–313

C

calling
    functions, 284
    instance methods, 353–355
call operator ( ), 284
cartograms, 465
Cartopy, 464–465
case sensitivity of variables, 209
catching exceptions when opening
    files, 342–343
categorical data, converting to
    numerical data, 610–612
catenary, 542
cells. *See specific types of cells (code,
    Markdown, output, and so on)*
chained assignment, 205–206
changing
    runtime configuration
        parameters, 574–576
    start date, 656
channels, defined, 16
checkpoints in notebooks, 109
choosing
    deep learning frameworks, 408
    geospatial libraries, 484–487
    image manipulation libraries, 410
    natural language processing
        libraries, 412
    plotting libraries, 450–456
    syntax styles in Jupyter Qt
        console, 53–54
circular dependencies, 309
class attributes, 350
classes, 347–348
    dataclasses, 361–362
        decorators, 362–364
        defining, 365–368
        optimizing, 372–373
        plotting with, 368–370
        post-initialization
            processing, 370–372
    defining, 349–352, 364–365
    docstrings for, 386–387
    inheritance, 355–359
    instance methods
        calling, 353–355
        defining, 352–353

instantiating objects, 353–355,
    357–358
    object control, 359–361
classification problems, 587. *See
    also* Palmer Penguins project
class modules, creating, 373–375
cleaning package cache, 48
clearing
    namespace in IPython console
        (Spyder), 76–77
    workspaces in JupyterLab, 157
CLI
    conda environments, 34–48
        cleaning package cache, 48
        creating, 36–37
        duplicating and sharing, 44–46
        list of commands, 35
        package management with,
            39–44
        removing, 47
        restoring, 46–47
        storage locations for, 37–39
    extensions, installing, 169–170
    JupyterLab, installing and
        launching, 142
    Jupyter Notebook, installing and
        launching, 96
    Jupyter Qt console, installing and
        launching, 52–53
    launching, 34
    Spyder, installing and launching,
        63–64
cloning. *See* duplicating conda
    environments
close( ) method, 326
closing
    files, 326, 329–330
    notebooks, 109
    shelves, 338
    workspaces, in JupyterLab, 157
Code Analysis pane (Spyder), 85–86,
    391–395
code blocks, 259–260
code cells, 101
    in Editor pane (Spyder), 81–83
    in JupyterLab, 150–152
    in Jupyter Notebook, 104–106

code repositories, sharing notebooks, 125–128
Colab, 131
collection-controlled loops, 264
color blindness, 578
columns, renaming in pandas, 590–591
column_stack() functions, 522
combining
    dictionaries, 252
    sequences into dictionaries, 248–249
    sets, 244
command line interface. *See* CLI
command mode keyboard shortcuts in Jupyter Notebook, 110–111
command palette in Jupyter Notebook, 112
commands (conda)
    environment management, 35
    package management, 39
command shell, defined, 2
comments, 210, 377, 378–382
    commenting-out code, 381
    inline, 380–381
    multiline, 380
    single-line, 379
Community tab (Anaconda Navigator), 17–18
comparison operators, 214–217
comprehensions, 271–274
computer vision libraries, 409
concrete paths, 320
conda, 8–9
    pip and, 24
    prompt, changing, 38–39
conda configuration file, 38
conda environment manager, purpose of, 22
conda environments
    in Anaconda Navigator, 15–17, 24–34
        backing up, 33
        creating, 25–26
        duplicating, 33
        package management in, 27–33
        removing, 34
    base environment, 23

with CLI, 34–48
    cleaning package cache, 48
    creating, 36–37
    duplicating and sharing, 44–46
    list of commands, 35
    package management with, 39–44
    removing, 47
    restoring, 46–47
    storage locations for, 37–39
defined, 9
JupyterLab, installing, 140–143
Jupyter Notebook, installing, 94–97
organization of, 22–23
Palmer Penguins project setup, 587–588
purpose of, 21
Spyder, installing, 66–68
conda info command, 23
conda package manager, purpose of, 23
condition-controlled loops, 264
conditions in flow control statements, 259
configuring Spyder interface, 64–66
consoles, adding in JupyterLab, 152–153
constants, 209, 295
container data types
    dictionaries, 246–256
    list of, 220
    lists, 229–238
    sets, 239–246
    tuples, 220–229
Contextily, 462–463
continue statement, 270
control statements. *See* flow control statements
converting strings to dates and times, 630–631
coordinated universal time, 633
copying
    cells between notebooks in JupyterLab, 158–159
    lists, 235–237
copy() method, 236
copy module, 237

correlations, quantifying, 608–609

corr() method, 608–609

counting in DataFrames, 600–601

count() method, 196

countplot() method, 597–601

cross_validate() method, 618

cross-validation, 616–620

cufflinks, 435

current date and time, retrieving, 626–627

current figure in Matplotlib, 546

current working directory, 317–319

custom extensions, creating in JupyterLab, 171

customizing widgets, 121–122

D

Dash, 433, 446–447

dashboards, 445–450
    Dash, 446–447
    Panel, 449–450
    Streamlit, 447–448
    Voilà, 448–449

Dask, 404, 416–418, 441

data
    loading in JSON format, 341
    requesting, 413–418
    saving in JSON format, 340

databases, 336

dataclasses, 361–362
    decorators, 362–364
    defining, 365–368
    optimizing, 372–373
    plotting with, 368–370
    post-initialization processing, 370–372

dataclass module, 361–362

DataFrames, 403, 585–586
    converting categorical data to numerical data, 610–612
    counting in, 600–601
    describing, 596–597
    displaying, 590–591
    duplicates in, 592
    missing values in, 592–596
    quantifying correlations, 608–609
    reindexing, 595–596

data normalization with scikit-learn, 613–615

data serialization
    with json module, 339–342
    with pickle module, 334–336
    pickling vs. JSON, 334

datasets. *See also* DataFrames; Palmer Penguins project
    exploring, 596–609
    loading, 589–590
    in seaborn, 589
    size of, 451
    for training algorithms, 612–613

Datashader, 441–443

data structures. *See* arrays

data types
    accessing, 185
    Boolean, 215
    built-in, 184–185
    containers, list of, 219–220
    in datetime module, 627
    dictionaries, 246–256
    floats, 186–189
    integers, 185–186
    lists, 229–238
    in NumPy, 495
    sets, 239–246
    strings, 189–200
    tuples, 220–229
    type casting, 186–187

data visualization
    dashboards, 445–450
        Dash, 446–456
        Panel, 449–450
        Streamlit, 447–448
        Voilà, 448–449
    geospatial libraries
        Bokeh, 484
        Cartopy, 464–465
        choosing, 484–487
        folium, 470–473
        GeoPandas, 460–464
        Geoplot, 465–467
        GeoViews, 476–479
        ipyleaflet, 473–476
        KeplerGL, 479–481
        list of, 459

data visualization *(continued)*
    geospatial libraries *(continued)*
        Plotly, 467–469
        purpose of, 458
        pydeck, 481–484
    plotting libraries, 419–420
        Altair, 429–430
        Bokeh, 430–456
        choosing, 450–456
        Datashader, 441–443
        HoloViews, 436–441
        for InfoVis, 421
        Matplotlib, 422–456, 537–538
        Mayavi, 443–445
        pandas plotting API, 428–456
        Plotly, 431–436
        seaborn, 424–456
    radial visualization, 605–608
    types of, 420–445
data wrangling, 413
date offsets (pandas), 636, 645
date_range() method, 640–642
date ranges in pandas, 640–642
DatetimeIndex class (pandas), 637
datetime() method, 627–628
datetime module, 626
    converting strings to dates and
        times, 630–631
    current date and time, 626–627
    data types, 627
    durations, 627–628
    formatting dates and times,
        628–630
    plotting with, 632–633
    timestamps, 627–628
    time zones, 633–636
date times (pandas), 636
dbm module, 336
Debugger pane (Spyder), 90
debugging, defined, 3, 90
decimal module, 186
decorators, 362–364
decrementing arrays, 528
deepcopy() method, 237
deep learning frameworks, 406–407
default key values, creating, 252–253
default parameters, 287–288

defining
    classes, 349–352, 364–365
    code cells
        in Editor pane (Spyder), 81–83
        in JupyterLab, 150–152
        in Jupyter Notebook, 104–106
    dataclasses, 365–368
    functions, 284–285
    instance methods, 352–353
def keyword, 284
deleting. *See* removing
delimiters in split() method, 198
dependencies
    circular, 309
    defined, 8, 23
depth of recursion, 298
derived classes, 355
describe() method, 597
describing DataFrames, 596–597
de-serialization, 334
designing functions, 298–299
dict() function, 248
dictionaries, 246–256
    combining, 252
    combining sequences into,
        248–249
    creating, 247–248
    creating empty, 249
    functions and methods for, 249–250
    key-value pairs
        adding, 251
        removing, 252
    key values
        creating default, 252–253
        retrieving, 251
    printing, 254–255
    retrieving contents, 250–251
    reverse lookups, 253–254
    sorting, 254
dictionary comprehensions, 273–274
difference() method, 243
differences between sets, finding, 243
dimension of arrays, 492–494
directories
    current working directory, 317–319
    for Jupyter notebooks, 98–100,
        144–145
    for Spyder project files, 69–72

directory paths
    absolute and relative, 319–320
    naming, 316
    normalizing, 319
    os module, 317–319
    `pathlib` module, 320–322
    `shutil` module, 322–324
`dir()` function, 306
disabling logging, 280
displaying
    DataFrames, 590–591
    image files in JupyterLab, 153–154
division operator (/), 176
docstrings, 74, 377, 382–395
    for classes, 386–387
    formats for, 392
    for functions and methods, 387–388
    for modules, 384–386
    in Spyder Code Analysis pane, 391–395
    updating, 388–391
`doctest` module, 388–391
documentation, 377–378
    comments, 378–382
    docstrings, 382–395
    with JupyterLab text editor, 164–165
`dot()` function, 528
dot notation, 180
dot product for matrices, 527–528
downloading notebooks, 123–125
downsampling time series in pandas, 650–656
`drop_duplicates()` method, 592
`dropna()` method, 594
`dstack()` function, 522
dtypes. *See* data types
dunder (double underscore) methods, 351, 365
`duplicated()` method, 592
duplicates
    in datasets, 592
    finding, 243–244
duplicating conda environments
    in Anaconda Navigator, 33
    with CLI, 44–46
durations in `datetime` module, 627–628

dynamic typing, 184, 211

E

Earth Engine, 483–484
edit mode keyboard shortcuts in Jupyter Notebook, 111–112
Editor pane (Spyder), 78–84
    autocompleting text, 84
    defining code cells, 81–83
    setting run configuration, 83–84
    writing programs with, 78–81
`elif` clause, 260–262
`else` clause, 259, 260–262
embedding widgets, 122
empty dictionaries, creating, 249
`empty()` function, 503
enabling extensions in Jupyter Notebook, 113–115
end-of-line (EOL) markers, 329
`enumerate()` function, 269
environment files, creating, 44–46
environments. *See* conda environments
Environments tab (Anaconda Navigator), 15–17
error messages, 182–183
errors, ignoring, 277–278
escape sequences, 190–191
events, handling with widgets, 120–121
exceptions
    catching when opening files, 342–343
    handling, 274–278
        ignoring errors, 277–278
        `try` and `except` statements, 274–276
    raising, 182–183, 274, 276–277
`except` statement, 274–276
exploring
    datasets, 596–609
    simulations in JupyterLab, 154–155
exponent operator (**), 177
expressions
    assigning variables, 204
    defined, 176
    generator, 302
    mathematical. *See* mathematical expressions
    ternary, 262

extend() method, 231

extensibility, defined, 3

eXtensible Markup Language (XML), 343

Extension Manager, 166–169

extensions
    in JupyterLab
        creating custom, 171
        installing and managing with CLI, 169–170
        installing and managing with Extension Manager, 166–169
        list of, 165–166
    in Jupyter Notebook, 113–115

F

facet plots, 435

ffill() method, 648

Figure object (Matplotlib), 538–539
    methods for, 559–560

File menu (Anaconda Navigator), 18–19

files. *See also* text files
    binary files, accessing, 335
    closing, 326, 329–330
    opening, catching exceptions, 342–343

filling missing values in datasets, 594–595

fillna() method, 594, 650

filter() function, 300

finding
    differences between sets, 243
    duplicates in sets, 243–244
    list index, 233–234
    missing values in datasets, 593
    packages
        in Anaconda Navigator, 27–30
        with CLI, 40–42

flattening arrays, 519–520

flatten() method, 519

float() function, 187

floats, 186–189
    converting to/from integers, 186–187
    rounding, 187–189

floor division operator (//), 176

flow control statements, 258
    if statement, 258–264
    loops, 264–274

flow of execution, 258
    functions and, 292–297
    tracing, 278–281

folders. *See* directories

folium, 470–473

formats for docstrings, 392

formatting
    dates and times, 628–630
    strings, 192–194, 206

for statement, 267–269

frequencies for time series, 641

From Data to Viz website, 456

fromkeys() method, 253

frozenset() function, 245–246

frozensets, creating, 245–246

fruitful functions, 285

f-strings, 192–194, 206

full() function, 503

FuncAnimation class (Matplotlib), 569, 571–573

functions, 283. *See also* methods
    assigning variables, 204–205
    base condition, 297
    built-in, 290–292
    calling, 284
    creating, 253
    creating arrays
        arange() function, 499–501
        array() function, 497
        linspace() function, 501–502
        list of, 494
        prefilled, 502–504
    default parameters, 287–288
    defining, 284–285
    designing, 298–299
    for dictionaries, 249–250
    docstrings for, 387–388
    flow of execution, 292–297
    generators, 300–303
    global variables, 294–295
    lambda, 299–300
    for lists, 230–231
    main(), 295–297

math module, 179–181
namespaces and scopes, 293–294
naming, 290
in NumPy
  aggregate, 531–533
  pseudorandom numbers, 533
  universal, 529–531
parameters and arguments,
  285–286
positional and keyword
  arguments, 286–287
recursion, 297–298
returning values, 289
for sets, 241–242
for tuples, 222

G

garbage collection, 203
generator expressions, 302
generators, 300–303
Gensim, 412
GeoPandas, 460–464
Geoplot, 465–467
geospatial data, 457–458
geospatial libraries
  Bokeh, 484
  Cartopy, 464–465
  choosing, 484–487
  folium, 470–473
  GeoPandas, 460–464
  Geoplot, 465–467
  GeoViews, 476–479
  ipyleaflet, 473–476
  KeplerGL, 479–481
  list of, 459
  Plotly, 467–469
  purpose of, 458
  pydeck, 481–484
GeoViews, 476–479
GeoVis, 420
getcwd() function, 309
get_dummies() method, 600, 611
get() method, 251
Gist, sharing notebooks, 125–128
GitHub, sharing notebooks, 125–128
global scope, 293–294
global statement, 295

global variables, 294–295
globular clusters, 144
Google Earth Engine, 483–484
GridSearchCV class (scikit-learn),
  620–622
GridSpec class, creating multipanel
  displays, 549–555, 564–567
groupby() method, 600

H

handling
  events with widgets, 120–121
  exceptions, 274–278
    ignoring errors, 277–278
    try and except statements,
      274–276
hash() function, 240
hash symbol (#) , 378
hash tables, 239
HDF5 (Hierarchical Data Format), 343
head() method, 591, 596, 651
heatmaps in seaborn, 608–609
helper libraries, 413–418
Help menu (Jupyter Notebook),
  109–110
Help pane (Spyder), 72–74
Heroku, 448
Hierarchical Data Format (HDF5), 343
History pane in IPython console
  (Spyder), 77
HoloViews, 436–441
HoloViz, 441
home() method, 322
Home tab (Anaconda Navigator), 13–14
hsplit() function, 523
hstack() function, 522
HTML (HyperText Markup
  Language), 413
HTTP (HyperText Transfer
  Protocol), 413
hvPlot, 440–441, 462
hyperparameters, 616, 620–622

I

identities of variables, 202–203
IDEs (integrated development
  environments), defined, 2, 3

idxmax() method, 618, 653
if statement, 258–264
    Boolean operators, 263–264
    code blocks, 259–260
    elif clause, 260–262
    else clause, 259, 260–262
    ternary expressions, 262
ignoring errors, 277–278
image files, displaying in JupyterLab, 153–154
image manipulation libraries, 409
images, adding in notebooks, 107–109
immutability of strings, 197
importing
    modules, 179, 304–306
    packages, 589
incrementing arrays, 528
indexes
    of list items, finding, 233–234
    in series, 584–585
indexing
    arrays
        Boolean indexing, 515–517
        multidimensional arrays, 511–514
        1D arrays, 507–509
        2D arrays, 509–511
    time series in pandas, 646–647
index() method, 233–234
InfoVis plotting libraries, 420
    Altair, 429–430
    Bokeh, 430–431
    Datashader, 441–443
    HoloViews, 436–441
    list of, 421
    Matplotlib, 422–423
    pandas plotting API, 428–429
    Plotly, 431–436
    seaborn, 424–428
inheritance, 355–359
initialization methods, 351
__init__() method, 351
inline comments, 380–381
in operator, 196
input() function, 213–214
inserting list items, 232
insert() method, 232
insetting plots, 567–568

insignificant variables, 212
inspecting modules, 306–307
installing
    Anaconda
        on Linux, 12–13
        on macOS, 11–12
        space requirements, 9
        on Windows, 9–11
    extensions
        with CLI, 169–170
        with Extension Manager, 166–169
        in Jupyter Notebook, 113
    ipywidgets, 115–116, 170–171
    JupyterLab, 140–143
        with Anaconda Navigator, 140–142
        with CLI, 142
    Jupyter Notebook, 94–97
        with Anaconda Navigator, 95
        with CLI, 96
    Jupyter Qt console
        with Anaconda Navigator, 51–52
        with CLI, 52–53
    packages
        in Anaconda Navigator, 27–30
        with CLI, 40–42
        with conda and pip, 24
    RISE extension, 132–133
    seaborn, 50
    Spyder
        with Anaconda Navigator, 62–63
        with CLI, 63–64
        for conda environments and packages, 66–68
instance attributes, 351
instance methods
    calling, 353–355
    defining, 352–353
instantiating objects, 353–355, 357–358
instantiation, 351
integers, 185–186
    converting to/from floats, 186–187
integrated development environments (IDEs), defined, 2, 3
interact class, creating widgets, 116–118

interactive class, creating widgets,
118–119
Interactive Python (IPython), 2
interfaces for Matplotlib, 539–540
object-oriented, 555–557
pyplot, 539–541
internment, 205–206
interpolate() method, 656–657
interpolation in pandas, 656–660
intersection() method, 243–244
int() function, 187
introspection, defined, 4
invoking, objects, 284
ipyleaflet, 473–476
IPython console (Spyder), 74–77
clearing namespace, 76–77
History pane, 77
kernels in, 76
output and plotting, 75
IPython (Interactive Python), 2
IPython notebooks. *See* notebooks
ipywidgets, installing, 115–116, 170–171
irregular time series in pandas,
656–660
isinstance() function, 185
isnull() method, 595
issubset() method, 245
issuperset() function, 245
items() method, 250–251, 339
iterables, 220

J

joining arrays, 521–522
join() method, 228
jointplot() method, 604–605
Jovian, 131
JSON (JavaScript Object Notation), 334
data
loading, 341
saving, 340
tuples, saving, 341–342
json.dumps() method, 254, 340
json module, 339–342
pickle vs., 334
Jupyter, 3
JupyterDash, 446
Jupyter-gmaps, 476

JupyterHub, 2, 131
JupyterLab, 94
code cells, defining, 150–152
consoles, adding, 152–153
defined, 2
extensions, 3
creating custom, 171
installing and managing
with CLI, 169–170
installing and managing
with Extension Manager,
166–169
list of, 165–166
image files, displaying, 153–154
installing, 140–143
with Anaconda Navigator,
140–142
with CLI, 142
vs. Jupyter Notebook, 140
left sidebar, 147–148
Markdown cells, 149–150
menu bar, 146–147
navigating, 145–146
notebooks
copying cells between,
158–159
creating, 148
naming, 149
opening multiple, 156
sharing, 171
project folders, creating, 144–145
purpose of, 139
simulations, exploring, 154–155
single document mode, 160
synchronized views, creating, 158
text editor
documentation with, 164–165
running scripts in notebooks,
163–164
running scripts in terminal,
162–163
writing scripts, 161–162
widgets, 170–171
workspace
clearing, 157
closing, 157
saving, 156–157

Jupyter Notebook. *See also* notebooks
  command palette, 112
  defined, 2
  extensions, 113–115
  Help menu, 109–110
  installing, 94–97
    with Anaconda Navigator, 95
    with CLI, 96
  vs. JupyterLab, 140
  keyboard shortcuts, 110–112
  navigating, 100–101
  Palmer Penguins project setup,
    587–588
  purpose of, 93–94
  widgets, 115
    creating manually, 119–120
    creating with interact class,
      116–118
    creating with interactive class,
      118–119
    customizing, 121–122
    embedding, 122
    handling events, 120–121
    installing ipywidgets, 115–116
Jupyter Notebook Viewer, sharing
  notebooks, 128–129
Jupyter Qt console
  defined, 2
  installing and launching
    with Anaconda Navigator,
      51–52
    with CLI, 52–53
  as interactive, 53
  keyboard shortcuts, 54–55
  multiline editing, 58–59
  printing and saving, 56–58
  purpose of, 49
  syntax highlighting, 53
  syntax styles, choosing, 53–54
  tab and kernel options, 55

K

KeplerGL, 479–481
Keras, 407–418
kernel density estimation (KDE), 604
Kernel menu (Jupyter Notebook),
  checking and running
  notebooks, 123

kernels
  defined, 4
  in IPython console (Spyder), 76
  in Jupyter Qt console, 55
  restarting, 310
keyboard shortcuts
  in Jupyter Notebook, 110–112
  in Jupyter Qt console, 54–55
keys() method, 250–251, 338
key-value pairs
  adding, 251
  creating default, 252–253
  removing, 252
  retrieving, 251
keyword arguments, 286–287
k-NN (*k*-Nearest Neighbor), 609–622
  converting categorical data to
    numerical data, 610–612
  normalizing data, 613–615
  optimizing
    with cross-validation, 616–620
    with GridSearchCV class,
      620–622
  prediction accuracy, 615–616
  running, 615–616
  training dataset, 612–613

L

lambda functions, 299–300
launching
  Anaconda Navigator, 13, 25
  CLI, 34
  JupyterLab
    with Anaconda Navigator,
      140–142
    with CLI, 142
  Jupyter Notebook
    with Anaconda Navigator, 95
    with CLI, 96
  Jupyter Qt console
    with Anaconda Navigator,
      51–52
    with CLI, 52–53
  Spyder
    with Anaconda Navigator,
      62–63
    with CLI, 63–64
    from Start menu, 64

lazy evaluation, 300
Learning tab (Anaconda Navigator), 17
left sidebar (JupyterLab), 147–148
len() function, 223–224
length of tuples, 223–224
libraries
    Beautiful Soup, 414
    for computer vision (image
        manipulation), 409
    dashboards
        Dash, 446–447
        list of, 446
        Panel, 449–450
        Streamlit, 447–448
        Voilà, 448–449
    Dask, 416–418
    deep learning frameworks, 406–407
    defined, 8
    geospatial libraries
        Bokeh, 484
        Cartopy, 464–465
        choosing, 484–487
        folium, 470–473
        GeoPandas, 460–464
        Geoplot, 465–467
        GeoViews, 476–479
        ipyleaflet, 473–476
        KeplerGL, 479–481
        list of, 459
        Plotly, 467–469
        purpose of, 458
        pydeck, 481–484
    helper libraries, 413–418
    Keras, 407
    list of, 400
    for machine learning, 404–406
    for natural language processing,
        411–412
    NLTK, 411–412
    NumPy, 401–418
    OpenCV, 409–410
    pandas, 403–404
    Pillow, 410
    plotting libraries, 419–420
        Altair, 429–456
        Bokeh, 430–456
        choosing, 450–456

        Datashader, 441–443
        HoloViews, 436–441
        Matplotlib, 422–456, 537–538
        Mayavi, 443–445
        pandas plotting API, 428–456
        Plotly, 431–436
        seaborn, 424–456
    PyTorch, 408
    for regular expressions, 415–416
    requests, 413–414
    scikit-image, 410
    scikit-learn, 404–407
    SciPy, 401
    spaCy, 412
    statsmodels, 406
    SymPy, 402–403
    TensorFlow, 407
line continuation character (\), 189
lineplot() method, 619
linspace() function, 501–502
Linux, Anaconda installation, 12–13
list comprehensions, 272–273
list() function, 230, 251
lists, 229–238
    adding items to, 231–232
    changing item values, 233
    checking for membership, 237–238
    converting data types to, 230
    copying, 235–237
    creating, 230
    finding index of items, 233–234
    functions and methods for,
        230–231
    inserting items, 232
    removing items, 232–233
    sorting, 234–235
load_dataset() method, 590
load() function, 534
loading
    data in JSON format, 341
    datasets, 589–590
localize() method, 634
local scope, 293–294
logging.disable() function, 280
logging levels, 279
logging module, 278–281
loop control statements, 269–271

loops, 264–274
    animating plots, 569–571
    `break` statement, 270
    `continue` statement, 270
    `for` statement, 267–269
    `pass` statement, 271
    replacing with comprehensions, 271–274
    `while` statement, 265–267

M

machine learning, 404
    with *k*-NN, 609–622
    libraries, 404–406
macOS, Anaconda installation, 11–12
magic commands
    defined, 57
    list of, 58
magic methods, 351, 365
`main()` function, 295–297
`make_column_transformer()` method, 614
`maketrans()` method, 199
manually creating widgets, 119–120
maps. *See* geospatial libraries
Markdown cells
    adding images in, 107–109
    adding text with, 102–104
    in JupyterLab, 149–150
markers, 470
mathematical expressions
    assignment operator, 177
    augmented assignment operators, 178
    defined, 176
    mathematical operators, 176–177
    `math` module, 179–181
    precedence, 178–179
`math` module, 179–181
Matplotlib, 422–456, 537–538
    with `datetime` module, 632–633
    interfaces for, 539–540
        object-oriented, 555–557
        pyplot, 539–541
    multipanel displays, creating with GridSpec class, 549–555, 564–567
    plots
        3D plots, 568–569
        animating, 569–573

        creating with object-oriented approach, 557–561
        creating with pyplot, 542–545
        insetting, 567–568
        styling, 573–580
        subplots, 545–549, 561–564
        terminology for, 538–539
matrix, 493
matrix dot product, 527–528
`max()` function, 224–225
maximum values of tuples, 224–225
`max()` method, 653
Mayavi, 443–445
`mean()` function, 531
`melt()` method, 619
membership operators, 196
memory allocation in NumPy, 496
menu bar (JupyterLab), 146–147
`meshgrid()` function, 502
`meshgrid()` method, 552
Method Resolution Order, 358
methods, 347. *See also* functions
    for dictionaries, 249–250
    docstrings for, 387–388
    for file objects, 326
    initialization, 351
    instance methods, 352–355, 353–355
    for lists, 230–231
    object-oriented
        for `Axes` objects, 560–561
        creating plots, 559
        for `Figure` objects, 559–560
    in os module, 317
    in pandas
        aggregation methods, 650
        handling missing values, 594
        I/O methods, 590
    in `pathlib` module, 321, 332
    in pyplot
        creating plots, 543
        manipulating plots, 544
    in seaborn
        `boxplot()`, 603–604
        `countplot()`, 597–601
        `jointplot()`, 604–605
        list of, 598
        `pairplot()`, 601–602

scatterplot(), 602–603
stripplot(), 603–604
for sets, 241–242
in shelve module, 338–339
in shutil module, 323
strings, 196–200
for tuples, 222
mgrid() function, 502
Microsoft Azure Notebooks, 131
min() function, 224–225
Miniconda, 9
minimum values of tuples, 224–225
MinMaxScaler() method, 614
missing values in datasets, 592–596
Modin, 404
modular approach for installation
    JupyterLab, 142–143
    Jupyter Notebook, 96–97
    Spyder, 66–68
modules, 283
    built-in, 311–313
    class modules, creating, 373–375
    copy, 237
    dataclass, 361–362
    datetime, 626–636
    dbm, 336
    decimal, 186
    defined, 8
    docstrings for, 384–386
    doctest, 388–391
    importing, 179, 304–306
    inspecting, 306–307
    json, 334, 339–342
    logging, 278–281
    math, 179–181
    naming, 310
    os, 317–319
    pathlib, 320–322, 332–333
    pickle, 334–336
    purpose of, 303–304
    re, 415–416
    shelve, 336–339
    shutil, 322–324
    stand-alone mode, 310–311
    string, 199
    writing, 307–310
modulo operator (%), 176

moving averages, 660–663
MRO (Method Resolution Order), 358
multidimensional arrays, 493
    indexing and slicing, 511–514
multiline comments, 380
multiline editing in Jupyter Qt console, 58–59
multipanel displays, creating with GridSpec class, 549–555, 564–567
multiple inheritance, 358
multiple notebooks, opening in JupyterLab, 156
mutability, 220
    hashability and, 240
    tuples and, 227

N

naive approach for installation
    JupyterLab, 140–142
    Jupyter Notebook, 94–96
    Spyder, 66
naive objects in datetime module, 633–636
namespaces
    clearing in IPython console (Spyder), 76–77
    in functions, 293–294
naming
    directory paths, 316
    functions, 290
    modules, 310
    notebooks
        in JupyterLab, 149
        in Jupyter Notebook, 101–102
    variables, 206–213
natural language processing, 411–412
Natural Language Tool Kit, 411–412
navigating
    JupyterLab, 145–146
    Jupyter Notebook, 100–101
Navigator. *See* Anaconda Navigator
nbextensions. *See* extensions
nbviewer, sharing notebooks, 128–129
ndarray class, 494. *See also* arrays
nested code blocks, 260
NLP (natural language processing), 411–412

NLTK (Natural Language Tool Kit), 411–412

normalizing
  pathnames, 319
  data with scikit-learn, 613–615

notebooks. *See also* Jupyter Notebook
  closing, 109
  code cells, defining, 104–106
  consoles, adding, 152–153
  copying cells between in JupyterLab, 158–159
  creating
    in Jupyter Notebook, 100–101
    in JupyterLab, 148
  images, adding, 107–109
  Markdown cells
    adding images, 107–109
    adding text, 102–104
  naming
    in JupyterLab, 149
    in Jupyter Notebook, 101–102
  opening multiple in JupyterLab, 156
  output cells, 106–107
  project folders, creating, 98–100, 144–145
  purpose of, 93–94
  running scripts in, 163–164
  saving, 109
  sharing, 122
    via Binder, 129–130
    checking and running from Kernel menu, 123
    with Colab, 131
    downloading, 123–125
    via GitHub and Gist, 125–128
    with Jovian, 131
    with JupyterHub, 131
    in JupyterLab, 171
    via Jupyter Notebook Viewer, 128–129
    with Microsoft Azure Notebooks, 131
    trusting, 131–132
  as slideshows
    creating, 133–136
    installing RISE extension, 132–133
    speaker notes, 136

synchronized views, creating, 158
  text, adding, 102–104
not in operator, 196
now() method, 626–627
np.eye() function, 503
np.savez() function, 534
Numba, 441
numerical data, converting categorical data to, 610–612
NumPy (Numerical Python), 401
  arrays
    accessing attributes, 504–506
    broadcasting, 526–527
    creating, 494–504
    describing with dimension and shape, 492–494
    flattening, 519–520
    incrementing and decrementing, 528
    indexing and slicing, 506–517
    joining, 521–522
    matrix dot product, 527–528
    printing, 497
    purpose of, 492
    reading and writing data, 533–535
    shaping, 518–519
    splitting, 522–524
    transposing, 520–521
    vectorization, 524–526
  data types, 495
  functions
    aggregate, 531–533
    pseudorandom numbers, 533
    universal, 529–531
  memory allocation, 496
  purpose of, 491–492

O

object-oriented approach (Matplotlib), 555–557
  multipanel displays, 564–567
  plots, 557–561
objects, 348
  controlling with objects, 359–361
  as instances, 351
  instantiating, 353–355, 357–358

invoking, 284
variables and, 202
1D arrays, 493, 507–509
ones() function, 502
OOP (object-oriented
    programming), 347
  classes
    defining, 349–352, 364–365
    docstrings for, 386–387
    inheritance, 355–359
    instance methods, 352–353
    instantiating objects, 353–355,
      357–358
  class modules, creating, 373–375
  dataclasses, 361–362
    decorators, 362–364
    defining, 365–368
    optimizing, 372–373
    plotting with, 368–370
    post-initialization processing,
      370–372
  object control, 359–361
  when to use, 348
OpenCV, 409–410
open() function, 325
opening
  files, catching exceptions, 342–343
  multiple notebooks in
    JupyterLab, 156
operators
  assignment, 177
  augmented assignment, 178
  Boolean, 263–264
  comparison, 214–217
  mathematical, 176–177
  membership, 196
  overloading
    in strings, 191–192
    in tuples, 227
    in variable assignment, 204
  precedence, 178–179
optimizing
  dataclasses, 372–373
  k-NN
    with cross-validation, 616–620
    with GridSearchCV class,
      620–622
os.chdir() method, 318

os.getcwd() method, 317
os module, 317–319
os.normpath() method, 319
os.path.join() method, 318
output cells in notebooks, 106–107
output in IPython console (Spyder), 75
overfitting, 616
overloading operators
  in strings, 191–192
  in tuples, 227
  in variable assignment, 204

P

package cache, 23
  cleaning, 48
packages
  in Anaconda Navigator, 15–17
  conda package manager, purpose
    of, 23
  defined, 8
  dependencies, defined, 23
  finding
    in Anaconda Navigator, 27–30
    with CLI, 40–42
  importing, 589
  installing
    in Anaconda Navigator, 27–30
    with CLI, 40–42
    with conda and pip, 24
  managing
    in Anaconda Navigator, 27–33
    with CLI, 39–44
  removing
    in Anaconda Navigator, 30–33
    with CLI, 42–44
  Spyder, installing, 66–68
  updating
    in Anaconda Navigator, 30–33
    with CLI, 42–44
pairplot() method, 601–602
pairplots, 424
Palmer Penguins project
  displaying DataFrames, 590–591
  duplicates in, 592
  exploring dataset, 596–609
  importing packages, 589
  loading dataset, 589–590
  missing values in, 592–596

Palmer Penguins project *(continued)*
    predictions with, 609–622
    purpose of, 586–587
    renaming columns, 590–591
    setup, 587–588
    steps in, 587
pandas, 403–404, 583
    aggregation methods, 650
    alternatives to, 404
    DataFrames, 585–586
        converting categorical data to
            numerical data, 610–612
        counting in, 600–601
        describing, 596–597
        displaying, 590–591
        duplicates in, 592
        missing values in, 592–596
        quantifying correlations,
            608–609
        reindexing, 595–596
    datasets, loading, 589–590
    plotting API, 428–456
    plotting syntax, 620
    radial visualization, 605–608
    resources for information, 623
    series, 584–585
    time series, 636–637
        changing start date, 656
        date offsets, 645
        date ranges, 640–642
        downsampling, 650–656
        indexing and slicing, 646–647
        interpolation, 656–660
        parsing data, 637–640
        resampling, 647–663
        sliding window functions,
            660–663
        time deltas, 644
        time spans, 642–644
        upsampling, 648–650
Pandas-Bokeh, 431
Panel, 449–450
parallel processing, 417
parameters
    default, 287–288
    of functions, 285–286
ParaView, 444
parse() method, 630

parsing time series data in pandas,
    637–640
pass statement, 271
pathlib module, 320–322
    reading and writing text files,
        332–333
pathnames, 316
    absolute and relative, 319–320
    normalizing, 319
    os module, 317–319
    pathlib module, 320–322
    shutil module, 322–324
Pattern, 412
period_range() method, 643
periods (pandas), creating, 642–644
pickled data, shelving, 336–338
pickle.dump() function, 335
pickle.load() function, 335
pickle module, 334–336
pie charts, 553
PIL (Python Image Library), 410
Pillow libraries, 410
pip, conda and, 24
plaintext files, 325. *See also* text files
Plotly, 431–436, 467–469
Plotly Express, 433–436, 467–469
plots
    with datetime module, 632–633
    in Matplotlib
        animating, 569–573
        creating with object-oriented
            approach, 557–561
        creating with pyplot, 542–545
        insetting, 567–568
        pyplot approach, 539–541
        styling, 573–580
        subplots, 545–549, 561–564
        terminology for, 538–539
        3D plots, 568–569
    types of, 452
plotting
    with dataclasses, 368–370
    in IPython console (Spyder), 75
plotting libraries, 419–420. *See*
        also geospatial libraries
    Altair, 429–456
    Bokeh, 430–456
    choosing, 450–456

Datashader, 441–443
HoloViews, 436–441
for InfoVis and SciVis, 421
Matplotlib, 422–456, 537–538
Mayavi, 443–445
pandas plotting API, 428–456
Plotly, 431–436
seaborn, 424–456
plt.pie() method, 553
plus sign (+)
addition operator, 176
string concatenation operator, 192
Polyglot, 412
pop() method, 232–233, 252
positional arguments, 286–287
postBuild files, 130
__post_init__ function, 370–372
post-initialization processing in
dataclasses, 370–372
power operator (**), 177
precedence, 178–179
predictions with *k*-NN, 609–622
predict() method, 615
prefilled arrays, creating, 502–504
print() function, 278
printing
arrays, 497
dictionaries, 254–255
in Jupyter Qt console, 56–58
tuples, 228–229
processes, 318
Profiler pane (Spyder), 89–90
profiling, defined, 4
project files in Spyder, 68–72
project folders, creating
in JupyterLab, 144–145
in Jupyter Notebook, 98–100
Project pane (Spyder), 72
projects, creating in Spyder, 316,
348–349. *See also* Palmer
Penguins project
prompt (conda), changing, 38–39
pseudorandom numbers in NumPy, 533
pure paths, 320
pydeck, 481–484
PyPI (Python Package Index), 24
pyplot, 539–541
multipanel displays, 549–555

plots, 542–545
subplots, 545–549
Python. *See also* flow control statements;
functions; libraries; modules;
variables
comments, 210
data types, 184–200
datetime module, 626
converting strings to dates
and times, 630–631
current date and time,
626–627
data types, 627
durations, 627–628
formatting dates and times,
628–630
plotting with, 632–633
timestamps, 627–628
time zones, 633–636
dictionaries, 246–256
documentation, 377–378
comments, 378–382
docstrings, 382–395
error messages, 182–183
exception handling, 274–278
lists, 229–238
mathematical expressions, 176–181
objects, 202
reserved keywords, 207–209
resources for information, 174
scientific ecosystem, 397–398
sets, 239–246
standard library, 303
tuples, 220–229
Python Data Analysis library.
See pandas
Python Image Library, 410
Python Package Index, 24
Python package management system.
See pip
PyTorch, 408
pytz library, 634

Q

Qt console. *See* Jupyter Qt console
Qt, defined, 4
quantifying correlations, 608–609
quotation marks for strings, 189–190

R

radial visualization, 605–608
radviz() method, 605–608
raise keyword, 276–277
raising exceptions, 182–183, 274, 276–277
random() function, 504
random numbers in NumPy, 533
range() function, 300
rank of arrays, 492
raster data, 457
ravel() function, 519
raw strings, 191
RcParams class (Matplotlib), 574
reading
    array data, 533–535
    text files, 325–329
        with pathlib, 332–333
readline() method, 328
readlines() method, 328
read() method, 327
recursion, 297–298
regular expressions (regex), 415–416
reindexing DataFrames, 595–596
relational operators. *See* comparison operators
relative paths, 319–320
re module, 415–416
remove() method, 233
removing
    conda environments
        in Anaconda Navigator, 34
        with CLI, 47
    key-value pairs, 252
    list items, 232–233
    missing values in datasets, 594–595
    packages
        in Anaconda Navigator, 30–33
        with CLI, 42–44
renaming columns in pandas, 590–591
replace() method, 198
replacing loops with comprehensions, 271–274
requesting data, 413–414
requests library, 413–414
resample() method, 647–650

resampling time series in pandas, 647–663
reserved keywords, 207–209
reset_index() method, 595
reshape() function, 518
reshaping arrays, 518–519
resizing multipanel displays, 554–555
resources for information on Python, 174
restarting kernels, 310
restoring conda environments with CLI, 46–47
retrieving current date and time, 626–627
return address of functions, 286
returning function values, 289
return keyword, 286
reverse lookups, 253–254
RISE extension, installing, 132–133
rmtree() method, 324
rolling() method, 661
round() function, 187–189
rounding floats, 187–189
row_stack() functions, 522
rstrip() method, 329
run configuration, setting in Editor pane (Spyder), 83–84
running
    k-NN, 615–616
    scripts
        in notebooks, 163–164
        in terminal, 162–163
runtime configuration parameters, changing, 574–576

S

Sankey diagrams, 466
save() function, 534
savez_compressed() function, 534
saving
    data in JSON format, 340
    in Jupyter Qt console, 56–58
    notebooks, 109
    tuples in JSON format, 341–342
    workspaces in JupyterLab, 156–157
scatterplot() method, 602–603
scatterplots, 552

scientific libraries. *See* libraries
Scientific Python Development IDE. *See* Spyder
scikit-image, 410
scikit-learn, 404–407, 584
    cross-validation, 616–620
    GridSearchCV class, 620–622
    k-NN, 609–622
    normalizing data, 613–615
    prediction accuracy, 615–616
    resources for information, 623
    training datasets, 612–613
SciPy, 401
SciPy stack, 400
SciVis plotting libraries, 420
    Mayavi, 443–445
scopes in functions, 293–294
scripts
    running
        in notebooks, 163–164
        in terminal, 162–163
        writing in JupyterLab text editor, 161–162
seaborn, 424–456, 584
    datasets
        loading, 590
        practice datasets, 589
    heatmaps, 608–609
    installing, 50
    methods
        boxplot(), 603–604
        countplot(), 597–601
        jointplot(), 604–605
        list of, 598
        pairplot(), 601–602
        scatterplot(), 602–603
        stripplot(), 603–604
    resources for information, 623
    wide-form and long-form data, 619
seek() method, 327
select_dtypes() method, 612
sequences, 220, 248–249
serialization
    with json module, 339–342
    with pickle module, 334–336
    pickling vs. JSON, 334
series, 584–585

set comprehensions, 274
setdefault() method, 252–253
set() function, 240
sets, 239–246
    combining, 244
    creating, 239–241
    differences between, 243
    duplicates in, 243–244
    frozensets, 245–246
    functions and methods for, 241–242
    supersets, 245
shape() function, 519, 520
shape of arrays, 492–494
shaping arrays, 518–519
shared package cache, 23
sharing
    conda environments
        with CLI, 44–46
    notebooks, 122
        via Binder, 129–130
        checking and running from Kernel menu, 123
        with Colab, 131
        downloading, 123–125
        via GitHub and Gist, 125–128
        with Jovian, 131
        with JupyterHub, 131
        in JupyterLab, 171
        via Jupyter Notebook Viewer, 128–129
        with Microsoft Azure Notebooks, 131
        trusting, 131–132
shell utilities (shutil) module, 322–324
shelve module, 336–339
shelve.open() method, 336
shelves, closing, 338
shelving pickled data, 336–338
shutil module, 322–324
sidebar (JupyterLab), 147–148
simulations, exploring in JupyterLab, 154–155
single document mode (JupyterLab), 160
single-line comments, 379
size of datasets, 451

slicing
    arrays
        multidimensional arrays,
            511–514
        1D arrays, 507–509
        2D arrays, 509–511
    strings, 194–196
    time series in pandas, 646–647
slideshows, notebooks as
    creating, 133–136
    installing RISE extension, 132–133
    speaker notes, 136
sliding window functions, 660–663
__slots__ class variable, 372–373
sorted() function, 254
sorting
    dictionaries, 254
    lists, 234–235
sort() method, 234–235
spaCy, 412
spatial indexing, 463
speaker notes for slideshows, 136
special consoles in Spyder, 77
special methods, 351, 365
specifications files, creating, 46
spines, 557
splat (*) operator, 226, 228
split() function, 523
SplitMap, 474
split() method, 198
splitting arrays, 522–524
Spyder
    Code Analysis pane, 85–86,
        391–395
    configuring interface, 64–66
    Debugger pane, 90
    defined, 2
    Editor pane, 78–84
        autocompleting text, 84
        defining code cells, 81–83
        setting run configuration,
            83–84
        writing programs with, 78–81
    Help pane, 72–74
    installing
        with Anaconda Navigator,
            62–63
        with CLI, 63–64

    for conda environments and
        packages, 66–68
    IPython console, 74–77
        clearing namespace, 76–77
        History pane, 77
        kernels in, 76
        output and plotting, 75
    launching
        with Anaconda Navigator,
            62–63
        with CLI, 63–64
        from Start menu, 64
    Profiler pane, 89–90
    project files and folders, 68–72
    Project pane, 72
    projects, creating, 316, 348–349
    purpose of, 61
    special consoles, 77
    Variable Explorer pane, 86–89
SQLite, 343
stacking, arrays, 521–522
stack overflow, 298
stand-alone mode for modules,
    310–311
standard library (Python), 303
start date, changing, 656
Start menu (Spyder), launching, 64
statements, 177
static typing, 184
statistical methods in NumPy, 531–533
statsmodels, 406
storage locations for conda
    environments, specifying,
    37–39
Streamlit, 447–448
strftime() method, 628–630
str() function, 190
string concatenation operator (+), 192
string module, 199
string replication operator (*), 192
strings, 189–200. *See also* text
    converting to dates and times,
        630–631
    escape sequences, 190–191
    formatting, 192–194, 206
    immutability of, 197
    interning, 205
    membership operators, 196

methods, 196–200
operator overloading, 191–192
quotation marks for, 189–190
raw, 191
slicing, 194–196
stripplot() method, 603–604
stripplots, 425
strptime() method, 631
structured arrays, 535
style files, creating, 576–578
style sheets, applying, 578–580
style.use() method, 577
styling plots, 573–580
   runtime configuration parameters,
     574–576
   style files, 576–578
   style sheets, 578–580
subclasses, 355
subplot() method, 545–549
subplot_mosaic() method, 566
subplots in Matplotlib
   object-oriented approach, 561–564
   pyplot, 545–549
subplots() method, 561–564
subtraction operator (-), 176
superclasses, 355
super() function, 358–359
supersets, 245
swapaxes() function, 521
swapping array axes, 520–521
SymPy, 402–403
synchronized views, creating in
   JupyterLab, 158
syntactic sugar, 363–364
syntax highlighting, 53
syntax styles, choosing in Jupyter Qt
   console, 53–54

T

tabs in Jupyter Qt console, 55
TensorFlow, 407
tensors, 493
terminal
   defined, 4
   running scripts in, 162–163
terminal window. *See* CLI
ternary expressions, 262

text. *See also* strings
   adding in notebooks, 102–104
   autocompleting in Editor pane
     (Spyder), 84
TextBlob, 412
text editor, JupyterLab
   documentation with, 164–165
   running scripts
     in notebooks, 163–164
     in terminal, 162–163
   writing scripts, 161–162
text editor, Spyder. *See* Editor pane,
   (Spyder)
text files
   closing, 329–330
   reading, 325–329
     with pathlib, 332–333
   writing to, 330–331
     with pathlib, 332–333
text() method, 551
threads, 417
3D arrays. *See* multidimensional arrays
3D plots, 568–569
thresholding, 515
tight_layout() method, 547, 554
tile maps, 463
timedelta object, 627–628
timedelta_range() method, 644
time deltas (pandas), 636, 644
time series
   datetime module, 626
     converting strings to dates
       and times, 630–631
     current date and time,
       626–627
     data types, 627
     durations, 627–628
     formatting dates and times,
       628–630
     plotting with, 632–633
     timestamps, 627–628
     time zones, 633–636
   pandas, 636–637
     changing start date, 656
     date offsets, 645
     date ranges, 640–642
     downsampling, 650–656

time series *(continued)*

    pandas *(continued)*

        indexing and slicing, 646–647

        interpolation, 656–660

        parsing data, 637–640

        resampling, 647–663

        sliding window functions, 660–663

        time deltas, 644

        time spans, 636, 642–644

        upsampling, 648–650

time spans (pandas), 636, 642–644

Timestamp class (pandas), 637

timestamps

    in datetime module, 627–628

    in pandas, 637

time zones

    in datetime module, 633–636

    in pandas, 639–640

to_datetime() method, 637–640

tracebacks, 182

tracing flow of execution, 278–281

training datasets, 612–613

train_test_split() method, 612–613

translate() function, 199

transpose() function, 520

transposing arrays, 520–521

trusting notebooks, 131–132

try statement, 274–276

tuple() function, 221–222

tuples, 220–229

    converting data types to, 221–222

    creating, 221

    functions and methods for, 222

    length of, 223–224

    minimum and maximum values, 224–225

    mutability and, 227

    operator overloading, 227

    printing, 228–229

    saving in JSON format, 341–342

    unpacking, 225–227

2D arrays, 493, 509–511

type annotations, 366–367

type casting, 186–187

type() function, 185

type hints, 184, 366–367

tz_convert() method, 639

tz_localize() method, 639

U

unary universal functions in NumPy, 529–530

Unicode, 190

union() method, 244

universal functions in NumPy, 529–531

unpacking tuples, 225–227

update() method, 251

updating

    docstrings, 388–391

    packages

        in Anaconda Navigator, 30–33

        with CLI, 42–44

upsampling time series in pandas, 648–650

user input, 213–214

UTC (coordinated universal time), 633

V

Vaex, 404

value_counts() method, 659

values() method, 250–251

Variable Explorer pane (Spyder), 86–89

variables, 201

    assigning, 177, 203–206

    comparison operators, 214–217

    global, 294–295

    identities of, 202–203

    insignificant, 212

    naming, 206–213

    for user input, 213–214

vector data, 457

vectorization, 524–526

vectors, 493

visualizing data

    dashboards, 445–450

        Dash, 446–447

        Panel, 449–450

        Streamlit, 447–448

        Voilà, 448–449

    geospatial libraries

        Bokeh, 484

        Cartopy, 464–465

choosing, 484–487
folium, 470–473
GeoPandas, 460–464
Geoplot, 465–467
GeoViews, 476–479
ipyleaflet, 473–476
KeplerGL, 479–481
list of, 459
Plotly, 467–469
purpose of, 458
pydeck, 481–484
plotting libraries, 419–420
Altair, 429–430
Bokeh, 430–431
choosing, 450–456
Datashader, 441–443
HoloViews, 436–441
for InfoVis, 421
Matplotlib, 422–456, 537–538
Mayavi, 443–445
pandas plotting API, 428–456
Plotly, 431–436
seaborn, 424–456
radial visualization, 605–608
types of visualizations, 420–445
void functions, 285
Voilà, 448–449
vsplit() function, 523
vstack() function, 522

W

Wang, Peter, 443
web scraping, 413
while statement, 265–267
whitespace in mathematical
expressions, 179
widgets, 115
creating
with interact class, 116–118
with interactive class, 118–119
manually, 119–120

customizing, 121–122
embedding, 122
handling events, 120–121
installing ipywidgets, 115–116
in JupyterLab, 170–171
Windows, Anaconda installation, 9–11
with statement
closing files, 329–330
closing shelves, 338
workspaces (JupyterLab)
clearing, 157
closing, 157
saving, 156–157
writelines() method, 331
write() method, 330
writing
array data, 533–535
with Editor pane (Spyder), 78–81
modules, 307–310
scripts in JupyterLab text editor,
161–162
to text files, 330–331
with pathlib, 332–333

X

XML (eXtensible Markup
Language), 343

Y

YAML, 44, 343
yield statement, 301

Z

zero() function, 502
zip() function, 248

RESOURCES

Visit *https://nostarch.com/python-tools-scientists* for errata and more information.